Food Toxicology and Safety

Food toxicology deals with the existence of dangerous toxic components that food may contain, and various food-processing contaminants are formed during the production process as a direct result of traditional and emerging food-processing techniques. This book brings together food toxicology and food safety, linking them to several types of food processing. It addresses the gaps in how contaminants are formed at different stages of food processing, highlighting ongoing efforts related to food safety. In addition, it studies how processing technologies affect food products, explaining the advantages, disadvantages, and process operations of these techniques.

Sofia Agriopoulou has a Bachelors in Agriculture from Aristotle University of Thessaloniki, Greece (1992) and a Masters in Plant Production from Agricultural University of Athens, Greece (2000). She gained her Ph.D. in Food Science from University of Patras, Greece (2015). She has done postdoctoral research in Chemometrics in University Aix Marseille, France (2020). Dr. Agriopoulou is an assistant professor with a specialization in Postharvest Management of Plant-Based Foods – Mycotoxins, at the Department of Food Science and Technology at the University of Peloponnese, Greece. She has written 44 research papers and reviews and her research has been presented at 34 international and national conferences. She has also written 10 chapters in books. She is a reviewer for 38 scientific journals, including in *Food Chemistry*, *Food Control*, *Toxins*, *Foods*, *Mycotoxin Research*, *Sustainability*, *Molecules*, and *Coatings*. She has participated in three international and three national research programs.

Maria Tarapoulouzi comes from Cyprus. She has a BSc in Chemistry (2009), and an MSc in Environmental Engineering (2013) from the University of Cyprus, as well as a PGDp in Toxicology (2010) from the University of Birmingham, UK. In 2020, she gained her Ph.D. in Chemistry from the University of Cyprus focusing on spectroscopic characterization and authenticity of Cypriot traditional cheese Halloumi in combination with chemometrics. She is currently a postdoctoral researcher at the University of Cyprus, a chemometrics specialist, and a research collaborator at the State General Laboratory of Cyprus for the research project IsoDatabase. Dr. Tarapoulouzi has also served as visiting teaching staff at several academic institutions across Cyprus and has supervised nine undergraduate dissertations.

Contemporary Food Engineering

The **Contemporary Food Engineering Series** will consist of edited books addressing engineering processes related to food production. Classical unit operations in engineering applied to food manufacturing will be covered as well as such topics including the transport and storage of liquid and solid foods; heating, chilling and freezing of foods; mass transfer in foods; dehydration, thermal processing, non-thermal processing, extrusion, liquid food concentration, membrane processes and applications of membranes in food processing; packaging, cleaning, and sanitation. The books will be aimed at professional food scientists, academicians researching food engineering problems, and graduate level students.

Innovative Technologies in Seafood Processing edited by Yesim Ozogul

Advances in Vinegar Production edited by Argyro Bekatorou

Advances in Food Bioproducts and Bioprocessing Technologies edited by Monica Lizeth Chavez Gonzalez, Nagamani Balagurusamy, Christobal N. Aguilar

Advances in Noninvasive Food Analysis, edited by Muhammad Kashif Iqbal Khan

Agricultural, Biosystems, and Biological Engineering Education: Global Perspectives and Current Practice, edited by Umezuruike Linus Opara

Bioreactor Technology in Food Processing, edited by Rosane F. Schwan, V. K. Joshi, and Disney R. Dias

Innovative Technologies for Meat Processing, edited by Slim Smaoui and Tanmay Sarkar

Engineering Innovations in Sensory Science, edited by Mahendran Radhakrishnan and Anbarasan Rajan

Food Toxicology and Safety: Food Manufacturing Related Effects, edited by Sofia Agriopoulou and Maria Tarapoulouzi

For more information about this series, please visit: www.crcpress.com/Contemporary-Food-Engineering/book-series/CRCCONFOOENG

Food Toxicology and Safety

Food Manufacturing Related Effects

Edited by
Sofia Agriopoulou and Maria Tarapoulouzi

CRC Press
Taylor & Francis Group
Boca Raton London New York

CRC Press is an imprint of the
Taylor & Francis Group, an **informa** business

First edition published 2026
by CRC Press
2385 NW Executive Center Drive, Suite 320, Boca Raton FL 33431

and by CRC Press
4 Park Square, Milton Park, Abingdon, Oxon, OX14 4RN

CRC Press is an imprint of Taylor & Francis Group, LLC

ISBN: 978-1-032-68787-2 (hbk)
ISBN: 978-1-032-71232-1 (pbk)
ISBN: 978-1-032-71229-1 (ebk)

DOI: 10.1201/9781032712291

Typeset in Times
by Newgen Publishing UK

Contents

Contributors

Sofia Agriopoulou
Department of Food Science and Technology
University of the Peloponnese
Kalamata, Greece

Didem Peren Aykas Cinkilic
Department of Food Engineering, Faculty of Engineering, Adnan Menderes University,
Aydin 09100, Turkey

Sena Bakir
Recep Tayyip Erdogan University
Rize, Turkiye

Sana Benosmane
Département des sciences de la nature et de la vie
Université 8 Mai 1945 Guelma
Guelma, Algeria
and
Laboratory of Cellular Toxicology
Faculty of Sciences
Badji Mokhtar University
Annaba, Algeria

Ludmila Rosa Bergsten-Torralba
Laboratory of Cytotoxicity and Genotoxicity (SCG)
Pharmacology and Toxicology Department (DFT)
INCQS, FIOCRUZ
Rio de Janeiro, Brazil

Sofiane Boudalia
Département d'Écologie et Génie de l'Environnement
Université 8 Mai 1945
Guelma, Algeria
and
Laboratoire de Biologie
Eau et Environnement
Université 8 Mai 1945
Guelma, Algeria

Aissam Bousbia
Département d'Écologie et Génie de l'Environnement
Université 8 Mai 1945
Guelma, Algeria
and
Laboratoire de Biologie, Eau et Environnement
Université 8 Mai 1945
Guelma, Algeria

Antonio Eugenio Castro Cardoso de Almeida
Graduate Program in Sanitary Surveillance (PPGVS)
National Institute of Health Quality Control
Oswaldo Cruz Foundation (FIOCRUZ)
Rio de Janeiro, Brazil

Shazia Chohan
National Centre of Excellence in Analytical Chemistry
University of Sindh
Jamshoro, Pakistan

Carlos Adam Conte-Junior
Department of Biochemistry
Chemistry Institute
Federal University of Rio de Janeiro
Rio de Janeiro, Brazil

Teresa D'Amore
Laboratory of Preclinical and Translational Research, IRCCS CROB
Centro di Riferimento Oncologico della Basilicata
Rionero in Vulture, Italy

Peng Deng
State Key Laboratory of Food Science and Technology
Jiangnan University
Wuxi, Jiangsu, China

Monia Ennouri
Olive Tree Institute
University of Sfax
Sfax, Tunisia
and
Higher Institute of Applied Science and Technology of Mahdia
University of Monastir
Monastir, Tunisia

Yassine Gueroui
Département de Biologie
Université 8 Mai 1945
Guelma, Algeria

Safia Habila
Laboratoire de Pharmocologie et de Phytochimie
Université Mohammed Seddik Benyahia
Jijel, Algeria
and
Laboratoire de biotechnologie Santé et Environnement
Université Mohammed Seddik Benyahia
Jijel, Algeria

Zouheyr Hadri
Faculty of Sciences and Technology
Department of Agricultural Sciences
University of Relizane
Relizane, Algeria

Fatmagul Hamzaoglu
Recep Tayyip Erdogan University
Rize, Turkiye

Rachel Ann Hauser-Davis
Oswaldo Cruz Foundation (FIOCRUZ)
Rio de Janeiro, Brazil

Demetris Kafouris
State General Laboratory
Ministry of Health
Nicosia, Cyprus

Hüseyin Kara
National Centre of Excellence in Analytical Chemistry
University of Sindh
Jamshoro, Pakistan

Priyanka Kataria
Department of Food Science and Technology
Graphic Era Deemed University
Dehradun, Uttarakhand, India

Asma Lahouel
Laboratoire de Pharmocologie et de Phytochimie
Université Mohammed Seddik Benyahia
Jijel, Algeria

Yong Li
College of Food Science and Engineering
Shanxi Agricultural University
Taiyuan 030031, China

Najma Memon
National Centre of Excellence in Analytical Chemistry
University of Sindh
Jamshoro, Pakistan

Malha Oudir
Higher School of Food Science and Agri-Food Industry, ESSAIA
El Harrach, Alger, Algeria
and
Laboratoire des Procédés énergétiques et nanotechnologie
Université Saad Dahleb
Blida, Algeria

Raziye Nur Özçiçek
Department of Chemistry
Faculty of Natural Sciences
Selcuk University
Konya, Turkiye

Sunil Pareek
Department of Agriculture and Environmental Sciences
National Institute of Food Technology Entrepreneurship and Management
Kundli, Sonipat, Haryana, India

Hansika Sati
Department of Agriculture and Environmental Sciences
National Institute of Food Technology Entrepreneurship and Management
Kundli, Sonipat, Haryana, India

Júlia Scarpa-de-Souza
Instituto de Física da Universidade Federal Fluminense
Rio de Janeiro, Brazil

Devyani Shinde
Department of Agriculture and Environmental Sciences
National Institute of Food Technology Entrepreneurship and Management
Kundli, Sonipat, Haryana, India

Nikolaos Sidirokastritis
Laboratory of Food and Agricultural Industries Technology
Chemical Engineering Department
Aristotle University of Thessaloniki
Thessaloniki, Greece

Slim Smaoui
Laboratory of Microbial and Enzymatic Biotechnologies and Biomolecules
Center of Biotechnology of Sfax (CBS)
University of Sfax
Sfax, Tunisia

Sanam Iram Soomro
Department of Chemistry
Faculty of Natural Sciences
Selcuk University
Konya, Turkiye
and
Departments of Community Health Sciences
Agha Khan University
Karachi, Pakistan

Georgios Stavroulakis
State General Laboratory
Ministry of Health
Nicosia, Cyprus

Maria Tarapoulouzi
Department of Chemistry
Faculty of Pure and Applied Science
University of Cyprus
Nicosia, Cyprus
and
State General Laboratory
Ministry of Health
Nicosia, Cyprus

Patroklos Vareltzis
Laboratory of Food and Agricultural Industries Technology
Chemical Engineering Department
Aristotle University of Thessaloniki
Thessaloniki, Greece

Júlia Vianna de Pinho
Universidade Federal do Rio de Janeiro
Rio de Janeiro, Brazil

Maomao Zeng
State Key Laboratory of Food Science and Resources
Jiangnan University
Wuxi 214122, China

Anissa Zergui
National Center of Toxicology
and
Department of Pharmacy
Faculty of Medicine
University of Oran
Oran, Algeria

Introduction 1

Maria Tarapoulouzi and Sofia Agriopoulou

1.1 FOOD PROCESSING AND MANUFACTURING

Processed food has become an indispensable part of the human food chain. Processed foods are typically ready-to-eat or require minimal preparation, making them ideal for busy lifestyles. This convenience is a major reason for their widespread adoption, particularly in urban settings where time is often scarce. Processed foods have been integrated into various cultures and lifestyles, with many products catering to specific dietary needs, preferences, and restrictions. This adaptability makes them a staple in diverse diets around the world. Mass production and the use of inexpensive ingredients make processed foods more affordable than many fresh, whole foods (Kumar et al., 2022). This accessibility is crucial for populations with limited financial resources. Moreover, processed foods are integral to the global food supply chain. Processing techniques, such as canning, freezing, and drying, significantly extend the shelf life of food (Sridhar et al., 2021). In this context, processed foods can be transported over long distances and stored for long periods, making it possible to distribute food to regions where fresh produce is not readily available. In addition, the processed food industry is a significant contributor to the global economy, providing jobs and supporting related industries like agriculture, transportation, and retail (Rob and Cattaneo, 2021).

On the other hand, establishing traceability systems for processed food faces a different set of challenges compared to primary agro-food, because of the variety of raw materials, batch mixing, and resource transformation (Lebelo et al., 2021). The consumption of processed foods also raises concerns regarding health, particularly due to high levels of sugar, salt, and unhealthy fats found in many products (Fuhrman, 2018). Nonetheless, their role in feeding a growing global population and supporting modern lifestyles underscores their indispensable place in the human food chain.

DOI: 10.1201/9781032712291-1

1.2 FOOD TOXICOLOGY – CONTAMINATION IN THE FOOD PRODUCTION CHAIN

Chemical contaminants in food can generally be grouped into four main categories: natural toxins, environmental pollutants, agrochemical residues, and toxic substances formed during food processing, along with chemicals intentionally added to food (Lindeman et al., 2021). As seen in Figure 1.1, the harvesting and transportation processes are the initial phase of raw materials contamination, thus before food processing and manufacturing. Furthermore, there are several steps in which foods may be contaminated after food production. Food packaging receives the food product, therefore transportation as well as storage and distribution of packed food play a significant role in food quality. Food preparation under correct conditions is vital and it depends on consumers, and after that correct food disposal is crucial to influence assimilation of contaminants into the environment.

Food processing and manufacturing play crucial roles in ensuring the safety and quality of food products (Kotsanopoulos and Arvanitoyannis, 2017). Several factors during food processing can impact food quality, and the first-very important one is **contamination during processing**. Food contamination sources could be *biological, chemical, and physical*. It can occur at multiple stages of the production process, including raw material handling, equipment contact, and exposure to environmental contaminants. Therefore, it can involve the introduction of harmful bacteria, viruses, fungi, chemicals, and physical hazards, which can significantly degrade the safety and quality of the final product (Abdolshahi and Shokrollahi Yancheshmeh, 2020). **Chemical contamination,** which includes the use of contaminated water, as well as pesticides, cleaning agents, or cross-contact with allergens can introduce harmful chemicals into the food (Wang et al., 2023). **Cross-contamination** due to improper cleaning of equipment and utensils can lead to cross-contamination between batches of food, potentially introducing allergens or harmful microorganisms, as well as poor personal hygiene practices of workers can

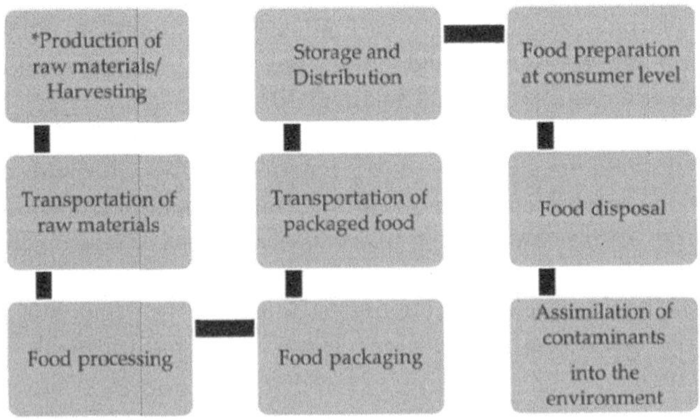

FIGURE 1.1 Key stages in food production, from farm to fork and ultimately to disposal (Lebelo et al., 2021).

be a source of contamination if proper handwashing and protective measures are not followed (Holah, 2023a). **Temperature control** is another important factor which may contain inadequate cooking or pasteurization; thus, if food is not cooked or pasteurized at the correct temperature for the required time, harmful microorganisms may survive, leading to food-borne illnesses. Furthermore, temperature abuse, in case of incorrect storage temperatures during processing and transportation can result in the growth of pathogens or spoilage microorganisms (Deak and Mohácsi-Farkas, 2023). **Packaging and storage**, e.g., contaminants from packaging materials, such as migration of chemicals from packaging to food, can affect food safety (Lebelo et al., 2021). Moreover, **incorrect storage conditions**, such as exposure to light, humidity, or temperature fluctuations, can compromise the safety and quality of the food (Mazzoleni et al., 2023). **Food fraud** in terms of intentional contamination or adulteration of food products for economic gain can pose serious safety risks. This includes the substitution or addition of substances for fraudulent purposes (Kelly et al., 2023). Also, **hygienic design of facilities** must be approved by authorities, as facility design and poorly designed processing facilities may harbor pathogens or be difficult to clean, thus increasing the risk of contamination (Holah, 2023b).

Especially, the categories of *fermented foods* or *alternatively processed foods* should be evaluated, in comparison with fresh foods (Voidarou et al., 2020). Cases of **microbial food-borne infection** have been reported in association with fresh cheese, sausages, fermented fish and fermented cereals (Ashaolu and Reale, 2020). The development of *functional foods containing probiotics* has experienced a great boom in the last years, reflected by the increasing number of novel products available at the market (Balthazar et al., 2022). Another risk involves **microbial food intoxications** due to mycotoxin contaminated raw materials, production of bacterial toxins or possible mycotoxin production by fungal inoculants (Agriopoulou et al., 2020; Smaoui et al., 2023). In addition, **toxic by-products** of fermentation may be produced, including ethyl carbamate and biogenic amines (Wang et al., 2023). From a food processing point of view, major risk enhancing factors are **the use of contaminated raw materials, lack of pasteurization, and use of poorly controlled natural fermentations** (Skowron et al., 2022).

It has to be emphasized that **heat-inducing reactions** during processing of food can generate several toxic compounds. For instance, acrylamide (AA) (responsible for multiorgan toxicity, *i.e.*, antifertility, carcinogenicity, neurotoxicity and teratogenicity) is largely derived from the heat-inducing reactions between the amino group of the amino acid asparagine and carbonyl groups of glucose and fructose in plant-derived foods including cereals, coffees, almonds, olives, potatoes, and sweet potatoes (Friedman, 2015).

Moreover, **canning**, as a preservation technique, is widely used to extend the shelf life as well as to maintain the quality of perishable foods. During the canning process, most of the microorganisms are killed, reducing their impact on food quality and safety. However, the presence of a range of undesirable chemical contaminants has been reported in canned foods and in relation to the canning process. Many chemical contaminants, including metals, polymeric contaminants and biogenic amine contaminants. They have various origins, including **migration from the can materials, formation during the canning process, or contamination during steps required prior to canning** (e.g., the disinfection step). Some other *can-packaged foods* (e.g. beverages or milk powder), which are not *canned foods* by definition, are also affected by contamination (Martínez and Carballo, 2021; Zheng et al., 2023).

To mitigate these risks, food processors and manufacturers should implement robust food safety management systems, including Good Manufacturing Practices (GMP), Hazard Analysis and Critical Control Points (HACCP), and regular monitoring and testing procedures. Effective measures, such as stringent hygiene practices, regular equipment maintenance, and thorough monitoring, are essential to minimize contamination and ensure the integrity of the food supply. Additionally, ongoing training and education for personnel are crucial to maintaining a strong food safety culture throughout the entire production process (Awuchi, 2023; Overbosch and Blanchard, 2023).

1.3 BOOK GOALS

The content of this book, "Food toxicology and safety: food manufacturing related effects" is of scientific interest for food chemistry, food toxicology and safety, as well as food engineering, but also has the potential to benefit human health. It is a comprehensive resource for understanding the complex interplay between food manufacturing processes and the safety of food. Future Trends in food safety are presented, such as emerging technologies and methods for enhancing food quality. The book goes through food safety of particular categories of contaminants, whose presence is related to food processing and manufacturing related effects. Those contaminants fell into the categories of endocrine disruptors, heavy metals, AA, hydroxymethylfurfural (HMF), polycyclic aromatic hydrocarbons (PAHs), N-nitroso compounds, aromatic amines, amides and nitrosamines. Moreover, dioxins or dioxin-like compounds, furans, 4(5)-methylimidazole, aromatic nitro compounds, trans fats, food additives, mycotoxins and agrochemicals are the main contaminants of interest in this book. Lastly, regulatory issues in terms of food processing and manufacturing related effects and food toxicology are also highlighted. Regulatory compliance may refer to adherence to standards, thus failure to comply with food safety regulations and industry standards can result in unsafe products (Overbosch and Blanchard, 2023). Therefore, the book closes with an overview of food safety regulations and standards whose use and application are crucial for food safety.

1.4 SUSTAINABILITY ASPECTS AND RESEARCH NEEDED IN THE FOOD-PROCESSING INDUSTRY

Sustainability in the food-processing industry, particularly concerning **sources of ingredients** and **contamination of food sources**, is a complex issue that requires careful consideration of environmental, economic, and health impacts. The sustainability aspects focus on ensuring that ingredients are sourced responsibly and that contamination is minimized to protect both the environment and public health (Onyeaka et al., 2024).

The **sourcing of ingredients** is an important step, starting from selecting suppliers and raw materials that meet environmental and social sustainability criteria. Development of certification schemes and standards that ensure sustainable sourcing (e.g., organic, fair trade, Rainforest Alliance) is important to take place (Meemken et al., 2021; Tallontire, 2023).

Assessment of the environmental impact of different agricultural practices, including deforestation, biodiversity loss, and soil degradation. In this context, exploration of regenerative agriculture practices that restore soil health and biodiversity while providing raw materials for food processing must be ensured (Kumar et al., 2022). Moreover, monoculture farming, where a single crop is grown extensively, can lead to soil degradation, loss of biodiversity, and increased pesticide use. Therefore, investigation of the long-term impacts of monoculture on soil health and ecosystem services is needed. In addition, exploring crop rotation, intercropping, and polyculture systems as sustainable alternatives can be studied, as well as development of resilient crop varieties that require fewer chemical inputs and are less susceptible to pests and diseases (Adedibu, 2023).

Furthermore, ensuring that the sourcing of ingredients is transparent and traceable is crucial for sustainability, as it allows companies to verify the environmental and social practices of their suppliers. The new aspect called blockchain and other digital technologies to enhance supply chain traceability may be implemented, as presented in Figure 1.2. Development of tools for evaluating the sustainability of supply chains, including carbon footprint, water usage, and labor conditions are important in this era (Mukherjee et al., 2022; Park and Li, 2021).

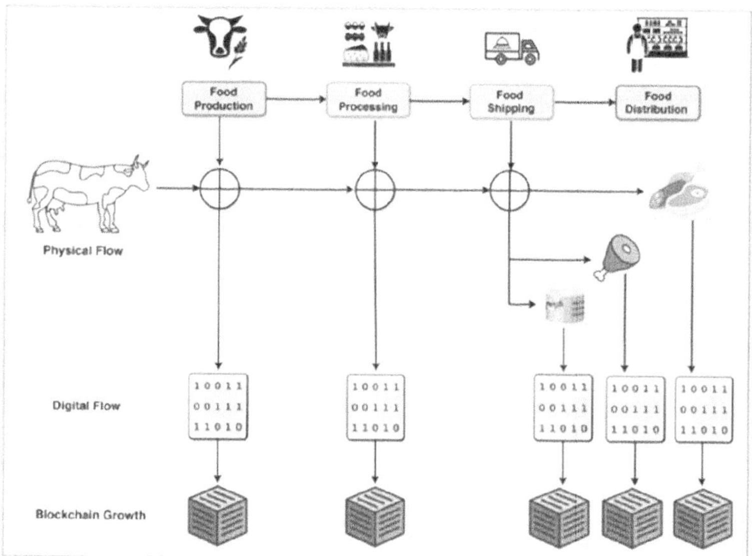

FIGURE 1.2 Blockchain model to enhance supply chain traceability (Torky and Hassanein, 2020).

In addition, climate change poses significant risks to the availability and quality of food ingredients, affecting crop yields, altering growing seasons, and increasing the prevalence of pests and diseases (Gomez-Zavaglia et al., 2020). Studying the resilience of different crops and ingredients to climate change and identifying those most at risk is of great importance. Development of climate-resilient crop varieties through breeding and biotechnology seems to be very useful. Exploration of alternative sourcing strategies and diversification of ingredient sources to mitigate the risks posed by climate change is also a valuable solution for future development (Saini, et al., 2020).

Advanced detection methods like use of (portable) biosensors, etc., for microbial and other contaminants in raw materials and processed foods should be considered (Eyvazi et al., 2021; Rodriguez et al., 2020). Furthermore, study of the cumulative effects of low-level exposure to heavy metals and toxins in food must be counted in risk assessment-related research (Acharya, 2024). Exploration of biodegradable and non-toxic packaging alternatives that reduce the risk of contamination must take place (Bose et al., 2023; Chaudhary et al., 2022). See Figure 1.3.

Exploring land management practices that reduce contamination risks, such as agroforestry, organic farming, and conservation agriculture should be addressed (Adedibu, 2023). Also, comprehensive risk assessment models to predict and prevent contamination events, including the combined effects of multiple contaminants must be applied. By addressing these sustainability aspects and focusing on targeted research, the food-processing industry can improve the sustainability of its ingredient sourcing and reduce the risk of contamination, ultimately leading to safer, more sustainable food systems before manufacturing and processing of food.

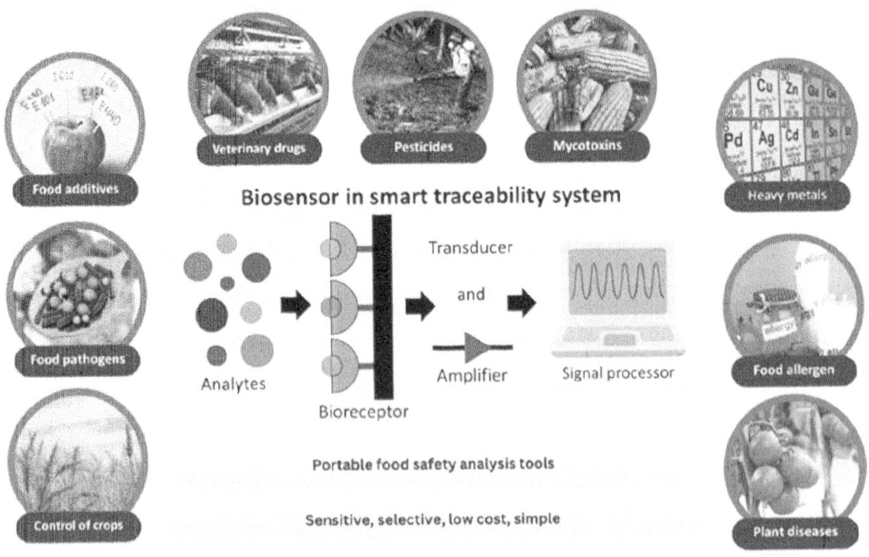

FIGURE 1.3 Portable biosensor system in food safety analysis (Meliana et al., 2024).

REFERENCES

Abdolshahi, A. and Shokrollahi Yancheshmeh, B., 2020. Food contamination. In: S. Sabuncuoglu (ed), *Mycotoxins and Food Safety* (pp. 5–16). Intech Open.

Acharya, S., 2024. Heavy metal contamination in food: Sources, impact, and remedy. In: Ogwu, M.C., Izah, S.C., Ntuli, N.R. (eds), *Food Safety and Quality in the Global South* (pp. 233–261). Springer Nature Singapore.

Adedibu, P.A., 2023. Ecological problems of agriculture: Impacts and sustainable solutions. *ScienceOpen Preprints*.

Agriopoulou, S., Stamatelopoulou, E. and Varzakas, T., 2020. Advances in occurrence, importance, and mycotoxin control strategies: Prevention and detoxification in foods. *Foods*, *9*(2), p. 137.

Ashaolu, T.J. and Reale, A., 2020. A holistic review on Euro-Asian lactic acid bacteria fermented cereals and vegetables. *Microorganisms*, *8*(8), p. 1176.

Awuchi, C.G., 2023. HACCP, quality, and food safety management in food and agricultural systems. *Cogent Food & Agriculture*, *9*(1), p. 2176280.

Balthazar, C.F., Guimarães, J.F., Coutinho, N.M., Pimentel, T.C., Ranadheera, C.S., Santillo, A., Albenzio, M., Cruz, A.G. and Sant'Ana, A.S., 2022. The future of functional food: Emerging technologies application on prebiotics, probiotics and postbiotics. *Comprehensive Reviews in Food Science and Food Safety*, *21*(3), pp. 2560–2586.

Bose, I., Roy, S., Pandey, V.K. and Singh, R., 2023. A comprehensive review on significance and advancements of antimicrobial agents in biodegradable food packaging. *Antibiotics*, *12*(6), p. 968.

Chaudhary, V., Punia Bangar, S., Thakur, N. and Trif, M., 2022. Recent advancements in smart biogenic packaging: Reshaping the future of the food packaging industry. *Polymers*, *14*(4), p. 829.

Deak, T. and Mohácsi-Farkas, C., 2023. Thermal treatment. In Food safety management (pp. 405–419). Academic Press.

Eyvazi, S., Baradaran, B., Mokhtarzadeh, A. and de la Guardia, M., 2021. Recent advances on development of portable biosensors for monitoring of biological contaminants in foods. *Trends in Food Science & Technology*, *114*, pp. 712–721.

Friedman, M., 2015. Acrylamide: Inhibition of formation in processed food and mitigation of toxicity in cells, animals, and humans. *Food & Function*, *6*(6), pp. 1752–1772.

Fuhrman, J., 2018. The hidden dangers of fast and processed food. *American Journal of Lifestyle Medicine*, *12*(5), pp. 375–381.

Gomez-Zavaglia, A., Mejuto, J.C. and Simal-Gandara, J., 2020. Mitigation of emerging implications of climate change on food production systems. *Food Research International*, *134*, p. 109256.

Holah, J., 2023a. Minimum hygienic design requirements for food processing factories. In Holah, J., Huub, L.M.L., Moerman, F. (eds), *Hygienic Design of Food Factories* (pp. 221–240). Woodhead Publishing.

Holah, J., 2023b. Principles of hygienic practice in food processing and manufacturing. In Veslemøy, A., Huub L., Motarjemi, Y. (eds), *Food Safety Management* (pp. 587–613). Academic Press.

Kelly, S.D., 2023. Emerging contaminants in the context of food fraud. In Knowles, M. E., Lucia E. A., Boobis, A.R., Popping, B. (eds), *Present Knowledge in Food Safety* (pp. 315–319). Academic Press.

Kotsanopoulos, K.V. and Arvanitoyannis, I.S., 2017. The role of auditing, food safety, and food quality standards in the food industry: A review. *Comprehensive Reviews in Food Science and Food Safety*, *16*(5), pp. 760–775.

Kumar, G.S., Kulkarni, M. and Rathi, N., 2022. Evolving food choices among the urban Indian middle-class: A qualitative study. *Frontiers in Nutrition*, *9*, p. 844413.

Kumar, R., Kumar, A. and Saikia, P., 2022. Deforestation and forests degradation impacts on the environment. In Singh, V.P., Yadav, S., Yadav, K.K., Yadava, R.N. (eds), *Environmental Degradation: Challenges and Strategies for Mitigation* (pp. 19–46). Springer International Publishing.

Lebelo, K., Malebo, N., Mochane, M.J. and Masinde, M., 2021. Chemical contamination pathways and the food safety implications along the various stages of food production: A review. *International Journal of Environmental Research and Public Health*, *18*(11), p. 5795.

Lindeman, B., Johansson, Y., Andreassen, M., Husøy, T., Dirven, H., Hofer, T., Knutsen, H.K., Caspersen, I.H., Vejrup, K., Paulsen, R.E. and Alexander, J., 2021. Does the food processing contaminant acrylamide cause developmental neurotoxicity? A review and identification of knowledge gaps. *Reproductive Toxicology*, *101*, pp. 93–114.

Martínez, S. and Carballo, J., 2021. Physicochemical, sensory and nutritional properties of foods affected by processing and storage. *Foods*, *10*(12), p. 2970.

Mazzoleni, S., Magni, S., Tretola, M., Luciano, A., Ferrari, L., Bernardi, C.E.M., Lin, P., Ottoboni, M., Binelli, A. and Pinotti, L., 2023. Packaging contaminants in former food products: Using Fourier Transform Infrared Spectroscopy to identify the remnants and the associated risks. *Journal of Hazardous Materials*, *448*, p. 130888.

Meemken, E.M., Barrett, C.B., Michelson, H.C., Qaim, M., Reardon, T. and Sellare, J., 2021. Sustainability standards in global agrifood supply chains. *Nature Food*, *2*(10), pp. 758–765.

Meliana, C., Liu, J., Show, P.L. and Low, S.S., 2024. Biosensor in smart food traceability system for food safety and security. *Bioengineered*, *15*(1), p. 2310908.

Mukherjee, A.A., Singh, R.K., Mishra, R. and Bag, S., 2022. Application of blockchain technology for sustainability development in agricultural supply chain: Justification framework. *Operations Management Research*, *15*(1), pp. 46–61.

Onyeaka, H., Ghosh, S., Obileke, K., Miri, T., Odeyemi, O.A., Nwaiwu, O. and Tamasiga, P., 2024. Preventing chemical contaminants in food: Challenges and prospects for safe and sustainable food production. *Food Control*, *155*, p. 110040.

Overbosch, P. and Blanchard, S., 2023. Principles and systems for quality and food safety management. In Veslemøy, A., Huub L., Motarjemi, Y. (eds), *Food Safety Management* (pp. 497–512). Academic Press.

Park, A. and Li, H., 2021. The effect of blockchain technology on supply chain sustainability performances. *Sustainability*, *13*(4), p. 1726.

Rob, V.O.S. and Cattaneo, A., 2021. Poverty reduction through the development of inclusive food value chains. *Journal of Integrative Agriculture*, *20*(4), pp. 964–978.

Rodriguez, R.S., O'Keefe, T.L., Froehlich, C., Lewis, R.E., Sheldon, T.R. and Haynes, C.L., 2020. Sensing food contaminants: advances in analytical methods and techniques. *Analytical Chemistry*, *93*(1), pp. 23–40.

Saini, P., Saini, P., Kaur, J.J., Francies, R.M., Gani, M., Rajendra, A.A., Negi, N., Jagtap, A., Kadam, A., Singh, C. and Chauhan, S.S., 2020. Molecular approaches for harvesting natural diversity for crop improvement. *Rediscovery of Genetic and Genomic Resources for Future Food Security*, pp. 67–169.

Sridhar, A., Ponnuchamy, M., Kumar, P.S. and Kapoor, A., 2021. Food preservation techniques and nanotechnology for increased shelf life of fruits, vegetables, beverages and spices: A review. *Environmental Chemistry Letters*, *19*, pp. 1715–1735.

Skowron, K., Budzyńska, A., Grudlewska-Buda, K., Wiktorczyk-Kapischke, N., Andrzejewska, M., Wałecka-Zacharska, E. and Gospodarek-Komkowska, E., 2022. Two faces of fermented foods—The benefits and threats of its consumption. *Frontiers in Microbiology*, *13*, p. 845166.

Smaoui, S., D'Amore, T., Tarapoulouzi, M., Agriopoulou, S. and Varzakas, T., 2023. Aflatoxins contamination in feed commodities: From occurrence and toxicity to recent advances in analytical methods and detoxification. *Microorganisms*, *11*(10), p. 2614.

Tallontire, A., 2023. Sustainability standards and certification for agriculture: An overview. In: Manning, L., (ed), Improving standards and certification in agri-food supply chains: Ensuring safety, sustainability and social responsibility. Burleigh Dodds. ISBN 9781801464512.

Torky, M. and Hassanein, A.E., 2020. Integrating blockchain and the internet of things in precision agriculture: Analysis, opportunities, and challenges. *Computers and Electronics in Agriculture*, *178*, p. 105476.

Voidarou, C., Antoniadou, M., Rozos, G., Tzora, A., Skoufos, I., Varzakas, T., Lagiou, A. and Bezirtzoglou, E., 2020. Fermentative foods: Microbiology, biochemistry, potential human health benefits and public health issues. *Foods*, *10*(1), p. 69.

Wang, Y., Liu, Y., Huang, X., Xiao, Z., Yang, Y., Yu, Q., Chen, S., He, L.I., Liu, A., Liu, S. and Zou, L., 2023. A review on mechanistic overview on the formation of toxic substances during the traditional fermented food processing. *Food Reviews International*, *39*(3), pp. 1275–1292.

Zheng, J., Tian, L. and Bayen, S., 2023. Chemical contaminants in canned food and can-packaged food: A review. *Critical Reviews in Food Science and Nutrition*, *63*(16), pp. 2687–2718.

2 Endocrine Disruptors and Safety

Food Processing and Manufacturing Related Effects

Sofiane Boudalia, Asma Lahouel, Malha Oudir, Safia Habila, Sana Benosmane, and Zouheyr Hadri

2.1 INTRODUCTION

Endocrine disruptors in food processing and manufacturing are a major concern due to their potential adverse effects on human health (C. Li et al., 2021). These contaminants can enter the food supply through various sources, such as polluted soil, water, and air, or during food processing. Studies have indicated that dietary exposure to endocrine disruptors poses serious human health risks (Boudalia et al., 2021) (Zhu et al., 2021). Moreover, many endocrine-disrupting chemicals tend to bioaccumulate, meaning they concentrate more in the tissues of organisms than in the surrounding environment. Furthermore, these chemicals can biomagnify, causing their concentrations to increase progressively in organisms higher up the food chain (Bertram et al., 2022).

Food products can become contaminated at multiple stages, including transportation (due to vehicle exhaust emissions or cross-contamination with other transported items), processing (from the use of disinfectants and cleaning agents), production (through contaminated water or heating processes that create acrylamides (AAs) and

 DOI: 10.1201/9781032712291-2

nitrosamines), and storage (from exposure to naphthalenes) (Rather et al., 2017). As a result, food processing techniques, such as boiling, steaming, and microwave-cooking can significantly impact the levels of endocrine disruptors in food products (de Paula et al., 2024) (Sundhar et al., 2023). For example, contamination occurs when endocrine disruptors migrate from packaging materials into food, particularly during heating, prolonged contact, repeated use of the packaging, or when the food is fatty or acidic (Muncke, 2021).

Several techniques can be employed to prevent, eliminate, or reduce contamination with endocrine disruptors in food. These methods include cleaning, peeling, stewing, using chemical additives, and incorporating natural plant ingredients such as polyphenols, which can help remove nitrites from food. Additionally, adsorption separation technology can also be utilized (C. Li et al., 2021) (Domingo, 2010). Research on food safety has underscored the significance of understanding how different processing methods impact the safety of food products (Olmedo et al., 2024). In summary, the reviewed literature highlights the crucial need to address endocrine disruptors' contamination in food processing and manufacturing to protect consumer health.

2.2 DEFINITION AND CONCEPTS

2.2.1 Endocrine Disruptor Definition

According to the literature, endocrine-disrupting (EDs) can be either natural, such as phytoestrogens and mycotoxins, or synthetic, including atmospheric pollutants, pesticides, detergents, plastic derivatives, varnishes, synthetic hormones, pharmaceuticals, and veterinary products (Darbre, 2019). These molecules exhibit distinct hormonal properties; for example, some can be estrogenic, like genistein and bisphenol A (Iwamoto et al., 2021), while others are anti-androgenic, such as vinclozolin (Boudalia et al., 2017).

In 1999, the European Commission (C.C. E, 1999) proposed the following definition: "An endocrine disruptor (ED) is an exogenous substance or mixture, altering the functions of the endocrine system and thereby inducing adverse health effects in an intact organism, its descendants or subpopulations". This definition has been expanded with the addition of the following definitions:

> A known endocrine disruptor is an exogenous substance or mixture that alters the function of the endocrine system and thus induces adverse effects on the health of an intact organism, descendants or a population.
>
> (WHO-UNEP, 2012)

In 2016, the European Commission defined endocrine disruptors as exogenous substances or mixtures that interfere with the functions of the endocrine system, leading to adverse health effects in an intact organism, its progeny, or specific subpopulations (Solecki et al., 2017). Over the past decade, the number of both natural and synthetic molecules with endocrine-disrupting properties has increased. In 2002, the European Commission published a report identifying 435 endocrine disruptors (C.C.E., 2002). As of today, TEDX (The Endocrine Disruption Exchange) lists 1,482 potential endocrine-disrupting chemicals (TEDX).

It should be noted that, for a comprehensive risk assessment of dietary endocrine disruptors, several critical factors must be considered. These include the potential for "cocktail effects" from exposure to multiple chemicals simultaneously, the relevance of doses to human exposure levels, and the timing and duration of exposure, whether *in utero*, postnatal, or across a lifetime. Additionally, the mode of exposure, such as dermal, oral, or inhalation, must be evaluated, along with the possibility of transgenerational effects, where adverse outcomes extend across multiple generations following a single gestational exposure. Finally, epigenetic mechanisms, which involve heritable but reversible changes in gene expression that do not alter the DNA sequence, should also be taken into account (Boudalia, 2017).

2.2.2 Endocrine Disruptors in Processed Foods and Sources of Exposure

In the recent years, dietary patterns have changed in most societies, as the fresh or minimally processed foods have been replaced by processed and ultra-processed foods. It is estimated that the processed foods contribute from 20% to 60% of total energy intake worldwide. Consumption is the highest in high-income countries but it is increasing in middle-income countries (Marino et al., 2021). The consumption of processed foods has been associated with obesity and metabolic disorders in both adults and children. The pediatric population is particularly vulnerable to these effects because exposure occurs during critical windows of development (Nardocci et al., 2021) (Khandpur et al., 2020). The harmful effects of processed foods are partly due to their high levels of sugar, fat, and salt, and low levels of proteins, fibers, and micronutrients. Additionally, they contain high levels of various molecules considered EDs (Tobias et al., 2021). Raw food materials may contain endocrine disruptors, such as mycotoxins, heavy metals and pesticides, as well as natural compounds like phytoestrogens, leading to their potential accumulation in the final products (C. Li et al., 2021). Furthermore, during processing, packaging, and storage, other molecules may be added to foods either intentionally, such as food additives, or unintentionally, mainly due to interactions with processing machines and packaging. In many cases, these molecules are considered disruptors and can negatively affect consumer health (de Paula et al., 2024).

2.2.2.1 Endocrine disruptor levels in food raw materials

✓ **Mycotoxins**

Including aflatoxins, ochratoxin A, fumonisins, trichothecenes, zearalenone, and patulin, are toxic compounds produced by fungal pathogens that represent a serious threat to food safety and consumer health (Khan et al., 2024). These toxins can contaminate a wide range of food crops globally, such as cereals (wheat, rice, corn, barley), fruits, lentils, almonds, peanuts, pistachios, walnuts, coffee, seeds, cotton, spices (paprika, pepper, ginger), and meat (Pleadin et al., 2019) (Rodrigues et al., 2012). In China, Ji et al. (2024) found that around 65.1% (69 out of 106) of the raw materials were contaminated with mycotoxins, with average levels between 3.66 and 321.8 µg/kg. Deoxynivalenol (DON) and tenuazonic acid (TeA) were the most common mycotoxins identified, based on their higher frequencies and concentrations in the samples. Additionally, more than two mycotoxins co-occurred in 53.8% (57 out of 106) of the raw materials. Wheat flour showed higher contamination levels and a wider variety of mycotoxins compared to other samples. In Namibia, cereal grains have been stored traditionally in silos for more than 12 months under poorly ventilated conditions. Analysis identified 30 mycotoxins, with around 42% of the samples contaminated by aflatoxins. Furthermore, 14% and 9% of samples surpassed the EU limits for AFB1 (2 µg/kg) and total aflatoxins (4 µg/kg), respectively. Over 50% of the samples contained five emerging mycotoxins: moniliformin, alternariolmethylether, tenuazonic acid, 3-nitropropionic acid (3-NPA), and enniatin B (Angula et al., 2024), which can affect consumers. Additionally, mycotoxins can contaminate human milk, with a median aflatoxin M1 (AFM1) level of 3.03 pg/mL found in breastmilk samples from Turkey (Memiş et al., 2021). They can also contaminate other animal-derived foods, such as raw cow milk (Tarannum et al., 2020), meat samples from Egypt (Abd-Elghany et al., 2015), and eggs (Herzallah, 2013) reaching the human food chain and also causing impairment of the health status of consumers.

✓ **Phytoestrogens**

Phytoestrogens are a diverse group of naturally occurring bioactive compounds found in approximately 300 plants from 16 different plant families. These compounds are structurally and functionally similar to mammalian estrogens. Among them, specific phenolics, terpenoids, and saponins are recognized as natural phytoestrogens, with phenolics being the most extensively studied. Phenolics are further categorized into flavonoids and non-flavonoids. Within the flavonoids, isoflavones like genistein and daidzein, some flavanones (eg., naringenin), flavonols (eg., kaempferol, quercetin), and coumestans are known for their endocrine activity. Non-flavonoids include lignans (eg., secoisolariciresinol, matairesinol, lariciresinol) and stilbenes (eg., resveratrol) (Stiefel et al., 2023) (Canivenc-Lavier et al., 2023). On a molecular and cellular level, phytoestrogens mimic the synthetic endocrine-disrupting compounds (Jedeon et al., 2014) (K. Jedeon et al., 2013) (T. A. Auxietre et al., 2014) (S. Boudalia et al., 2017), which suggests potential negative effects on consumers health (X. Wang et al., 2021) (Torrens-Mas et al., 2020). However, the data remains inconsistent (Patisaul et al., 2010).

Regarding phytoestrogen levels and consumption, earlier studies, including one by A. Lee et al. (2022), have shown that soybean consumption in France has recently increased, resulting in higher isoflavone exposure (genistein and daidzein). Based on 140 measurements across nine categories of soy-containing food items, the estimated isoflavone exposure is 6915 ± 1220 µg/day. Since 2005, isoflavone exposure among French individuals who do not consume soy has risen by a factor of 75. In Japan, older adults consume between 30 and 50 mg/day of isoflavones, while the per capita intake in the United States and Europe is only a few milligrams (Messina et al., 2022).

✓ **Pesticides**
Synthetic pesticides fall into four main categories: organochlorine pesticides, organophosphorus pesticides, carbamate pesticides, and pyrethroid pesticides (Boudebbouz, Boudalia, Boussadia, et al., 2022). Presently, organophosphates, such as malathion, chlorpyrifos, and dichlorvos, are widely used across the globe (Nappi et al., 2016). Pesticides are diverse chemicals capable of interacting with the endocrine system and acting as endocrine disruptors through non-monotonic dose-response relationships (T. A. Auxietre et al., 2014). Due to their lipophilic and somewhat persistent properties, organophosphorus pesticides tend to accumulate in the food chain, mainly in fatty tissues. Exposure to organochlorine pesticides has been associated with metabolic disorders, including obesity and type 2 diabetes, and other pathologies (Nappi et al., 2016). Furthermore, endocrine-disrupting effects have been identified in other pesticide and fungicide classes, such as triazoles, pyrethroids, and neonicotinoids (Kadic et al., 2024) (Andersen et al., 2022) (Suwannarin et al., 2021).

2.2.2.2 Endocrine disruptor levels during food processing

✓ **Food additives**
As defined by the European Union, food additives are "*any substance intentionally added to a food during the process of manufacture, processing, preparation, treatment, packaging, transport or storage for a technological purpose.*" (B. Sun et al., 2017). Food additives can be either synthetic or of natural sources. They are primarily used as sensory enhancers such as flavors, sweeteners and colorants, as conservative agents and stabilizers such as pH regulators and antioxidants; or as nutritive substances, such as vitamins and minerals (Gürtler, 2021).

The use of several food additives is currently under scrutiny, with some already banned due to their demonstrated toxicity in experimental and epidemiological studies. Cyclamate, for example, is an artificial sweetener banned in a few countries, including the USA, due to concerns about carcinogenesis (Roberts, 2016) However, it is still used in soft drinks, baked and processed foods, many desserts, gelatins, canned fruit, and as a tabletop sweetener in most of Europe (Ižaković et al., 2021). It has been associated with effects related to endocrine disruption, such as obesity, insulin resistance, and sterility in both epidemiological and experimental studies (Naik et al., 2023) (Czarnecka et al., 2021). Same effects have been linked to other artificial sweeteners such as saccharine and Sucralose (Hess et al., 2018) (Jiang et al., 2018) (Pałkowska-Goździk et al., 2018).

Food dyes, which are very common food additives in all kinds of processed foods, have been considered for long as neutral substances without health adverse effects. However, recent data keeps revealing their potential toxicity. Allura red, which is one of the most used dyes, is already banned in India and restricted in European countries due its potential carcinogenesis and implication in trigging autoimmune reactions and hyperactivity in children (Ravichandran et al., 2022) (Mishra, 2020). Tartrazine is another dye commonly used as food additive suspected to induce attention deficit hyperactivity disorder (ADHD) in children, among other toxic effects (Amin et al., 2018). It has been reported that tartrazine mimics estrogen effects *in vitro* (Axon et al., 2012) (Nasri et al.,

2021), but not *in vivo* (Meyer et al., 2017). However, a recent study reported that tartrazine altered levels of thyroidal and reproductive hormones in rats.

Nitrate and nitrites occur naturally in water and soil. They are as well food additives heavily used as antimicrobial agents and color enhancers in the chain of meat processing (Y. Zhang et al., 2023). In 2015, the WHO classified the processed meat as "carcinogenic" (Group 1) due high risk of colorectal carcinogenesis (D. Y. Lee et al., 2021). Moreover, nitrates as food additives were positively associated with breast cancer risk and nitrites as food additives were positively associated with prostate cancer risk (Chazelas et al., 2022). There are particularly, current concerns about the endocrine-disrupting effects of nitrates and nitrites as experimental data consistently suggest their interaction with thyroid hormones production and function (Poulsen et al., 2018) (Jannat et al., 2014).

Many other food additives such as butylated hydroxyanisole (BHA) and butylated hydroxytoluene (BHT) which are synthetic antioxidant phenols have been reported to induce endocrine disruption according to the experimental studies (X.-J. Zhang et al., 2023) (Z. Sun et al., 2021) (Anca Pop et al., 2018). Parabens particularly, are a class of preservatives and antimicrobial inhibitors that the scientific community has extensively studied and debated its endocrine-disrupting activity (Vandenberg et al., 2021). Parabens are aliphatic esters of p-hydroxybenzoic acid widely used in food, cosmetics, and pharmaceutical industries. They show an estrogen like activity; however, this activity is debated as weak compared to the endogenous estrogen activity. With the lack of conclusive and firm results supporting endocrine-disrupting effects of parabens, they are still used worldwide and considered safe in the range of the acceptable daily intake (ADI) of 10 mg/kg/day (Vandenberg et al., 2021) (Nowak et al., 2018).

✓ **AA and advanced glycated end products**

AA and advanced glycation end (AGEs) products originate from Maillard reaction (MR). MR is a spontaneous non-enzymatic reaction between the carbonyl groups of sugars and carbohydrates and free amino groups in amino acids under the influence of heat. MR is often desirable for the manufacturing of various food products such as bread, potato-based products, cocoa, and coffee in order to develop appealing color, crust texture, and appetizing aroma (Kathuria et al., 2023). Nevertheless, these foods contain high concentrations of Maillard products. High consumption of processed foods rich in AGEs and AA is raising great concerns as these products are implicated in the pathogenesis of several diseases such as cancer, diabetes, and neurodegenerative diseases (M. I. Khan et al., 2023) (Hosseini-Esfahani et al., 2023). Both AGEs and AA have been described as potential endocrine disruptors as they could interact with the activity of several hormones, including reproductive hormones, thyroidal hormones, and insulin (Matoso et al., 2019) (Ravichandran et al., 2019). They have been reported as well to induce reproductive toxicity in experimental animals (Matoso et al., 2019) (Merhi et al., 2020).

✓ **Heavy metals**

From literature, heavy metals may act as significant EDs, potentially leading to serious health effects. However, the mechanisms behind their endocrine disruption are not yet fully understood (D. Liu et al., 2023) (Iavicoli et al., 2009). We dedicated Chapter 3 to providing a comprehensive overview of critical data on the impact of food processing techniques on the presence of toxic metals.

Briefly, heavy metals, defined as elements with densities above 5 g/cm^3 and atomic weights between 63.5 and 200.6 g/mol, are toxic at low concentrations (Olmedo et al., 2024) (Meng et al., 2023). Essential trace metals like copper (Cu), zinc (Zn), and iron (Fe) can become toxic at abnormal levels, while others, such as arsenic (As), cadmium (Cd), and lead (Pb), are harmful even in small amounts (Malematja et al., 2023) (Rai et al., 2019). These metals are non-biodegradable and bioaccumulate in the environment (Zhuang et al., 2016) (F. Xu et al., 2017). Plant-based foods often become contaminated through polluted soil or irrigation, as observed in China, Morocco, Nigeria, and Bangladesh (Xiang et al., 2022) (Chaoua et al., 2019) (Orosun et al., 2023) (Hassan et al., 2024). Similarly, animal-based foods like fish, meat, and milk can be contaminated through exposure to polluted air, water, or feed, with studies reporting heavy metals in various regions, including Bangladesh and Spain (Aissioui et al., 2021) (Boudebbouz, Boudalia, Bousbia, et al., 2022) (Zergui et al., 2023) (Hamed Hammad Mohammed et al., 2020) (Bousbia et al., 2019) (Das Pinkey et al., 2024), (Forcada et al., 2023).

2.2.2.3 Endocrine disruptor levels during food packaging and storage process

The food packaging industry is one of the most dynamic industries that has evolved dramatically over the years and plays a key role in the modern economy. Currently, the expending demands on processed foods due the expansion of population and the changing in food habits, many materials and packaging products have been introduced to the market but without full consideration to their related risks on the public health (Mukherjee et al., 2021).

Today, it is well proven that several packaging materials contain harmful substances including well recognized EDs such as phthalates, bisphenols, and prefluoralkyls substances (de Paula et al., 2024). These substances are used as additives to ameliorate the physicochemical and mechanical properties of packages, to increase its durability and to protect the organoleptic characteristics of food from possible interactions with packages (Alamri et al., 2021). Chemicals composing food contact materials including packages are not allowed to be added directly to foods. Yet, they have been constantly detected in foods either due primary contamination of foods or contamination from food contact materials during processing or through migration of chemicals from packages to the foods (de Paula et al., 2024). Migration of chemicals from packages to foods either by direct contact or simple diffusion depends on several factors. It depends mainly on food nature, its pH and fat composition, the chemical nature of the migrant, its solubility, and its concentration as well as the temperature of the system, the contact surface ratio and the contact duration. Migration of substances from packages to foods is raising serious concerns as it becomes an important contributor to exposure of the general population to EDs (Ong et al., 2022).

✓ **Phthalate esters**

Phthalate esters are among the most detected EDs in foods and food packages. They are a family of high molecular weight synthetic organic chemicals widely used as polymer catalyst and plasticizers for their ability to increase softness and plasticity of plastic food packages. On the other hand, low molecular weight phthalates are synthesized and used as emulsifiers in pharmaceutical and personal care products (Serrano et al., 2014). Phthalates are not covalently bonded to the polymer consequently they readily migrate from plastic packaging into food products during production, handling, heat treatment, transportation, storage, cooking, and consumption, thus resulting in human exposure through ingestion (Same et al., 2023) (Wittassek et al., 2008) (Wittassek et al., 2011). Yet, phthalate esters and particularly diethylhexyl phthalate (DEHP) and diisononyl phthalate (DINP) are detected ubiquitously in plastic packaged foods, particularly in beverages, bottled water and ready-to- eat and take away foods (Yang et al., 2019) (Sakhi et al., 2014) (Schecter et al., 2013). Actually, phthalates are not attached to the core of plastics, so they can leach easily to the packaged food (Serrano et al., 2014). Moreover, the migration of phthalates increases remarkably in foods rich in fat and with the increase of temperature and time of storage (Yang et al., 2019). The outdoor storage of bottled water is raising concerns as it correlates positively with the concentration of phthalate esters in several studies (X. Xu et al., 2020) (Zaki et al., 2018) (Jeddi et al., 2016). Although these concentrations do not exceed usually the specific migration limit and the maximum contaminant level, the health risk of cumulative chronic exposure could not be excluded (Schecter et al., 2013). In fact, the specific migration limits (SML) for these phthalates have been reduced: SML DEHP from 1.5 mg/kg to 0.6 mg/kg, SML DBP from 0.3 mg/kg to 0.12 mg/kg, SML BBP from 30 mg/kg to 6 mg/kg. The European Commission has also established a new Tolerable Daily Intake (TDI) for Di-n-butyl phthalate (DBP), Benzyl-butyl phthalate (BBP), Di(2-ethylhexyl) phthalate (DEHP), and Di-isononyl-phtalate (DINP) of 50 micrograms per kilogram of bodyweight (μg/kg bw) expressed as DEHP equivalent strength.

In fact, ultra-processed food consumption was associated with increased concentrations of urinary phthalate esters metabolites (Huang et al., 2021) (Rudel et al., 2011). Phthalate esters are well identified endocrine disruptors that have been associated with breast cancer, diabetes, obesity, and heart diseases in epidemiological studies (G. Liu et al., 2021) (Zhu et al., 2021) (Song et al., 2014). High urinary levels of PAEs metabolites have been associated recently with increased all-cause mortality (Beyer et al., 2022) what implies the need of a rigorous reassessment to the health outcomes of phthalate esters exposure.

✓ **Alkylphenols**

Just like phtalates, alkylphenols are widely used as plasticizers and antioxidants in plastic manufacturing. Due to their surfactant properties, alkylphenols are used also as emulsifiers and stabilizers in detergents and pesticides. Nonylphenol (NP), Octylphenol (OP), and Nonylphenol Ethoxylates (NOE) are the most detected alkylphenols in the environment and households (Snedeker, 2014). These specific compounds are well characterized as estrogen active molecules and endocrine disruptors (Acir et al., 2018). Even that they are not allowed in NP, OP, and NOE have been detected consistently in foods, mainly as a result of primary contamination. Moreover, tris(nonylphenyl) phosphite (TNPP) and alkylphenol ethoxylates (APEO) are used as stabilizers and antioxidant in packaging polymers and papers. NP and OP may be present and migrate from packages to foods as impurities or breakdown products of TNPP or APEO (Snedeker,

2014). Polyvinyl Chloride (PVC) cling films and high-impact polystyrene (HIPS) have been reported to contain high concentrations of NP comparing to polycarbonate and vinyl acetate packaging (Fernandes et al., 2008) (Bradley, 2010). Moreover, the migration of NP to foods packaged in PVC films and HIPS containing high concentrations of NP has been demonstrated experimentally even at a short time of storage at 4 days and 20 ℃ temperature (Snedeker, 2014) (Bradley, 2010). Experimental tests revealed as well the potential of NP and OP to migrate to bottled water depending on the type of polymer package (Guart et al., 2011). NP and OP have been reported as well in different kinds of packaged foods and in bottled water along with phthalates and bisphenol A (Baute-Pérez et al., 2022) (Fasano et al., 2015) (Amiridou et al., 2011).

✓ **Bisphenol A**

Bisphenol A (BPA) is a chemical used widely in producing plastics and polymers; it enters in the manufacturing of several products, including plastic food packaging materials. BPA is used as well as a coating agent in metal canes to prevent corrosion and in paper food packaging (de Paula et al., 2024). BPA is a wide spread pollutant in the environments, so it could be a primary contaminant in foods. Moreover, migration of BPA from packages is a major source of contamination. The highest concentrations of BPA were found in packaged products (18.68 μg/kg) when compared with unpackaged foods (1.5 μg/kg) (Vilarinho et al., 2019). All the types of packaging implicating BPA have shown a degree of its migration into foods (Vilarinho et al., 2019). However, it seems that canned foods contain the highest concentrations of BPA compared to other types of packaged foods (González et al., 2020). Several types of canned sea foods have been reported to contain important concentrations of PBA. Somme analyzed samples contained BPA at a concentration that even exceeded the SLM (0.05 mg/kg) specified by the European Union (Lin, 2022). Furthermore, the European Commission has reduced the SML of BPA from 0.6 mg/kg to 0.05 mg/kg for plastic FCMs, varnishes, and coatings. TDI has been also reduced from 40 μg/kg body weight per day to 0.2 ng BPA/kg bw per day (Deng et al., 2024) and EFSA report 2022.

BPA is a confirmed ED because of its estrogenic activity (Katia Jedeon et al., 2013). In animal models BPA showed signs of productive toxicity, neurotoxicity, immunotoxicity, and cancer development (Ma et al., 2019) (S. Boudalia et al., 2014) while exposure to PBA has been associated to diabetes and insulin resistance obesity, and hypertension (Boudalia et al., 2021) (Rancière et al., 2015). The most serious BPA effects have been manifested as neurodevelopmental alterations in children of mothers exposed to this agent during pregnancy and breastfeeding (X. Wang et al., 2023). The migration of BPA to infant food formulations packaged in canes with BPA basing epoxy resin coating has been identified as a major source of infant exposure to BPA. In 2012, FDA has amended its regulation to no longer provide for the use of BPA based materials in baby bottles, sippy cups, and infant formula packaging (Thoene et al., 2018). However, the use of BPA in infant's formula packaging is still common in developing countries (S Boudalia et al., 2016).

✓ **Per and polyfluoroalkyl substances**

Per and polyfluoroalkyl substances (PFAS) are a large group of fully or partially fluoroated hydrophobic alkyl chains of variable size with a final hydrophilic group

(Ramírez Carnero et al., 2021). PFAS are used as surfactants in food packaging, and in food processing for their non-stick and grease, oil, and water-resistant properties (Heo et al., 2014). Indeed, PFAS concentrations ranging from 3 to 750 ng/g were reported in various food contact materials, including fast-food packaging, popcorn bags, paper and cardboard bags, wrapping packages, pizza boxes, ice cream tubs, pastry trays, cups, cream containers, sandwiches, muffin packaging, baking paper, plates, and even in eco-friendly food packaging (Miralles et al., 2023) (Scholes et al., 2024) (Simonetti et al., 2024). PFAS were also detected in the virgin plastic materials at 3.84% (Tumu et al., 2024). These PFAS can migrate from the food contact materials into food products and beverages, during preparation, handling, transportation, storage, cooking, and consumption, resulting in human exposure through ingestion (Ramírez Carnero et al., 2021) (Scholes et al., 2024) (Simonetti et al., 2024). The migration values were found to be between 0.86 and 22.3 μg/kg in paper and cardboard (Vera et al., 2024). In parallel, PFAS were detected in several food products, notably food of animal origin including seafood, freshwater fish, salmon, dairy products, eggs, and meat, at concentrations above tolerable weekly intake (TWI) of 4.4 ng/kg body weight established by the European Food Safety Authority in 2020 (European Union, 2022). Furthermore, PFAS were also found in instant rice at concentrations ranging from 0.021 to 0.253 ng/g (S. H. Hwang et al., 2024), in ready-to-eat food at concentrations varied between 0.014 and 0.391 ng/g, in frozen vegetables at concentrations ranging from 0.009 to 0.166 ng/g and in fresh vegetables at 0.007 to 0.296 ng/g (Piva et al., 2023). PFAS were even detected in drinking water at concentrations ranging from 2.3 to 7135 ng/L (Pelch et al., 2023).

The main concern about PFAS besides their toxicity is their persistence and resistance to degradation in the environment. Thus, both PFOS and PFOA are no longer used in food packaging. However, due their high persistency in the environment, they are still constantly detected at high concentrations in different food products, including processed foods such as frozen meat and sea foods (Authority, 2020) (Jeong et al., 2019).

Short chain PFAS ($C \leq 7$) have been proposed as alternatives to long chain PFAS as these molecules showed less toxicity and less persistency in the environment. However, the short chain PFAS tend to migrate from packages to foods more than long chain PFAS (Dueñas-Mas et al., 2023) (Zabaleta et al., 2020). Moreover, some of the short chain PFAS and their products of degradation have shown the same persistence and toxic effects, including endocrine disruption as long chain PFAS (Brendel et al., 2018). Recycled packages particularly contain high concentrations of these residues since they remain in the fibers of the recycled papers and plastics. In several processed food samples, bonzophenoes have been detected at concentrations that exceeded the European specific migration limit set at 0.6 mg/kg (Snedeker, 2014) (Bugey et al., 2013) (Jung et al., 2013) which could be worrisome to the health of the population.

✓ **Butylated hydroxyanisole (BHA) and butylated hydroxytoluene (BHT)**

Butylated hydroxyanisole (BHA) and butylated hydroxytoluene (BHT) are among the most widely used synthetic phenolic antioxidants in the food industry as food additives to increase shelf life and improve the quality, freshness, taste, and texture of consumer products (Galal et al., 2024) (Felter et al., 2021). These two compounds are similar in nature and use but differ in their endocrine disruption potential; BHA is being evaluated as an endocrine disruptor, while BHT is considered non-carcinogenic, and there are too many uncertainties to assess its effects on health and the environment. Existing *in vitro* studies indicate that BHA has a weak estrogenic effect as well as anti-androgenic properties, while an *in vivo* study revealed it to have anti-estrogenic properties (A. Pop et al., 2013) (Ousji et al., 2020). Despite the large number of scientific articles describing the controversial effects of BHT and its analogues using *in vitro* and *in vivo* animal models, very few studies have described these synthetic antioxidants in humans (Zhang et al., 2018) (W. Wang et al., 2019). Regarding the endocrine-disrupting effect, there is currently not enough data available to draw a conclusion about the safety of BHA and BHT. See Figure 2.1 and Table 2.1.

2.3 EFFECTS OF FOOD PROCESSING TECHNIQUES ON ENDOCRINE DISRUPTORS CONCENTRATIONS

Food processing techniques can significantly affect the concentration of EDs in the final product. These techniques can either reduce or increase the concentration of EDs, depending on the type of process involved, such as heating, fermentation, or packaging.

✓ **Washing, soaking**

From previous studies, traditional cooking methods have proven effective in reducing isoflavone content. For instance, water soaking reduced isoflavone levels in soy-textured proteins by over 70%. Compared to commercial foods, isoflavone content was found to be 20, 2.6, 4.5, and 9.8 times lower in "homemade" soy juice, tofu, tempeh, and miso, respectively (Bensaada et al., 2024), which is related to the water solubility of glycosylated isoflavones (Fernandez-Lopez et al., 2016).

Some pesticides are considered as endocrine disruptors (T.-A. Auxietre et al., 2014) (Boudebbouz, Boudalia, Boussadia, et al., 2022), and various processing methods can influence their concentrations in foods. For instance, washing and soaking have been shown to reduce pesticide levels and promote their degradation. For example, washing okra in tap water for 5 minutes can reduce chlorfenapyr levels by 90% and acetamiprid levels by 48% (Hanafi et al., 2016).

✓ **Fermentation**

Fermented foods are produced through desired microbial growth and enzymatic conversion of food components to enhance taste and nutritional quality (Leksir et al., 2019). However, poorly controlled fermentation can result in contamination by mycotoxigenic fungi and mycotoxins (Tian et al., 2022). From previous studies, it has been reported that traditional fermented soybean products and cheese present a food

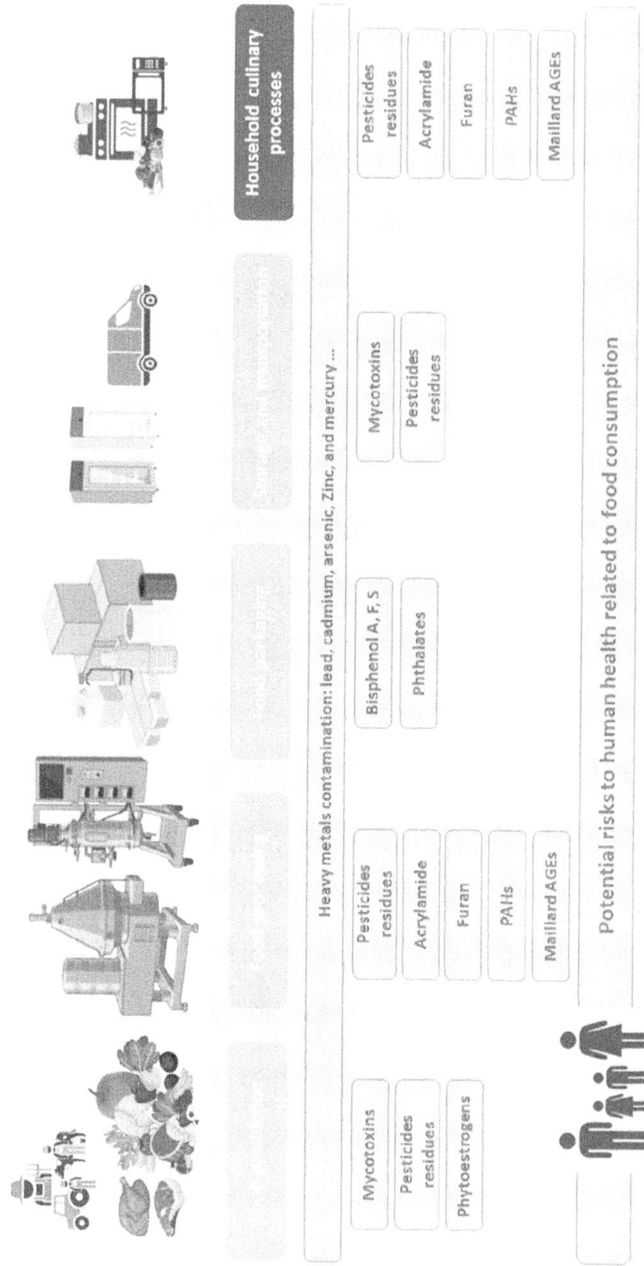

FIGURE 2.1 EDs during the food production process. (Photograph by the author.)

TABLE 2.1 Chemical names, applications, and documented effects of specific compounds with potential endocrine activity

NAMES	*SOURCE/ APPLICATION*	*EFFECTS ON HUMAN HEALTH*	*REFERENCE*
Raw material			
Mycotoxin	Produced by certain species of molds (microscopic fungi)	Liver cancer, nephropathy, fertility problem, neurotoxic, Immune suppressive	Balwan et al. (2023)
Phytoestrogens	Raw material	Uterine fibroids	Sridevi et al. (2021)
Pesticides	Raw material	Neurological effects, cancer, reproduction effects, respiratory problems, diabetes effects, fetal death, allergic reaction, infertility, hypospadias, birth effects	M. F. Ahmad et al. (2024)
Food processing			
Heavy metals	Raw material, food processing, packaging, and storage and transportation	Neurotoxicity, nephrotoxicity, cancer, cardiovascular toxicity, hepato toxicity, immuno toxicity, skin toxicity, repro toxicity	Mitra et al. (2022)
Food additives	Food processing, packaging, and storage and transportation	Cardiac arrhythmias, headache, nausea, vomiting, weakness, shortness of breath, diarrhea, fever, changes in skin, hair and nails, and birth defects, gut microbiota	Cao et al. (2020) Mary (2023)
Acrylamide and Advanced glycated end products	Maillard reaction	Age-related illnesses, diabetes mellitus, and autoimmune/ inflammatory rheumatic diseases, systemic sclerosis, and psoriasis	M. I. Khan et al. (2023)

TABLE 2.1 (Continued)

NAMES	*SOURCE/ APPLICATION*	*EFFECTS ON HUMAN HEALTH*	*REFERENCE*
Food packaging			
Phthalates	Plastic packaging	Disruption of androgenic biosynthesis Variation in the androgenic index Increased LH/ testosterone ratio overweight, obesity, asthma. Pregnancy loss, preterm birth and breast cancer, affect endocrine and respiratory system for children	Arrigo et al. (2023) Y. Wang et al. (2021)
Bisphenol A and analogs	Plastic packaging, baby bottles, sippy cups, and infant formula packaging Coatings of some paper-based products	Endocrine disruption (neuro-, repro-, and metabo-disruption)	S. Boudalia et al. (2014), (2021) Andújar et al. (2019)

safety risk due to the presence of high levels of mycotoxins (Chin et al., 2024) (Izzo et al., 2022).

Additionally, fermentation can influence the levels of phytoestrogens in processed foods. For instance, it has been noted that fermented soymilk contains fewer total isoflavones compared to its unfermented counterpart. During fermentation, the amounts of glucoside, malonylglucoside, and acetylglucoside isoflavones are greatly reduced, while the content of aglycone isoflavones-which are the most potent form of phytoestrogens-significantly increases (C. E. Hwang et al., 2016) (Chien et al., 2006).

Microbial activity during fermentation can degrade certain EDs, such as DBP, thereby reducing their concentration. However, this effect is dependent on the specific microorganisms involved, such as *Lactobacillus plantarum*, *Leuconostoc mesenteroides*, and *Lactobacillus acidophilus*, as well as the fermentation conditions (L. Zhao et al., 2021).

✓ **Heat treatment**

During food processing, heat treatment can affect the level of food contamination, for example, the high temperature generated by mechanical processing of cereals, or during dry heating (baking, roasting, frying, (pressure) cooking, and extrusion cooking), might promote destroy the structure of mycotoxins and reduce their content (Schaarschmidt et al., 2021).

Thermal processing methods have also been reported to reduce isoflavone content in both decorticated and whole pulses. In Indian households, common practices include soaking or germination, followed by either pressure or open pan cooking. Chickpeas, in particular, required the longest cooking time, with pressure cooking reducing their isoflavone content by up to 63%, while open pan cooking decreased it by as much as 90% (Deorukhkar et al., 2021).

Thermal processing techniques, such as blanching, oven drying, boiling, frying, and stir-frying, have generally been effective in reducing pesticide residues in food products. For instance, blanching has been shown to completely eliminate residues in tomatoes and significantly reduce them in eggplants, peppers, and cowpeas (Arowolo et al., 2022) (Naman et al., 2022) (Bajwa et al., 2014).

Similarly, Fierens et al. (2012) reported that phthalate concentrations in foods decreased after cooking. The authors emphasized that cooking should be considered when assessing human dietary exposure, as it significantly affects phthalate levels in foods. However, in general, heating can accelerate the migration of EDs from packaging materials and processing equipment into food, thereby increasing their concentration (Abbasi et al., 2024). From previous studies, pasteurization and sterilization at high temperatures can degrade some EDs, thereby reducing their concentration. However, the migration of bisphenol A from drinking water bottles can increase at elevated temperatures of 35°C and 45°C (Abdulazeez et al., 2024).

✓ **Storage methods**

According to the literature, refrigeration tends to slow down the degradation of pesticides, leading to higher residue levels over time compared to storage at room temperature. In contrast, higher temperatures or heat treatment can accelerate pesticide breakdown. For example, Liang et al. (2012) found that storing cucumbers at 4°C for 48 hours reduced levels of trichlorfon, dimethoate, dichlorvos, fenitrothion, and chlorpyrifos by 60.9%, 80.4%, 83.3%, 66.7%, and 70.4%, respectively. When stored at 25°C for the same duration, these pesticides were reduced by 90.2%, 82.6%, 97.6%, 88.9%, and 71.6%, respectively. Similarly, F. Zhao et al. (2019), reported that a pesticide degraded by 12% after six months at -20°C, whereas the same level of degradation was achieved within just one week at 4°C.

Additionally, the migration of EDs into food is influenced by several factors, primarily temperature and storage duration. Đurić et al. (2024) recorded the highest concentrations of bis(2-ethylhexyl) phthalate (DEHP) after six months of storage in salmon meat packed in polyamide polyethylene (253.56 μg/kg), particularly in fatty foods. Moreover, 10 to 30 minutes of ultrasound treatment, storage temperatures, UV radiation for 1 to 6 hours, and prolonged storage duration all increase the migration of bisphenol A from polyethylene terephthalate (PET) in drinking water bottles in Turkey (Abdulazeez et al., 2024).

2.4 METHODS FOR REDUCING AND REMOVING ENDOCRINE DISRUPTORS FROM FOOD

To remove endocrine disruptors, advanced technologies for effectively removing these disruptors from water systems, food matrices, and the air include aeration, ultraviolet treatment, oxidation, chlorination, coagulation, filtration, biodegradation, and membrane processes (Tapia-orozco et al., 2016). Table 2.2 summarizes methods involved in removing EDs.

2.4.1 Adsorption

Adsorption is a surface process involving the transfer of a substance to the boundary between two distinct phases. The substance that attaches to a surface is called the adsorbate, while the solid material to which it adheres is known as the adsorbent. Adsorption can be categorized into two types based on the interaction between adsorbates and adsorbents: physisorption and chemisorption. Physisorption is related to weak intermolecular forces, such as electrostatic and van der Waals forces, whereas chemisorption entails a chemical reaction between the adsorbing substance and the surface of the adsorbent (W. T. Vieira et al., 2020).

Adsorption stands out as the most efficient and economical method for removing contaminants. However, ongoing research and development of new adsorbents remains a dynamic field within adsorption science and technology. Activated carbon (AC) has been extensively used to eliminate various EDs due to several advantages: its high surface area and pore structure, ability to regenerate and reuse, low initial investment cost, effectiveness at low concentrations, suitability for both batch and continuous processes, hydrophobic surface properties, and versatility enhanced through various activation methods (Ayaliew et al., 2022). Recently, a study conducted by Westrup et al. (2021) and colleagues investigated the use of electrospun poly (butyleneadipate-co-terephthalate) (PBAT) fibers as an adsorbent material for removing EDs. The PBAT microfibers demonstrated high adsorption capacity for three contaminants: estrone, 17β-estradiol, and 17α-ethynylestradiol.

Additionally, various types of AC and biosorbents are capable of removing many classes of pesticides. The pH and temperature conditions are also noted, as they are typically kept constant when assessing the monolayer adsorption capacity (Ighalo et al., 2021). On the other hand, the metal ions can bioaccumulate in the major systems of living organisms and cause hazardous impacts on plants, animals, and humans. To mitigate this issue, adsorption (physisorption and chemisorption) techniques are applied (Ayob et al., 2021). In numerous instances, AC is employed as an adsorption agent in water treatment plants to capture heavy metals, owing to its microporous structure and extensive surface area (Maftouh et al., 2023) (Vedenyapina et al., 2021).

TABLE 2.2 Multiple methods for removing EDs

TECHNOLOGY	*TYPE OF COMPOUND*	*EDS*	*EFFECT ON EDS*	*REFERENCES*
Adsorption	Biochar from mushroom	17α-ethinylestradiol and progesterone	High adsorption	R. A. L. Vieira et al. (2022)
	Powdered and granular activated carbon	Estrogens estrone, 17β-estradiol, estriol, 17α-ethinylestradiol, and bisphenol A, nonylphenol	Removal effectiveness	Marcelo et al. (2017)
	F400 activated carbon	Xenoestrogen Bisphenol A (BPA)	Removal effect	Hernández-Abreu et al. (2020)
	Diaion SP207, Diaion SP825, Dowex L493, Active Carbon	Phenol 2,3,4, Phenol 2,4, clofibric methyl ester, Clofibrate, Clofibric Acid, Triclosan, Carbamazepine, E1, E2 and EE2	Good performance for removal of EDs from water	Solak et al. (2014)
Microfiltration	Modified and unmodified polyethersulfone (PES)	17β-estradiol (E2)	Modified PES: adsorption of E2 >60%. Unmodified PES: adsorption of E2 up to 30%	Niavarani et al. (2021)
	Microfiltration	Bisphenol A (BPA)	MF removed 24% of BPA	Zielińska et al. (2016)
	PTFE membrane modified using ZIF-8 MOFs	Progesterone	95% of removal	Ragab et al. (2016)
	Polyacrylonitrile (PAN) modified	Chromium and cadmium	Removal capacity	X. Liu et al. (2020)
Nanofiltration	Thin-film nanocomposite (TFN) with functionalized graphene oxide	Amitriptylene HCl (ATT HCl) and Bisphenol-A (BPA), Acetaminophen (ACT), and Caffeine (CFN)	100% removal for BPA, 90% for CFN and ATT HCl, and 80% removal for ACT	Baig et al. (2024)

	Nanofiltration	Bisphenol A (BPA)	NF removed up to thrice more BPA compared to MF	Zielińska et al. (2016)
	sulfonated graphene oxide, Polyaniline nanoparticles	Heavy metals	Rejection of Cu, Cr, and Cd	Mahmoud et al. (2023)
Ultrafiltration	Micellar Enhanced Ultrafiltration (MEUF)	Heavy metals	Removal of Heavy metals	Yaqub et al. (2019)
	Carbon nanotube–ultrafiltration composite membrane	Steroid hormones	Removal to 50–75%	Nguyen et al. (2021)
	Hybrid Polystyrene Nanoparticle-Ultrafiltration	Estrone	Adsorption of estrone	Akanyeti et al. (2017)
Reverse osmosis	Reverse osmosis membrane	Bisphenol A (BPA),	A greater BPA removal efficiency compared to UV/H_2O_2	Moreira et al. (2019)
	polyamide-based RO membranes	Pesticides	Low and high rejection of small and uncharged secondary amide pesticides	Fujioka et al. (2020)
	RO FilmTec™ XLE membranes	Pesticides (atrazine, simazine, isoproturon, metolachlor ESA, 2,4-D, and chlorothalonil)	Rejection rate >97%	Rodríguez-Alegre et al. (2024)
Advanced oxidation	Fenton/Photo-Fenton + UV	Pesticide: Atrazine	Higher Higer removal	Soto-Verjel et al. (2022)
	TiO2/UV photocatalysis	Pesticides in aqueous solution	Removal of pesticides	
	Ozone, Ozone/TiO_2 photocatalysis, Ozone/H_2O_2, UV	EE2, caffeine, progesterone, Bisphenol An, PPCPs and steroids 17β-estradiol, 17α -ethinylestradiol and estriol	High removal efficiency	Cesaro et al. (2016)

(continued)

TABLE 2.2 (Continued)

TECHNOLOGY	*TYPE OF COMPOUND*	*EDS*	*EFFECT ON EDS*	*REFERENCES*
Bioremediation	Fungal: *Tinea versicolor*	Phenols, parabens and phthalate	All removal	Pezzella et al. (2017)
	Microbial enzymes	NPs and BPA	High removal rate	Hashim et al. (2018)
Phytoremediation	Growing native weeds and grasses	Heavy metals	High accumulation of Fe, Zn, Cu, Mn, Ni, and Pb	Chandra et al. (2018)
	Phytodegradation, phytostimulation, phytovolatilization, phytoextraction, phytostabilization, and rhizofiltration: By whole plant, root, or shoot	Insecticides, pharmaceutical compounds, Heavy metals, organic compounds	Each technique removed specific EDs	Kristanti et al. (2023)

Among endocrine disruptors, bisphenol A and ethinylestradiol are known for their effects on the human endocrine system (Cunha et al., 2016) (Mg et al., 2022). In 2017, Hemidouche et al. (2017), found that granular activated carbon and Chemviron Filtrasorb recorded a high adsorption capacity of bisphenol A and ethinylestradiol. On the other hand, hydrochars derived from lignocellulosic biomass have garnered significant attention across various disciplines, particularly in adsorption. These hydrothermally produced materials prove highly efficient in adsorbing diverse pollutants and offer a cost-effective, sustainable source for producing ACs with extensive surface areas through subsequent activation processes (Qureshi et al., 2020). Furthermore, it has been reported that modified biochar has the ability to remove various types of EDs, and it is also possible to regenerate them (Dixit et al., 2023).

2.4.2 Membrane Technology

Membrane technology, which involves a phase-changing process, is an appealing alternative to conventional treatment methods. These membranes can be constructed from diverse materials, endowing them with unique physicochemical characteristics. This versatility enables them to efficiently eliminate a broad array of endocrine disruptors (Katibi, Yunos, Man, Aris, bin Mohd Nor, et al., 2021).

Membrane processes use various materials, such as ceramics, polymers, and zeolites, to filter high-molecular-weight solutes by employing hydrostatic pressure. The removal process operates through three main mechanisms: size exclusion, charge repulsion, or physicochemical interactions. Typically, these mechanisms interact, but one often dominates. Therefore, the filtration characteristics, removal efficiency, and contaminant rejection depend on the type of membrane used, the specific compounds, and the conditions they are exposed to. Factors such as pore size, hydrophobicity, functional groups, and pKa of both the membrane materials and contaminants, as well as the quality of the water treated, all play a role in influencing the removal process (W. T. Vieira et al., 2020). Membrane filtration techniques such as microfiltration, ultrafiltration, nanofiltration, and reverse osmosis exploit the specific physicochemical properties of the membrane material to effectively block a broad range of endocrine disruptors (Pironti et al., 2021).

2.4.2.1 Microfiltration

Microfiltration (MF) is one of the earliest commercially practiced pressure-driven membrane technologies. It effectively removes micrometer-sized particles, such as suspended solids, major pathogens, large bacteria, proteins, and yeast cells, using physical separation. MF membranes have pore diameters ranging from 0.1 to 5 μm, allowing them to separate particles larger than 0.1 μm with an open membrane structure. This broad pore size range makes MF suitable for diverse applications, including pharmaceuticals, wastewater treatment, food processing, desalination, and biotechnology (Anis et al., 2019). A new scientific experiment published in 2024 demonstrated that polyethersulfone (PES) membranes modified with urethane exhibit high removal efficiency for 17β-estradiol, estriol, bisphenol A, estrone, ethinylestradiol, and equilin. The adsorption loading of 17β-estradiol on the functionalized PES membrane was 6.7 ± 0.7 mg/m^2, representing a five-fold increase compared to the unmodified PES membrane (Niavarani et al., 2024). Notably, the addition of synthetic organic layers to microfiltration membranes increases their ability to remove different classes of endocrine disruptors. An investigation conducted in 2018 demonstrated that microfiltration membranes with synthetic organic fouling layers, such as calcium humate and calcium alginate, exhibited high adsorption of estriol, 17β-estradiol, 17α-ethynylestradiol, and 4-nonylphenol through mechanisms of increased hydrophobicity and thickness (Xue et al., 2018). However, membrane separation techniques such as microfiltration, ultrafiltration, nanofiltration, and reverse osmosis have shown the highest efficiency in separating heavy metals (Yunchu et al., 2023).

2.4.2.2 Nanofiltration and ultrafiltration

Nanofiltration (NF) membranes offer superior rejection of multivalent ions and organic molecules while maintaining a higher flux, making NF an ideal choice for high-performance water treatment applications. The use of NF membrane technology is growing, encompassing not only the removal of heavy metals and pesticides from groundwater but also the separation of organic compounds in wastewater, industrial process water recycling, and brackish water softening. NF membrane technology is sustainable, energy-efficient, and provides high separation efficiency, making it suitable for applications in various industrial sectors such as food, medicine, chemicals, and biotechnology, in addition to wastewater treatment (A. Ahmad et al., 2023).

To remove EDs from wastewater, nanofiltration membranes intercalated with hydrophilic MoS_2 nanosheets can be utilized. These MoS_2 nanosheet-intercalated membranes significantly improve ED rejection, with water/ED selectivity nearly six times higher than that of the control membrane (Dai et al., 2021). Furthermore, nanofiltration has the potential to remove various classes of pesticides and estrogenic compounds, such as atrazine, dimethoate, simazine, isoproturon, and prometryn, with removal rates ranging from 15.5% to over 95%, depending on the pesticide type. For example, membrane bioreactors are often used in combination with nanofiltration or reverse osmosis to enhance the removal of contaminants such as bisphenol A and alkylphenols (Kamaz et al., 2019) (Besha et al., 2017) (Pironti et al., 2021).

Ultrafiltration is a membrane separation technology that lies between microfiltration and nanofiltration, targeting molecules with molecular weights of 500 to 500,000 Da and pore sizes of 0.001 to 0.1 μm, operating at pressures between 0.1 and 0.8 MPa (X. Li et al., 2018). Due to the low molar mass of some contaminants, size exclusion is often ineffective, making adsorption the primary removal mechanism, depending on the interaction between the adsorbate and membrane. For instance, polyvinylidene fluoride membranes removed 78% of diethylhexyl phthalate but only 13% of carbamazepine (W. T. Vieira et al., 2020). Ultrafiltration's efficiency can be enhanced by combining it with technologies like advanced oxidation processes. A hybrid ultrafiltration-ozonation system removed up to 99% of endocrine disruptors such as bisphenol A, compared to 46% with ultrafiltration alone and 70% with ozonation alone (Pironti et al., 2021) (Si et al., 2018).

2.4.3 Reverse Osmosis (RO)

RO effectively removes up to 99% of EDs and other contaminants, producing potable water. RO membrane performance is influenced by structural and surface characteristics, with modern membranes incorporating nanostructured additives like graphene derivatives to enhance properties (Abdullah et al., 2023). Advances in RO design focus on improved control, reduced energy consumption, and optimization. Combining RO with other techniques boosts ED rejection. For instance, a polyamide thin-film composite (PA TFC) membrane enhanced with titania nanotubes (TNT) achieved 89.05% bisphenol A (BPA) and 97.89% caffeine rejection (N. A. Ahmad et al., 2022).

Reverse osmosis effectively removes synthetic steroid constituents from aqueous solutions, making it a viable option for treating water and effluents to improve potability and reduce toxicity risks to both the environment and human health. A study demonstrated a 95.6% retention rate of ethinylestradiol and levonorgestrel hormones using reverse osmosis (Vanni et al., 2023). Additionally, an excellent BPA rejection rate of 98% was achieved when reverse osmosis was combined with nanofiltration (Yüksel et al., 2013). Similarly, a removal efficiency of 97% was reported when microfiltration (MF) and RO were used together to treat endocrine disruptors (Katibi, Yunos, Man, Aris, bin Mohd Nor, et al., 2021).

2.4.4 Advanced Oxidation

Advanced oxidation processes (AOPs) generate highly reactive radicals to degrade pollutants at near-ambient conditions. These methods typically produce hydroxyl radicals using oxidants like ozone and hydrogen peroxide, combined with light, catalysts (Fe^{2+}, Fe^{3+}, TiO_2), ultrasonic energy, or heat. Examples include Fenton's reagent, photo-Fenton, UV-assisted peroxidation, and various O_3/TiO_2 systems. AOPs generally consume less energy than direct oxidation methods (Ameta et al., 2018). The photo-Fenton process, for instance, has shown 95–100% removal efficiency for pharmaceuticals, hormones, phenols, pesticides, and PPCP compounds (Hu et al., 2018) (Kohantorabi et al., 2019) (Pironti et al., 2021) (Raji et al., 2021).

From literature, Zong et al. (2021) evaluated five MnO_2 phases for oxidizing EDs like estrone (E1), ethynylestradiol (EE2), and bisphenol A (BPA) in methanol solutions. They found that δ-MnO_2 had the highest reactivity, achieving complete E1 removal at pH 3 within 120 minutes. Another study by Ziembowicz et al. (2021) explored Fenton-like oxidation of dibutyl phthalate (DBP) in landfill leachate, finding that sediments doped with 5% Fe, Cu, and Mn were most effective, achieving 95% DBP degradation at pH 3.0.

Ozonation is widely used across various industries as a powerful oxidizing agent. In the dairy industry, it is employed to sanitize milk production equipment, while in water treatment processes; it is utilized to eliminate residual organic pollutants such as pesticides, herbicides, antibiotics, and detergents. Notably, studies have examined and confirmed the oxidation of bisphenol A (BPA) through ozonation (Alsager et al., 2021). Ozone can break down organic pollutants either directly or indirectly by producing a highly reactive oxidizing agent known as the hydroxyl radical (Katibi, Yunos, Man, Aris, Nor, et al., 2021).

Photocatalytic oxidation is an effective technique for breaking down endocrine disruptors. Titanium dioxide (TiO_2) is typically used as the standard photocatalyst. When supported on activated carbon and exposed to ultraviolet light, TiO_2 has demonstrated high degradation efficiency for estrogens E_2 and EE_2, achieving over 90% removal within just 6 minutes. Certain pesticides and herbicides, such as atrazine, 2,4-dichlorophenoxyacetic acid (2,4-D), DDT, and lindane, are also classified as EDs. In the photocatalytic oxidation process, complete detoxification of atrazine was achieved using In S-TiO_2-rGO as the catalyst. Under visible light, atrazine was nearly fully degraded with In, S-TiO_2-rGO, demonstrating an efficiency 80 times greater than that of undoped TiO_2 (R. Wang et al., 2020).

2.4.5 Bioremediation and Phytoremediation

Bioremediation involves the use of living microorganisms to break down environmental pollutants or prevent pollution. It is a method for eliminating contaminants, such as EDs, thereby restoring natural conditions and preventing further environmental degradation (Sardrood et al., 2013).

Microbes can naturally inhabit areas contaminated by EDs or be introduced artificially to these sites to reduce the harmful effects of these pollutants. This can be done through either in-situ or ex-situ methods. As the terms suggest, in-situ bioremediation involves treating contaminants directly at the contaminated site without excavation or relocation, whereas ex-situ methods involve transporting the contaminated soil or water to a different location for treatment away from the original site. In-situ techniques are particularly effective in breaking down pollutants such as EDs (Antony et al., 2022).

Despite limited exploration, research shows that microbial communities, enzymes, and fungi hold promise in converting EDs into safer end products (Moukhtari et al., 2023). Aerobic bacteria, using electron acceptors like nitrate and manganese iron sulfate, produce enzymes such as lipases, peroxidases, and oxygenases under stress. These enzymes break down EDs like bisphenols and phthalates, using them as a carbon source and converting them into less toxic compounds. Key bacteria in ED remediation include *Pseudomonas sp.*, *Rhodococcus sp.*, *Sphingomonas sp.*, *Achromobacter sp.*, *Alcaligenes sp.*, *Burkholderia sp.*, and *Ralstonia sp* (Antony et al., 2022) (Thacharodi et al., 2023).

Microalgae can utilize substances like pesticides, hydrocarbons, endocrine disruptors, and cyanides as carbon and nitrogen sources. Their cell wall carbohydrates also absorb toxic chemicals from wastewater, making microalgae effective in reducing nitrogen, phosphorus, heavy metals, and other pollutants. Proper management of microalgal bioremediation is key, as it can remove nutrients from various sources, including domestic and industrial wastewater, agricultural runoff, and livestock farms (Hammed et al., 2016).

Phytoremediation uses hyperaccumulator plants and rhizospheric microorganisms to stabilize, transfer, or degrade pollutants in soil and water. The five main methods are phytodegradation, phytofiltration, phytoextraction, phytostabilization, and phytovolatilization. Plants that rapidly grow, produce substantial biomass, and accumulate heavy metals in their shoots without toxicity are effective for remediating contaminated soils and agro-ecosystems (Nedjimi, 2021).

Recent reports indicate that plants such as *Eclipta alba*, *Rumex dentatus*, *Cammelina benghalensis*, *Ricinus communis*, *Phragmites communis*, and *Solanum nigrum* have the capability to accumulate substantial amounts of metals, including Cd, Pd, Ni, Fe, and Cu, in their roots, shoots, and leaves (Sharma et al., 2020). It was observed that subsurface flow constructed wetlands containing *Scirpus grossus* exhibited enhanced efficiency in treating water contaminated with paracetamol and ibuprofen. They achieved impressive micropollutant removal rates of 94% and 99. Moreover, there was a noticeable increase in rhizobacterial count (from 17×10^3 to 29×10^3 CFU/ml), suggesting that extended retention periods promoted more effective interaction between plants and microbes. For instance, Microbial breakdown of pharmaceuticals and personal care products benefits from the combined effects of increased production of humic acids, which stimulate root growth and enhance adsorption sites, secretion of plant exudates (such as carbohydrates, amino acids, and organic acids), and radial oxygen loss. Studies have indicated that these interactions in the plant rhizosphere contribute to improved removal efficiency (S. Singh et al., 2024). As part of phytoremediation, Rhizospheric bacterial communities including *Bacillus thuringiensis*, *Bacillus cereus*, *Achromobacter denitrificans*, and *Bacillus subtilis* demonstrated elevated accumulation of metals (Fe, Cu, Zn, Mn, Pb, Ni, Cd, Cr) in *Cannabis sativa* shoot, root, and subsequently in their leaves (K. Singh et al., 2023).

2.5 CONCLUDING REMARKS AND CONCLUSIONS

The reviewed literature suggests that food can become contaminated with endocrine disruptors during the production of raw materials, both from crops and animal products, as well as throughout food processing. The toxicity of these disruptors in food is influenced by multiple factors, including the natural conditions during production and practices related to storage, processing, packaging, and cooking methods.

To reduce the impact of endocrine disruptors' contamination in food production and processing, several strategies are employed. Thorough post-harvest handling, such as extensive washing, can greatly decrease surface-level contamination. Proper cooking methods can further reduce the levels of these disruptors by changing their bioavailability and lowering their concentrations in the final product. Additionally, it is essential to improve monitoring and control measures throughout the entire food production chain, from sourcing to packaging. This involves using advanced detection technologies to identify and mitigate contamination risks early.

Additionally, advancements in food processing technologies should focus on innovations that reduce contamination from endocrine disruptors. Food 4.0 technologies can help prevent and mitigate contamination in the food supply chain through various methods: precision agriculture using IoT and AI to manage soil and water quality, smart processing, and packaging technologies with AI-driven detection systems to identify and eliminate contaminated products, blockchain for transparent traceability to quickly recall affected items, advanced sensors for real-time monitoring, robotics to minimize human error in handling, and data analytics for informed decision-making. Exploring alternative materials for equipment that come into contact with food can also reduce the risk of endocrine disruptors leaching during processing. Promoting sustainable agricultural practices can mitigate the uptake of endocrine disruptors in crops and livestock, addressing contamination at its source. Education and awareness campaigns for both producers and consumers are crucial in fostering an understanding of the risks associated with endocrine disruptors in food and encouraging best practices to minimize exposure. Lastly, ongoing research into the long-term health effects of chronic low-level exposure to endocrine disruptors will inform the development of more precise mitigation strategies and regulatory frameworks.

REFERENCES

Abbasi, F., De-la-Torre, G. E., KalantarHormozi, M. R., Schmidt, T. C., & Dobaradaran, S., 2024. A review of endocrine disrupting chemicals migration from food contact materials into beverages. *Chemosphere*, 355, 141760. https://doi.org/10.1016/j.chemosphere.2024.141760

Abd-Elghany, S. M., & Sallam, K. I., 2015. Rapid determination of total aflatoxins and ochratoxins A in meat products by immuno-affinity fluorimetry. *Food Chemistry*, 179, 253–256. https://doi.org/10.1016/j.foodchem.2015.01.140

Abdulazeez, Z. M., Yazici, F., & Aksoy, A., 2024. Influence of UV light, ultrasound, and heat treatment on the migration of bisphenol A from polyethylene terephthalate bottle into the food simulant. *Food Chemistry*, 439, 138162. https://doi.org/10.1016/j.foodchem.2023.138162

Abdullah, M. S., Goh, P. S., Ismail, A. F., & Hasbullah, H., 2023. The treatment of endocrine-disruptive chemicals in wastewater through asymmetric reverse osmosis membranes: A review. *Symmetry*, 15(5), 1049. https://doi.org/10.3390/sym15051049

Acir, I.-H., & Guenther, K., 2018. Endocrine-disrupting metabolites of alkylphenol ethoxylates – A critical review of analytical methods, environmental occurrences, toxicity, and regulation. *Science of the Total Environment*, 635, 1530–1546. https://doi.org/10.1016/j.scitotenv.2018.04.079

Ahmad, A., & Alshammari, M. B., 2023. *Nanofiltration Membrane for Water Purification* (Springer ed.). https://doi.org/10.1007/978-981-19-5315-6

Ahmad, M. F., Ahmad, F. A., Alsayegh, A. A., Zeyaullah, M., AlShahrani, A. M., Muzammil, K., Saati, A. A., Wahab, S., Elbendary, E. Y., Kambal, N., Abdelrahman, M. H., & Hussain, S., 2024. Pesticides impacts on human health and the environment with their mechanisms of action and possible countermeasures. *Heliyon*, 10(7), e29128. https://doi.org/10.1016/j.heliyon.2024.e29128

Ahmad, N. A., Goh, P. S., Azman, N., Ismail, A. F., Hasbullah, H., Hashim, N., Kerisnan Krishnan, N. D., Yahaya, N. K. E. M., Mohamed, A., Mohamed Yusoff, M. A., Karim, J., & Abdullah, N. S., 2022. Enhanced removal of endocrine-disrupting compounds from wastewater using reverse osmosis membrane with Titania nanotube-constructed nanochannels. *Membranes*, 12(10), 958. https://doi.org/10.3390/membranes12100958

Aissioui, S., Poirier, L., Amara, R., & Ramdane, Z., 2021. Concentrations of lead, cadmium, and mercury in Mullus barbatus barbatus (L.) from the Algerian coast and health risks associated to its consumption. *Regional Studies in Marine Science*, 47, 101959. https://doi.org/10.1016/j.rsma.2021.101959

Akanyeti, İ., Kraft, A., & Ferrari, M. C., 2017. Hybrid polystyrene nanoparticle-ultrafiltration system for hormone removal from water. *Journal of Water Process Engineering*, 17, 102–109. https://doi.org/10.1016/j.jwpe.2017.02.014

Alamri, M. S., Qasem, A. A. A., Mohamed, A. A., Hussain, S., Ibraheem, M. A., Shamlan, G., Alqah, H. A., & Qasha, A. S., 2021. Food packaging's materials: A food safety perspective. *Saudi Journal of Biological Sciences*, 28(8), 4490–4499. https://doi.org/10.1016/j.sjbs.2021.04.047

Alsager, O. A., & Alnajrani, M. N., 2021. Effect of various reagents on the decomposition of endocrine disruptors by ozonation: A detailed study. *Journal of Water Process Engineering*, 40(December 2020), 101814–101814. https://doi.org/10.1016/j.jwpe.2020.101814

Ameta, R., K. Chohadia, A., Jain, A., & Punjabi, P. B., 2018. Chapter 3–Fenton and photo-Fenton processes. In S. C. Ameta & R. Ameta (Eds.), *Advanced Oxidation Processes for Waste Water Treatment* (pp. 49–87): Academic Press. https://doi.org/10.1016/B978-0-12-810499-6.00003-6

Amin, K. A., & Al-Shehri, F. S., 2018. Toxicological and safety assessment of tartrazine as a synthetic food additive on health biomarkers: A review. *African Journal of Biotechnology*, 17(6), 139–149. https://doi.org/10.5897/AJB2017.16300

Amiridou, D., & Voutsa, D., 2011. Alkylphenols and phthalates in bottled waters. *Journal of Hazardous Materials*, 185(1), 281–286. https://doi.org/10.1016/j.jhazmat.2010.09.031

Andersen, H. R., David, A., Freire, C., Fernández, M. F., D'Cruz, S. C., Reina-Pérez, I., Fini, J.-B., & Blaha, L., 2022. Pyrethroids and developmental neurotoxicity–A critical review of epidemiological studies and supporting mechanistic evidence. *Environmental Research*, 214, 113935. https://doi.org/10.1016/j.envres.2022.113935

Andújar, N., Gálvez-Ontiveros, Y., Zafra-Gómez, A., Rodrigo, L., Álvarez-Cubero, M. J., Aguilera, M., Monteagudo, C., & Rivas, A., 2019. Bisphenol A analogues in food and their hormonal and obesogenic effects: A review. *Nutrients*, 11(9), 2136. https://doi.org/10.3390/nu11092136

Angula, M., Ishola, A., Tjiurutue, M., Sulyok, M., Krska, R., Ezekiel, C. N., & Misihairabgwi, J., 2024. Mycotoxins in stored cereals from rural households in central northern Namibia. *Food Control*, 163, 110532. https://doi.org/10.1016/j.foodcont.2024.110532

Anis, S. F., Hashaikeh, R., & Hilal, N., 2019. Microfiltration membrane processes: A review of research trends over the past decade. *Journal of Water Process Engineering*, 32(September), 100941. https://doi.org/10.1016/j.jwpe.2019.100941

Antony, S., Antony, S., Rebello, S., George, S., Biju, D. T., R, R., Madhavan, A., Binod, P., Pandey, A., Sindhu, R., & Awasthi, M. K., 2022. Bioremediation of endocrine disrupting chemicals–Advancements and challenges. *Environmental Research*, 213(June), 113509–113509. https://doi.org/10.1016/j.envres.2022.113509

Arowolo, O. K., Gbaye, O. A., & Aborisade, A. T., 2022. Effect of household processing methods on insecticide residues in two tomato cultivars from the market in Ondo State, Nigeria. *Toxicology and Environmental Health Sciences*, 14(4), 371–378. https://doi.org/10.1007/s13530-022-00151-8

Arrigo, F., Impellitteri, F., Piccione, G., & Faggio, C., 2023. Phthalates and their effects on human health: Focus on erythrocytes and the reproductive system. *Comparative Biochemistry and Physiology Part C: Toxicology & Pharmacology*, 270, 109645. https://doi.org/10.1016/j.cbpc.2023.109645

Authority, E. F. S., 2020. Outcome of a public consultation on the draft risk assessment of perfluoroalkyl substances in food. *EFSA Supporting Publications*, 17(9), 1931E. https://doi.org/10.2903/sp.efsa.2020.EN-1931

Auxietre, T.-A., Dumontier, M.-F., Balguy, I., Frapart, Y., Canivenc-Lavier, M.-C., Berges, R., Boudalia, S., Auger, J., Corvol, M.-T., & Savouret, J.-F., 2014. Sub-NOAEL amounts of vinclozolin and xenoestrogens target rat chondrogenesis in vivo. *Biochimie*, 99, 169–177. https://doi.org/10.1016/j.biochi.2013.12.001

Auxietre, T. A., Dumontier, M. F., Balguy, I., Frapart, Y., Canivenc-Lavier, M. C., Berges, R., Boudalia, S., Auger, J., Corvol, M. T., & Savouret, J. F., 2014. Sub-NOAEL amounts of vinclozolin and xenoestrogens target rat chondrogenesis in vivo. *Biochimie*, 99, 169–177. https://doi.org/10.1016/j.biochi.2013.12.001

Axon, A., May, F. E. B., Gaughan, L. E., Williams, F. M., Blain, P. G., & Wright, M. C., 2012. Tartrazine and sunset yellow are xenoestrogens in a new screening assay to identify modulators of human oestrogen receptor transcriptional activity. *Toxicology*, 298(1), 40–51. https://doi.org/10.1016/j.tox.2012.04.014

Ayaliew, A., Berhe, S., Hailu, G., & Gashaw, A., 2022. Heliyon removal of endocrine disrupters from the contaminated environment: Public health concerns, treatment strategies and future perspectives–A review. *Heliyon*, 8(December 2019), e09206. https://doi.org/10.1016/j.heliyon.2022.e09206

Ayob, S., Othman, N., Ali Hamood Altowayti, W., Khalid, F. S., Bakar, N. A., Tahir, M., & Soedjono, E. S., 2021. A review on adsorption of heavy metals from wood-industrial wastewater by oil palm waste. *Journal of Ecological Engineering*, 22(3), 249–265. https://doi.org/10.12911/22998993/132854

Baig, N., & Matin, A., 2024. Incorporating functionalized graphene oxide into diethylene triamine-based nanofiltration membranes can improve the removal of emerging organic micropollutants. *Journal of Colloid and Interface Science*, 676(July), 657–669. https://doi.org/10.1016/j.jcis.2024.06.180

Bajwa, U., & Sandhu, K. S., 2014. Effect of handling and processing on pesticide residues in food–A review. *Journal of Food Science and Technology*, 51(2), 201–220. https://doi.org/10.1007/s13197-011-0499-5

Balwan, W. K., Saba, N., & Kour, S., 2023. Study of impact of mycotoxins on the human health. *Scholar Bulletin*, 9(2), 19–23. https://doi.org/10.36348/sb.2023.v09i02.003

Baute-Pérez, D., Santana-Mayor, Á., Herrera-Herrera, A. V., Socas-Rodríguez, B., & Rodríguez-Delgado, M. Á., 2022. Analysis of alkylphenols, bisphenols and alkylphenol ethoxylates in microbial-fermented functional beverages and bottled water: Optimization of a dispersive liquid-liquid microextraction protocol based on natural hydrophobic deep eutectic solvents. *Food Chemistry,* 377, 131921. https://doi.org/10.1016/j.foodchem.2021.131921

Bensaada, S., Peruzzi, G., Cubizolles, L., Denayrolles, M., & Bennetau-Pelissero, C., 2024. Traditional and domestic cooking dramatically reduce estrogenic isoflavones in Soy foods. *Foods*, 13(7), 999. https://doi.org/10.3390/foods13070999

Bertram, M. G., Gore, A. C., Tyler, C. R., & Brodin, T., 2022. Endocrine-disrupting chemicals. *Current Biology*, 32(13), R727–R730. https://doi.org/10.1016/j.cub.2022.05.063

Besha, A. T., Gebreyohannes, A. Y., Tufa, R. A., Bekele, D. N., Curcio, E., & Giorno, L., 2017. Removal of emerging micropollutants by activated sludge process and membrane bioreactors and the effects of micropollutants on membrane fouling: A review. *Journal of Environmental Chemical Engineering*, 5(3), 2395–2414. https://doi.org/10.1016/j.jece.2017.04.027

Beyer, A., Schorgg, P., Karavasiloglou, N., Sarwar, S., Rohrmann, S., Bärnighausen, T., Cassidy, A., Connolly, L., & Kühn, T., 2022. Urinary phthalate concentrations and mortality risk: A population-based study. *Environmental Research,* 214, 113927. https://doi.org/10.1016/j.envres.2022.113927

Boudalia, S., 2017. Endocrine disruptors: History and difficulties in predicting effects. *Insights Nutrients Metabolics*, 1(2), 1–2

Boudalia, S., Belloir, C., Miller, M.-L., & Canivenc-Lavier, M.-C., 2017. Early endocrine disruptors exposure acts on 3T3-L1 differentiation and endocrine activity. *BioImpacts*, 7(2), 83–89. https://doi.org/10.15171/bi.2017.11

Boudalia, S., Berges, R., Chabanet, C., Folia, M., Decocq, L., Pasquis, B., Abdennebi-Najar, L., & Canivenc-Lavier, M. C., 2014. A multi-generational study on low-dose BPA exposure in Wistar rats: Effects on maternal behavior, flavor intake and development. *Neurotoxicology and Teratology*, 41, 16–26. https://doi.org/10.1016/j.ntt.2013.11.002

Boudalia, S., Bousbia, A., Boumaaza, B., Oudir, M., & Canivenc Lavier, M. C., 2021. Relationship between endocrine disruptors and obesity with a focus on bisphenol A: A narrative review. *BioImpacts*, 11(4), 289–300. https://doi.org/10.34172/bi.2021.33

Boudalia, S., & Oudir, M., 2016. Bisphenol-A: Legislation in industrials countries and in Algeria. *Research Journal of Environmental Toxicology*, 10, 189–192. https://doi.org/10.3923/rjet.2016.189.192

Boudebbouz, A., Boudalia, S., Bousbia, A., Gueroui, Y., Boussadia, M. I., Chelaghmia, M. L., Zebsa, R., Affoune, A. M., & Symeon, G. K., 2022. Determination of heavy metal levels and health risk assessment of raw cow milk in Guelma Region, Algeria. *Biological Trace Element Research*, 201, 1704–1716. https://doi.org/10.1007/s12011-022-03308-1

Boudebbouz, A., Boudalia, S., Boussadia, M. I., Gueroui, Y., Habila, S., Bousbia, A., & Symeon, G. K., 2022. Pesticide residues levels in raw cow's milk and health risk assessment across the globe: A systematic review. *Environmental Advances*, 9, 100266. https://doi.org/10.1016/j.envadv.2022.100266

Bousbia, A., Boudalia, S., Gueroui, Y., Ghebache, R., Amrouchi, M., Belase, B., Meguelati, S., Belkheir, B., Benidir, M., & Chelaghmia, M., 2019. Heavy metals concentrations in raw cow milk produced in the different livestock farming types in Guelma province (Algeria): Contamination and risk assessment of consumption. *JAPS: Journal of Animal & Plant Sciences*, 29(2), 386–395.

Bradley, E., 2010. *Nonylphenol in Food Contact Plastics and Migration into Foods.* FS Agency, Editor. The Food and Environment Research Agency.

Brendel, S., Fetter, É., Staude, C., Vierke, L., & Biegel-Engler, A., 2018. Short-chain perfluoroalkyl acids: Environmental concerns and a regulatory strategy under REACH. *Environmental Sciences Europe*, 30(1), 9. https://doi.org/10.1186/s12302-018-0134-4

Bugey, A., Janin, Y., Edder, P., & Bieri, S., 2013. Targeted multidimensional gas chromatography using a heart-cutting device and cryogenic focusing for the determination of benzophenone derivatives in foodstuffs. *Analytical and Bioanalytical Chemistry*, 405(12), 4177–4185. https://doi.org/10.1007/s00216-013-6808-2

Canivenc-Lavier, M.-C., & Bennetau-Pelissero, C., 2023. Phytoestrogens and health effects. *Nutrients*, 15(2), 317. https://doi.org/10.3390/nu15020317

Cao, Y., Liu, H., Qin, N., Ren, X., Zhu, B., & Xia, X., 2020. Impact of food additives on the composition and function of gut microbiota: A review. *Trends in Food Science & Technology*, 99, 295–310. https://doi.org/10.1016/j.tifs.2020.03.006

C.C.E, 1999. Stratégie communautaire concernant les perturbateurs endocriniens: une série de substances suspectées d'influer sur le système hormonal des hommes et des animaux. In COM (Ed.), (Vol. 706). Bruxelles: Commission des Communautés Européennes.

C.C.E., 2002. *Endocrine Disruptors: Study on Gathering Information on 435 Substances with Insufficient Data* (pp. 279). Bruxelles: Commission des Communautés Européennes.

Cesaro, A., & Belgiorno, V., 2016. Removal of endocrine disruptors from urban wastewater by Advanced Oxidation Processes (AOPs): A review. *The Open Biotechnology Journal*, 10(1), 151–172. https://doi.org/10.2174/1874070701610010151

Chandra, R., Kumar, V., Tripathi, S., & Sharma, P., 2018. Heavy metal phytoextraction potential of native weeds and grasses from endocrine-disrupting chemicals rich complex distillery sludge and their histological observations during in-situ phytoremediation. *Ecological Engineering*, 111(November 2017), 143–156. https://doi.org/10.1016/j.ecoleng.2017.12.007

Chaoua, S., Boussaa, S., El Gharmali, A., & Boumezzough, A., 2019. Impact of irrigation with wastewater on accumulation of heavy metals in soil and crops in the region of Marrakech in Morocco. *Journal of the Saudi Society of Agricultural Sciences*, 18(4), 429–436. https://doi.org/10.1016/j.jssas.2018.02.003

Chazelas, E., Pierre, F., Druesne-Pecollo, N., Esseddik, Y., Szabo de Edelenyi, F., Agaesse, C., De Sa, A., Lutchia, R., Gigandet, S., Srour, B., Debras, C., Huybrechts, I., Julia, C., Kesse-Guyot, E., Allès, B., Galan, P., Hercberg, S., Deschasaux-Tanguy, M., & Touvier, M., 2022. Nitrites and nitrates from food additives and natural sources and cancer risk: Results from the NutriNet-Santé cohort. *International Journal of Epidemiology*, 51(4), 1106–1119. https://doi.org/10.1093/ije/dyac046

Chien, H.-L., Huang, H.-Y., & Chou, C.-C., 2006. Transformation of isoflavone phytoestrogens during the fermentation of soymilk with lactic acid bacteria and bifidobacteria. *Food Microbiology*, 23(8), 772–778. https://doi.org/10.1016/j.fm.2006.01.002

Chin, X. H., Elhalis, H., Chow, Y., & Liu, S. Q., 2024. Enhancing food safety in soybean fermentation through strategic implementation of starter cultures. *Heliyon*, 10(2), e25007. https://doi.org/10.1016/j.heliyon.2024.e25007

Cunha, D. L. D., Silva, S. M. C. D., Bila, D. M., Oliveira, J. L. D. M., Sarcinelli, P. D. N., & Larentis, A. L., 2016. Regulation of the synthetic estrogen 17 α-ethinylestradiol in water bodies in Europe, the United States, and Brazil Regulamentação do estrogênio sintético 17 α-etinilestradiol em matrizes aquáticas na Europa, Estados Unidos e Brasil Regulación del es. *Cad. Saúde Pública, Rio de Janeiro*, 32(3), 1–12

Czarnecka, K., Pilarz, A., Rogut, A., Maj, P., Szymańska, J., Olejnik, Ł., & Szymański, P., 2021. Aspartame-true or false? Narrative review of safety analysis of general use in products. *Nutrients*, 13(6), 1957. https://doi.org/10.3390/nu13061957

Dai, R., Han, H., Wang, T., Li, X., & Wang, Z., 2021. Enhanced removal of hydrophobic endocrine disrupting compounds from wastewater by nanofiltration membranes intercalated with hydrophilic MoS2 nanosheets: Role of surface properties and internal nanochannels. *Journal of Membrane Science*, 628(January), 119267–119267. https://doi.org/10.1016/j.memsci.2021.119267

Darbre, P. D., 2019. The history of endocrine-disrupting chemicals. *Current Opinion in Endocrine and Metabolic Research*, 7, 26–33. https://doi.org/10.1016/j.coemr.2019.06.007

Das Pinkey, P., Nesha, M., Bhattacharjee, S., Chowdhury, M. A. Z., Fardous, Z., Bari, L., & Koley, N. J., 2024. Toxicity risks associated with heavy metals to fish species in the transboundary River–Linked Ramsar conservation site of Tanguar Haor, Bangladesh. *Ecotoxicology and Environmental Safety*, 269, 115736. https://doi.org/10.1016/j.ecoenv.2023.115736

de Paula, L. C. P., & Alves, C., 2024. Food packaging and endocrine disruptors. *Jornal de Pediatria*, 100, S40–S47. https://doi.org/10.1016/j.jped.2023.09.010

Deng, X., Liang, S., Tang, Y., Li, Y., Xu, R., Luo, L., Wang, Q., Zhang, X., & Liu, Y., 2024. Adverse effects of bisphenol A and its analogues on male fertility: An epigenetic perspective. *Environmental Pollution*, 345, 123393. https://doi.org/10.1016/j.envpol.2024.123393

Deorukhkar, A., & Ananthanarayan, L., 2021. Effect of thermal processing methods on flavonoid and isoflavone content of decorticated and whole pulses. *Journal of Food Science and Technology*, 58(2), 465–473. https://doi.org/10.1007/s13197-020-04555-7

Dixit, A., & Ahammed, M. M., 2023. Use of modified biochar for removal of endocrine disrupting compounds from water and wastewater: A review. *ioresource Technology Reports*, 23, 101519–101519. https://doi.org/10.1016/j.biteb.2023.101519

Domingo, J. L., 2010. Influence of cooking processes on the concentrations of toxic metals and various organic environmental pollutants in food: A review of the published literature. *Critical Reviews in Food Science and Nutrition*, 51(1), 29–37. https://doi.org/10.1080/10408390903044511

Dueñas-Mas, M. J., Ballesteros-Gómez, A., & de Boer, J., 2023. Determination of several PFAS groups in food packaging material from fast-food restaurants in France. *Chemosphere*, 339, 139734. https://doi.org/10.1016/j.chemosphere.2023.139734

Đurić, B., Kartalović, B., Habschied, K., Novakov, N., Vranešević, J., Brkić, B., & Mastanjević, K., 2024. Effects of packaging material type, storage time and lipid content on phthalate migration in smoked fish meat. *Applied Sciences*, 14(4), 1660. https://doi.org/10.3390/app14041660

European Union., 2022. Commission Recommendation (EU) 2022/1431 of 24 August 2022 on the monitoring of perfluoroalkyl substances in food. *Official Journal of the European Union*, 221, 105–109

Fasano, E., Cirillo, T., Esposito, F., & Lacorte, S., 2015. Migration of monomers and plasticizers from packed foods and heated microwave foods using QuEChERS sample preparation and gas chromatography/mass spectrometry. *LWT–Food Science and Technology*, 64(2), 1015–1021. https://doi.org/10.1016/j.lwt.2015.06.066

Felter, S. P., Zhang, X., & Thompson, C., 2021. Butylated hydroxyanisole: Carcinogenic food additive to be avoided or harmless antioxidant important to protect food supply? *Regulatory Toxicology and Pharmacology*, 121, 104887. https://doi.org/10.1016/j.yrtph.2021.104887

Fernandes, A. R., Rose, M., & Charlton, C., 2008. 4-Nonylphenol (NP) in food-contact materials: Analytical methodology and occurrence. *Food Additives & Contaminants: Part A*, 25(3), 364–372. https://doi.org/10.1080/02652030701564548

Fernandez-Lopez, A., Lamothe, V., Delample, M., Denayrolles, M., & Bennetau-Pelissero, C., 2016. Removing isoflavones from modern soyfood: Why and how? *Food Chemistry*, 210, 286–294. https://doi.org/10.1016/j.foodchem.2016.04.126

Fierens, T., Vanermen, G., Van Holderbeke, M., De Henauw, S., & Sioen, I., 2012. Effect of cooking at home on the levels of eight phthalates in foods. *Food and Chemical Toxicology*, 50(12), 4428–4435. https://doi.org/10.1016/j.fct.2012.09.004

Forcada, S., Menéndez Miranda, M., Stevens, F., Royo, L. J., Fernández Pierna, J. A., Baeten, V., & Soldado, A., 2023. Industrial impact on sustainable dairy farms: Essential elements, hazardous metals and polycyclic aromatic hydrocarbons in forage and cow's milk. *Heliyon*, 9(10). https://doi.org/10.1016/j.heliyon.2023.e20977

Fujioka, T., Kodamatani, H., Yujue, W., Yu, K. D., Wanjaya, E. R., Yuan, H., Fang, M., & Snyder, S. A., 2020. Assessing the passage of small pesticides through reverse osmosis membranes. *Journal of Membrane Science*, 595, 117577–117577. https://doi.org/10.1016/j.memsci.2019.117577

Galal, S. A. B., Madhat Mousa, M., Elzanfaly, E. S., Hussien, E. M., Amer, E. A. H., & Zaazaa, H. E., 2024. Quantitative analysis of residual butylated hydroxytoluene and butylated hydroxyanisole in Salmo salar, milk, and butter by liquid chromatography–tandem mass spectrometry. *Food Chemistry*, 453, 139653. https://doi.org/10.1016/j.foodchem.2024.139653

González, N., Cunha, S. C., Ferreira, R., Fernandes, J. O., Marquès, M., Nadal, M., & Domingo, J. L., 2020. Concentrations of nine bisphenol analogues in food purchased from Catalonia (Spain): Comparison of canned and non-canned foodstuffs. *Food and Chemical Toxicology*, 136, 110992. https://doi.org/10.1016/j.fct.2019.110992

Guart, A., Bono-Blay, F., Borrell, A., & Lacorte, S., 2011. Migration of plasticizersphthalates, bisphenol A and alkylphenols from plastic containers and evaluation of risk. *Food Additives & Contaminants: Part A*, 28(5), 676–685. https://doi.org/10.1080/19440049.2011.555845

Gürtler, R., 2021. Risk assessment of food additives. In F.-X. Reichl & M. Schwenk (Eds.), *Regulatory Toxicology* (pp. 1323–1337). Cham: Springer International Publishing. https://doi.org/10.1007/978-3-030-57499-4_116

Hamed Hammad Mohammed, H., Jin, G., Ma, M., Khalifa, I., Shukat, R., Elkhedir, A. E., Zeng, Q., & Noman, A. E., 2020. Comparative characterization of proximate nutritional compositions, microbial quality and safety of camel meat in relation to mutton, beef, and chicken. *LWT*, 118, 108714. https://doi.org/10.1016/j.lwt.2019.108714

Hammed, A. M., Prajapati, S. K., Simsek, S., & Simsek, H., 2016. Growth regime and environmental remediation of microalgae. *Algae*, 31(3), 189–204. https://doi.org/10.4490/algae.2016.31.8.28

Hanafi, A., Elsheshetawy, H. E., & Faied, S. F., 2016. Reduction of pesticides residues on okra fruits by different processing treatments. *Journal für Verbraucherschutz und Lebensmittelsicherheit*, 11(4), 337–343. https://doi.org/10.1007/s00003-016-1054-0

Hashim, N., Kassim, M., Yusof, N., & Sharifuddin, S., 2018. Bioremediation of endocrine disruptive chemicals: The power of microbial enzymes. *Asian Journal of Biotechnology and Bioresource Technology*, 2(4), 1–9. https://doi.org/10.9734/ajb2t/2017/39092

Hassan, J., Rajib, M. M. R., Khan, M. N.-E. A., Khandaker, S., Zubayer, M., Ashab, K. R., Kuba, T., Marwani, H. M., Asiri, A. M., Hasan, M. M., Islam, A., Rahman, M. M., & Awual, M. R., 2024. Assessment of heavy metals accumulation by vegetables irrigated with different stages of textile wastewater for evaluation of food and health risk. *Journal of Environmental Management*, 353, 120206. https://doi.org/10.1016/j.jenvman.2024.120206

Hemidouche, S., Assoumani, A., Favier, L., Simion, A. I., Grigoras, C. G., Wolbert, D., Gavrila, L., Nationale, E., & Chimie, S. D. (2017, 2017). *Removal of some Endocrine Disruptors via Adsorption on Activated Carbon*. Sinaia, Romania: E-Health and Bioengineering Conference (EHB). https://doi.org/10.1109/EHB.2017.7995448

Heo, J.-J., Lee, J.-W., Kim, S.-K., & Oh, J.-E., 2014. Foodstuff analyses show that seafood and water are major perfluoroalkyl acids (PFAAs) sources to humans in Korea. *Journal of Hazardous Materials*, 279, 402–409. https://doi.org/10.1016/j.jhazmat.2014.07.004

Hernández-Abreu, A. B., Álvarez-Torrellas, S., Águeda, V. I., Larriba, M., Delgado, J. A., Calvo, P. A., & García, J., 2020. Enhanced removal of the endocrine disruptor compound Bisphenol A by adsorption onto green-carbon materials. Effect of real effluents on the adsorption process. *Journal of Environmental Management*, 266(December 2019). https://doi.org/10.1016/j.jenvman.2020.110604

Herzallah, S. M., 2013. Aflatoxin B1 residues in eggs and flesh of laying hens fed Aflatoxin B1 contaminated diet. *American Journal of Agricultural and Biological Sciences*, 8(2). https://doi.org/10.3844/ajabssp.2013.156.161

Hess, E. L., Myers, E. A., Swithers, S. E., & Hedrick, V. E., 2018. Associations between nonnutritive sweetener intake and metabolic syndrome in adults. *Journal of the American College of Nutrition*, 37(6), 487–493. https://doi.org/10.1080/07315724.2018.1440658

Hosseini-Esfahani, F., Beheshti, N., Nematollahi, A., Koochakpoor, G., verij-Kazemi, S., Mirmiran, P., & Azizi, F., 2023. The association between dietary acrylamide intake and the risk of type 2 diabetes incidence in the Tehran lipid and glucose study. *Scientific Reports*, 13(1), 8235. https://doi.org/10.1038/s41598-023-35493-x

Hu, C., Huang, D., Zeng, G., Cheng, M., Gong, X., Wang, R., Xue, W., Hu, Z., & Liu, Y., 2018. The combination of Fenton process and Phanerochaete chrysosporium for the removal of bisphenol A in river sediments: Mechanism related to extracellular enzyme, organic acid and iron. *Chemical Engineering Journal*, 338(August 2017), 432–439. https://doi.org/10.1016/j.cej.2018.01.068

Huang, Y.-C., Huang, P.-R., Lo, Y.-T. C., Sun, C.-W., Pan, W.-H., Wang, S.-L., & Huang, H.-B., 2021. Food processing and phthalate exposure: The nutrition and health survey in Taiwan (1993–1996 and 2005–2008). [Original Research]. *Frontiers in Nutrition*, 8. https://doi.org/10.3389/fnut.2021.766992

Hwang, C. E., Haque, M. A., Lee, J. H., Ahn, M. J., Lee, H. Y., Lee, B. W., Lee, Y.-Y., Lee, C., Kim, B. J., & Park, J.-Y., 2016. Changes in phytoestrogen contents and antioxidant activities during fermentation of soybean-powder milks prepared from different soybean cultivars by Lactobacillus plantarum P1201. *Korean Journal of Microbiology*, 52(2), 202–211. https://doi.org/10.7845/kjm.2016.5049

Hwang, S. H., Ryu, S. Y., Seo, D., Kim, S. Y., Lee, J., & Cho, Y. S., 2024. Development of a method using QuEChERS and LC–MS/MS for analysis of per-and polyfluoroalkyl substances in rice matrix. *Food Chemistry*, 445, 138687. https://doi.org/10.1016/j.foodchem.2024.138687

Iavicoli, I., Fontana, L., & Bergamaschi, A., 2009. The effects of metals as endocrine disruptors. *Journal of Toxicology and Environmental Health, Part B*, 12(3), 206–223. https://doi.org/10.1080/10937400902902062

Ighalo, J. O., Adeniyi, A. G., & Adelodun, A. A., 2021. Recent advances on the adsorption of herbicides and pesticides from polluted waters: Performance evaluation via physical attributes. *Journal of Industrial and Engineering Chemistry*, 93, 117–137. https://doi.org/10.1016/j.jiec.2020.10.011

Iwamoto, M., Masuya, T., Hosose, M., Tagawa, K., Ishibashi, T., Suyama, K., Nose, T., Yoshihara, E., Downes, M., Evans, R. M., & Matsushima, A., 2021. Bisphenol A derivatives act as novel coactivator-binding inhibitors for estrogen receptor β. *Journal of Biological Chemistry*, 297(5), 101173. https://doi.org/10.1016/j.jbc.2021.101173

Ižaković, M., Ačkar, Đ., & Šubarić, D., 2021. Commonly used artificial sweeteners in Europe. *Hrana u zdravlju i bolesti: znanstveno-stručni časopis za nutricionizam i dijetetiku*, 10(1), 24–34.

Izzo, L., Mikušová, P., Lombardi, S., Sulyok, M., & Ritieni, A., 2022. Analysis of mycotoxin and secondary metabolites in commercial and traditional Slovak cheese samples. *Toxins*, 14(2), 134. https://doi.org/10.3390/toxins14020134

Jannat, M., Fatimah, R., & Kishida, M., 2014. Nitrate (NO3−) and nitrite (NO2−) are endocrine disruptors to downregulate expression of tyrosine hydroxylase and motor behavior through conversion to nitric oxide in early development of zebrafish. *Biochemical and Biophysical Research Communications*, 452(3), 608–613. https://doi.org/10.1016/j.bbrc.2014.08.114

Jeddi, M. Z., Rastkari, N., Ahmadkhaniha, R., & Yunesian, M., 2016. Endocrine disruptor phthalates in bottled water: Daily exposure and health risk assessment in pregnant and lactating women. *Environmental Monitoring and Assessment*, 188(9), 534. https://doi.org/10.1007/s10661-016-5502-1

Jedeon, K., De la Dure-Molla, M., Brookes, S. J., Loiodice, S., Marciano, C., Kirkham, J., Canivenc-Lavier, M.-C., Boudalia, S., Bergès, R., Harada, H., Berdal, A., & Babajko, S., 2013. Enamel defects reflect perinatal exposure to bisphenol A. *Amercian Journal of Pathology*, 183(1), 108–118. https://doi.org/10.1016/j.ajpath.2013.04.004

Jedeon, K., Marciano, C., Loiodice, S., Boudalia, S., Canivenc Lavier, M. C., Berdal, A., & Babajko, S., 2014. Enamel hypomineralization due to endocrine disruptors. *Connect Tissue Research*, 55(1), 43–47. https://doi.org/10.3109/03008207.2014.923857

Jeong, Y.-J., Bang, S., Kim, J., Chun, S.-H., Choi, S., Kim, J., Chung, M.-S., Kang, G. J., Kang, Y.-W., Kim, J., Kho, Y., Joo, Y., & Lee, K.-W., 2019. Comparing levels of perfluorinated compounds in processed marine products. *Food and Chemical Toxicology*, 126, 199–210. https://doi.org/10.1016/j.fct.2019.01.028

Ji, X., Xiao, Y., Yang, W., Wei, W., Lyu, W., Wang, X., & Yang, H., 2024. Comprehensive identification and risk assessment of regulated and emerging mycotoxins in infant foods and related raw materials and risk management advice: A case study of an infant food company in China. *Food Research International*, 187, 114304. https://doi.org/10.1016/j.foodres.2024.114304

Jiang, J., Qi, L., Wei, Q., & Shi, F., 2018. Effects of daily exposure to saccharin sodium and rebaudioside A on the ovarian cycle and steroidogenesis in rats. *Reproductive Toxicology*, 76, 35–45. https://doi.org/10.1016/j.reprotox.2017.12.006

Jung, T., Simat, T. J., Altkofer, W., & Fügel, D., 2013. Survey on the occurrence of photo-initiators and amine synergists in cartonboard packaging on the German market and their migration into the packaged foodstuffs. *Food Additives & Contaminants: Part A*, 30(11), 1993–2016. https://doi.org/10.1080/19440049.2013.837586

Kadic, A., Oles, P., Fischer, B. C., Reetz, A. E., Sylla, B. S., Feiertag, K., Ritz, V., Heise, T., Marx-Stoelting, P., Tralau, T., Renko, K., & Solano, M. d. L. M., 2024. In vitro and in vivo investigation of a thyroid hormone system-specific interaction with triazoles. *Scientific Reports*, 14(1), 6503. https://doi.org/10.1038/s41598-024-55019-3

Kamaz, M., Wickramasinghe, S. R., Eswaranandam, S., Zhang, W., Jones, S. M., Watts, M. J., & Qian, X., 2019. Investigation into micropollutant removal from wastewaters by a membrane bioreactor. *International Journal of Environmental Research and Public Health*, 16(8), article 1363. https://doi.org/10.3390/ijerph16081363

Kathuria, D., Hamid, Gautam, S., & Thakur, A., 2023. Maillard reaction in different food products: Effect on product quality, human health and mitigation strategies. *Food Control*, 153, 109911. https://doi.org/10.1016/j.foodcont.2023.109911

Katibi, K. K., Yunos, K. F., Man, H. C., Aris, A. Z., bin Mohd Nor, M. Z., & binti Azis, R. S., 2021. Recent advances in the rejection of endocrine-disrupting compounds from water using membrane and membrane bioreactor technologies: A review. *Polymers*, 13(3), 1–52. https://doi.org/10.3390/polym13030392

Katibi, K. K., Yunos, K. F., Man, H. C., Aris, A. Z., Nor, M. Z. M., Azis, R. S., & Umar, A. M., 2021. Contemporary techniques for remediating endocrine-disrupting compounds in various water sources: Advances in treatment methods and their limitations. *Polymers*, 13(19), 1–46. https://doi.org/10.3390/polym13193229

Khan, M. I., Ashfaq, F., Alsayegh, A. A., Hamouda, A., Khatoon, F., Altamimi, T. N., Alhodieb, F. S., & Beg, M. M. A., 2023. Advanced glycation end product signaling and metabolic complications: Dietary approach. *World Journal of Diabetes*, 14(7), 995. https://dx.doi.org/10.4239/wjd.v14.i7.995

Khan, R., Anwar, F., & Ghazali, F. M., 2024. A comprehensive review of mycotoxins: Toxicology, detection, and effective mitigation approaches. *Heliyon*, 10(8), e28361. https://doi.org/10.1016/j.heliyon.2024.e28361

Khandpur, N., Neri, Daniela A., Monteiro, C., Mazur, A., Frelut, M.-L., Boyland, E., Weghuber, D., & Thivel, D., 2020. Ultra-processed food consumption among the paediatric population: An overview and call to action from the European Childhood obesity Group. *Annals of Nutrition and Metabolism*, 76(2), 109–113. https://doi.org/10.1159/000507840

Kohantorabi, M., Giannakis, S., Gholami, M. R., Feng, L., & Pulgarin, C., 2019. A systematic investigation on the bactericidal transient species generated by photo-sensitization of natural organic matter (NOM) during solar and photo-Fenton disinfection of surface waters. *Applied Catalysis B: Environmental,* 244, 983–995. https://doi.org/10.1016/j.apcatb.2018.12.012

Kristanti, R. A., Tirtalistyani, R., Tang, Y. Y., Thao, N. T. T., Kasongo, J., & Wijayanti, Y., 2023. Phytoremediation mechanism for emerging pollutants: A review. *Tropical Aquatic and Soil Pollution*, 3(1), 88–108. https://doi.org/10.53623/tasp.v3i1.222

Lee, A., Bensaada, S., Lamothe, V., Lacoste, M., & Bennetau-Pelissero, C., 2022. Endocrine disruptors on and in fruits and vegetables: Estimation of the potential exposure of the French population. *Food Chemistry,* 373, 131513. https://doi.org/10.1016/j.foodchem.2021.131513

Lee, D. Y., Lee, S. Y., Jo, C., Yoon, Y., Jeong, J. Y., & Hur, S. J., 2021. Effect on health from consumption of meat and meat products. *Journal of Animal Science and Technology*, 63(5), 955–976. https://doi.org/10.5187/jast.2021.e101

Leksir, C., Boudalia, S., Moujahed, N., & Chemmam, M., 2019. Traditional dairy products in Algeria: Case of Klila cheese. *Journal of Ethnic Foods*, 6(1), 7. https://doi.org/10.1186/s42779-019-0008-4

Li, C., Li, C., Yu, H., Cheng, Y., Xie, Y., Yao, W., Guo, Y., & Qian, H., 2021. Chemical food contaminants during food processing: Sources and control. *Critical Reviews in Food Science and Nutrition*, 61(9), 1545–1555. https://doi.org/10.1080/10408398.2020.1762069

Li, X., Jiang, L., & Li, H., 2018. Application of ultrafiltration technology in water treatment. *IOP Conference Series: Earth and Environmental Science*, 186(3), article 012009. https://doi.org/10.1088/1755-1315/186/3/012009

Liang, Y., Wang, W., Shen, Y., Liu, Y., & Liu, X. J., 2012. Effects of home preparation on organophosphorus pesticide residues in raw cucumber. *Food Chemistry*, 133(3), 636–640. https://doi.org/10.1016/j.foodchem.2012.01.016

Lin, N., 2022. Migration of Bisphenol A and Its Related Compounds in Canned Seafood and Dietary Exposure Estimation. Oxford University Press (OUP). https://doi.org/10.1093/fqsafe/fyac006

Liu, D., Shi, Q., Liu, C., Sun, Q., & Zeng, X., 2023. Effects of endocrine-disrupting heavy metals on human health. *Toxics*, 11(4), 322. https://doi.org/10.3390/toxics11040322

Liu, G., Cai, W., Liu, H., Jiang, H., Bi, Y., & Wang, H., 2021. The association of bisphenol A and phthalates with risk of breast cancer: A meta-analysis. *International Journal of Environmental Research and Public Health*, 18(5), 2375. https://doi.org/10.3390/ijerph18052375

Liu, X., Jiang, B., Yin, X., Ma, H., & Hsiao, B. S., 2020. Highly permeable nanofibrous composite microfiltration membranes for removal of nanoparticles and heavy metal ions. *Separation and Purification Technology*, 233(631), 1–35. https://doi.org/10.1016/j.seppur.2019.115976

Ma, Y., Liu, H., Wu, J., Yuan, L., Wang, Y., Du, X., Wang, R., Marwa, P. W., Petlulu, P., Chen, X., & Zhang, H., 2019. The adverse health effects of bisphenol A and related toxicity mechanisms. *Environmental Research*, 176, 108575. https://doi.org/10.1016/j.envres.2019.108575

Maftouh, A., El Fatni, O., El Hajjaji, S., Jawish, M. W., & Sillanpää, M., 2023. Comparative review of different adsorption techniques used in heavy metals removal in water. *Biointerface Research in Applied Chemistry*, 13(4), 1–19. https://doi.org/10.33263/briac134.397

Mahmoud, A. E. D., & Mostafa, E., 2023. Nanofiltration membranes for the removal of heavy metals from aqueous solutions: Preparations and applications. *Membranes*, 13(9), article 789. https://doi.org/10.3390/membranes13090789

Malematja, E., Manyelo, T. G., Sebola, N. A., Kolobe, S. D., & Mabelebele, M., 2023. The accumulation of heavy metals in feeder insects and their impact on animal production. *Science of the Total Environment*, 885, 163716. https://doi.org/10.1016/j.scitotenv.2023.163716

Marcelo, A. N., Kamila, O. G., & Grace, C., 2017. Determination and removal of endocrine disruptors in wastewater by activated carbon. *Journal of Civil Engineering and Architecture*, 11(7), 653–662. https://doi.org/10.17265/1934-7359/2017.07.003

Marino, M., Puppo, F., Del Bo', C., Vinelli, V., Riso, P., Porrini, M., & Martini, D., 2021. A systematic review of worldwide consumption of ultra-processed foods: Findings and criticisms. *Nutrients*, 13(8), 2778. https://doi.org/10.3390/nu13082778

Mary, M. M., 2023. Food additives: Recent trends in the food sector. In A. Dr. Muhammad Sajid & K. Mr. Waseem (Eds.), *Health Risks of Food Additives – Recent Developments and Trends in Food Sector (pp. Ch. 1)*. Rijeka: IntechOpen. https://doi.org/10.5772/intechopen.109484

Matoso, V., Bargi-Souza, P., Ivanski, F., Romano, M. A., & Romano, R. M., 2019. Acrylamide: A review about its toxic effects in the light of Developmental Origin of Health and Disease (DOHaD) concept. *Food Chemistry*, 283, 422–430. https://doi.org/10.1016/j.foodchem.2019.01.054

Memiş, E. Y., Yalçın, S. S., & Yalçın, S., 2021. Mycotoxin carry-over in breast milk and weight of infant in exclusively-breastfed infants. *Archives of Environmental & Occupational Health*, 76(6), 313–318. https://doi.org/10.1080/19338244.2020.1828242

Meng, R., Zhu, Q., Long, T., He, X., Luo, Z., Gu, R., Wang, W., & Xiang, P., 2023. The innovative and accurate detection of heavy metals in foods: A critical review on electrochemical sensors. *Food Control*, 150, 109743. https://doi.org/10.1016/j.foodcont.2023.109743

Merhi, Z., Du, X. Q., & Charron, M. J., 2020. Perinatal exposure to high dietary advanced glycation end products affects the reproductive system in female offspring in mice. *Molecular Human Reproduction*, 26(8), 615–623. https://doi.org/10.1093/molehr/gaaa046

Messina, M., Mejia, S. B., Cassidy, A., Duncan, A., Kurzer, M., Nagato, C., Ronis, M., Rowland, I., Sievenpiper, J., & Barnes, S., 2022. Neither soyfoods nor isoflavones warrant classification as endocrine disruptors: A technical review of the observational and clinical data. *Critical Reviews in Food Science and Nutrition*, 62(21), 5824–5885. 10.1080/10408398.2021.1895054

Meyer, S. K., Probert, P. M. E., Lakey, A. F., Axon, A. R., Leitch, A. C., Williams, F. M., Jowsey, P. A., Blain, P. G., Kass, G. E. N., & Wright, M. C., 2017. Hepatic effects of tartrazine (E 102) after systemic exposure are independent of oestrogen receptor interactions in the mouse. *Toxicology Letters*, 273, 55–68. https://doi.org/10.1016/j.toxlet.2017.03.024

Mg, R., Girigoswami, A., Chakraborty, S., & Girigoswami, K., 2022. Bisphenol A–An overview on its effect on health and environment. *Biointerface Research in Applied Chemistry*, 12(1), 105–119. https://doi.org/10.33263/BRIAC121.105119

Miralles, P., Beser, M. I., Sanchís, Y., Yusà, V., & Coscollà, C., 2023. Determination of 21 per-and poly-fluoroalkyl substances in paper-and cardboard-based food contact materials by ultra-high-performance liquid chromatography coupled to high-resolution mass spectrometry. *Analytical Methods*, 15(12), 1559–1568.

Mishra, D., 2020. Food colors and associated oxidative stress in chemical carcinogenesis. In S. Chakraborti, B. K. Ray & S. Roychowdhury (Eds.), *Handbook of Oxidative Stress in Cancer: Mechanistic Aspects* (pp. 1–14). Singapore: Springer Singapore. https://doi.org/10.1007/978-981-15-4501-6_182-1

Mitra, S., Chakraborty, A. J., Tareq, A. M., Emran, T. B., Nainu, F., Khusro, A., Idris, A. M., Khandaker, M. U., Osman, H., Alhumaydhi, F. A., & Simal-Gandara, J., 2022. Impact of heavy metals on the environment and human health: Novel therapeutic insights to counter the toxicity. *Journal of King Saud University–Science*, 34(3), 101865. https://doi.org/10.1016/j.jksus.2022.101865

Moreira, C. G., Moreira, M. H., Silva, V. M. O. C., Santos, H. G., Bila, D. M., & Fonseca, F. V., 2019. Treatment of bisphenol A (BPA) in water using UV/H2O2 and reverse osmosis (RO) membranes: Assessment of estrogenic activity and membrane adsorption. *Water Science and Technology*, 80(11), 2169–2178. https://doi.org/10.2166/wst.2020.024

Moukhtari, F. E., Martín-Pozo, L., & Zafra-Gómez, A., 2023. Strategies based on the use of microorganisms for the elimination of pollutants with endocrine-disrupting activity in the environment. *Journal of Environmental Chemical Engineering*, 11(1), article 109268. https://doi.org/10.1016/j.jece.2023.109268

Mukherjee, R., Pandya, P., Baxi, D., & Ramachandran, A. V., 2021. Endocrine disruptors–Food' for thought. *Proceedings of the Zoological Society*, 74(4), 432–442. https://doi.org/10.1007/s12595-021-00414-1

Muncke, J., 2021. Tackling the toxics in plastics packaging. *PLOS Biology*, 19(3), e3000961. https://doi.org/10.1371/journal.pbio.3000961

Naik, A. Q., Zafar, T., & Shrivastava, V. K., 2023. The impact of non-caloric artificial sweetener aspartame on female reproductive system in mice model. *Reproductive Biology and Endocrinology*, 21(1), 73. https://doi.org/10.1186/s12958-023-01115-4

Naman, M., Masoodi, F. A., Wani, S. M., & Ahad, T., 2022. Changes in concentration of pesticide residues in fruits and vegetables during household processing. *Toxicology Reports*, 9, 1419–1425. https://doi.org/10.1016/j.toxrep.2022.06.013

Nappi, F., Barrea, L., Di Somma, C., Savanelli, M. C., Muscogiuri, G., Orio, F., & Savastano, S., 2016. Endocrine aspects of environmental "Obesogen" pollutants. *International Journal of Environmental Research and Public Health*, 13(8), 765. https://doi.org/10.3390/ijerph13080765

Nardocci, M., Polsky, J. Y., & Moubarac, J.-C., 2021. Consumption of ultra-processed foods is associated with obesity, diabetes and hypertension in Canadian adults. *Canadian Journal of Public Health*, 112(3), 421–429. https://doi.org/10.17269/s41997-020-00429-9

Nasri, A., & Pohjanvirta, R., 2021. In vitro estrogenic, cytotoxic, and genotoxic profiles of the xenoestrogens 8-prenylnaringenine, genistein and tartrazine. *Environmental Science and Pollution Research*, 28(22), 27988–27997. https://doi.org/10.1007/s11356-021-12629-y

Nedjimi, B., 2021. Phytoremediation: A sustainable environmental technology for heavy metals decontamination. *SN Applied Sciences*, 3(3), 1–19. https://doi.org/10.1007/s42452-021-04301-4

Nguyen, M. N., Trinh, P. B., Burkhardt, C. J., & Schäfer, A. I., 2021. Incorporation of single-walled carbon nanotubes in ultrafiltration support structure for the removal of steroid hormone micropollutants. *Separation and Purification Technology*, 264(January), 1–13. https://doi.org/10.1016/j.seppur.2021.118405

Niavarani, Z., Breite, D., Prager, A., Abel, B., & Schulze, A., 2021. Estradiol removal by adsorptive coating of a microfiltration membrane. *Membranes*, 11(2), 1–13. https://doi.org/10.3390/membranes11020099

Niavarani, Z., Breite, D., Yasir, M., Sedlarik, V., Prager, A., Schönherr, N., Abel, B., Gläser, R., & Schulze, A., 2024. Removal of endocrine disrupting chemicals from water through urethane functionalization of microfiltration membranes via electron beam irradiation. *Frontiers of Environmental Science and Engineering*, 18(45). https://doi.org/10.1007/s11783-024-1805-6

Nowak, K., Ratajczak–Wrona, W., Górska, M., & Jabłońska, E., 2018. Parabens and their effects on the endocrine system. *Molecular and Cellular Endocrinology*, 474, 238–251. https://doi.org/10.1016/j.mce.2018.03.014

Olmedo, P., Chen, R., Grau-Perez, M., Nachman, K. E., Gil, F., Rule, A. M., & Navas-Acien, A., 2024. Manufacturing practices as potential sources of cadmium and other relevant metals in processed meat products. Risk and nutritional assessment. *Food Control*, 158, 110225. https://doi.org/10.1016/j.foodcont.2023.110225

Ong, H.-T., Samsudin, H., & Soto-Valdez, H., 2022. Migration of endocrine-disrupting chemicals into food from plastic packaging materials: An overview of chemical risk assessment, techniques to monitor migration, and international regulations. *Critical Reviews in Food Science and Nutrition*, 62(4), 957–979. https://doi.org/10.1080/10408398.2020.1830747

Orosun, M. M., Nwabachili, S., Alshehri, R. F., Omeje, M., Alshdoukhi, I. F., Okoro, H. K., Ogunkunle, C. O., Louis, H., Abdulhamid, F. A., Osahon, S. E., Mohammed, A. U., Ehinlafa, E. O., Yunus, S. O., & Ife-Adediran, O., 2023. Potentially toxic metals in irrigation water, soil, and vegetables and their health risks using Monte Carlo models. *Scientific Reports*, 13(1), 21220. https://doi.org/10.1038/s41598-023-48489-4

Ousji, O., & Sleno, L., 2020. Identification of in vitro metabolites of synthetic phenolic antioxidants BHT, BHA, and TBHQ by LC-HRMS/MS. *International Journal of Molecular Sciences*, 21(24), 9525. https://doi.org/10.3390/ijms21249525

Pałkowska-Goździk, E., Bigos, A., & Rosołowska-Huszcz, D., 2018. Type of sweet flavour carrier affects thyroid axis activity in male rats. *European Journal of Nutrition*, 57(2), 773–782. https://doi.org/10.1007/s00394-016-1367-x

Patisaul, H. B., & Jefferson, W., 2010. The pros and cons of phytoestrogens. *Frontiers in Neuroendocrinology*, 31(4), 400–419. https://doi.org/10.1016/j.yfrne.2010.03.003

Pelch, K. E., McKnight, T., & Reade, A., 2023. 70 analyte PFAS test method highlights need for expanded testing of PFAS in drinking water. *Science of The Total Environment*, 876, 162978

Pezzella, C., Macellaro, G., Sannia, G., Raganati, F., Olivieri, G., Marzocchella, A., Schlosser, D., & Piscitelli, A., 2017. Exploitation of Trametes versicolor for bioremediation of endocrine disrupting chemicals in bioreactors. *PLoS One*, 12(6), 1–12. https://doi.org/10.1371/journal.pone.0178758

Pironti, C., Ricciardi, M., Proto, A., Bianco, P. M., Montano, L., & Motta, O., 2021. Endocrine-disrupting compounds: An overview on their occurrence in the aquatic environment and human exposure. *Water (Switzerland)*, 13(10), 1–32. https://doi.org/10.3390/w13101347

Piva, E., Fais, P., Ioime, P., Forcato, M., Viel, G., Cecchetto, G., & Pascali, J. P., 2023. Per-and polyfluoroalkyl substances (PFAS) presence in food: Comparison among fresh, frozen and ready-to-eat vegetables. *Food Chemistry*, 410, 135415

Pleadin, J., Frece, J., & Markov, K., 2019. Chapter Eight–Mycotoxins in food and feed. In F. Toldrá (Ed.), *Advances in Food and Nutrition Research* (Vol. 89, pp. 297–345): Academic Press. https://doi.org/10.1016/bs.afnr.2019.02.007

Pop, A., Drugan, T., Gutleb, A. C., Lupu, D., Cherfan, J., Loghin, F., & Kiss, B., 2018. Estrogenic and anti-estrogenic activity of butylparaben, butylated hydroxyanisole, butylated hydroxytoluene and propyl gallate and their binary mixtures on two estrogen responsive cell lines (T47D-Kbluc, MCF-7). *Journal of Applied Toxicology*, 38(7), 944–957. https://doi.org/10.1002/jat.3601

Pop, A., Kiss, B., & Loghin, F., 2013. Endocrine disrupting effects of butylated hydroxyanisole (BHA–E320). *Clujul Medical*, 86(1), 16–20

Poulsen, R., Cedergreen, N., Hayes, T., & Hansen, M., 2018. Nitrate: An environmental endocrine disruptor? A review of evidence and research needs. *Environmental Science & Technology*, 52(7), 3869–3887. https://doi.org/10.1021/acs.est.7b06419

Qureshi, U. A., Hameed, B. H., & Ahmed, M. J., 2020. Adsorption of endocrine disrupting compounds and other emerging contaminants using lignocellulosic biomass-derived porous carbons: A review. *Journal of Water Process Engineering*, 38(March), 101380–101380. https://doi.org/10.1016/j.jwpe.2020.101380

Ragab, D., Gomaa, H. G., Sabouni, R., Salem, M., Ren, M., & Zhu, J., 2016. Micropollutants removal from water using microfiltration membrane modified with ZIF-8 metal organic frameworks (MOFs). *Chemical Engineering Journal*, 300, 273–279. https://doi.org/10.1016/j.cej.2016.04.033

Rai, P. K., Lee, S. S., Zhang, M., Tsang, Y. F., & Kim, K.-H., 2019. Heavy metals in food crops: Health risks, fate, mechanisms, and management. *Environment International*, 125, 365–385. https://doi.org/10.1016/j.envint.2019.01.067

Raji, M., Mirbagheri, S. A., Ye, F., & Dutta, J., 2021. Nano zero-valent iron on activated carbon cloth support as Fenton-like catalyst for efficient color and COD removal from melanoidin wastewater. *Chemosphere*, 263, 127945–127945. https://doi.org/10.1016/j.chemosphere.2020.127945

Ramírez Carnero, A., Lestido-Cardama, A., Vazquez Loureiro, P., Barbosa-Pereira, L., Rodríguez Bernaldo de Quirós, A., & Sendón, R., 2021. Presence of Perfluoroalkyl and Polyfluoroalkyl Substances (PFAS) in Food Contact Materials (FCM) and its migration to food. *Foods*, 10(7), 1443. https://doi.org/10.3390/foods10071443

Rancière, F., Lyons, J. G., Loh, V. H. Y., Botton, J., Galloway, T., Wang, T., Shaw, J. E., & Magliano, D. J., 2015. Bisphenol A and the risk of cardiometabolic disorders: A systematic review with meta-analysis of the epidemiological evidence. *Environmental Health*, 14(1), 46. https://doi.org/10.1186/s12940-015-0036-5

Rather, I. A., Koh, W. Y., Paek, W. K., & Lim, J., 2017. The sources of chemical contaminants in food and their health implications. [Mini Review]. *Frontiers in Pharmacology*, 8. https://doi.org/10.3389/fphar.2017.00830

Ravichandran, G., Lakshmanan, D. K., Arunachalam, A., & Thilagar, S., 2022. Food obesogens as emerging metabolic disruptors; A toxicological insight. *The Journal of Steroid Biochemistry and Molecular Biology*, 217, 106042. https://doi.org/10.1016/j.jsbmb.2021.106042

Ravichandran, G., Lakshmanan, D. K., Raju, K., Elangovan, A., Nambirajan, G., Devanesan, A. A., & Thilagar, S., 2019. Food advanced glycation end products as potential endocrine disruptors: An emerging threat to contemporary and future generation. *Environment International*, 123, 486–500. https://doi.org/10.1016/j.envint.2018.12.032

Roberts, A., 2016. The safety and regulatory process for low calorie sweeteners in the United States. *Physiology & Behavior*, 164, 439–444. https://doi.org/10.1016/j.physbeh.2016.02.039

Rodrigues, I., & Naehrer, K., 2012. A three-year survey on the worldwide occurrence of mycotoxins in feedstuffs and feed. *Toxins*, 4(9), 663–675. https://doi.org/10.3390/toxins4090663

Rodríguez-Alegre, R., Pérez Megías, L., Sanchis, S., Andecochea Saiz, C., & You, X., 2024. Nanofiltration & reverse osmosis technical assessment for pesticides removal. *Discover Environment*, 2(1). https://doi.org/10.1007/s44274-024-00075-9

Rudel, R. A., Gray, J. M., Engel, C. L., Rawsthorne, T. W., Dodson, R. E., Ackerman, J. M., Rizzo, J., Nudelman, J. L., & Brody, J. G., 2011. Food packaging and Bisphenol A and Bis(2-Ethyhexyl) phthalate exposure: Findings from a dietary intervention. *Environmental Health Perspectives*, 119(7), 914–920. https://doi.org/10.1289/ehp.1003170

Sakhi, A. K., Lillegaard, I. T. L., Voorspoels, S., Carlsen, M. H., Løken, E. B., Brantsæter, A. L., Haugen, M., Meltzer, H. M., & Thomsen, C., 2014. Concentrations of phthalates and bisphenol A in Norwegian foods and beverages and estimated dietary exposure in adults. *Environment International*, 73, 259–269. https://doi.org/10.1016/j.envint.2014.08.005

Same, O. S., Nobosse, P., Ngea, G. L. N., Piveteau, C., Lemdani, M., Kamga, R., & Deprez, B., 2023. Migration study of phthalates from non-food plastic containers used in food preservation. *Heliyon*, 9(9). https://doi.org/10.1016/j.heliyon.2023.e20002

Sardrood, B. P., Goltapeh, E. M., & Varma, A., 2013. An introduction to bioremediation. In E. M. Goltapeh, Y. R. Danesh & A. Varma (Eds.), *Fungi as Bioremediators* (pp. 3–27). Berlin, Heidelberg: Springer Berlin Heidelberg. https://doi.org/10.1007/978-3-642-33811-3_1

Schaarschmidt, S., & Fauhl-Hassek, C., 2021. The fate of mycotoxins during secondary food processing of maize for human consumption. *Comprehensive Reviews in Food Science and Food Safety*, 20(1), 91–148. https://doi.org/10.1111/1541-4337.12657

Schecter, A., Lorber, M., Guo, Y., Wu, Q., Yun, S. H., Kannan, K., Hommel, M., Imran, N., Hynan, L. S., Cheng, D., Colacino, J. A., & Birnbaum, L. S., 2013. Phthalate concentrations and dietary exposure from food purchased in New York State. *Environmental Health Perspectives*, 121(4), 473–479. https://doi.org/10.1289/ehp.1206367

Scholes, R. C., Hart-Cooper, W., Glenn, G. M., & Orts, W. J., 2024. *Poly-and Perfluorinated Alkyl Substances in Food Packaging Materials Food Packaging Materials: Current Protocols* (pp. 99–114): Springer.

Serrano, S. E., Braun, J., Trasande, L., Dills, R., & Sathyanarayana, S., 2014. Phthalates and diet: A review of the food monitoring and epidemiology data. *Environmental Health*, 13(1), 43. https://doi.org/10.1186/1476-069X-13-43

Sharma, P., Singh, S. P., Pandey, S., Thanki, A., & Singh, N. K., 2020. 2–Role of potential native weeds and grasses for phytoremediation of endocrine-disrupting pollutants discharged from pulp paper industry waste. In V. C. Pandey & V. Singh (Eds.), *Bioremediation of Pollutants* (pp. 17–37): Elsevier. https://doi.org/10.1016/B978-0-12-819025-8.00002-8

Si, X., Hu, Z., & Huang, S., 2018. Combined process of ozone oxidation and ultrafiltration as an effective treatment technology for the removal of endocrine-disrupting chemicals. *Applied Sciences (Switzerland)*, 8(8), article 1240. https://doi.org/10.3390/app8081240

Simonetti, G., Riccardi, C., Pomata, D., Acquaviva, L., Fricano, A., Buiarelli, F., Senofonte, M., & Di Filippo, P., 2024. Studies of potential migration of hazardous chemicals from sustainable food contact Materials. *Foods*, 13(5), 645

Singh, K., Tripathi, S., & Chandra, R., 2023. Bacterial assisted phytoremediation of heavy metals and organic pollutants by Cannabis sativa as accumulator plants growing on distillery sludge for ecorestoration of polluted site. *Journal of Environmental Management*, 332(January), 117294–117294. https://doi.org/10.1016/j.jenvman.2023.117294

Singh, S., Pant, A., Dutta, K., Rani, R., Vithanage, M., & Daverey, A., 2024. Phytoremediation of pharmaceuticals and personal care products using the constructed wetland. *Environmental Chemistry and Ecotoxicology*, 6(April), 104–116. https://doi.org/10.1016/j.enceco.2024.04.001

Snedeker, S. M., 2014. Benzophenone UV-Photoinitiators used in food packaging: Potential for human exposure and health risk considerations. In S. M. Snedeker (Ed.), *Toxicants in Food Packaging and Household Plastics: Exposure and Health Risks to Consumers* (pp. 151–176). London: Springer London. https://doi.org/10.1007/978-1-4471-6500-2_6

Solak, S., Vakondios, N., Tzatzimaki, I., Diamadopoulos, E., Arda, M., Kabay, N., & Yüksel, M., 2014. A comparative study of removal of endocrine disrupting compounds (EDCs) from treated wastewater using highly crosslinked polymeric adsorbents and activated carbon. *Journal of Chemical Technology and Biotechnology*, 89(6), 819–824. https://doi.org/10.1002/jctb.4315

Solecki, R., Kortenkamp, A., Bergman, A., Chahoud, I., Degen, G. H., Dietrich, D., Greim, H., Hakansson, H., Hass, U., Husoy, T., Jacobs, M., Jobling, S., Mantovani, A., Marx-Stoelting, P., Piersma, A., Ritz, V., Slama, R., Stahlmann, R., van den Berg, M., Zoeller, R. T., & Boobis, A. R., 2017. Scientific principles for the identification of endocrine-disrupting chemicals: A consensus statement. *Archives of Toxicology*, 91(2), 1001–1006. https://doi.org/10.1007/s00204-016-1866-9

Song, Y., Hauser, R., Hu, F. B., Franke, A. A., Liu, S., & Sun, Q., 2014. Urinary concentrations of bisphenol A and phthalate metabolites and weight change: A prospective investigation in US women. *International Journal of Obesity*, 38(12), 1532–1537. https://doi.org/10.1038/ijo.2014.63

Soto-Verjel, J., Maturana, A. Y., & Villamizar, S. E., 2022. Advanced catalytic oxidation coupled to biological systems to treat pesticidecontaminated water: A review on technological trends and future challenges. *Water Science & Technology*, 85(4), 1263–1294. https://doi.org/10.2166/wst.2021.642

Sridevi, V., Naveen, P., Karnam, V. S., Reddy, P. R., & Arifullah, M., 2021. Beneficiary and adverse effects of phytoestrogens: A potential constituent of plant-based diet. *Current Pharmaceutical Design*, 27(6), 802–815. https://doi.org/10.2174/13816128269992009171547 47

Stiefel, C., & Stintzing, F., 2023. Endocrine-active and endocrine-disrupting compounds in food–Occurrence, formation and relevance. *NFS Journal*, 31, 57–92. https://doi.org/10.1016/j.nfs.2023.03.004

Sun, B., & Wang, J., 2017. Food Additives. In John (Ed.), Food Safety in China (pp. 186–200): Wiley & Sons. https://doi.org/10.1002/9781119238102.ch12

Sun, Z., Gao, R., Chen, X., Liu, X., Ding, Y., Geng, Y., Mu, X., Liu, T., Li, F., Wang, Y., & He, J., 2021. Exposure to butylated hydroxytoluene compromises endometrial decidualization during early pregnancy. *Environmental Science and Pollution Research*, 28(31), 42024–42036. https://doi.org/10.1007/s11356-021-13720-0

Sundhar, S., Shakila, R. J., Shalini, R., Aanand, S., & Jayakumar, N., 2023. Impact of thermal cooking processes on organochlorine pesticide residues (OCPs) in the edible green seaweed, Ulva lactuca, and associated human health risk assessment. *Journal of Food Composition and Analysis*, 121, 105370. https://doi.org/10.1016/j.jfca.2023.105370

Suwannarin, N., Prapamontol, T., Isobe, T., Nishihama, Y., Hashimoto, Y., Mangklabruks, A., Pantasri, T., Chantara, S., Naksen, W., & Nakayama, S. F., 2021. Exposure to organophosphate and neonicotinoid insecticides and its association with steroid hormones among male reproductive-age farmworkers in Northern Thailand. *International Journal of Environmental Research and Public Health*, 18(11), 5599. https://doi.org/10.3390/ijerph18115599

Tapia-orozco, N., Ibarra-cabrera, R., Tecante, A., Gimeno, M., Parra, R., & Garcia-arrazola, R., 2016. Removal strategies for endocrine disrupting chemicals using cellulose-based materials as adsorbents: A review. *Biochemical Pharmacology*, 4(4), 3122–3142. https://doi.org/10.1016/j.jece.2016.06.025

Tarannum, N., Nipa, M. N., Das, S., & Parveen, S., 2020. Aflatoxin M1 detection by ELISA in raw and processed milk in Bangladesh. *Toxicology Reports*, 7, 1339–1343. https://doi.org/10.1016/j.toxrep.2020.09.012

TEDX. The Endocrine Disruption Exchange. List of potential endocrine disruptors, the endocrine disruptor exchange. Retrieved July 7, 2024, from www.endocrinedisruption.com/endocrine.TEDXList.overview.php.

Thacharodi, A., Hassan, S., Hegde, T. A., Thacharodi, D. D., Brindhadevi, K., & Pugazhendhi, A., 2023. Water a major source of endocrine-disrupting chemicals: An overview on the occurrence, implications on human health and bioremediation strategies. *Environmental Research*, 231(P1), 116097–116097. https://doi.org/10.1016/j.envres.2023.116097

Thoene, M., Rytel, L., Nowicka, N., & Wojtkiewicz, J., 2018. The state of bisphenol research in the lesser developed countries of the EU: A mini-review. *Toxicology Research*, 7(3), 371–380. https://doi.org/10.1039/c8tx00064f

Tian, F., Woo, S. Y., Lee, S. Y., Park, S. B., Im, J. H., & Chun, H. S., 2022. Mycotoxins in soybean-based foods fermented with filamentous fungi: Occurrence and preventive strategies. *Comprehensive Reviews in Food Science and Food Safety*, 21(6), 5131–5152. https://doi.org/10.1111/1541-4337.13032

Tobias, D. K., & Hall, K. D., 2021. Eliminate or reformulate ultra-processed foods? Biological mechanisms matter. *Cell Metabolism*, 33(12), 2314–2315. https://doi.org/10.1016/j.cmet.2021.10.005

Torrens-Mas, M., & Roca, P., 2020. Phytoestrogens for cancer prevention and treatment. *Biology*, 9(12), 427. https://doi.org/10.3390/biology9120427

Tumu, K., Vorst, K., & Curtzwiler, G., 2024. Understanding intentionally and non-intentionally added substances and associated threshold of toxicological concern in post-consumer polyolefin for use as food packaging materials. *Heliyon*, 10(1). https://doi.org/10.1016/j.heliyon.2023.e23620

Vandenberg, L. N., & Bugos, J., 2021. Assessing the public health implications of the food preservative propylparaben: Has this chemical been safely used for decades. *Current Environmental Health Reports*, 8(1), 54–70. https://doi.org/10.1007/s40572-020-00300-6

Vanni, D. C., Silvestre, W. P., & Baldasso, C., 2023. Study of the efficiency of the removal of steroid hormones from water by reverse osmosis. *Scientia cum Industria*, 11(1), e231104–e231104. https://doi.org/10.18226/23185279.e231104

Vedenyapina, M. D., Kurmysheva, A. Y., Kulaishin, S. A., & Kryazhev, Y. G., 2021. Adsorption of heavy metals on activated carbons (A review). *Solid Fuel Chemistry*, 55(2), 83–104. https://doi.org/10.3103/s0361521921020099

Vera, P., Canellas, E., Dreolin, N., Goshawk, J., & Nerín, C., 2024. The analysis of the migration of per and poly fluoroalkyl substances (PFAS) from food contact materials using ultrahigh performance liquid chromatography coupled to ion-mobility quadrupole time-of-flight mass spectrometry (UPLC-IMS-QTOF). *Talanta*, 266, 124999.

Vieira, R. A. L., Pickler, T. B., Segato, T. C. M., Jozala, A. F., & Grotto, D., 2022. Biochar from fungiculture waste for adsorption of endocrine disruptors in water. *Scientific Reports*, 12(1), 6507. https://doi.org/10.1038/s41598-022-10165-4

Vieira, W. T., de Farias, M. B., Spaolonzi, M. P., da Silva, M. G. C., & Vieira, M. G. A., 2020. Removal of endocrine disruptors in waters by adsorption, membrane filtration and biodegradation. A review. *Environmental Chemical Letters*, 18, 1113–1143. https://doi.org/10.1007/s10311-020-01000-1

Vilarinho, F., Sendón, R., van der Kellen, A., Vaz, M. F., & Silva, A. S., 2019. Bisphenol A in food as a result of its migration from food packaging. *Trends in Food Science & Technology*, 91, 33–65. https://doi.org/10.1016/j.tifs.2019.06.012

Wang, R., Ma, X., Liu, T., Li, Y., Song, L., Tjong, S. C., Cao, L., Wang, W., Yu, Q., & Wang, Z., 2020. Degradation aspects of endocrine disrupting chemicals: A review on photocatalytic processes and photocatalysts. *Applied Catalysis A: General*, 597(November 2019), 117547–117547. https://doi.org/10.1016/j.apcata.2020.117547

Wang, W., & Kannan, K., 2019. Quantitative identification of and exposure to synthetic phenolic antioxidants, including butylated hydroxytoluene, in urine. *Environment International*, 128, 24–29. https://doi.org/10.1016/j.envint.2019.04.028

Wang, X., Ha, D., Yoshitake, R., Chan, Y. S., Sadava, D., & Chen, S., 2021. Exploring the biological activity and mechanism of xenoestrogens and phytoestrogens in cancers: Emerging methods and concepts. *International Journal of Molecular Sciences*, 22(16), 8798. https://doi.org/10.3390/ijms22168798

Wang, X., Luo, Z.-C., Du, O., Zhang, H.-J., Fan, P., Ma, R., Chen, Y., Wang, W., Zhang, J., & Ouyang, F., 2023. The association between maternal urinary Bisphenol A levels and neurodevelopment at age 2 years in Chinese boys and girls: A prospective cohort study. *Ecotoxicology and Environmental Safety*, 264, 115413. https://doi.org/10.1016/j.ecoenv.2023.115413

Westrup, J. L., Bertoldi, C., Cercena, R., Dal-Bó, A. G., Soares, R. M. D., & Fernandes, A. N., 2021. Adsorption of endocrine disrupting compounds from aqueous solution in poly(butyleneadipate-co-terephthalate) electrospun microfibers. *Colloids and Surfaces A: Physicochemical and Engineering Aspects*, 611, 125800–125800. https://doi.org/10.1016/j.colsurfa.2020.125800

WHO-UNEP, 2012. World Health Organization, 2012. United Nations Environment Programme (WHO-UNEP). In: Bergman, A., Heindel, J. J., Jobling, S., Kidd, K. A., Zoeller, R. T. (Eds.), *State of the Science of Endocrine Disrupting Chemicals*. WHO-UNEP.

Wittassek, M., & Angerer, J., 2008. Phthalates: Metabolism and exposure. *International Journal of Andrology*, 31(2), 131–138

Wittassek, M., Koch, H. M., Angerer, J., & Brüning, T., 2011. Assessing exposure to phthalates–The human biomonitoring approach. *Molecular Nutrition & Food Research*, 55(1), 7–31

Xiang, M., Ma, J., Cheng, J., Lei, K., Li, F., Shi, Z., & Li, Y., 2022. Collaborative evaluation of heavy metal pollution of soil-crop system in the southeast of Yangtze River Delta, China. *Ecological Indicators*, 143, 109412. https://doi.org/10.1016/j.ecolind.2022.109412

Xu, F., Liu, Z., Cao, Y., Qiu, L., Feng, J., Xu, F., & Tian, X., 2017. Assessment of heavy metal contamination in urban river sediments in the Jiaozhou Bay catchment, Qingdao, China. *CATENA*, 150, 9–16. https://doi.org/10.1016/j.catena.2016.11.004

Xu, X., Zhou, G., Lei, K., LeBlanc, G. A., & An, L., 2020. Phthalate esters and their potential risk in PET bottled water stored under common conditions. *International Journal of Environmental Research and Public Health*, 17(1), 141

Xue, W., Xiao, K., Liang, P., & Huang, X., 2018. Roles of membrane and organic fouling layers on the removal of endocrine disrupting chemicals in microfiltration. *Journal of Environmental Sciences (China)*, 72, 176–184. https://doi.org/10.1016/j.jes.2018.01.004

Yang, J., Song, W., Wang, X., Li, Y., Sun, J., Gong, W., & Sun, C., 2019. Migration of phthalates from plastic packages to convenience foods and its cumulative health risk assessments. *Food Additives & Contaminants: Part B*, 12(3), 151–158. https://doi.org/10.1080/19393210.2019.1574909

Yaqub, M., & Lee, S. H., 2019. Heavy metals removal from aqueous solution through micellar enhanced ultrafiltration: A review. *Environmental Engineering Research*, 24(3), 363–375. https://doi.org/10.4491/eer.2018.249

Yüksel, S., Kabay, N., & Yüksel, M., 2013. Removal of bisphenol A (BPA) from water by various nanofiltration (NF) and reverse osmosis (RO) membranes. *Journal of Hazardous Materials*, 263, 307–310. https://doi.org/10.1016/j.jhazmat.2013.05.020

Yunchu, L., Ziyan, H., Chenxi, L., Yan, Y., & Zhu, S., 2023. Exploration of treatment technology for heavy metal wastewater. *E3S Web of Conferences*, 393(03031), 1–6

Zabaleta, I., Blanco-Zubiaguirre, L., Baharli, E. N., Olivares, M., Prieto, A., Zuloaga, O., & Elizalde, M. P., 2020. Occurrence of per- and polyfluorinated compounds in paper and board packaging materials and migration to food simulants and foodstuffs. *Food Chemistry*, 321, 126746. https://doi.org/10.1016/j.foodchem.2020.126746

Zaki, G., & Shoeib, T., 2018. Concentrations of several phthalates contaminants in Egyptian bottled water: Effects of storage conditions and estimate of human exposure. *Science of the Total Environment*, 618, 142–150. https://doi.org/10.1016/j.scitotenv.2017.10.337

Zergui, A., Boudalia, S., & Joseph, M. L., 2023. Heavy metals in honey and poultry eggs as indicators of environmental pollution and potential risks to human health. *Journal of Food Composition and Analysis*, 119, 105255. https://doi.org/10.1016/j.jfca.2023.105255

Zhang, R., Li, C., Li, Y., Cui, X., & Ma, L. Q., 2018. Determination of 2,6-di-tert-butyl-hydroxytoluene and its transformation products in indoor dust and sediment by gas chromatography–mass spectrometry coupled with precolumn derivatization. *Science of the Total Environment*, 619–620, 552–558. https://doi.org/10.1016/j.scitotenv.2017.11.115

Zhang, X.-J., Diao, M.-N., & Zhang, Y.-F., 2023. A review of the occurrence, metabolites and health risks of butylated hydroxyanisole (BHA). *Journal of the Science of Food and Agriculture*, 103(13), 6150–6166. https://doi.org/10.1002/jsfa.12676

Zhang, Y., Zhang, Y., Jia, J., Peng, H., Qian, Q., Pan, Z., & Liu, D., 2023. Nitrite and nitrate in meat processing: Functions and alternatives. *Current Research in Food Science*, 6, 100470. https://doi.org/10.1016/j.crfs.2023.100470

Zhao, F., Liu, J., Han, B., & Luo, J., 2019. Investigation and validation of Detection of storage stability of difenoconazole residue in Mango. *Journal of Food Quality*, 2019(1), 5641643. https://doi.org/10.1155/2019/5641643

Zhao, L., Pan, X., Yang, Q., Zhao, R., & Li, X., 2021. The ability of lactic acid bacteria strains to remove di-n-butyl phthalate in simulated food matrices. *International Journal of Food Science & Technology*, 56(2), 553–562. https://doi.org/10.1111/ijfs.14741

Zhu, X., Yin, T., Yue, X., Liao, S., Cheang, I., Zhu, Q., Yao, W., Lu, X., Shi, S., Tang, Y., Zhou, Y., Li, X., & Zhang, H., 2021. Association of urinary phthalate metabolites with cardiovascular disease among the general adult population. *Environmental Research,* 202, 111764. https://doi.org/10.1016/j.envres.2021.111764

Zhuang, W., Liu, Y., Chen, Q., Wang, Q., & Zhou, F., 2016. A new index for assessing heavy metal contamination in sediments of the Beijing-Hangzhou Grand Canal (Zaozhuang Segment): A case study. *Ecological Indicators,* 69, 252–260. https://doi.org/10.1016/j.ecolind.2016.04.029

Zielińska, M., Bułkowska, K., Cydzik-Kwiatkowska, A., Bernat, K., & Wojnowska-Baryła, I., 2016. Removal of bisphenol A (BPA) from biologically treated wastewater by microfiltration and nanofiltration. *International Journal of Environmental Science and Technology*, 13(9), 2239–2248. https://doi.org/10.1007/s13762-016-1056-6

Ziembowicz, S., Kida, M., Koszelnik, P., Czarnota, J., & Miąsik, M., 2021. Fenton-like degradation of di-n-butyl phthalate in landfill leachate by endogenous catalysts or iron, copper and manganese loaded bottom sediments. *Environmental Nanotechnology, Monitoring and Management*, 16(May), article 100551. https://doi.org/10.1016/j.enmm.2021.100551

Zong, W., Guo, Z., Wu, M., Yi, X., Zhou, H., Jing, S., Zhan, J., Liu, L., & Liu, Y., 2021. Synergistic multiple active species driven fast estrone oxidation by δ-MnO2 in the existence of methanol. *Science of the Total Environment*, 761, 143201. https://doi.org/10.1016/j.scitotenv.2020.143201

Heavy Metals and Safety

Food Processing and Manufacturing Related Effects

3

Sofiane Boudalia, Safia Habila, Zouheyr Hadri, Asma Lahouel, Anissa Zergui, Yassine Gueroui, and Aissam Bousbia

3.1 INTRODUCTION

Heavy metals in food processing and manufacturing are a major concern due to their potential adverse effects on human health. These contaminants can enter the food supply through various sources, such as polluted soil, water, and air, or during food processing. Studies have indicated that dietary exposure to heavy metals poses serious human health risks (Boudebbouz et al., 2021). For instance, vegetables irrigated with treated wastewater have been found to accumulate heavy metals, which can be hazardous to health (Qureshi et al., 2016). The extent of health dangers linked to toxic metals hinges on how much of these metals are present in specific environments and the duration of exposure. Many organs can be affected by heavy metal toxicity, including the skin, nervous system, liver, kidneys, and heart.

DOI: 10.1201/9781032712291-3

(Mitra et al., 2022) Food processing techniques can influence the levels of heavy metals in food products (Ho et al., 2021). For example, ultrasonic treatment has been investigated as a method to decontaminate milk from heavy metals like lead, mercury, and arsenic without compromising its quality. The use of low-frequency ultrasound has shown promise in reducing heavy metal contamination in milk, especially in areas with poor environmental conditions (Porova et al., 2014). Manufacturing practices in the food industry are critical in managing dietary heavy metal contamination (Lee et al., 2019). Research on food safety has underscored the significance of understanding how different processing methods impact the safety of food products (Olmedo et al., 2024). In summary, the reviewed literature highlights the crucial need to address heavy metal contamination in food processing and manufacturing to protect consumer health.

3.2 DEFINITION AND CONCEPTS

3.2.1 Heavy Metals Definition

Heavy metals are defined as metallic elements with a density higher than 5 g/cm^3 and atomic weights between 63.5 and 200.6 g/mol, exhibiting toxic effects at low concentrations (Olmedo et al., 2024) (Meng et al., 2023). Some of these metals, such as copper (Cu), zinc (Zn), chromium (Cr), nickel (Ni), cobalt (Co), molybdenum (Mo), and iron (Fe), are essential trace elements necessary for biological processes, but they can be toxic at very low or very high concentrations (Malematja et al., 2023). Others, like aluminum (Al), arsenic (As), cadmium (Cd), mercury (Hg), tin (Sn), and lead (Pb) (also called xenobiotics), have no biological role and are considered harmful, causing toxic effects even at very low concentrations (Rai et al., 2019). A common characteristic of all heavy metals is their non-biodegradability in the environment and their tendency to bioaccumulate (Zhuang et al., 2016) (Xu et al., 2017).

3.2.2 Sources of Heavy Metals

The presence of heavy metals in the environment and/or food could be due to natural sources or to the human activity.

✓ **Natural sources**

The most important natural sources of heavy metals are physical and chemical weathering, metallic minerals and parent rocks, volcanic eruptions and erosion, sea-salt sprays, forest fires, biogenic sources, sulfates, sulfides, phosphates, organic compound (Oladoye et al., 2022) (Saidon et al., 2024);

✓ **Anthropogenic sources**

The anthropogenic sources of heavy metals are depending on human activities such as industrial activities (power plants, coal combustion, industry metallurgical, automobile repair facility, chemical plants, batteries, industries, pharmaceuticals and electroplating etc.), road traffic (exhaust particles from vehicles, tire wear particles, damaged road surface particles and wear particles from brake linings), domestic emissions (sewage sludge municipal), fossil fuels combustion and agricultural activity (use of pesticides, fertilizers, and biosolids) (Lee et al., 2019) (Yousuf et al., 2022) (Machate, 2023).

3.3 ANALYSIS METHODS AND TOXICITY OF HEAVY METALS

From the literature, several studies have reported the use of various analytical methods to detect the presence of heavy metals in food. These methods include inductively coupled plasma optical emission spectrometry (ICP-OES) (Luis et al., 2015) (Sneddon et al., 2017) (Fathabad et al., 2018) flame atomic absorption spectrometry (F-AAS) (Akinyele et al., 2015) (Abbasi et al., 2020) (Adjei-Mensah et al., 2021) (Hassan et al., 2024) (Boudebbouz et al., 2022), inductively coupled plasma mass spectrometry (ICP-MS) (N. Khan et al., 2014) (Mohamed et al., 2020) (Oladeji et al., 2023) (Zeng et al., 2023) (Zergui et al., 2024), Graphite Furnace Atomic Absorption Spectrometry (GF-AAS) (de Oliveira et al., 2017); High-resolution Continuum Source Graphite Furnace Atomic Absorption; Spectrometry (HR-CS-GFAAS) (Leao et al., 2016) (Zhong et al., 2016) (Tinas et al., 2018); Anion Exchange Chromatography Coupled to Inductively Coupled Plasma Mass Spectrometry (AEC-ICP MS); Microwave Induced Plasma Optical Emission Spectrometry-(MIP OES) (Song et al., 2022); and Cold Vapor Atomic Absorption Spectroscopy (CVAAS) (Román-Ochoa et al., 2021). These techniques can detect heavy metals even at very low concentrations.

From the environment, humans can be exposed to heavy metals through inhalation, skin contact, or ingestion, particularly in workplaces and through the consumption of contaminated water and food. Several authors N. Shaheen et al. (2016) Mitra et al. (2022) have reported that the persistence and accumulation of heavy metals in the environment are responsible for their toxic effects, leading to various health issues. Heavy metals can affect living organisms by interacting with important protein and enzyme sites, including SH groups. They can compete with other essential metals for cofactors in regulatory enzymes, preventing them from functioning normally and fulfilling their metabolic roles (Balali-Mood et al., 2021) (Witkowska et al., 2021). Additionally, heavy metals can generate oxygen-free radicals and disrupt cellular processes, especially within the membrane system (Briffa et al., 2020) (Samet et al., 2020).

Prolonged consumption of foodstuffs containing heavy metals can lead to chronic accumulation; resulting in various toxic effects (see Table 3.1).

TABLE 3.1 Toxic effects of chronic exposure to low doses of heavy metals

HEAVY METALS	*PERMISSIBLE LIMITS (WHO) (PPM)*	*TOXIC EFFECT*	*REFERENCE*
Cr	20	Prolonged exposure to chromium disrupts male reproductive function and may contribute to cancers of the lung, larynx, kidney, testicles, bone, bladder, and thyroid. It can cause numerous physiological dysfunctions, including cutaneous, renal, neurological, and gastrointestinal diseases.	Balali-Mood et al. (2021) Pavesi et al. (2020)
Al	0.005 to 0.2	Chronic aluminum exposure can lead to neurological diseases, including Alzheimer's disease, Parkinson's disease, autism spectrum disorder, multiple sclerosis, and dialysis encephalopathy. It can also cause severe osteomalacia, hypoparathyroidism, microcytic anemia, cholestatic hepatotoxicity, and the inhibition of the kidney enzyme 25-hydroxyvitamin D-1 alpha-hydroxylase.	Bryliński et al. (2023) Bondy et al. (2022)
As	0.01	Chronic exposure to arsenic increases the risk of skin, lung, kidney, and liver cancer. It can also cause systemic health issues, including cardiovascular, dermatological, neurological, hepatobiliary, renal, gastrointestinal, and respiratory diseases.	Palma-Lara et al. (2020) Kesici (2016) K. M. Khan et al. (2020)
Cd	0.3	Cadmium is recognized as a carcinogen (causing breast, lung, prostate, nasopharynx, pancreas, and kidney cancers), mutagen, and endocrine disruptor. Chronic exposure can lead to anemia, emphysema, osteoporosis, renal problems, anosmia, and chronic rhinitis. It can also have a depressive effect. Cadmium may play a role in the development of diseases related to the central nervous system (CNS), such as Alzheimer's disease (AD), Parkinsonism, and Parkinson's disease (PD).	Genchi et al. (2020) Galvez-Fernandez et al. (2021) Charkiewicz et al. (2023)
Cu	40	Exposure to copper can be linked to various diseases, including Wilson's disease, hepatic disorders (cirrhosis, hepatitis, gastroenteritis), neurodegenerative diseases (Alzheimer's and Parkinson's), anemia, metabolic syndrome, cardiovascular diseases, and cancer.	Gromadzka et al. (2020) Taylor et al. (2020) Maghool et al. (2023)

TABLE 3.1 (Continued)

HEAVY METALS	*PERMISSIBLE LIMITS (WHO) (PPM)*	*TOXIC EFFECT*	*REFERENCE*
Hg	2	Chronic mercury exposure is associated with neurological disorders, including loss of balance, memory issues, difficulty speaking, tremors, sleeplessness, and neuromuscular effects. It can also cause congenital abnormalities and perinatal death. Additionally, it can lead to renal and fetal toxicity, dental and adrenal issues, diabetes, hypertension, anemia, and changes in the genetic, enzyme, and immunological systems.	Duan et al. (2020) Marrugo-Madrid et al. (2022) Petrova et al. (2020) Zhang et al. (2021)
Ni	1.5	Chronic exposure to nickel can cause several health issues, including cardiovascular disease, contact dermatitis, lung fibrosis, asthma, and respiratory tract cancer.	Genchi et al. (2020) Klein et al. (2022) Renu et al. (2021)
Pb	5	Chronic lead exposure can impair neurobehavioral development, cause learning problems, speech and language difficulties, and poor attention span, and lead to neurodegenerative diseases. It can also limit fertility, cause fetal death, spontaneous miscarriages, cardiovascular damage, and anemia, and have mutagenic or carcinogenic effects.	Shah-Kulkarni et al. (2016) Rajendra Sukhadeorao (2020)
Zn	60	Excess zinc concentration in the body can lead to neurological disorders such as Alzheimer's and Parkinson's disease. It can also interfere with copper absorption, leading to low copper status, which consequently impairs immune function.	Fan et al. (2024) Plum et al. (2010)

3.4 FOOD CONTAMINATION BY HEAVY METALS

Foods, either fresh or processed, are mainly contaminated with heavy metals (Abbasi et al., 2020). These metals first contaminate food via contact with the environment.

However, the technology processing can increase the chances of food contamination with heavy metals.

3.4.1 Environmental Contamination of Food

Heavy metals can enter plant-based foods through irrigation with polluted water, contaminated soil, or the application of pesticides and fertilizers (Rai et al., 2019). In animal-based foods, metal contamination primarily occurs through the consumption of contaminated water or feed (Zhou et al., 2019).

3.4.1.1 Contamination of plant-based food

Polluted soil is the primary cause of heavy metal contamination in crops. The content of toxic metals in vegetable samples was reported high in many regions across the globe. Xiang et al. (2022) detected elevated levels of heavy metals (As, Cu, Cr, Hg, Ni, and Pb) in rice, wheat, soybean, lettuce, potato, and orange cultivated on polluted soil in China. On the other hand, Sithole et al. (2022) detected concentrations of Cd, Zn, and Cr that exceeded the permissible limit in edibles parts of mushroom grown on mine polluted soil in South Africa.

The use of wastewater for irrigation can be a potential source of vegetable contamination with heavy metals. Throughout the world, several authors reported a high contamination of food trough irrigation, for instance, Chaoua et al. (2019) found high concentrations of Zn, Cu, Pb, and Cd in edible plants such as broad bean, durum wheat, soft wheat, oat, nettle, broadleaf, plantain, alfalfa, and mallow irrigated with wastewater in a study conducted in Marrakech (Morocco). Similarly, Orosun et al. (2023) reported a significant level of As, Cd, Cr, and Pb in Spinach, and Cabbage irrigated by contaminated water in Nigeria: in the other hand, high levels of Cd, Cr, Cu, Ni, Pb, and Zn were detected in red amaranth, radish, Indian spinach, cauliflower, and tomato irrigated by textile wastewater in Bangladesh (Hassan et al., 2024).

3.4.1.2 Contamination of animal-based food

The contamination of animal-based food can occur in several ways, such as inhaling polluted air or ingesting of contaminated water and plants (Hassan Emami et al., 2023). Numerous studies worldwide have reported the presence of heavy metals in animal-based foods such as fish, red meat, chicken meat, eggs, and raw milk (Aissioui et al., 2021) (Boudebbouz et al., 2022) (Zergui et al., 2023) (Hamed Hammad Mohammed et al., 2020) (Bousbia et al., 2019).

In a recent study carried out in Bangladesh, heavy metals such as Cd, Pb, Cr, Cu, Mn, and Zn were detected in eight fish species (Das Pinkey et al., 2024), meanwhile, Ni, Cd, Cu, Cr, and Mn were present in meat samples collected from commercial markets in Pakistan (Ahmad, 2016). Similarly, Scivicco et al. (2024), reported greater bioaccumulation of heavy metals in fish samples (*Diplodus annularis*) collected from the Gulf of Naples in Italy. On the other hand, As, Cd, Pb, and Hg were detected in five fish species in Cameroon (Manz et al., 2023). The contamination of raw milk by heavy metals has been extensively investigated. Forcada et al. (2023) detected the presence of As, Pb, and Hg in Spain, but in concentrations below the permissible limit. In another study conducted in Iran, Shahbazi et al. (2016) detected low levels of Pb, Cd, Zn, and Cu in raw milk. On the other hand, in raw milk collected from industrial areas in China, Su et al. (2021) reported ranges of Cr, As, Cd, and Pb levels at 0.54–10.61, 0.10–1.49, 0.02–0.39, and ND–15.22 μg/kg, respectively.

3.5 EFFECTS OF FOOD PROCESSING TECHNIQUES ON HEAVY METAL CONCENTRATIONS

The aim of food processing is to transform raw foods into marketable, nutritious food products with improved physicochemical properties (Chellaram et al., 2014). This process includes many operations, such as fermentation, sterilization, pasteurization, cooking, baking, roasting, toasting, frying, boiling, grilling, canning, and packaging (Ghoshal, 2018). These operations are often essential components of food production. However, as shown in Table 3.2, they can also be potential sources of contamination by various chemical compounds, including heavy metals (Hembrom et al., 2020).

TABLE 3.2 Effects of food processing techniques on heavy metal concentrations

FOOD	*PROCESSING TECHNIQUE*	*EFFECTS ON HEAVY METALS CONTENT*	*REFERENCES*
Oilseed	Extracting oils	Reduce heavy metals	Lee et al. (2019)
Noodles	Boiling		
Tea	Infusing tea		
Meats	Boiling, grilling	Reduce Cd, Ni, Co Increase Cu	Bushuty et al. (2018)
Fish	Grilling, frying and steaming	Increase heavy metals	Kazemi, Esmaeilbeigi, Ansari, et al. (2022)
Fresh yam, cocoyam, cassava, and plantain	Frying and boiling Roasting	Reduce As and Pb The least reduction in the heavy metal contaminations	Adjei-Mensah et al. (2021)
Coffee and yogurt	Packaging and cups (Polypropylene)	Migration of Pb, Cd, Hg, Cr, and Sb in foods	Eti et al. (2023)
Soft drinks and water	Bottle (Polyethylene Terephthalate)		
Take-out food	Food containers (Plastic)	Sb transfer from the food containers to food	Han et al. (2024)

(*continued*)

TABLE 3.2 (Continued)

FOOD	*PROCESSING TECHNIQUE*	*EFFECTS ON HEAVY METALS CONTENT*	*REFERENCES*
Smoked fish	Washing three times with distilled water	Decrease in metal concentration	Inobeme et al. (2020)
White fish	Grilling, boiling and micorwaving	Decrease concentration of Pb, Cr and Cd	Mehdipour et al. (2018)
Carrots	Steaming	High concentration of Cu	Diyarov et al. (2022)
Rice	Fermentation	Reduction of Cd levels in rice (over 80%)	Zhai et al. (2019)
Cocoa beans	Fermentation	Reduction of Pb and Ni concentrations	Millena et al. (2023)
UHT dairy milk in Sao	Pasteurization and sterilization	Increase of Pb, Hg, and Ni	Ribeiro Sant'Ana et al. (2021)
Marine fish	Boiling	Reduction of As, Cr, Ni and Pb	Kazemi, Esmaeilbeigi, Sahebi, et al. (2022)
Mussels and shrimp	Frying	Increase of Na, Mg, Ca, Cu, Fe, Zn, Pb, Cd, and Hg	Anna et al. (2013)
Sardine and hake	Grilling	Increase of As, Hg, and Pb concentrations	Perelló et al. (2008)
Meat	Grilling	Increase of Cu and Fe levels	Joyce et al. (2016)
Fruits and vegetables	Canning	The concentrations of Pb, Cr, Ni, As, and Cd exceeded the permissible limits set by different health organizations	Massadeh et al. (2018)
Rice	Cooking	Cooking significantly reduced As (42.3%) and Pb in rice (42.9%)	Sharafi et al. (2019)

3.5.1 Heavy Metal Concentrations Through Fermentation

Fermented foods are defined as "foods or beverages produced through controlled microbial growth and the conversion of food components through enzymatic action" (Dimidi et al., 2019). Fermented foods offer numerous benefits for human health, including the reduction of blood cholesterol levels, improvement of immunity, protection against pathogens, provision of beneficial bacteria, and combating diabetes, obesity, allergies, and atherosclerosis. They also improve nutrient availability and digestibility (Şanlier et al., 2019). It has been reported that the fermentation process can reduce heavy metal concentrations in food due to the ability of microorganisms to remove heavy metals from foodstuffs (Siripongvutikorn et al., 2016) (Massoud et al., 2019). For example, Zhai et al. (2019) observed a reduction in Cd levels in rice after applying the fermentation process using the bacteria *Lactobacillus plantarum*. Similarly, Millena et al. (2023) noted a reduction in Pb and Ni concentrations in cocoa beans. Moreover, fermentation improved the bioaccessibility of Cd, Cu, Zn, and Mn in mantou by promoting greater hydrolysis of heavy metal-phytate complexes during longer fermentation periods (Li et al., 2023).

3.5.2 Heavy Metal Concentrations During Pasteurization and Sterilization

Pasteurization, also known as heat treatment, inactivates enzymes and kills heat-sensitive microorganisms. It is one of the most crucial phases in preservation and is indispensable for food safety (Azizi-Lalabadi et al., 2023). Although effective in reducing most biological contaminants in food, these techniques are not effective in eliminating chemical contaminants, such as heavy metals (Alinezhad et al., 2024). In fact, they can potentially increase the concentration of heavy metals in the final product due to evaporation during pasteurization (Alinezhad et al., 2024).

Numerous studies worldwide have confirmed the presence of heavy metals in pasteurized food, including dairy products. For instance, Feizi et al. (2022) detected Cd and Zn at concentrations higher than the permissible limit in different pasteurized milk sold in Iran. Ribeiro Sant'Ana et al. (2021) identified Hg, Pb, Se, and Ni in UHT dairy milk in Sao Luís Brazil. In contrast, Christophoridis et al. (2019) detected Cu, Mn, Zn, Cd, Pb, and Se at very low concentrations in cheese sold in Greece. With the exception of Cr, the estimated daily intake (EDI) of the heavy metals present in cow raw milk and pasteurized milk was determined to be below the permissible limit in Bangladesh. Based on the Target Carcinogenic Risk values for As and Pb, it appears that children who consume milk may be particularly susceptible to carcinogenic health risks (Hasan et al., 2022). Furthermore, the concentration of Cd determined in the sterilized milk in Libya was also below the limits permitted by FAO/WHO (Abdeljalil et al., 2021).

3.5.3 Heavy Metal Concentrations During Cooking

Cooking or thermal processing, such as boiling, frying, grilling, and baking, is essential for households and commercial establishments (e.g., hotels, hostels, restaurants, roadside stores) (Davis et al., 2018). It is one of the most commonly used processing operations globally to improve the nutritional quality of food. Cooking prolongs storage time, improves digestibility and sensory characteristics, destroys microorganisms, deactivates anti-nutrients or toxic substances, and reduces the metal content of raw products (Hajeb et al., 2014) (McGowan et al., 2017) (Rondanelli et al., 2017). The cooking process can influence heavy metal levels in meals. It may also adversely affect food quality and contribute to contamination through the use of contaminated water or the leaching of metals from cooking utensils and equipment (Hajeb et al., 2014).

From the literature, the impact of cooking methods on the heavy metal content of Iraqi rice available in the local market was investigated. The results indicate that soaking, washing, and cooking rice significantly reduced arsenic concentration (Abu-Almaaly, 2020). In Japan, this method decreased arsenic levels in rice by approximately 20 times, from 0.2104 to 0.01032 mg/kg (Phrukphicharn et al., 2021). Mohapatra et al. (2021) conducted a separate investigation involving the cooking of whole sorghum grain in excessive water, fermentation under submerged conditions, steaming post-fermentation, and a process of fermentation, steaming, and flaking. The study observed an increase in the content of calcium (Ca), magnesium (Mg), and sodium (Na) during the fermentation process. Additionally, there was a significant decrease in the levels of heavy metals such as arsenic (As), lead (Pb), titanium (Ti), and barium (Ba). Most cooking methods had minimal influence on the metal content in the prepared meat. For instance, the Pb content in raw chicken meat, initially at 0.30 mg/kg, decreased to 0.25, 0.18, and 0.10 mg/kg in grilled, marinated, and simmered chicken, respectively (Elsharawy, 2018).

✓ **Heavy metal concentrations during boiling**

Boiling is a popular method where food is cooked in liquid at a high temperature in a pot on a heated stove. Boiling causes alterations in food items through the transfer of heat and chemical reactions, resulting in a soft, tasty, and easily digestible meal (Ncobela et al., 2017) (Mehmood et al., 2020) (Yang et al., 2023). It has been reported that boiling raw food can reduce heavy metal concentrations in meals by causing these metals to leach into the dissolved salts in the water. Boiling has been found to significantly decrease cadmium levels by 33.3% in crabs and 58.8% in shrimps, as well as lead levels by 60% in shrimps (Abd-Elghany et al., 2020). Additionally, a decrease in arsenic (from 1.50 ± 0.21 to 1.00 ± 0.27 mg/kg), chromium (from 0.19 ± 0.02 to 0.11 ± 0.04 mg/kg), nickel (from 0.11 ± 0.012 to 0.03 ± 0.004 mg/kg), and lead (from 0.34 ± 0.02 to 0.32 ± 0.14 mg/kg) was recorded in boiled marine fish (*Thunnus tonggol*). However, mercury levels increased from 2.34 ± 0.11 to 3.56 ± 0.89 mg/kg (Kazemi, Esmaeilbeigi, Ansari, et al., 2022). Adjei-Mensah et al. (2021) reported a decrease in manganese, lead, arsenic, and mercury levels by 30.7%, 71.3%, 50%, and 59.4%, respectively, in cassava crops; by 65.6%, 59.5%, 76.4%, and 21.1%, respectively, in cocoyam crops; by 30.2%, 93.9%, 50%, and 34.9%, respectively, in plantain; and by 86%, 99%, 88.2%, and 92.7%, respectively, in yam crops. In contrast, an increase in heavy metal concentration due to the evaporation loss of water has been reported in cooked seafood (Chera-Anghel et al., 2023), noodles and infused tea (Lee et al., 2019). Onyeka et al. (2018) observed an increase in concentrations of Al, Pb, Cu, Fe, and Ni in cooked meals (vegetable, fleshy, tubers; cereals and legume) in Nigeria as compared to uncooked meals, attributed to the leaching of metals from cooking utensils.

✓ **Heavy metal concentrations during frying**

Frying is a popular food processing technique that involves immersing raw food in heated oil at a temperature of 180°C, which generates several chemical reactions, including thermal oxidation, polymerization, and hydrolysis (Gertz, 2014). These reactions enhance the sensory characteristics of food, such as flavor, color, and texture (Nayak et al., 2016).

From previous studies, predicting the effects of frying on heavy metal levels in food has proven to be challenging. Various factors, such as the initial concentration of heavy metals in the way of preparing food could impact the reduction or rise in these elements in food (Phrukphicharn et al., 2021).

Frying techniques can reduce the levels of heavy metals in fried food by causing the metals to leach into the oil. This reduction in heavy metal levels has been observed in shrimp (Ulaganathan et al., 2022), meat (Pistón et al., 2020), and crops (Adjei-Mensah et al., 2021). However, in some cases, heavy metal concentrations in fried food may increase due to water loss, which has a "concentrating" effect (Bassey et al., 2014). For instance, Gheisari et al. (2016) detected an increase in the concentrations of Pb and Cd in fried shrimp compared to fresh samples. Similarly, Bassey et al. (2014) showed a significant increase in the concentrations of Cd, Pb, Hg, Mn, and Zn in fried fish compared to raw fish. Another study conducted in 2016 found that the concentration of Cr had decreased in grass carp when cooked using the deep-fried method (Khodanazary et al., 2016). Moreover, Anna et al. (2013) found notable rises in levels of macro and microelements, along with heavy metals (Na, Mg, Ca, Cu, Fe, Zn, Pb, Cd, and Hg), except for K and Mg, in mussels and shrimp.

✓ **Heavy metal concentrations during grilling**

Grilling is a cooking method that uses direct high heat, typically from below, to quickly cook food, enhancing its flavor and creating a desirable brown crust (Suleman et al., 2020). The grilling process can lead to an increase in heavy metal concentration due to water loss through evaporation in various foods, including fish (Bassey et al., 2014). The interaction between the grilling material and the food may also cause an increase in heavy metal concentrations, such as As, Hg, and Pb in grilled sardine and hake, as well as Cu and Fe in meat (Perelló et al., 2008) (Joyce et al., 2016). However, Ersoy (2011), reported no significant differences in Pb and As concentrations between raw and grilled fish. On the other hand, a reduction in heavy metal concentrations has been noticed in grilled fish, meat, crab, and shrimp. This reduction could be attributed to the loss of heavy metals as a result of protein denaturation during cooking, which can destroy the disulfide bridges between the cysteine groups containing the metals (Abd-Elghany et al., 2020) (Mieiro et al., 2016). Metal reduction varies depending on the source. For instance, grilling significantly reduces cadmium levels in shrimps, whereas a notable reduction in lead levels was observed in crabs (Abd-Elghany et al., 2020). Furthermore, grilling of tilapia fish reduced the concentration of heavy metals compared to raw samples (M Ibrahim et al., 2020).

3.5.4 Heavy Metal Concentrations During After Packaging

Food packaging is an essential component of the food industry, used to keep food and beverages hygienic and ensure their safety. The most commonly used materials for packaging foods belong to the following classes: metals, glass, paper, and polymer containers (plastic) (Teck Kim et al., 2014) (Borah et al., 2019). Despite its usefulness for food safety, packaging can be a potential source of exposure to several toxic chemicals, including heavy metals (Noureddine El Moussawi et al., 2019) (Alamri et al., 2021).

✓ **Heavy metals concentrations in canned food**

"Canned food" can be defined as pasteurized food packaged in metal cans and glass jars (Kowalska et al., 2020). Canned foods, such as canned fruits, canned vegetables, canned meat, canned fish, and canned drinks, are important sources of minerals, proteins, carbohydrates, vitamins, and trace elements (Massadeh et al., 2018). During canned food preservation, the packaging material can interact with the food components, potentially transferring its contents, including heavy metals (Sn, Fe, Cd, Pb), into the food due to leaching from the packaging material (Parkar et al., 2014) (Olmedo et al., 2024).

In general, heavy metal contamination in canned food depends on the materials used in the canning process. For instance, in food canned in metals, heavy metal contamination is mostly caused by the corrosion of metallic packaging, which is usually composed of tin plate, chromium-coated steel, or aluminum (Alzahrani et al., 2017) (Alamri et al., 2021). Worldwide, several studies have reported heavy metal contamination in various types of canned food, including canned sardines in Egypt (Sabala et al., 2024), canned fruits and vegetables sold in Jordan markets (Massadeh et al., 2018), and canned meat in Iraq (Hussein et al., 2021).

Glass jar packaging is utilized for better-quality products and allows consumers to clearly see the contents, thus improving their trust in product quality (Almeida et al., 2023). In contrast to food packaged in metal cans, heavy metals such as Cd and Pb are rarely detected in food packed in glass jars. However, the risk of contamination increases with the reutilization of recycled jars (Alamri et al., 2021).

✓ **Heavy metals concentrations in plastic packed food**

In recent years, plastic has been widely employed as a food-contact packaging material in the food industry. It helps maintain the quality, freshness, and aroma of food during storage and transportation (Eti et al., 2023). Plastic packaging is composed of multiple polymers, including polyethylene (both low and high density), polypropylene, polyvinyl chloride, and polystyrene, along with numerous additives such as plasticizers, antioxidants, stabilizers, curing agents, and coloring agents (Groh et al., 2019) (Tajeddin et al., 2020). Additionally, metals such as cobalt, manganese, nickel, lead, cadmium, antimony, and chromium are used as inert fillers, pigments for color, and stabilizers to improve the physicochemical properties of plastics (Ranta-Korpi et al., 2014) (Klöckner et al., 2021) (Eti et al., 2023). These metals can also function as biocides, antimicrobial agents, lubricants, and flame retardants (Turner et al., 2021). The interaction between food and plastic can lead to the transfer of packaging components, including heavy metals, into the food products. Numerous studies have shown the migration of heavy metals from plastic packaging. For example, Eti et al. (2023) demonstrated that heavy metals can migrate from yogurt containers. Similarly, S. Khan et al. (2015) reported the leaching of Pb, Cu, Ni, Zn, Mn, Cr, and Cd from locally produced food containers used in India. Other studies have also detected the presence of heavy metals in food packaged in plastic. De Toni et al. (2017) found lead (Pb) and nickel (Ni) in coffee packaged in plastic. In various countries, papers and paperboards were utilized for packaging purposes. In India, certain samples were found to contain levels of heavy metals such as Al, As, Ba, Cr, Co, Ni, Pb, and V that exceeded the permitted concentrations (Sood et al., 2019). Paper, paperboard, corrugated board wrappers, and other related food packaging materials were gathered from various fast-food restaurants in Istanbul for analysis of heavy metals. The study revealed that the migration of Pb into the food exceeded authorized limits. However, other metals such as Cd, Zn, Cr, Cu, and Ni migrated into the food as well, but their concentrations remained within permissible limits (Mertoğlu-Elmas et al., 2018). Abdel-Rahman et al. (2019) also detected the presence of Cu, Fe, Mn, and Ni in soft drink sold plastic bottles in Egypt.

3.6 METHODS FOR REDUCING AND REMOVING HEAVY METALS FROM FOOD

Research in the field of heavy metal decontamination in food products has made significant progress in recent years, with a growing focus on developing cost-effective and efficient methods (Scutarașu et al., 2023). As reported previously, various factors can influence the presence of heavy metal levels in food, spanning natural conditions during production, storage, packaging, post-harvest preparation, and cooking techniques. Crops grown in polluted soils or environments tend to have elevated metal content. Post-harvest practices such as washing typically remove metal pollutants, while cooking can either reduce or augment the metal content (Deepmallya, 2021). Various approaches have been developed to mitigate or reduce the presence of heavy metals during the food manufacturing process. Among these approaches, nanotechnology has emerged as a promising solution due to its potential to specifically target contaminants and enhance the effectiveness of detoxification and degradation processes (Huang et al., 2017). However, despite these potential benefits, many challenges remain, particularly concerning cost and the feasibility of large-scale implementation. In addition, concerns about the safety and environmental impact of nanomaterials must be thoroughly examined (Scutarașu et al., 2023). Table 3.3 provides an overview of promising and cost-effective approaches to improve heavy metal decontamination in food products.

3.6.1 Washing, Soaking and Cooking

Contamination of the environment, soil, and groundwater by toxic heavy metals, along with concerns about food safety, is a global issue today. A recent study found that soaking the rice for two hours was the most effective method for reducing heavy metal concentrations, with overnight soaking also proving successful in the removal of aluminum (Al), cadmium (Cd), and lead (Pb) (Al-Harbi et al., 2023). Additionally, among arsenic (As), cadmium (Cd), and lead (Pb), the highest reduction during rice washing was observed for Pb (60%), followed by As (40%), Zn (32%), and Cd (27%). Furthermore, washing can lead to a greater loss of Pb, Cd, and As from the aleurone layer and outer layer of the endosperm in rice (Mohammadi et al., 2021).

TABLE 3.3 Methods for reducing metal content

METHODS	*PRINCIPLE*	*EFFECTS*	*SUBSTRATE*	*REFERENCE*
Chemical precipitation	The administration of certain coagulants, such as lime, alum, and organic polymers.	The use of sulfur precipitation agents is effective for significantly removing several heavy metals, such as Cd^{2+}, Cr^{2+}, Co^{2+}, le Cu^{2+}, le Hg^{2+}, and others.	Water	Goswami et al. (2022)
	Chemical reagents like $Ca(OH)_2$–calcium hydroxide or NaOH–sodium hydroxide.	Remove Cu^{2+} ions from aqueous solutions. This procedure is easy to implement and feasible for most metals, with low costs.	Water	Pohl (2020)
Electrochemical methods	Ion exchange, electrooxidation, electrochemical reduction, and electroflotation.	Remove food contaminants where ion exchange demonstrates excellent selectivity for heavy metals.	Raw material and food products	Zhu et al. (2019)
Electro dialysis	Effect of the electric potential of semipermeable membranes (bipolar membranes).	This method is easy to implement, feasible for most types of metals, and effective for large volumes of samples.	Dairy industry	Zhu et al. (2019) Bazinet (2005)
Coagulation/ flocculation process	Compounds such as aluminum, ferrous sulfate, and ferric chloride can be used.	Effective methods for removing certain metals, such as Pb, Zn, and Fe.	Water and wastewater	Pang et al. (2011) Zhu et al. (2019)
Flotation methods	The effects of different single-chain surfactants, obtained by reacting cysteine with octanoyl (C8), decanoyl (C10), and dodecanoyl (C12) chloride, were investigated for their use in the ion flotation removal of low levels of arsenic, mercury, lead, cadmium, and chromium ions from aqueous solutions.	Removal of heavy metal ions from water using ion flotation.	Aqueous solution	Taseidifar et al. (2017)

Photo catalytic removal methods	Photocatalytic reduction of metal ions is considered a promising method for their removal due to its flexible operation, use of light energy, minimal need for additional chemicals, and absence of sludge production.	This technique enables the simultaneous removal of metals and by-products that have low toxicity.	Wastewater	Liu et al. (2024)
Adsorption process	Adsorption-based processes can be effective for reducing contamination. For this purpose, nanoporous carbon-based adsorbents and adsorbents derived from certain agricultural wastes can be used.	Adsorption methods are simple, cost-effective, and efficient for most metals, utilizing abundant and inexpensive biosorbents.	Aqueous solution	Raji et al. (2023) Goswami et al. (2022)
Phytoremediation	Phytoremediation is the use of plants to remove pollutants from the environment. Typically, a plant is classified as a hyperaccumulator if it can accumulate more than 1,000 mg/kg of Cu, Co, Cr, Ni, or Pb, or more than 10,000 mg/kg of Mn or Zn.	In 2011, more than 400 plant species were identified as metal hyperaccumulators. The best hyperaccumulators vary depending on the region's climate and soil types.	Heavy metals content in the environment	Krämer (2010) Grabowska (2011)
Use of organic amendments	Effective reduction of the bioavailability of heavy metals in soils can be achieved by using organic amendments such as compost and farmyard manure.	Organic matter is important for reducing metal levels, especially in acidic soils with low cation exchange capacity (Cd, Zn, and Ni).	Soil	Brown et al. (2003) Erdogan et al. (2007)
Use of inorganic amendments	Mineral amendments include: Al-montmorillonite, clinoptilolite, diammonium phosphate, ferrous sulfate, hydroxyapatite, lime, manganese oxide, red mud, and synthetic zeolites.	Inorganic amendments, unlike organic amendments, are more effective at reducing the bioavailability of metals.	Soil	Bolan et al. (2003)

A processing aid is a substance used to enhance specific technical aspects of food manufacturing. Examples include antimicrobials in meat processing, filter aids for oils and beverages, and enzymes for bread dough. Filter aids like diatomaceous earth (DE) and perlite, as well as bentonite clay for beverage clarification, can contain various elements, including heavy metals, due to their mineral origins. Washing with a 9.7% hydrochloric acid (HCl) solution effectively reduces lead (Pb) and arsenic (As) levels in DE, perlite, and cellulose filter aids (Redan, 2020). Moreover, spinach washed with citric acid solution (10%) had the ability to reduce residues of mercury, lead, zinc, and arsenic using 10% citric acid: 23%, 28%, 54%, and 22%, respectively. In comparison, treatment with tap water also resulted in reductions of 7%, 7%, 15%, and 6% for the same elements (Amir et al., 2019).

3.6.2 Adsoption

Standard approaches to address heavy metals adsorption include chemical precipitation, oxidation, ultrafiltration, ion exchange, reverse osmosis, electrodialysis, nanofiltration, and chelation within treatment systems. These methods are used in various applications, such as the food industry, pharmaceutical industry, and wastewater treatment (Mathew et al., 2023) (Verma et al., 2021).

Biosorption: Biosorption is a method employing biological materials or components as adsorbents to remove toxic metals and non-metal ions, as well as small particulates, from wastewater originating from various industries. The biological materials employed as adsorbents in this biosorption process are called biosorbents. These biosorbents can be categorized as natural, biological, and waste-oriented (Faizan et al., 2020).

Biochar, composed of carbon, stands out as an excellent adsorbent for recovering heavy metals from polluted areas. The effective sorption of heavy metals in modified biochar relies on key factors such as pyrolysis temperature variation, redox complexation, and the introduction of organic substances (Sarker et al., 2022). Additionally, activated carbon was used in batch experiments to assess its efficacy in removing lead, cadmium, nickel, chromium, and zinc from water. Among the tested heavy metals, nickel exhibited the highest removal percentages across all concentrations, with removal rates decreasing as the concentration of the heavy metal increased (Karnib et al., 2014).

Biomaterial: The comparison of outcomes using different biomaterials indicated that *S. cerevisiae* stands out as a distinctive biomaterial for metal biosorption. Given its abundance as a by-product of the fermentation industry, *Saccharomyces cerevisiae* finds extensive use in beverage and food production. Numerous studies have demonstrated its capability to eliminate toxic metals, recover precious metals, and purify aqueous solutions containing radioactive elements. *S. cerevisiae* has been proven effective in removing various metals such as lead (Pb), cadmium (Cd), zinc (Zn), mercury (Hg), nickel (Ni), arsenic (As), chromium (Cr), gold (Au), platinum (Pt), and uranium (U) (Massoud et al., 2019). Heba et al. (2017) evaluated the efficiency of *Enterococcus faecium* on the Cd and Pb removal from the milk. The reduction of Cd ranged from 79.4 to 82.8%, while Pb had lower reduction rate (from 72 to 75%). Additionally, the biomass of algae was used as a wastewater treatment method to remove ions of Cu, Pb, Cd, and Zn. They determined that the efficiency of algae biomass in adsorbing harmful metals from wastewater is attributed to its functional groups (Utomo et al., 2016).

Membrane filtration: Adsorption depends on the physical and chemical interactions between heavy metal ions and adsorbents. The unique structural properties of adsorptive membranes have been shown to be highly effective in removing heavy metals from wastewater (Vo et al., 2020). Ceramic membranes are classified as porous materials used in industry. In water and wastewater treatment, ceramic membranes offer several benefits compared to polymer-based filtration systems. These advantages include high-temperature stability, resistance to fouling, and minimal maintenance requirements, ultimately resulting in reduced lifecycle costs (A. N. Mohamad et al., 2023). Empirical data indicate that polymer membranes are successful in eliminating a diverse array of contaminants, encompassing suspended pollutants, organic substances, as well as organic and inorganic compounds, including heavy metal ions (Wieszczycka et al., 2017).

Dietary fiber: The primary functional groups of dietary fiber (DF), including hydroxyl (-OH), carboxyl (-COOH), carbonyl (C=O), and amino ($-NH_2$), carry a negative charge and can engage in interactions with heavy metals and other ions. These interactions may take place through processes such as chelation, complexation, ion exchange, or adsorption, influencing the bioavailability and toxicity of heavy metals in the body (Raji et al., 2023). Potentially, the adsorption capacity is enhanced by the electrostatic attraction between positively charged metal ions and the negatively charged functional groups of DF (S. M. Shaheen et al., 2019). The functional groups of dietary fiber create stable complexes with metal ions through the process of chelation. Specifically, carboxyl groups found in fibers like pectin, hemicellulose, and alginate have the capability to chelate heavy metals such as Pb and Cd (Chen et al., 2020). A fiber's capacity to bind chemical components is influenced by its chemical structure and content, the type of metal involved, and the source of the fiber. The process of adsorption is intricate and can occur through three distinct mechanisms: physical adsorption, chemisorption, or mechanical sorption. Physical adsorption is driven by temperature-dependent van der Waals forces, while mechanical adsorption is determined by the porosity, crystallinity, and ability of the adsorbent to entrap elements within its spatial structure. Chemisorption is associated with the presence of phenolic groups from lignin and carboxyl groups (-COOH) from uronic acids in the fiber matrix (Raji et al., 2023).

3.6.3 Agri-Food Wastes for Heavy Metals Removal

Agricultural and food industry wastes appear to offer a viable solution for the bioremediation of heavy metals due to their notable ability to effectively bind these metals (Mathew et al., 2023). Recently, soybean oil residues were used to remove cadmium, zinc, and lead ions from aqueous solutions. Discontinuous experiments were conducted to investigate the effect of initial concentration (100–300 ppm), solution pH (1–5), contact time (1–60 min), adsorbent amount (0.02–2 g) on the uptake of cadmium, zinc, and lead ions by soybean oil residues. According to the results, the best pH for adsorption of metal ions is about 3–5, and the equilibrium time for cadmium ions is 40 minutes, for zinc ions is 20 minutes, and for lead ions is 10 minutes (Mahmudiono et al., 2022). Furthermore, adsorbents derived from broccoli stalks demonstrated superior effectiveness in removing heavy metals from aqueous solutions. The broccoli stalk sample exhibited the highest adsorption capacities for all metallic adsorbates, with capacities of 4.34 mg/g for Cd^{2+}, 3.02 mg/g for Zn^{2+}, 3.97 mg/g for Ni^{2+}, and 7.21 mg/g for Cu^{2+} (Landin-Sandoval et al., 2020).

3.6.4 Nanotechnology

Nanomaterials possess distinct properties, including selectivity and small size, which make them effective absorbers of heavy metals. Nanomaterials produced from microbes and fungi have been shown to mitigate heavy metal contamination either through enzyme-based processes or by absorption mechanisms (Kumar et al., 2023). Nanotechnology has the capability to adsorb toxic substances through processes such as ion exchange, ion precipitation, and simple adsorption. Nanoadsorbents, which include carbon-based, metal-based, and polymeric types, are utilized in different categories for the removal of heavy metals from wastewater (Parvin et al., 2019). In addition, pulsed electric field is a promising new method for pasteurization that, when implemented correctly, has the potential to extend the shelf life (A. Mohamad et al., 2020). This method can provoke the accumulation of heavy metals ions, resulting in a higher concentration of metals (Pankiewicz et al., 2017). Moreover, microwave heating can be used for food pasteurization and also to help reduce the concentration of heavy metals. Guney et al. (2021) reported that infrared-assisted microwave applications reduced the concentrations of Al and Ni in milk. In contrast, heating food containing citric acid and packaged in plastic composed of recycled polyethylene terephthalate using microwaves stimulates the migration of lead, cadmium, chromium, nickel, and antimony (Whitt et al., 2016).

3.7 CONCLUDING REMARKS AND CONCLUSIONS

In summary, the reviewed literature highlights that heavy metals contamination can occur during the production phase of raw materials, whether crops or animal products, or during food processing. The toxicity of heavy metals in food is influenced by a variety of factors, including natural conditions during production, as well as practices related to storage, processing, packaging, and cooking techniques.

To mitigate the effects of heavy metal contamination in food production and processing, several strategies are used. First, the implementation of rigorous post-harvest handling measures, such as thorough washing, can significantly reduce surface-level contamination. Additionally, employing proper cooking techniques can further mitigate heavy metal levels by altering their bioavailability and reducing overall concentrations in the final food product. Beyond these practices, enhancing monitoring and control measures throughout the entire food production chain, from sourcing to packaging, is crucially important. This includes adopting advanced detection technologies to identify and mitigate contamination risks early on.

Moreover, advancements in food processing technologies should prioritize innovations that minimize heavy metal contamination. Food 4.0 technologies can prevent and mitigate heavy metal contamination in the food supply chain through precision agriculture using IoT and AI to manage soil and water quality, smart processing and packaging technologies equipped with AI-driven detection systems to identify and remove contaminated products, blockchain for transparent traceability to swiftly recall affected items, advanced sensors for real-time monitoring, robotics to reduce human error in handling, and data analytics for informed decision-making. In addition, the exploration of alternative materials for equipment that comes into contact with food,, which can reduce the risk of metal leaching during processing. Furthermore, promoting sustainable agricultural practices can mitigate heavy metal uptake in crops and livestock, thereby addressing contamination at its source. Education and awareness campaigns targeting both producers and consumers play a pivotal role in fostering understanding of the risks associated with heavy metals in food and promoting best practices to minimize exposure. Lastly, continued research into the long-term health effects of chronic low-level exposure to heavy metals will guide the development of more targeted mitigation strategies and regulatory frameworks.

REFERENCES

Abbasi, H., Shah, M. H., Mohiuddin, M., Elshikh, M. S., Hussain, Z., Alkahtani, J., Ullah, W., Alwahibi, M. S., & Abbasi, A. M., 2020. Quantification of heavy metals and health risk assessment in processed fruits' products. *Arabian Journal of Chemistry*, 13(12), 8965–8978. https://doi.org/10.1016/j.arabjc.2020.10.020

Abd-Elghany, S. M., Zaher, H. A., Elgazzar, M. M., & Sallam, K. I., 2020. Effect of boiling and grilling on some heavy metal residues in crabs and shrimps from the Mediterranean Coast at Damietta region with their probabilistic health risk assessment. *Journal of Food Composition and Analysis*, 93, 103606. https://doi.org/10.1016/j.jfca.2020.103606

Abdel-Rahman, G. N., Ahmed, M. B. M., Sabry, B. A., & Ali, S. S. M., 2019. Heavy metals content in some non-alcoholic beverages (carbonated drinks, flavored yogurt drinks, and juice drinks) of the Egyptian markets. *Toxicology Reports*, 6, 210–214. https://doi.org/10.1016/j.toxrep.2019.02.010

Abdeljalil, N. A., Benkhayal, A. A., Ahmida, M., El Sharaa, I. A., & Ahmida, N. H., 2021. Determination of lead and cadmium in different types of milk samples collected from different markets in Benghazi-Libya. *International Journal of Agriculture Innovations and Research*, 9(4), 253–264.

Abu-Almaaly, R. A., 2020. Effect of cooking method on the content of heavy metals in rice that available in local market. *Plant Archives*, 20, 2976–2981

Adjei-Mensah, R., Ofori, H., Tortoe, C., Torgbor Johnson, P.-N., Aryee, D., & Kofi Frimpong, S., 2021. Effect of home processing methods on the levels of heavy metal contaminants in four food crops grown in and around two mining towns in Ghana. *Toxicology Reports*, 8, 1830–1838. https://doi.org/10.1016/j.toxrep.2021.11.001

Ahmad, F., 2016. Analysis of raw meat for heavy metals and bacterial contamination and comparison of antibiotic susceptibility of isolated bacteria: Contamination of raw meat sold in commercial markets. *Proceedings of the Pakistan Academy of Sciences: B. Life and Environmental Sciences*, 53(1), 13–20

Aissioui, S., Poirier, L., Amara, R., & Ramdane, Z., 2021. Concentrations of lead, cadmium, and mercury in Mullus barbatus barbatus (L.) from the Algerian coast and health risks associated to its consumption. *Regional Studies in Marine Science*, 47, 101959. https://doi.org/10.1016/j.rsma.2021.101959

Akinyele, I. O., & Shokunbi, O. S., 2015. Comparative analysis of dry ashing and wet digestion methods for the determination of trace and heavy metals in food samples. *Food Chemistry*, 173, 682–684. https://doi.org/10.1016/j.foodchem.2014.10.097

Al-Harbi, H. F., Al-Mohaimeed, A. M., & El-Tohamy, M. F., 2023. Assessment of essential elements and heavy metals in Saudi Arabian rice samples underwent various processing methods. *Open Chemistry*, 21(1). https://doi.org/10.1515/chem-2022-0328

Alamri, M. S., Qasem, A. A. A., Mohamed, A. A., Hussain, S., Ibraheem, M. A., Shamlan, G., Alqah, H. A., & Qasha, A. S., 2021. Food packaging's materials: A food safety perspective. *Saudi Journal of Biological Sciences*, 28(8), 4490–4499. https://doi.org/10.1016/j.sjbs.2021.04.047

Alinezhad, Z., Hashemi, M., & Tavakoly Sany, S. B., 2024. Concentration of heavy metals in pasteurized and sterilized milk and health risk assessment across the globe: A systematic review. *PLoS One*, 19(2), e0296649. https://doi.org/10.1371/journal.pone.0296649

Almeida, C., Ceballos-Santos, S., Laso, J., Margallo, M., Aldaco, R., & Marques, A., 2023. Contribution of glass jar packaging to the environmental assessment of canned seafood products: Albacore tuna (Thunnus alalunga) and Atlantic chub mackerel (Scomber colias) as case studies. *Journal of Cleaner Production,* 420, 138366. https://doi.org/10.1016/j.jclepro.2023.138366

Alzahrani, H. R., Kumakli, H., Ampiah, E., Mehari, T., Thornton, A. J., Babyak, C. M., & Fakayode, S. O., 2017. Determination of macro, essential trace elements, toxic heavy metal concentrations, crude oil extracts and ash composition from Saudi Arabian fruits and vegetables having medicinal values. *Arabian Journal of Chemistry*, 10(7), 906–913. https://doi.org/10.1016/j.arabjc.2016.09.012

Amir, R. M., Randhawa, M. A., Sajid, M. W., Nadeem, M., Ahmad, A., & Wattoo, F. M., 2019. Evaluation of various soaking agents as a novel tool for heavy metal residues mitigation from spinach. *Food Science and Technology*, 39(1), 176–180. https://doi.org/10.1590/fst.00118

Anna, C., & Kamila, S., 2013. Effect of processing treatments (frozen, frying) on contents of minerals in tissues of 'frutti di mare'. *International Journal of Food Science & Technology*, 48(2), 238–245. https://doi.org/10.1111/j.1365-2621.2012.03179.x

Azizi-Lalabadi, M., Moghaddam, N. R., & Jafari, S. M., 2023. 9–Pasteurization in the food industry. In S. M. Jafari (Ed.), *Thermal Processing of Food Products by Steam and Hot Water* (pp. 247–273): Woodhead Publishing. https://doi.org/10.1016/B978-0-12-818616-9.00009-2

Balali-Mood, M., Naseri, K., Tahergorabi, Z., Khazdair, M. R., & Sadeghi, M., 2021. Toxic mechanisms of five heavy metals: Mercury, lead, chromium, cadmium, and arsenic. [Review]. *Frontiers in Pharmacology*, 12, article 643972. https://doi.org/10.3389/fphar.2021.643972

Bassey, F. I., Oguntunde, F. C., Iwegbue, C. M. A., Osabor, V. N., & Edem, C. A., 2014. Effects of processing on the proximate and metal contents in three fish species from Nigerian coastal waters. *Food Science & Nutrition*, 2(3), 272–281. https://doi.org/10.1002/fsn3.102

Bazinet, L., 2005. Electrodialytic phenomena and their applications in the dairy industry: A review. *Critical Reviews in Food Science and Nutrition*, 45(4), 307–326.

Bolan, N. S., & Duraisamy, V., 2003. Role of inorganic and organic soil amendments on immobilisation and phytoavailability of heavy metals: A review involving specific case studies. *Soil Research*, 41(3), 533–555.

Bondy, S. C., & Campbell, A., 2022. Aluminum and neurodegenerative disease. In R. M. Kostrzewa (Ed.), *Handbook of Neurotoxicity* (pp. 2283–2303). Cham: Springer International Publishing. https://doi.org/10.1007/978-3-031-15080-7_231

Borah, H., & Dutta, U., 2019. 1–Trends in beverage packaging. In A. M. Grumezescu & A. M. Holban (Eds.), *Trends in Beverage Packaging* (pp. 1–19): Academic Press. https://doi.org/10.1016/B978-0-12-816683-3.00001-3

Boudebbouz, A., Boudalia, S., Bousbia, A., Gueroui, Y., Boussadia, M. I., Chelaghmia, M. L., Zebsa, R., Affoune, A. M., & Symeon, G. K., 2022. Determination of heavy metal levels and health risk assessment of raw cow milk in Guelma Region, Algeria. *Biological Trace Element Research*, 201, 1704–1716. https://doi.org/10.1007/s12011-022-03308-1

Boudebbouz, A., Boudalia, S., Bousbia, A., Habila, S., Boussadia, M. I., & Gueroui, Y., 2021. Heavy metals levels in raw cow milk and health risk assessment across the globe: A systematic review. *Science of the Total Environment*, 751, 141830. https://doi.org/10.1016/j.scitotenv.2020.141830

Bousbia, A., Boudalia, S., Gueroui, Y., Ghebache, R., Amrouchi, M., Belase, B., Meguelati, S., Belkheir, B., Benidir, M., & Chelaghmia, M., 2019. Heavy metals concentrations in raw cow milk produced in the different livestock farming types in Guelma province (Algeria): Contamination and risk assessment of consumption. *JAPS: Journal of Animal & Plant Sciences*, 29(2), 386–395

Briffa, J., Sinagra, E., & Blundell, R., 2020. Heavy metal pollution in the environment and their toxicological effects on humans. *Heliyon*, 6(9), e04691. https://doi.org/10.1016/j.heliyon.2020.e04691

Brown, S., Chaney, R. L., Hallfrisch, J. G., & Xue, Q., 2003. Effect of biosolids processing on lead bioavailability in an urban soil. *Journal of Environmental Quality*, 32(1), 100–108

Bryliński, Ł., Kostelecka, K., Woliński, F., Duda, P., Góra, J., Granat, M., Flieger, J., Teresiński, G., Buszewicz, G., Sitarz, R., & Baj, J., 2023. Aluminium in the human brain: Routes of penetration, toxicity, and resulting complications. *International Journal of Molecular Sciences*, 24(8), 7228. https://doi.org/10.3390/ijms24087228

Bushuty, D. H. E., & Shanshan, N. M., 2018. Effect of cooking methods (Boiling-grilling) and processing using some natural foods on the concentrations of heavy metals in fresh meats. *Journal of Home Economics*, 28, 131–151.

Chaoua, S., Boussaa, S., El Gharmali, A., & Boumezzough, A., 2019. Impact of irrigation with wastewater on accumulation of heavy metals in soil and crops in the region of Marrakech in Morocco. *Journal of the Saudi Society of Agricultural Sciences*, 18(4), 429–436. https://doi.org/10.1016/j.jssas.2018.02.003

Charkiewicz, A. E., Omeljaniuk, W. J., Nowak, K., Garley, M., & Nikliński, J., 2023. Cadmium toxicity and health effects–A brief summary. *Molecules*, 28(18), 6620. https://doi.org/10.3390/molecules28186620

Chellaram, C., Murugaboopathi, G., John, A. A., Sivakumar, R., Ganesan, S., Krithika, S., & Priya, G., 2014. Significance of nanotechnology in food industry. *APCBEE Procedia*, 8, 109–113. https://doi.org/10.1016/j.apcbee.2014.03.010

Chen, J., Huang, H., Chen, Y., Xie, J., Song, Y., Chang, X., Liu, S., Wang, Z., Hu, X., & Yu, Q., 2020. Effects of fermentation on the structural characteristics and in vitro binding capacity of soluble dietary fiber from tea residues. *LWT,* 131, 109818. https://doi.org/10.1016/j.lwt.2020.109818

Chera-Anghel, I.-A., & Stefan-van Staden, R.-I., 2023. Extraction of heavy metals by cooking/preserving of seafood, tuna and poultry from Romania–A source of contamination with heavy metals. *Food Chemistry*, 407, 135158. https://doi.org/10.1016/j.foodchem.2022.135158

Christophoridis, C., Kosma, A., Evgenakis, E., Bourliva, A., & Fytianos, K., 2019. Determination of heavy metals and health risk assessment of cheese products consumed in Greece. *Journal of Food Composition and Analysis*, 82, 103238. https://doi.org/10.1016/j.jfca.2019.103238

Das Pinkey, P., Nesha, M., Bhattacharjee, S., Chowdhury, M. A. Z., Fardous, Z., Bari, L., & Koley, N. J., 2024. Toxicity risks associated with heavy metals to fish species in the Transboundary River–Linked Ramsar conservation site of Tanguar Haor, Bangladesh. *Ecotoxicology and Environmental Safety*, 269, 115736. https://doi.org/10.1016/j.ecoenv.2023.115736

Davis, B., Lockwood, A., Pantelidis, I. S., & Alcott, P., 2018. *Food and Beverage Management*: Routledge.

de Oliveira, T. M., Augusto Peres, J., Lurdes Felsner, M., & Cristiane Justi, K., 2017. Direct determination of Pb in raw milk by graphite furnace atomic absorption spectrometry (GF AAS) with electrothermal atomization sampling from slurries. *Food Chemistry*, 229, 721–725. https://doi.org/10.1016/j.foodchem.2017.02.143

De Toni, L., Tisato, F., Seraglia, R., Roverso, M., Gandin, V., Marzano, C., Padrini, R., & Foresta, C., 2017. Phthalates and heavy metals as endocrine disruptors in food: A study on pre-packed coffee products. *Toxicology Reports*, 4, 234–239. https://doi.org/10.1016/j.toxrep.2017.05.004

Deepmallya, G., 2021. Impact of processing techniques on heavy metal concentration in food products: A review. *International Journal of Creative Research*, 9(5), 508–519.

Dimidi, E., Cox, S. R., Rossi, M., & Whelan, K., 2019. Fermented foods: Definitions and characteristics, impact on the gut microbiota and effects on gastrointestinal health and disease. *Nutrients*, 11(8), 1806. https://doi.org/10.3390/nu11081806

Diyarov, A. E., Nursapina, N. A., Matveyeva, I. V., & Ponomarenko, O. I., 2022. Effect of food processing method on heavy metals content. *Chemical Bulletin of Kazakh National University*, 105(2), 14–19. https://doi.org/10.15328/cb1247

Duan, P., Khan, S., Ali, N., Shereen, M. A., Siddique, R., Ali, B., Iqbal, H. M. N., Nabi, G., Sajjad, W., & Bilal, M., 2020. Biotransformation fate and sustainable mitigation of a potentially toxic element of mercury from environmental matrices. *Arabian Journal of Chemistry*, 13(9), 6949–6965. https://doi.org/10.1016/j.arabjc.2020.06.041

Elsharawy, N., 2018. Control cadmium, copper, iron, and lead residues in chicken meat and their offal. *International Journal of Current Research*, 10(7), 71646–71651. https://doi.org/10.24941/ijcr.31397.07.2018

Erdogan, S., Baysal, A., Akba, O., & Hamamci, C., 2007. Interaction of metals with humic acid isolated from oxidized coal. *Polish Journal of Environmental Studies*, 16(5), 671–675.

Ersoy, B., 2011. Effects of cooking methods on the heavy metal concentrations of the African catfish (Clarias Gariepinus). *Journal of Food Biochemistry*, 35(2), 351–356. https://doi.org/10.1111/j.1745-4514.2010.00386.x

Eti, S. A., Islam, M. S., Shourove, J. H., Saha, B., Ray, S. K., Sultana, S., Ali Shaikh, M. A., & Rahman, M. M., 2023. Assessment of heavy metals migrated from food contact plastic packaging: Bangladesh perspective. *Heliyon*, 9(9), e19667. https://doi.org/10.1016/j.heliyon.2023.e19667

Faizan, A., & Sadaf, Z., 2020. Potential use of agro food wastes as biosorbents in the removal of heavy metals. In N. Aurel (Ed.), *Emerging Contaminants* (pp. Ch. 8). Rijeka: IntechOpen. https://doi.org/10.5772/intechopen.94175

Fan, Y.-G., Wu, T.-Y., Zhao, L.-X., Jia, R.-J., Ren, H., Hou, W.-J., & Wang, Z.-Y., 2024. From zinc homeostasis to disease progression: Unveiling the neurodegenerative puzzle. *Pharmacological Research*, 199, 107039. https://doi.org/10.1016/j.phrs.2023.107039

Fathabad, A. E., Shariatifar, N., Moazzen, M., Nazmara, S., Fakhri, Y., Alimohammadi, M., Azari, A., & Mousavi Khaneghah, A., 2018. Determination of heavy metal content of processed fruit products from Tehran's market using ICP- OES: A risk assessment study. *Food and Chemical Toxicology*, 115, 436–446. https://doi.org/10.1016/j.fct.2018.03.044

Feizi, R., Hamidi, F., Jaafarzadeh, N., Ghahrchi, M., & zafarzadeh, A., 2022. Determination and health risk assessment of heavy metals (Pb, Cd, Cu and Zn) in different brands of pasteurized milk. *International Journal of Environmental Analytical Chemistry*, 102(18), 6892–6903. https://doi.org/10.1080/03067319.2020.1817908

Forcada, S., Menéndez Miranda, M., Stevens, F., Royo, L. J., Fernández Pierna, J. A., Baeten, V., & Soldado, A., 2023. Industrial impact on sustainable dairy farms: Essential elements, hazardous metals and polycyclic aromatic hydrocarbons in forage and cow's milk. *Heliyon*, 9(10), article 20977. https://doi.org/10.1016/j.heliyon.2023.e20977

Galvez-Fernandez, M., Grau-Perez, M., Garcia-Barrera, T., Ramirez-Acosta, S., Gomez-Ariza, J. L., Perez-Gomez, B., Galan-Labaca, I., Navas-Acien, A., Redon, J., Briongos-Figuero, L. S., Dueñas-Laita, A., Perez-Castrillon, J. L., Tellez-Plaza, M., & Martin-Escudero, J. C., 2021. Arsenic, cadmium, and selenium exposures and bone mineral density-related endpoints: The HORTEGA study. *Free Radical Biology and Medicine*, 162, 392–400. https://doi.org/10.1016/j.freeradbiomed.2020.10.318

Genchi, G., Carocci, A., Lauria, G., Sinicropi, M. S., & Catalano, A., 2020. Nickel: Human health and environmental toxicology. *International Journal of Environmental Research and Public Health*, 17(3), 679. https://doi.org/10.3390/ijerph17030679

Gertz, C., 2014. Fundamentals of the frying process*. *European Journal of Lipid Science and Technology*, 116(6), 669–674. https://doi.org/10.1002/ejlt.201400015

Gheisari, E., Raissy, M., & Rahimi, E., 2016. *The Effect of Different Cooking Methods on Lead and Cadmium Contents of Shrimp and Lobster. Journal of Food Biosciences and Technology*, 6(2), 53–58.

Ghoshal, G., 2018. Chapter 2–Emerging food processing technologies. In A. M. Grumezescu & A. M. Holban (Eds.), *Food Processing for Increased Quality and Consumption* (pp. 29–65): Academic Press. https://doi.org/10.1016/B978-0-12-811447-6.00002-3

Goswami, A. K., & Baroliya, P. K., 2022. *Heavy Metals Adsorption: Low-Cost Adsorbents.*Berlin, Boston: Walter de Gruyter GmbH & Co KG. DOI: https://doi.org/10.1515/9783110774658

Grabowska, I., 2011. Reduction of heavy metals transfer into food. *Polish Journal of Environmental Studies*, 20(3), 635–642.

Groh, K. J., Backhaus, T., Carney-Almroth, B., Geueke, B., Inostroza, P. A., Lennquist, A., Leslie, H. A., Maffini, M., Slunge, D., Trasande, L., Warhurst, A. M., & Muncke, J., 2019. Overview of known plastic packaging-associated chemicals and their hazards. *Science of the Total Environment*, 651, 3253–3268. https://doi.org/10.1016/j.scitotenv.2018.10.015

Gromadzka, G., Tarnacka, B., Flaga, A., & Adamczyk, A., 2020. Copper Dyshomeostasis in neurodegenerative diseases–Therapeutic implications. *International Journal of Molecular Sciences,* 21(23), 9259. https://doi.org/10.3390/ijms21239259

Guney, B., & Gokmen, S., 2021. Effects of different heat treatment and radiation (microwave and infrared) sources on minerals and heavy metal contents of cow's milk. *Journal of Food Processing and Preservation*, 45(1), e15084. https://doi.org/10.1111/jfpp.15084

Hajeb, P., Sloth, J. J., Shakibazadeh, S., Mahyudin, N. A., & Afsah-Hejri, L., 2014. Toxic elements in food: Occurrence, binding, and reduction approaches. *Comprehensive Reviews in Food Science and Food Safety*, 13(4), 457–472. https://doi.org/10.1111/1541-4337.12068

Hamed Hammad Mohammed, H., Jin, G., Ma, M., Khalifa, I., Shukat, R., Elkhedir, A. E., Zeng, Q., & Noman, A. E., 2020. Comparative characterization of proximate nutritional compositions, microbial quality and safety of camel meat in relation to mutton, beef, and chicken. *LWT*, 118, 108714. https://doi.org/10.1016/j.lwt.2019.108714

Han, Y., Ryu, K., Song, N., Seo, J., Kang, I., Chung, H.-J., & Park, R., 2024. Potential migration and health risks of heavy metals and metalloids in take-out food containers in South Korea. *International Journal of Environmental Research and Public Health*, 21(2), 139. https://doi.org/10.3390/ijerph21020139

Hasan, G., Kabir, M., & Miah, M., 2022. Determination of heavy metals in raw and pasteurized liquid milk of Bangladesh to assess the potential health risks. *Food Research*, 6(1), 233–237.

Hassan Emami, M., Saberi, F., Mohammadzadeh, S., Fahim, A., Abdolvand, M., Ali Ehsan Dehkordi, S., Mohammadzadeh, S., & Maghool, F., 2023. A review of heavy metals accumulation in red meat and meat products in the Middle East. *Journal of Food Protection*, 86(3), 100048. https://doi.org/10.1016/j.jfp.2023.100048

Hassan, J., Rajib, M. M. R., Khan, M. N.-E. A., Khandaker, S., Zubayer, M., Ashab, K. R., Kuba, T., Marwani, H. M., Asiri, A. M., Hasan, M. M., Islam, A., Rahman, M. M., & Awual, M. R., 2024. Assessment of heavy metals accumulation by vegetables irrigated with different stages of textile wastewater for evaluation of food and health risk. *Journal of Environmental Management*, 353, 120206. https://doi.org/10.1016/j.jenvman.2024.120206

Heba, H., & Sherin, N., 2017. Assessment of heavy metals residues in some food stuffs and its biocontrol by probiotic strain Enterococcus facium (E980) in an experimental model. *Alexandria Journal of Veterinary Sciences*, 52(1), 87–96. https://doi.org/10.5455/ajvs.253280

Hembrom, S., Singh, B., Gupta, S. K., & Nema, A. K., 2020. A comprehensive evaluation of heavy metal contamination in foodstuff and associated human health risk: A global perspective. In P. Singh, R. P. Singh & V. Srivastava (Eds.), *Contemporary Environmental Issues and Challenges in Era of Climate Change* (pp. 33–63). Singapore: Springer Singapore. https://doi.org/10.1007/978–981-32-9595-7_2

Ho, K. K. H. Y., & Redan, B. W., 2021. Impact of thermal processing on the nutrients, phytochemicals, and metal contaminants in edible algae. *Critical Reviews in Food Science and Nutrition*, 62(2), 508–526. https://doi.org/10.1080/10408398.2020.1821598

Huang, B., Qi, C., Yang, Z., Guo, Q., Chen, W., Zeng, G., & Lei, C., 2017. Pd/Fe3O4 nanocatalysts for highly effective and simultaneous removal of humic acids and Cr (VI) by electro-Fenton with H2O2 in situ electro-generated on the catalyst surface. *Journal of Catalysis*, 352, 337–350.

Hussein, S. A., Rathi, M. H., & Kadhim, T. J., 2021. Determination of the levels of lead and cadmium in canned fish and meat, imported to the local markets of Diyala Province, Iraq. *IOP Conference Series: Earth and Environmental Science*, 790(1), 012043. https://doi.org/10.1088/1755-1315/790/1/012043

Inobeme, A., Ajai, A., Eziukwu, C., Obingwa, P., Okonkwo, S., & Ekwoba, L., 2020. Effect of cooking methods on heavy metals content of food. *Journal of Xidian University*, 14(8), 704–714. https://doi.org/10.37896/jxu14.8/076

Joyce, K., Emikpe, B., Asare, D., Asenso, T., Yeboah, R., Jarikre, T., & Jagun, J., 2016. Effects of different cooking methods on heavy metals level in fresh and smoked game meat. *Journal of Food Processing & Technology*, 7(9), 9–11. https://doi.org/10.4172/2157-7110.1000617

Karnib, M., Kabbani, A., Holail, H., & Olama, Z., 2014. Heavy metals removal using activated carbon, silica and silica activated carbon composite. *Energy Procedia,* 50, 113–120. https://doi.org/10.1016/j.egypro.2014.06.014

Kazemi, A., Esmaeilbeigi, M., Ansari, A., Asl, A. G., & Mohammadzadeh, B., 2022. Alterations and health risk assessment of the environmental concentration of heavy metals in the edible tissue of marine fish (Thunnus tonggol) consumed by different cooking methods. *Regional Studies in Marine Science,* 53, 102361. https://doi.org/10.1016/j.rsma.2022.102361

Kazemi, A., Esmaeilbeigi, M., Sahebi, Z., & Ansari, A., 2022. Health risk assessment of total chromium in the qanat as historical drinking water supplying system. *Science of The Total Environment,* 807, 150795. https://doi.org/10.1016/j.scitotenv.2021.150795

Kesici, G. G., 2016. Arsenic ototoxicity. *Journal of Otology*, 11(1), 13–17. https://doi.org/10.1016/j.joto.2016.03.001

Khan, K. M., Chakraborty, R., Bundschuh, J., Bhattacharya, P., & Parvez, F., 2020. Health effects of arsenic exposure in Latin America: An overview of the past eight years of research. *Science of the Total Environment*, 710, 136071. https://doi.org/10.1016/j.scitotenv.2019.136071

Khan, N., Jeong, I. S., Hwang, I. M., Kim, J. S., Choi, S. H., Nho, E. Y., Choi, J. Y., Park, K. S., & Kim, K. S., 2014. Analysis of minor and trace elements in milk and yogurts by inductively coupled plasma-mass spectrometry (ICP-MS). *Food Chemistry*, 147, 220–224. https://doi.org/10.1016/j.foodchem.2013.09.147

Khan, S., & Khan, A. R., 2015. Contamination of toxic heavy metal in locally made plastic food packaging containers. *Global Journal of Science Frontiers Research B Chemistry*, 15(1), 19–24.

Khodanazary, A., Golgolipour, S., & Ghanemi, K., 2016. Effect of different cooking methods on changes of free fatty acids, thiobarbitoric acid, heavy metal (Ni, Cr, Co, Cd, Pb) and sensory properties of grass carp (Ctenopharyngodonidella). *Iranian Food Science and Technology Research Journal*, 12(5), 630–638. https://doi.org/10.22067/ifstrj.v12i5.43025

Klein, C. B., & Costa, M., 2022. *Nickel Handbook on the Toxicology of Metals* (pp. 615–637). San Diego, USA: Elsevier.

Klöckner, P., Reemtsma, T., & Wagner, S., 2021. The diverse metal composition of plastic items and its implications. *Science of the Total Environment,* 764, 142870. https://doi.org/10.1016/j.scitotenv.2020.142870

Kowalska, G., Pankiewicz, U., & Kowalski, R., 2020. Determination of the level of selected elements in Canned meat and fish and risk assessment for consumer health. *Journal of Analytical Methods in Chemistry*, 2020(1), 2148794. https://doi.org/10.1155/2020/2148794

Krämer, U., 2010. Metal hyperaccumulation in plants. *Annual Review of Plant Biology,* 61, 517–534.

Kumar, A., Krishan, B., Shivani, Dhiman, S., & Sharma, A., 2023. Nanotechnology for bioremediation of heavy metals. In M. P. Shah (Ed.), *Modern Approaches in Waste Bioremediation: Environmental Microbiology* (pp. 515–526). Cham: Springer International Publishing. https://doi.org/10.1007/978-3-031-24086-7_24

Landin-Sandoval, V. J., Mendoza-Castillo, D. I., Bonilla-Petriciolet, A., Aguayo-Villarreal, I. A., Reynel-Avila, H. E., & Gonzalez-Ponce, H. A., 2020. Valorization of agri-food industry wastes to prepare adsorbents for heavy metal removal from water. *Journal of Environmental Chemical Engineering*, 8(5), 104067. https://doi.org/10.1016/j.jece.2020.104067

Leao, D. J., Junior, M. M. S., Brandao, G. C., & Ferreira, S. L. C., 2016. Simultaneous determination of cadmium, iron and tin in canned foods using high-resolution continuum source graphite furnace atomic absorption spectrometry. *Talanta,* 153, 45–50. https://doi.org/10.1016/j.talanta.2016.02.023

Lee, J.-G., Hwang, J.-Y., Lee, H.-E., Kim, T.-H., Choi, J.-D., & Gang, G.-J., 2019. Effects of food processing methods on migration of heavy metals to food. *Applied Biological Chemistry*, 62(1), 64. https://doi.org/10.1186/s13765-019-0470-0

Li, L., Liu, Y., Ippolito, J. A., Xing, W., Zuo, Q., & Wang, F., 2023. Fermentation affects heavy metal bioaccessibility in Chinese mantou. *Environmental Science and Pollution Research*, 30(20), 59013–59026. https://doi.org/10.1007/s11356-023-26727-6

Liu, J., Zhang, Z., Fan, Z., Tang, X., & Zhong, Q., 2024. Strategies for boosting the photocatalytic reduction of toxic metal ions: Progress and prospects. *Journal of Water Process Engineering*, 57, 104683. https://doi.org/10.1016/j.jwpe.2023.104683

Luis, G., Rubio, C., Revert, C., Espinosa, A., González-Weller, D., Gutiérrez, A. J., & Hardisson, A., 2015. Dietary intake of metals from yogurts analyzed by inductively coupled plasma optical emission spectrometry (ICP-OES). *Journal of Food Composition and Analysis*, 39, 48–54. https://doi.org/10.1016/j.jfca.2014.11.013

M Ibrahim, S., A El-Sherif, S., Abdel-Ghafour, S., S Abouzied, A.-R., M Rabea, H., A El-Lahamy, A., M El-Sayed, H., & S Abo–Zeid, K., 2020. Effect of location and grilling process on heavy metals concentration in muscles of different fish species, Egypt. *Egyptian Journal of Aquatic Biology and Fisheries*, 24(6), 15–24. https://doi.org/10.21608/ejabf.2020.109083

Machate, D. J., 2023. Anthropogenic hyperactivity for natural resources increases heavy metals concentrations in the environment: Toxicity of healthy food and cancer risks estimated. *Journal of Trace Elements and Minerals*, 4, 100057. https://doi.org/10.1016/j.jtemin.2023.100057

Maghool, F., Emami, M. H., Alipour, R., Mohammadzadeh, S., Sereshki, N., Dehkordi, S. A. E., Fahim, A., Tayarani-Najaran, Z., Sheikh, A., Kesharwani, P., & Sahebkar, A., 2023. Rescue effect of curcumin against copper toxicity. *Journal of Trace Elements in Medicine and Biology*, 78, 127153. https://doi.org/10.1016/j.jtemb.2023.127153

Mahmudiono, T., Bokov, D., Widjaja, G., Konstantinov, I. S., Setiyawan, K., Abdelbasset, W. K., Majdi, H. S., Kadhim, M. M., Kareem, H. A., & Bansal, K., 2022. Removal of heavy metals using food industry waste as a cheap adsorbent. *Food Science and Technology*, 42, article 111721. https://doi.org/10.1590/fst.111721

Malematja, E., Manyelo, T. G., Sebola, N. A., Kolobe, S. D., & Mabelebele, M., 2023. The accumulation of heavy metals in feeder insects and their impact on animal production. *Science of the Total Environment*, 885, 163716. https://doi.org/10.1016/j.scitotenv.2023.163716

Manz, J. C. K., Nsoga, J. V. F., Diazenza, J. B., Sita, S., Bakana, G. M. B., Francois, A., Ndomou, M., Gouado, I., & Mamonekene, V., 2023. Nutritional composition, heavy metal contents and lipid quality of five marine fish species from Cameroon coast. *Heliyon*, 9(3), e14031. https://doi.org/10.1016/j.heliyon.2023.e14031

Marrugo-Madrid, S., Pinedo-Hernández, J., Paternina-Uribe, R., Marrugo-Negrete, J., & Díez, S., 2022. Health risk assessment for human exposure to mercury species and arsenic via consumption of local food in a gold mining area in Colombia. *Environmental Research*, 215, 113950. https://doi.org/10.1016/j.envres.2022.113950

Massadeh, A. M., & Al-Massaedh, A. A. T., 2018. Determination of heavy metals in canned fruits and vegetables sold in Jordan market. *Environmental Science and Pollution Research*, 25(2), 1914–1920. https://doi.org/10.1007/s11356-017-0611-0

Massoud, R., Hadiani, M. R., Hamzehlou, P., & Khosravi-Darani, K., 2019. Bioremediation of heavy metals in food industry: Application of Saccharomyces cerevisiae. *Electronic Journal of Biotechnology*, 37, 56–60. https://doi.org/10.1016/j.ejbt.2018.11.003

Mathew, S., Soans, J. C., Rachitha, R., Shilpalekha, M. S., Gowda, S. G. S., Juvvi, P., & Chakka, A. K., 2023. Green technology approach for heavy metal adsorption by agricultural and food industry solid wastes as bio-adsorbents: A review. *Journal of Food Science and Technology*, 60(7), 1923–1932. https://doi.org/10.1007/s13197-022-05486-1

McGowan, L., Caraher, M., Raats, M., Lavelle, F., Hollywood, L., McDowell, D., Spence, M., McCloat, A., Mooney, E., & Dean, M., 2017. Domestic cooking and food skills: A review. *Critical Reviews in Food Science and Nutrition*, 57(11), 2412–2431. https://doi.org/10.1080/10408398.2015.1072495

Mehdipour, S. Z., Shokrzadeh, M., Khanzadi, S., & Shahsavani, D., 2018. Effects of cooking methods on the concentrations of lead, chromium, and cadmium in Whitefish (Rutilus frissi kutum) from the Caspian Sea, Iran. *Iranian Journal of Chemistry and Chemical Engineering*, 37(1), 141–147. https://doi.org/10.30492/ijcce.2018.26649

Mehmood, A., & Zeb, A., 2020. Effects of different cooking techniques on bioactive contents of leafy vegetables. *International Journal of Gastronomy and Food Science*, 22, 100246. https://doi.org/10.1016/j.ijgfs.2020.100246

Meng, R., Zhu, Q., Long, T., He, X., Luo, Z., Gu, R., Wang, W., & Xiang, P., 2023. The innovative and accurate detection of heavy metals in foods: A critical review on electrochemical sensors. *Food Control*, 150, 109743. https://doi.org/10.1016/j.foodcont.2023.109743

Mertoğlu-Elmas, G., & Çınar, G., 2018. Toxic metals in paper and paperboard food packagings. *BioResources*, 13(4), 7560–7580.

Mieiro, C. L., Coelho, J. P., Dolbeth, M., Pacheco, M., Duarte, A. C., Pardal, M. A., & Pereira, M. E., 2016. Fish and mercury: Influence of fish fillet culinary practices on human risk. *Food Control*, 60, 575–581. https://doi.org/10.1016/j.foodcont.2015.09.006

Millena, C. G., Balonzo, A. R. R., Rentoy, J. R., Ruivivar, S. S., & Bobiles, S. C., 2023. Effect of fermentation stages on the nutritional and mineral bioavailability of cacao beans (Theobroma cacao L.). *Journal of Food Composition and Analysis,* 115, 104886. https://doi.org/10.1016/j.jfca.2022.104886

Mitra, S., Chakraborty, A. J., Tareq, A. M., Emran, T. B., Nainu, F., Khusro, A., Idris, A. M., Khandaker, M. U., Osman, H., Alhumaydhi, F. A., & Simal-Gandara, J., 2022. Impact of heavy metals on the environment and human health: Novel therapeutic insights to counter the toxicity. *Journal of King Saud University – Science*, 34(3), 101865. https://doi.org/10.1016/j.jksus.2022.101865

Mohamad, A., Shah, N. N. A. K., Sulaiman, A., Mohd Adzahan, N., & Aadil, R. M., 2020. Impact of the pulsed electric field on physicochemical properties, fatty acid profiling, and metal migration of goat milk. *Journal of Food Processing and Preservation*, 44(12), e14940. https://doi.org/10.1111/jfpp.14940

Mohamad, A. N., Aziz, F., Yusof, N., Jaafar, J., Wan Salleh, W. N., Jye, L. W., & Ismail, A. F., 2023. Recent advances in heavy metal removal by thin film nanocomposite membrane. *Water Supply*, 23(9), 3635–3659. https://doi.org/10.2166/ws.2023.204

Mohamed, R., Zainudin, B. H., & Yaakob, A. S., 2020. Method validation and determination of heavy metals in cocoa beans and cocoa products by microwave assisted digestion technique with inductively coupled plasma mass spectrometry. *Food Chemistry,* 303, 125392. https://doi.org/10.1016/j.foodchem.2019.125392

Mohammadi, F., Marti, A., Nayebzadeh, K., Hosseini, S. M., Tajdar-oranj, B., & Jazaeri, S., 2021. Effect of washing, soaking and pH in combination with ultrasound on enzymatic rancidity, phytic acid, heavy metals and coliforms of rice bran. *Food Chemistry,* 334, 127583. https://doi.org/10.1016/j.foodchem.2020.127583

Mohapatra, D., Patel, A. S., Kar, A., Deshpande, S. S., & Tripathi, M. K., 2021. Effect of different processing conditions on essential minerals and heavy metal composition of sorghum grain. *Journal of Food Processing and Preservation*, 45(1), e14909. https://doi.org/10.1111/jfpp.14909

Nayak, P. K., Dash, U., Rayaguru, K., & Krishnan, K. R., 2016. Physio-chemical changes during repeated frying of cooked oil: A review. *Journal of Food Biochemistry*, 40(3), 371–390. https://doi.org/10.1111/jfbc.12215

Ncobela, C. N., Kanengoni, A. T., Hlatini, V. A., Thomas, R. S., & Chimonyo, M., 2017. A review of the utility of potato by-products as a feed resource for smallholder pig production. *Animal Feed Science and Technology,* 227, 107–117. https://doi.org/10.1016/j.anifeedsci.2017.02.008

Noureddine El Moussawi, S., Ouaini, R., Matta, J., Chébib, H., Cladière, M., & Camel, V., 2019. Simultaneous migration of bisphenol compounds and trace metals in canned vegetable food. *Food Chemistry,* 288, 228–238. https://doi.org/10.1016/j.foodchem.2019.02.116

Oladeji, O. M., Aasa, O. A., Adelusi, O. A., & Mugivhisa, L. L., 2023. Assessment of heavy metals and their human health risks in selected spices from South Africa. *Toxicology Reports,* 11, 216–220. https://doi.org/10.1016/j.toxrep.2023.09.008

Oladoye, P. O., Olowe, O. M., & Asemoloye, M. D., 2022. Phytoremediation technology and food security impacts of heavy metal contaminated soils: A review of literature. *Chemosphere,* 288, 132555. https://doi.org/10.1016/j.chemosphere.2021.132555

Olmedo, P., Chen, R., Grau-Perez, M., Nachman, K. E., Gil, F., Rule, A. M., & Navas-Acien, A., 2024. Manufacturing practices as potential sources of cadmium and other relevant metals in processed meat products. Risk and nutritional assessment. *Food Control,* 158, 110225. https://doi.org/10.1016/j.foodcont.2023.110225

Onyeka, E., & Ibeawuchi, O., 2018. Metal concentrations of water and food cooked in various cooking pots. *International Journal of Innovative Research and Development*, 7(6), 182–195. https://doi.org/10.24940/ijird/2018/v7/i6/JUN18093

Orosun, M. M., Nwabachili, S., Alshehri, R. F., Omeje, M., Alshdoukhi, I. F., Okoro, H. K., Ogunkunle, C. O., Louis, H., Abdulhamid, F. A., Osahon, S. E., Mohammed, A. U., Ehinlafa, E. O., Yunus, S. O., & Ife-Adediran, O., 2023. Potentially toxic metals in irrigation water, soil, and vegetables and their health risks using Monte Carlo models. *Scientific Reports*, 13(1), 21220. https://doi.org/10.1038/s41598-023-48489-4

Palma-Lara, I., Martínez-Castillo, M., Quintana-Pérez, J. C., Arellano-Mendoza, M. G., Tamay-Cach, F., Valenzuela-Limón, O. L., García-Montalvo, E. A., & Hernández-Zavala, A., 2020. Arsenic exposure: A public health problem leading to several cancers. *Regulatory Toxicology and Pharmacology,* 110, 104539. https://doi.org/10.1016/j.yrtph.2019.104539

Pang, F. M., Kumar, P., Teng, T. T., Mohd Omar, A. K., & Wasewar, K. L., 2011. Removal of lead, zinc and iron by coagulation–flocculation. *Journal of the Taiwan Institute of Chemical Engineers*, 42(5), 809–815. https://doi.org/10.1016/j.jtice.2011.01.009

Pankiewicz, U., Sujka, M., Kowalski, R., Mazurek, A., Włodarczyk-Stasiak, M., & Jamroz, J., 2017. Effect of pulsed electric fields (PEF) on accumulation of selenium and zinc ions in Saccharomyces cerevisiae cells. *Food Chemistry*, 221, 1361–1370. https://doi.org/10.1016/j.foodchem.2016.11.018

Parkar, J., & Rakesh, M., 2014. Leaching of elements from packaging material into canned foods marketed in India. *Food Control*, 40, 177–184. https://doi.org/10.1016/j.foodcont.2013.11.042

Parvin, F., Rikta, S. Y., & Tareq, S. M., 2019. 8 – Application of nanomaterials for the removal of heavy metal from wastewater. In A. Ahsan & A. F. Ismail (Eds.), *Nanotechnology in Water and Wastewater Treatment* (pp. 137–157): Elsevier. https://doi.org/10.1016/B978-0-12-813902-8.00008-3

Pavesi, T., & Moreira, J. C., 2020. Mechanisms and individuality in chromium toxicity in humans. *Journal of Applied Toxicology*, 40(9), 1183–1197. https://doi.org/10.1002/jat.3965

Perelló, G., Martí-Cid, R., Llobet, J. M., & Domingo, J. L., 2008. Effects of various cooking processes on the concentrations of arsenic, cadmium, mercury, and lead in foods. *Journal of Agricultural and Food Chemistry*, 56(23), 11262–11269. https://doi.org/10.1021/jf802411q

Petrova, M. V., Ourgaud, M., Boavida, J. R. H., Dufour, A., Tesán Onrubia, J. A., Lozingot, A., & Heimbürger-Boavida, L.-E., 2020. Human mercury exposure levels and fish consumption at the French Riviera. *Chemosphere*, 258, 127232. https://doi.org/10.1016/j.chemosphere.2020.127232

Phrukphicharn, A., Sripanyaphol, S., & Kanisornmongkolchai, T., 2021. Effects of cooking process on the level of heavy metal accumulation in vegetables: Literature review. *International Journal of Environment, Agriculture and Biotechnology*, 6(5), 133–141. https://doi.org/10.22161/ijeab.65.18

Pistón, M., Suárez, A., Bühl, V., Tissot, F., Silva, J., & Panizzolo, L., 2020. Influence of cooking processes on Cu, Fe, Mn, Ni, and Zn levels in beef cuts. *Journal of Food Composition and Analysis*, 94, 103624. https://doi.org/10.1016/j.jfca.2020.103624

Plum, L. M., Rink, L., & Haase, H., 2010. The essential toxin: Impact of zinc on human health. *International Journal of Environmental Research and Public Health*, 7(4), 1342–1365. https://doi.org/10.3390/ijerph7041342

Pohl, A., 2020. Removal of heavy metal ions from water and wastewaters by sulfur-containing precipitation agents. *Water, Air, & Soil Pollution*, 231(10), 503.

Porova, N., Botvinnikova, V., Krasulya, O., Cherepanov, P., & Potoroko, I., 2014. Effect of ultrasonic treatment on heavy metal decontamination in milk. *Ultrasonics Sonochemistry*, 21(6), 2107–2111. https://doi.org/10.1016/j.ultsonch.2014.03.029

Qureshi, A. S., Hussain, M. I., Ismail, S., & Khan, Q. M., 2016. Evaluating heavy metal accumulation and potential health risks in vegetables irrigated with treated wastewater. *Chemosphere*, 163, 54–61. https://doi.org/10.1016/j.chemosphere.2016.07.073

Rai, P. K., Lee, S. S., Zhang, M., Tsang, Y. F., & Kim, K.-H., 2019. Heavy metals in food crops: Health risks, fate, mechanisms, and management. *Environment International*, 125, 365–385. https://doi.org/10.1016/j.envint.2019.01.067

Rajendra Sukhadeorao, D., 2020. Lead: Toxicological profile, pollution aspects and remedial solutions. In C. Pipat (Ed.), *Lead Chemistry* (pp. Ch. 3). Rijeka: IntechOpen. https://doi.org/10.5772/intechopen.93095

Raji, Z., Karim, A., Karam, A., & Khalloufi, S., 2023. A review on the heavy metal adsorption capacity of dietary fibers derived from agro-based wastes: Opportunities and challenges for practical applications in the food industry. *Trends in Food Science & Technology*, 137, 74–91. https://doi.org/10.1016/j.tifs.2023.05.004

Ranta-Korpi, M., Vainikka, P., Konttinen, J., Saarimaa, A., & Rodriguez, M., 2014. *Ash Forming Elements in Plastics and Rubbers* (p. 131). Espoo, Finland: VTT Technical Research Centre of Finland.

Redan, B. W., 2020. Processing aids in food and beverage manufacturing: Potential source of elemental and trace metal contaminants. *Journal of Agricultural and Food Chemistry*, 68(46), 13001–13007. https://doi.org/10.1021/acs.jafc.9b08066

Renu, K., Chakraborty, R., Myakala, H., Koti, R., Famurewa, A. C., Madhyastha, H., Vellingiri, B., George, A., & Valsala Gopalakrishnan, A., 2021. Molecular mechanism of heavy metals (Lead, Chromium, Arsenic, Mercury, Nickel and Cadmium)–Induced hepatotoxicity–A review. *Chemosphere*, 271, 129735. https://doi.org/10.1016/j.chemosphere.2021.129735

Ribeiro Sant'Ana, M. A., de Carvalho, T. C., & da Silva, I. F., 2021. Concentration of heavy metals in UHT dairy milk available in the markets of São Luís, Brazil, and potential health risk to children. *Food Chemistry*, 346, 128961. https://doi.org/10.1016/j.foodchem.2020.128961

Román-Ochoa, Y., Choque Delgado, G. T., Tejada, T. R., Yucra, H. R., Durand, A. E., & Hamaker, B. R., 2021. Heavy metal contamination and health risk assessment in grains and grain-based processed food in Arequipa region of Peru. *Chemosphere*, 274, 129792. https://doi.org/10.1016/j.chemosphere.2021.129792

Rondanelli, M., Daglia, M., Meneghini, S., Di Lorenzo, A., Peroni, G., Faliva, M. A., & Perna, S., 2017. Nutritional advantages of sous-vide cooking compared to boiling on cereals and legumes: Determination of ashes and metals content in ready-to-eat products. *Food Science & Nutrition*, 5(3), 827–833. https://doi.org/10.1002/fsn3.469

Sabala, R. F., Zakaria, A. I., Imre, K., Morar, A., Herman, V., & Sallam, K. I., 2024. Potential cancer risks associated with the consumption of raw, salted, and canned sardine contaminated by mercury, arsenic, lead, and cadmium in Egypt. *Journal of Food Composition and Analysis*, 134, 106516. https://doi.org/10.1016/j.jfca.2024.106516

Saidon, N. B., Szabó, R., Budai, P., & Lehel, J., 2024. Trophic transfer and biomagnification potential of environmental contaminants (heavy metals) in aquatic ecosystems. *Environmental Pollution*, 340, 122815. https://doi.org/10.1016/j.envpol.2023.122815

Samet, J. M., Chen, H., Pennington, E. R., & Bromberg, P. A., 2020. Non-redox cycling mechanisms of oxidative stress induced by PM metals. *Free Radical Biology and Medicine*, 151, 26–37. https://doi.org/10.1016/j.freeradbiomed.2019.12.027

Şanlier, N., Gökcen, B. B., & Sezgin, A. C., 2019. Health benefits of fermented foods. *Critical Reviews in Food Science and Nutrition*, 59(3), 506–527. https://doi.org/10.1080/10408398.2017.1383355

Sarker, A., Kim, J.-E., Islam, A. R. M. T., Bilal, M., Rakib, M. R. J., Nandi, R., Rahman, M. M., & Islam, T., 2022. Heavy metals contamination and associated health risks in food webs—A review focuses on food safety and environmental sustainability in Bangladesh. *Environmental Science and Pollution Research*, 29(3), 3230–3245. https://doi.org/10.1007/s11356-021-17153-7

Scivicco, M., Cacciola, N. A., Esposito, F., Squillante, J., Ariano, A., Borriello, L., Cirillo, T., & Severino, L., 2024. Heavy metals in fishes from the Tyrrhenian Sea and risk assessment. *Journal of Food Composition and Analysis*, 128, 106027. https://doi.org/10.1016/j.jfca.2024.106027

Scutaraşu, E. C., & Trincă, L. C., 2023. Heavy metals in foods and beverages: Global situation, health risks and reduction methods. *Foods*, 12(18), 3340

Shah-Kulkarni, S., Ha, M., Kim, B.-M., Kim, E., Hong, Y.-C., Park, H., Kim, Y., Kim, B.-N., Chang, N., Oh, S.-Y., Kim, Y. J., Lee, B., & Ha, E.-H., 2016. Neurodevelopment in early childhood affected by prenatal lead exposure and iron intake. *Medicine*, 95(4). https://doi.org/10.1097/MD.0000000000002508

Shahbazi, Y., Ahmadi, F., & Fakhari, F., 2016. Voltammetric determination of Pb, Cd, Zn, Cu and Se in milk and dairy products collected from Iran: An emphasis on permissible limits and risk assessment of exposure to heavy metals. *Food Chemistry*, 192, 1060–1067. https://doi.org/10.1016/j.foodchem.2015.07.123

Shaheen, N., Irfan, N. M., Khan, I. N., Islam, S., Islam, M. S., & Ahmed, M. K., 2016. Presence of heavy metals in fruits and vegetables: Health risk implications in Bangladesh. *Chemosphere*, 152, 431–438. https://doi.org/10.1016/j.chemosphere.2016.02.060

Shaheen, S. M., Niazi, N. K., Hassan, N. E. E., Bibi, I., Wang, H., Tsang, Daniel C. W., Ok, Y. S., Bolan, N., & Rinklebe, J., 2019. Wood-based biochar for the removal of potentially toxic elements in water and wastewater: A critical review. *International Materials Reviews*, 64(4), 216–247. https://doi.org/10.1080/09506608.2018.1473096

Sharafi, K., Yunesian, M., Nodehi, R. N., Hossein Mahvi, A., Pirsaheb, M., & Nazmara, S., 2019. The reduction of toxic metals of various rice types by different preparation and cooking processes – Human health risk assessment in Tehran households, Iran. *Food Chemistry*, 280, 294–302. https://doi.org/10.1016/j.foodchem.2018.12.060

Siripongvutikorn, S., Asksonthong, R., & Usawakesmanee, W., 2016. Evaluation of harmful heavy metal (Hg, Pb and Cd) reduction using Halomonas elongata and Tetragenococcus halophilus for protein hydrolysate product. *Functional Foods in Health and Disease*, 6(4), 195–205. https://doi.org/10.31989/ffhd.v6i4.240

Sithole, S. C., Agboola, O. O., Mugivhisa, L. L., Amoo, S. O., & Olowoyo, J. O., 2022. Elemental concentration of heavy metals in oyster mushrooms grown on mine polluted soils in Pretoria, South Africa. *Journal of King Saud University – Science*, 34(2), 101763. https://doi.org/10.1016/j.jksus.2021.101763

Sneddon, E. J., Hardaway, C. J., Sneddon, J., Boggavarapu, K., Tate, A. S., Tidwell, S. L., Gary, D. P., & Douvris, C., 2017. Determination of selected metals in rice and cereal by inductively coupled plasma-optical emission spectrometry (ICP-OES). *Microchemical Journal*, 134, 9–12. https://doi.org/10.1016/j.microc.2017.04.009

Song, Y., Guo, F., Zeng, P., Liu, J., Wang, Y., & Cheng, H., 2022. Simultaneous measurements of Cr, Cd, Hg and Pb species in ng L−1 levels by interfacing high performance liquid chromatography and inductively coupled plasma mass spectrometry. *Analytica Chimica Acta*, 1212, 339935. https://doi.org/10.1016/j.aca.2022.339935

Sood, S., & Sharma, C., 2019. Levels of selected heavy metals in food packaging papers and paperboards used in India. *Journal of Environmental Protection*, 10(3), 360–368. https://doi.org/10.4236/jep.2019.103021

Su, C., Gao, Y., Qu, X., Zhou, X., Yang, X., Huang, S., Han, L., Zheng, N., & Wang, J., 2021. The occurrence, pathways, and risk assessment of heavy metals in raw milk from industrial areas in China. *Toxics*, 9(12), 320. https://doi.org/10.3390/toxics9120320

Suleman, R., Wang, Z., Aadil, R. M., Hui, T., Hopkins, D. L., & Zhang, D., 2020. Effect of cooking on the nutritive quality, sensory properties and safety of lamb meat: Current challenges and future prospects. *Meat Science*, 167, 108172. https://doi.org/10.1016/j.meatsci.2020.108172

Tajeddin, B., & Arabkhedri, M., 2020. Chapter 16 – Polymers and food packaging. In M. A. A. AlMaadeed, D. Ponnamma & M. A. Carignano (Eds.), *Polymer Science and Innovative Applications* (pp. 525–543): Elsevier. https://doi.org/10.1016/B978-0-12-816808-0.00016-0

Taseidifar, M., Makavipour, F., Pashley, R. M., & Rahman, A. F. M. M., 2017. Removal of heavy metal ions from water using ion flotation. *Environmental Technology & Innovation*, 8, 182–190. https://doi.org/10.1016/j.eti.2017.07.002

Taylor, A. A., Tsuji, J. S., Garry, M. R., McArdle, M. E., Goodfellow, W. L., Adams, W. J., & Menzie, C. A., 2020. Critical review of exposure and effects: Implications for setting regulatory health criteria for ingested copper. *Environmental Management*, 65(1), 131–159. https://doi.org/10.1007/s00267-019-01234-y

Teck Kim, Y., Min, B., & Won Kim, K., 2014. Chapter 2 – General characteristics of packaging materials for food system. In J. H. Han (Ed.), *Innovations in Food Packaging (Second Edition)* (pp. 13–35). San Diego: Academic Press. https://doi.org/10.1016/B978-0-12-394601-0.00002-3

Tinas, H., Ozbek, N., & Akman, S., 2018. Determination of lead in flour samples directly by solid sampling high resolution continuum source graphite furnace atomic absorption spectrometry. *Spectrochimica Acta Part B: Atomic Spectroscopy*, 140, 73–75. https://doi.org/10.1016/j.sab.2017.12.002

Turner, A., & Filella, M., 2021. Hazardous metal additives in plastics and their environmental impacts. *Environment International*, 156, 106622. https://doi.org/10.1016/j.envint.2021.106622

Ulaganathan, A., Robinson, J. S., Rajendran, S., Geevaretnam, J., Pandurangan, P., & Durairaj, S., 2022. Effect of different thermal processing methods on potentially toxic metals in the seafood, Penaeus vannamei, and the related human health risk assessment. *Journal of Food Composition and Analysis*, 105, 104259. https://doi.org/10.1016/j.jfca.2021.104259

Utomo, H. D., Tan, K. X. D., Choong, Z. Y. D., Yu, J. J., Ong, J. J., & Lim, Z. B., 2016. Biosorption of heavy metal by algae biomass in surface water. *Journal of Environmental Protection*, 7(11), 1547–1560. https://doi.org/10.4236/jep.2016.711128

Verma, B., Balomajumder, C., Sabapathy, M., & Gumfekar, S. P., 2021. Pressure-driven membrane process: A review of advanced technique for heavy metals remediation. *Processes*, 9(5), 752. https://doi.org/10.3390/pr9050752

Vo, T. S., Hossain, M. M., Jeong, H. M., & Kim, K., 2020. Heavy metal removal applications using adsorptive membranes. *Nano Convergence*, 7(1), 36. https://doi.org/10.1186/s40580-020-00245-4

Whitt, M., Brown, W., Danes, J. E., & Vorst, K. L., 2016. Migration of heavy metals from recycled polyethylene terephthalate during storage and microwave heating. *Journal of Plastic Film & Sheeting*, 32(2), 189–207. https://doi.org/10.1177/8756087915590

Wieszczycka, K., & Staszak, K., 2017. Polymers in separation processes. *Physical Sciences Reviews*, 2(5). https://doi.org/10.1515/psr-2016-0127

Witkowska, D., Słowik, J., & Chilicka, K., 2021. Heavy metals and human health: Possible exposure oathways and the competition for protein binding sites. *Molecules*, 26(19), 6060. https://doi.org/10.3390/molecules26196060

Xiang, M., Ma, J., Cheng, J., Lei, K., Li, F., Shi, Z., & Li, Y., 2022. Collaborative evaluation of heavy metal pollution of soil-crop system in the southeast of Yangtze River Delta, China. *Ecological Indicators,* 143, 109412. https://doi.org/10.1016/j.ecolind.2022.109412

Xu, F., Liu, Z., Cao, Y., Qiu, L., Feng, J., Xu, F., & Tian, X., 2017. Assessment of heavy metal contamination in urban river sediments in the Jiaozhou Bay catchment, Qingdao, China. *CATENA,* 150, 9–16. https://doi.org/10.1016/j.catena.2016.11.004

Yang, Z., Zhang, Y., Wu, Y., & Ouyang, J., 2023. Factors influencing the starch digestibility of starchy foods: A review. *Food Chemistry*, 406, 135009. https://doi.org/10.1016/j.foodchem.2022.135009

Yousuf, S., Donald, A., & Hassan, A., 2022. A review on particulate matter and heavy metal emissions; impacts on the environment, detection techniques and control strategies. *MOJ Ecology Environmental Science*, 7(1), 1–5.

Zeng, X., Liu, D., Wu, Y., Zhang, L., Chen, R., Li, R., Gu, W., Zhang, L., Liu, C., & Sun, Q., 2023. Heavy metal risk of disposable food containers on human health. *Ecotoxicology and Environmental Safety*, 255, 114797. https://doi.org/10.1016/j.ecoenv.2023.114797

Zergui, A., Boudalia, S., & Joseph, M. L., 2023. Heavy metals in honey and poultry eggs as indicators of environmental pollution and potential risks to human health. *Journal of Food Composition and Analysis,* 119, 105255. https://doi.org/10.1016/j.jfca.2023.105255

Zergui, A., Kerdoun, M. A., & Boudalia, S., 2024. Trace elements in tea in Ouargla, Algeria and health risk assessment. *Food Additives & Contaminants: Part B*, 17(2), 110–121. https://doi.org/10.1080/19393210.2024.2304233

Zhai, Q., Guo, Y., Tang, X., Tian, F., Zhao, J., Zhang, H., & Chen, W., 2019. Removal of cadmium from rice by Lactobacillus plantarum fermentation. *Food Control,* 96, 357–364. https://doi.org/10.1016/j.foodcont.2018.09.029

Zhang, Y., Song, Z., Huang, S., Zhang, P., Peng, Y., Wu, P., Gu, J., Dutkiewicz, S., Zhang, H., Wu, S., Wang, F., Chen, L., Wang, S., & Li, P., 2021. Global health effects of future atmospheric mercury emissions. *Nature Communications*, 12(1), 3035. https://doi.org/10.1038/s41467-021-23391-7

Zhong, W.-S., Ren, T., & Zhao, L.-J., 2016. Determination of Pb (Lead), Cd (Cadmium), Cr (Chromium), Cu (Copper), and Ni (Nickel) in Chinese tea with high-resolution continuum source graphite furnace atomic absorption spectrometry. *Journal of Food and Drug Analysis*, 24(1), 46–55. https://doi.org/10.1016/j.jfda.2015.04.010

Zhou, X., Zheng, N., Su, C., Wang, J., & Soyeurt, H., 2019. Relationships between Pb, As, Cr, and Cd in individual cows' milk and milk composition and heavy metal contents in water, silage, and soil. *Environmental Pollution*, 255, 113322. https://doi.org/10.1016/j.envpol.2019.113322

Zhu, Y., Fan, W., Zhou, T., & Li, X., 2019. Removal of chelated heavy metals from aqueous solution: A review of current methods and mechanisms. *Science of the Total Environment*, 678, 253–266.

Zhuang, W., Liu, Y., Chen, Q., Wang, Q., & Zhou, F., 2016. A new index for assessing heavy metal contamination in sediments of the Beijing-Hangzhou Grand Canal (Zaozhuang Segment): A case study. *Ecological Indicators*, 69, 252–260. https://doi.org/10.1016/j.ecolind.2016.04.029

Acrylamide and Safety

Food Processing and Manufacturing-Related Effects

4

Shazia Chohan, Sanam Iram Soomro, Raziye Nur Özçiçek, Hüseyin Kara, and Najma Memon

4.1 INTRODUCTION

During processing, food commodities go through a variety of heat treatments, which include methods like baking, pasteurization, frying, and roasting. The primary goals of these heat treatments are to ensure that the food remains stable for longer periods and to achieve the desired taste and texture. The application of heat, such as caramelization, lipid oxidation, and the Maillard reaction to food can trigger different reactions. This results in the production of various compounds that can have beneficial effects, like antimicrobial, antioxidant, and antiallergenic properties, as well as undesirable effects, such as cytotoxicity, carcinogenicity, and mutagenic responses. Acrylamide (AA) is one such compound produced during food processing that has gained significant attention because of its toxicity and frequent presence in processed foods (Borrelli and Fogliano 2005, van Boekel et al. 2010, Capuano and Fogliano 2011).

 DOI: 10.1201/9781032712291-4

Acrylamide (AA) is colorless and odorless crystalline solid (Krishnakumar and Visvanathan 2014a) that find applications in water purification, cosmetics, and tunnel construction. The presence of acrylamide in food products was discovered by researchers from the Swedish National Food Authority and the University of Stockholm in April 2002 (Sharp 2003). Originally, the researchers were investigating the influence of acrylamide exposure on tunnel workers following an accidental spillage of a grouting agent. To their surprise, they discovered that the control group, who had never been exposed to acrylamide, had measurable blood levels of the acrylamide-haemoglobin adduct. They concluded from this finding that heating foods high in carbohydrates, particularly potatoes, over 120°C (248°F) produced acrylamide. Further examination of food products obtained from stores in Stockholm confirmed acrylamide presence in various carbohydrate-rich foods, while meat products had significantly reduced acrylamide levels (Ali Zaber 2022).

The Swedish report quickly garnered attention and concern worldwide due to probable adverse health effects. As a result, an expert consultation on the "Health Implications of Acrylamide in Food" was held in June 2002 by the Food and Agricultural Organization (FAO) and the World Health Organization (WHO) (Sirot et al. 2012). Most often, acrylamide is an industrial chemical that is used to make polyacrylamide, grouting agents, concrete binders, soil treatment as a conditioner, paper manufacturing, ore processing, water treatment as a flocculant, and gel electrophoresis in various research. Acrylamide's chemical and toxicological characteristics have been well researched. (Kusnin et al. 2015, Taeymans et al. 2004). High-dose animal studies, mainly conducted on rodents, have shown that acrylamide is a genotoxin and cancer-causing compound. In humans, it acts as a neurotoxin, causing peripheral neuropathy. Acrylamide was categorized as Group 2A-"probably carcinogenic to humans" by the International Agency for Research on Cancer (IARC) based on these animal studies" (IARC 1994). AA can induce tumours in various organs and cause reproductive and neurotoxic damage. Following food ingestion, the digestive system quickly absorbs AA, which is then transferred to various tissues. Glycinamide (GA), which is more reactive towards DNA and proteins, is the result of the liver's metabolism of AA. When ordinary foods are cooked at high temperatures and processed by Maillard reactions, free asparagine (ASN) and reducing sugars are the key ingredients that create AA. The margin of exposure (MOE) criterion is used to measure the hazard of dietary AA exposure. The MOE for dietary AA exposure in various global zones is below 10,000, and in some cases, below 200. While some research has linked the use of AA to an increased risk of human cancer, the majority of investigations have not shown a strong correlation. The consultation identified numerous knowledge gaps and recommended more investigation in areas where there were knowledge gaps. This sparked an unparalleled global collaborative effort that is still going strong today. The delay was caused in part by a lack of toxicological information on human exposure to low concentrations of acrylamide found in most food items (Lineback et al. 2012). This consultation identified several knowledge gaps and recommended further research in those areas (FAO/WHO 2006).

This chapter focuses on significant advancements in our understanding of this significant problem and draws attention to critical future issues.

4.2 OCCURRENCE AND FORMATION OF ACRYLAMIDE IN PROCESSED FOOD

Food products that possess high values of acrylamide include potato chips, bread, biscuits, cereals, coffee, baby food, fruits, vegetables, meat, fish, popcorn, cocoa goods, green tea, oilseeds, condiments, sauces, and handmade potato products. Dietary AA exposures were assessed by the EFSA's 2015 Panel on Contaminants in the Food Chain (CONTAM) to be between 0.4 and 1.9 μg/kg body weight per day and between 0.6 and 3.4 μg/kg per day. The diets strong in carbohydrates, such as French fries, potato chips, and bakery goods, have greater amounts of AA (400 ppb–7000 ppb), followed by diets high in protein (5–400 ppb) (Chain 2015). The levels of AA in various food products from different countries are summarized in Table 4.1, which is acrylamide incidence statistics from 22 countries between 2002 and 2004 found in the Summary Report of the 64th Meeting of the Joint Food and Agriculture Organization of the United Nations/World Health Organization Expert Committee on Food Additives (www.who.int/). Out of the entire 6,752 analytical findings, 21.9% came from North America, 8.9% came from Asia, 1.6% came from the Pacific, and 67.6% came from Europe. The data provided offers insights into the level of acrylamide found in various food items, highlighting the potential exposure levels associated with various cooking practices. Notably, processed cereals and pasta, such as those toasted, fried, or grilled, have a higher mean concentration of 123 μg/kg compared to their raw or boiled counterparts, which have a concentration of 15 μg/kg. Breads and rolls, as well as pastry and biscuits, have elevated levels of 446 μg/kg and 350 μg/kg, respectively. Depending on how they are prepared, potatoes exhibit a range of concentrations, with baked potatoes at 169 μg/kg and potato crisps (US 5 chips) at a significantly higher concentration of 752 μg/kg. Coffee products, particularly extracts and decaffeinated versions, also contribute to acrylamide exposure, with concentrations reaching 1100 μg/kg and 668 μg/kg, respectively. These findings emphasize the importance of understanding acrylamide levels in different foods and adopting cooking practices that minimize its formation, promoting overall food safety (Maan et al. 2022, Soares 2015).

Acrylamide is not intentionally added to food; instead, it emerges during the heat processing of certain food items. Studies suggest that the heating of food plays an important role in the formation of acrylamide. This compound is generated in a diverse range of foods, especially those rich in carbohydrates (reducing sugars), when subjected to frying, baking, or roasting at temperatures exceeding 120°C. Notably, acrylamide formation in potato fries occurs at temperatures about 120°C, particularly in conditions of reduced moisture content and prolonged heating. The conditions are illustrated in Figure 4.1 where fundamental pathways leading to acrylamide formation in foods are depicted. Acrylamide is produced by a variety of pathways in connection with the Maillard reactions system found in food items. The ASN pathway is the main channel through which acrylamide is formed.

Exploring the pathways for acrylamide formation in processed food

The main process by which acrylamide is formed in food is the Maillard reaction, which occurs when reducing carbohydrates like glucose and fructose combine with free

TABLE 4.1 Amount of acrylamide found in processed food from summary report of acrylamide incidence statistics from 22 countries

FOOD CATEGORY	*FOOD ITEM*	*MEAN CONCENTRATION (MG/KG)*	*MAXIMUM (MG/KG)*
Cereals and cereal-based products	Cereals and pasta (raw and boiled)	15	47
	Cereals and pasta (processed: toasted, fried, grilled)	123	820
	Breads and rolls	446	3436
	Pastry and biscuits (US 5 cookies)	350	7834
	Breakfast cereals	96	1346
	Pizza	33	763
Fish and seafood (breaded, fried, baked)		25	233
Meat and offal (coated, cooked, fried)		19	313
Milk and dairy products		5.8	36
Nuts and oilseeds		84	1925
	Potato purees/mashed/boiled	16	69
	Potato baked	169	1270
	Potato crisps (US 5 chips)	752	4080
	Potato chips (US 5 French fries)	334	5312
	Potato chips, croquettes (frozen, not ready-to-serve)	110	750
	Coffee (brewed), ready-to-drink	13	116
	Coffee (ground, instant or roasted, not brewed)	288	1291
	Coffee extracts	1100	4948
	Coffee decaffeinate	668	5399
	Coffee substitutes	845	7300
	Cocoa products	220	909
	Green tea (roasted)	306	660
Sugars and honey	Mainly chocolate	24	112

(continued)

TABLE 4.1 (Continued)

FOOD CATEGORY	*FOOD ITEM*	*MEAN CONCENTRATION (MG/KG)*	*MAXIMUM (MG/KG)*
Fruits & Vegetables	Raw, boiled and canned	4.2	25
	Processed (toasted, baked, fried, grilled)	59	202
	Fruits (fresh)	<1	10
	Fruits (processed: dried, fried)	131	770
Miscellaneous	Alcoholic beverages (beer, gin, wine)	6.6	46
	Condiments and sauces	71	1168
	Infant formula	<5	15
	Baby food (canned, jarred)	22	121
	Baby food (dry powder)	16	73
	Baby food (biscuits, rusks, etc.)	181	1217
	Dried food	121	1184

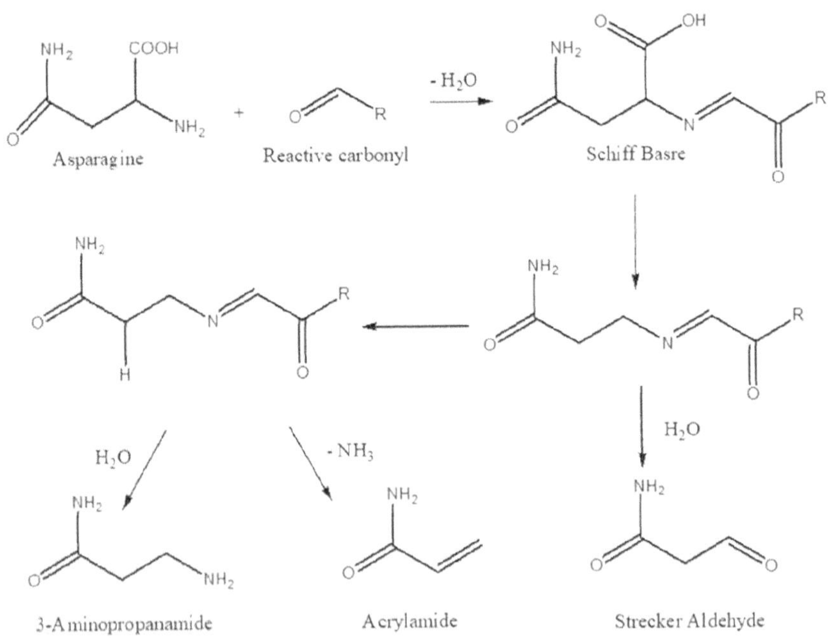

FIGURE 4.1 Mechanism that is hypothesized to produce acrylamide in food that has been heated.

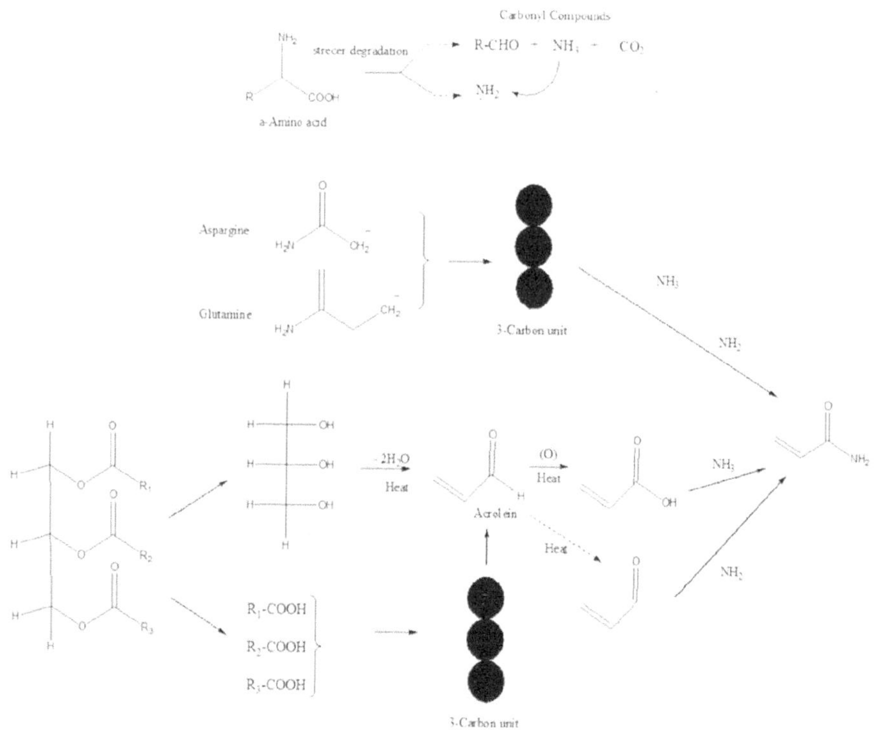

FIGURE 4.2 Hypothesized mechanism by which an amino acid and a lipid interact to form acrylamide.

amino acids like ASN. This non-enzymatic browning reaction occurs during baking or frying, with the optimal combination of carbohydrates, lipids, and proteins contributing to the desired color, flavor, and aroma. Asparagine, abundant in potatoes, necessitates carbohydrates for acrylamide formation glucose and fructose content significantly influences the possibility of acrylamide production. In rye varieties, the concentration of free ASN predominantly determines acrylamide formation, while in cooked flours and doughs, ASN is a major limiting factor. Potatoes are mainly limited by reducing sugars, while ASN plays a more substantial role in cereal products.

Although the primary routes for acrylamide formation involve ASN and reducing sugars, other suggested pathways include acrolein and ammonia (refer to Figure 4.2). In the absence of ASN, acrolein and ammonia can contribute to acrylamide formation, particularly in foods having more lipid. Triglycerides, or lipids, degrade at high temperatures is known for producing acrolein and acrylic acid. Additionally, the thermal decomposition of amino acids with ammonia results in the formation of acrylamide. It has been shown that some amino acids, such as glutamine, cysteine, and aspartic acid, can partially synthesize acrylamide. However, there is a suggestion that the process might not be as essential for the formation of acrylamide in food. (Krishnakumar and Visvanathan 2014b, Mottram, Wedzicha, and Dodson 2002, Stadler et al. 2002a, Maan et al. 2022). See Figure 4.2.

4.2.1 Human Health Effects of Acrylamide (AA)

Acrylamide (AA) is a potent neurotoxic substance, with laboratory animal studies indicating no-observable adverse effect levels (NOAEL) of 0.2–0.5 µg/kg/day and a lowest-observable-adverse effect level (LOAEL) of 2 µg/kg/day. The average dietary intake is estimated by the WHO to be 0.001 mg/kg/day. Both human occupational exposure and animal research have demonstrated the neurotoxic consequence of AA, include skeletal muscle weakening, hindlimb foot splay, and ataxia on regular exposure. Research has also shown that AA can cause BV-2 microglial cells to undergo apoptosis and mitochondrial dysfunction by inducing apoptosis in rat primary astrocytes. Numerous research projects have investigated the neurotoxic consequences of AA in work environments, such as a railroad tunnel construction project in southwest Sweden, where workers developed impaired nerve function due to exposure to a gel containing AA (Doerge et al. 2005, Zamani et al. 2017).

According to the IARC, acrylamide is categorized as a "probable human carcinogen" although there is limited information on its carcinogenic impacts in humans. AA shares structural similarities with carcinogenic compounds like vinyl carbamate and acrylonitrile. Prolonged exposure to drinking water with high levels of AA has been associated with the development of multiple tumours in both males and females. Research has suggested that AA may raise a postmenopausal woman's chance of developing various malignancies, including kidney and breast cancers. However, findings regarding the association between dietary AA and cancer hazard have been contradictory, with some studies demonstrating a positive correlation between dietary AA and renal cell cancer, while others have found no such association with bladder and prostate cancer risk. Studies have also investigated the AA's genotoxicity and that of glycidamide, one of its main metabolites, have demonstrated that AA may damage DNA at dosages of 10, 20, and 30 mg/kg (Doerge et al. 2005, Eisenbrand 2020).

Studies on reproductive toxicity of AA in laboratory animals have not shown evidence of reproductive toxicity in humans. It was shown that the NOAEL for AA reproductive harm in rats was 2–5 µg/kg/day. This is 2,000 times greater than expected dietary intakes and four times greater than the levels needed for neurotoxicity. About 0.5–10 mg/kg of AA has been shown by administration to decrease epididymal sperm reserves and cause histopathologic lesions in the testes of treated rats. Male rats injected with AA experienced decreased testosterone and prolactin concentrations. Additionally, AA was found to be toxic to the reproductive system of female mice, causing corpora lutea, organ weights, and body weights to decline (Ali et al. 1983, Sakamoto and Hashimoto 1986, Wang et al. 2010, Wei et al. 2014).

Research done by Jin et al. discovered that AA also exhibits immunotoxicity, although this aspect is less reported compared to its neurotoxicity and reproductive toxicity. The study demonstrated that AA led to decreased body weight, lymphocyte counts, thymus and spleen weights, and pathological changes in lymph glands, spleen, and thymus gland. Additionally, there was a significant reduction in the proportion of T cells and natural killer (NK) cells (Chain 2015, Semla et al. 2017, Timmermann et al. 2021).

4.3 EXPOSURE OF ACRYLAMIDE AND RISK ASSESSMENT

Acrylamide, a potentially harmful substance, has been linked to genetic and reproductive issues. A Swedish study in 2002 found that high-temperature processed carbohydrates, such as potato chips, bread, and French fries, may have high levels of AA. This has led to government and international investigations into AA in food products. Many food processors have taken steps to reduce AA levels. Acrylamide is present in a variety of food items, including potato crisps, bread, biscuits, cereals, coffee, baby foods, fruits, vegetables, seafood, pork, popcorn, oilseeds, green tea, cocoa items, sauces, spices, and handmade potato items. The study was conducted by Mesias et al., found average levels of AA in Spain's renowned food establishments adhered to EFSA standards. Nevertheless, 13.5% of the samples exceeded the European Commission's specified threshold for ready-to-eat French fries. The amount of AA found in food can vary between countries, and the main sources of AA are potato products, coffee, and bakery items. Direct exposure to AA usually occurs through the consumption of carbohydrate-rich foods, including crisps, potato chips, roasted cereals, and bread. Additionally, indirect exposure to AA can occur through food packaging that contains polyacrylamide residues. Potato products alone contribute to 50% of human utilization of AA, with bakery products like bread accounting for 20% of this total. Adults typically consume between 0.3–0.6 µg/kg/day of AA and the average intake of children is a little bit higher at 0.4–0.6 µg/kg. In the United States, foods with high levels of AA make up 38% of the total daily energy intake, and these foods provide 47% of the daily iron intake and 42% of the daily folate intake. Smokers tend to expose themselves to higher levels of AA compared to non-smokers. According to a study, men consume 0.36 µg of AA/kg of body weight on average per day, while women have slightly lower values at 0.33 µg AA/kg/day. For men and women, respectively, consumption of coffee raises these levels to 0.49 and 0.46 µg AA/kg/day (Svensson et al. 2003, Dybing 2007). Men between the ages of 16 and 30 have the greatest levels of AA. For individuals 7 to 19 years old, the average daily consumption of AA in Germany is 0.3 µg AA/kg/day, while in the Netherlands, the same intake is 0.71 µg AA/kg/day. In Brazil, the estimated mean intake of AA from foods is 0.12 µg/kg/day for individuals aged 11–17 years. These differences can be partially explained by variations in dietary patterns across different regions (Arisseto et al. 2009)

In 2011, JECFA (the Joint FAO/WHO Expert Committee on Food Additives) evaluated a MOE methodology to evaluate whether exposure to AA poses a danger to humans at specific levels. The MOE approach involves estimating the variation in the dosage that leads to a low cancer prevalence and the estimated human exposure. By utilizing this approach, risk managers can assess how closely human exposures align with levels known to cause observable effects in laboratory animals or humans. This information guides the implementation of appropriate regulatory or non-regulatory involvements to mitigate potential risks. Further research focusing on biological mechanisms is necessary to enhance our understanding of the dose-response effects of cancer caused by AA at low doses, which align with human dietary intakes. A report in 2010, mentioned physiological-based toxicokinetic model (PBTK) for the assessment of internal dosimetry of AA. The findings indicated that chronic rodent carcinogenicity studies were primarily indicative of hormonal dysregulation as a key factor of AA and/or glycidamide's carcinogenic operation. The scientists determined the cancer risk values of glycidamide and AA; and then compared them with the predicted internal exposures of humans based on daily consumption of either 4μg or 1μg of AA/kg of body weight (high consumption) or 1μg of AA/kg of body weight (medium consumption). For AA, the estimated MOE values were similar to those that JECFA had previously published (Sweeney et al. 2010, Tardiff et al. 2010). In 2011, JECFA conducted a review using preliminary data from the NTP bioassay to determine AA's MOE. The findings revealed that dietary exposure to AA increased the risk of developing male mice with Harderian gland tumours and benign mammary tumours in female rats. These amounts of MOE were consistent with the values obtained in the previous evaluation conducted in 2006. On the basis of comprehensive new NTP data, cancer bioassays, PBPK modeling, epidemiology research, as well as revised evaluations of dietary exposure, the committee reaffirmed the earlier assessment. The Committee acknowledged the potential health concerns associated with AA, which is known to possess both genotoxic and carcinogenic properties (Rudén 2004, Dybing et al. 2005, Xu et al. 2014, Zamani et al. 2017).

Risk management

Acrylamide risk management in foods has involved a voluntary collaboration between national regulatory organizations and AA-containing food producers. In 2002, Germany implemented a concept to reduce the level of AA in foods to the greatest extent possible. This concept involves categorizing food products into specific groups and establishing a signal value for the lower values of AA within the top 10% of each group. There is a limit of 1,000 μg/kg for the signal value. Annually, the results are assessed, and if there is a significant decrease in AA, the signal value is lowered accordingly. When products exceed the established signal value, producers are contacted, and discussions are held to explore methods of minimizing AA. While this system has been successful in reducing AA content in certain products, there are instances where AA levels in some products have increased. Member states have been requested by the European Commission (EC) to maintain the amount of AA present in foods that contain higher amounts of the substance and significantly contribute to dietary intake over a three-year period from 2007 to 2009. The results of this monitoring revealed that only three out of the 22 food groups showed a trend of decreasing AA content, while six groups indicated no change, while two groups exhibited an increase. To address this issue, the EC published indicative levels of AA using 10 different food categories. The values are not safety thresholds, rather, these serve as indicators to identify when further investigation is necessary to determine the reasons for exceeding the indicative values. The report resulting from this research was revised by December 31, 2012, to assess whether additional measures need to be implemented. While most countries advise consumers to adhere recommendations on behalf of a balanced diet, the available data is currently insufficient to warrant significant changes in these recommendations (Renn 2003, Slayne and Lineback 2005, Sarion et al. 2021).

4.4 INDUSTRIAL METHODS FOR THE ASSESSMENT OF ACRYLAMIDE IN FOODS

A variety of techniques have been developed in recent years for recognizing the AA monomer, particularly in biological fluids, sugar, field crops, and non-cooked meals. Most are conventional methods based on gas chromatography (GC) or high performance liquid chromatography (HPLC) (Ouyang et al.) Because of the notable level of AA in commonly consumed in bakery products mostly such as potato chips, French fries, bread, crackers, morning cereals, roasted coffee, and coffee replacements, need the use of accurate and extremely sensitive analytical procedures to ensure reliable quantitative data across different food items. The first, the LC-MS method, for analyzing AA in processed foods, was introduced by Rosen and Hellenas in 2002. The method used isotope dilution liquid chromatography mass spectrometry (LC-MS) to validate initial findings obtained in Sweden using GC-MS. The process involved extracting the analyte by using water, followed by centrifugation and solid-phase extraction using materials like Multimode cartridges, filtration through a 0.22 μm syringe filter. In last, centrifugation was carried out by using a spin filter. Chromatographic separation was achieved using HPLC on carbon black, hydrophilic reversed phase, or ion exchange columns. However, the process faced challenges due to AA's low molecular weight and weak retention with conventional LC reversed-phase sorbents. To confirm the analyte, the three-stage mass spectrometer was used to identify several different mass transitions that exhibit characteristics. It was suggested that protein precipitation might also be included in sample preparation (Rosén and Hellenäs 2002). Over time, a variety of methods have been discovered to determine the AA monomer. Particularly in food items such as sugar, field crops, and mushrooms, as well as in biological fluids like water. In the past, TLC and GC were the most often utilized techniques (Christova-Bagdassarian et al. 2012). Nevertheless, contemporary analytical methods must adhere to European standards for effectiveness and produce dependable data suitable for consumption evaluation. There are predominantly two methods to analyze AA in food: one relies on gas chromatography mass spectroscopy (GC-MS), either way around derivatization, while the other utilizes tandem liquid chromatography mass spectrometry (HPLC-MS-MS) (Wenzl et al. 2003). The commonly used procedures for determining AA for coffee and its products using GC-MS chromatography are often inadequate because co-eluting interferences are present along with the analyte. To address this issue, various methods have been developed for analyzing AA in chocolate powder, cocoa, coffee, and coffee substitutes. These methods typically involve sample preparation with an additional solid-phase extraction (SPE) clean-up step for elimination of residual matrix components. In these methods, samples are homogenized with C18 sorbent, followed by manual loading onto an SPE column (C18 Isolute), and elution of AA with water. The eluted AA is then derivatized through bromination and quantitatively determined with the help of GC-MS (GC-MS/MS) (Zhang et al. 2006). These approaches exhibit a linear range of 170–5,000 μg/kg, a limit of detection (LOD) was about 40 μg/kg, and a recovery range was between 68 to 86%. Recently, new methods have been developed utilizing modern techniques, allowing for the analysis of AA at much lower concentrations, at 0.1 µg/kg. Solid-phase microextraction (SPME) combined with gas chromatography and positive chemical ionization tandem mass spectrometry (GC-PCI-MS-MS) has been applied to analyze potato-based samples including chips and crisps for AA content (Lee et al. 2007).

4.5 INDUSTRIAL PRACTICES ASSOCIATED WITH ACRYLAMIDE FORMATION

Acrylamide is predominantly generated industrially to produce polyacrylamide that has been utilized as a flocking agent which promotes the aggregation of colloids and other suspended particles in liquids, creating a floc and dry strength agent in the paper-making process. Besides, this AA is unintentionally produced during industrial food processing. The Millard reaction is believed to be mainly responsible for the production of AA, however, its production depends on various factors which include the content of reducing sugars and ASN in source food material. The following section is devoted to discussing the factors that contribute to the production of AA during industrial food processing.

Maillard reaction: The Maillard reaction is already discussed where it was mentioned that carbonyl source like reducing sugars and ASN, an amide amino acid, undergoes condensation that is the key source of AA generation in processed foods. Furthermore, the sugar-asparagine adduct N-glycosylasparagine produces significant quantities of AA, indicating that the Maillard reaction's early phases are a substantial source of AA (Stadler et al. 2002b). The ASN route of Maillard reaction is currently recognized as a main precursor to AA. In contrast to the previously stated processes, the ASN pathway shows a major role for AA synthesis than either acrolein or acrylic acid. As a result, the ASN pathway becomes the main mechanism through which AA is produced when the Maillard reaction occurs. Moreover, 3-aminopropionamide, or decarboxylated ASNs, can produce AA despite the lack of reducing sugars (Zyzak et al. 2003). Current findings from model reaction systems indicate that acrolein and ASN together may, in some cases, produce a significant quantity of AA. This highlights the essential role that acrolein plays in the synthesis of AA in lipid-rich diets (Yasuhara et al. 2003). Other hypothesized sources of AA in food include heated gluten and its pyrolytic production (Claus, Weisz, et al. 2006) However, it is unknown how AA could be formed from protein. Nonetheless, it has been demonstrated that dried fruit like dates and prunes contains some AA (EFSA Panel on Contaminants in the Food Chain 2015). Furthermore, the studies have uncovered novel pathways, identifying acrylic acid in model systems as a potential source of AA. Given the close relationship between AA formation and food composition, several factors including the availability of raw materials (cultivars), potential seasonal fluctuations, fertilization methods, farming techniques, harvest timing, storage conditions, etc., are responsible for AA production (Taeymans et al. 2005). These factors are described as follows:

4.5.1 Effect Of Source/Raw Material Composition on the Generation of AA

Any industry that processes food finds its success largely in the selection of its raw materials. For frying and baking, ingredients at minimal concentrations of reducing sugars and ASN are usually favored to reduce the formation of AA. The essential way to select the appropriate cultivar, is backed by suitable agronomic techniques by maintaining ideal storage conditions. Studies have indicated that AA production is initiated by ASN and reducing sugars, with varying amounts of these precursors present in various crops. The reducing sugars and ASN amount in potatoes varies between the range of 97 to 2,550 mg/kg. The most abundant free amino compound in potato tubers is reducing sugars and ASN, usually making up about one-third of the entire pool of free amino acids (Elmore et al. 2007). Because of the high level of reducing sugars and ASN, it was initially believed that sugar levels would be the primary factor influencing the production of AA. The research was conducted by Shepherd et al. (2010) on segregated breeding population shows that the variance in AA formation potential was equally influenced by ASN and sugar concentrations. On the other hand, Elmore et al. (2007) found that ASN, as a proportion of the total free amino acid pool, was the decisive factor in AA formation. Shojaee-Aliabadi et al. (2013) conducted experiments on potato cultivars for low level of AA in potato chips, revealed the Sante variety potatoes, showed the greatest AA level (8,825 μg/kg) when the reducing sugar content was maximum. Claus, Schreiter, et al. (2006) investigated how cereal variability affected precursors and AA concentrations in yeast-leavened bread using nine wheat, two spelt, and two rye cultivars. The amount of ASN in wheat varied between 8.7 and 24.9 mg/kg, while the AA amount increased proportionately from 14 to 74 μg/kg. In another study, the impact of species of coffee (Espresso), roast level, and draught period of AA generation was demonstrated by Alves et al. (2010) using matrix solid-phase dispersion with GC-MS. In 30 milliliters of medium-roasted espresso, the mean concentration of AA was 1.16 ± 0.25 μg for samples of pure Arabica and 2.31 ± 0.43 μg for Robusta samples.

These studies suggest that reducing sugars and ASN content together plays a crucial role in the production of AA. However, a suitable selection of foods that contain low content may be preferred for the industrial production of foods. Any agricultural practice that increases the ASN content would eventually impact the production of AA during food processing. For example, *the use of fertilizers has a significant influence on the AA formation*. The concentration of free ASN is the key factor that determines the capacity for AA production in potatoes, wheat, and rye (Postles et al. 2013). ASN may build up to large concentrations in plants in situations when there is a surplus of nitrogen available and protein synthesis is restricted. This implies that in situations where plants are incapable of storing nitrogen as protein, they use free ASN as a nitrogen reserve (Lea et al. 2007). Higher nitrogen levels in wheat grains were shown to significantly increase ASN levels, which Lerner et al. attributed to improved utilization. The amount of AA in bread increased fourfold, from 10.6 to 55.6 mg/kg, when comparing the maximum dosage (220 kg N/ha) to no fertilizer treatment (0 kg N/ha) (Lerner et al. 2006). In 2010, Whittaker and colleagues demonstrated that the mineral composition of tubers, influenced by both soil mineral content and pH, impacted the expression of

ASN and reducing sugars. The study concluded that the location of cultivation significantly influenced the mineral composition of the tubers as a result, reducing sugar contents were inversely related to potassium and calcium levels, while they were positively associated with zinc and copper contents (Whittaker et al. 2010).

Another factor that can produce a high content of ASN is unfavorable meteorological conditions. A study revealed the influence of environmental factors on the accumulation of ASN in wheat and rye grains showcased its findings by examining six wheat varieties grown across multiple locations in the UK during two harvest periods, in addition to a varied assortment of rye cultivars cultivated in Hungary, UK, France, and Poland, gathered between 2005 and 2007 (Curtis et al. 2009, Curtis et al. 2010). The production of cereals is greatly hampered by the sprouting of cereal grains, which is brought on by unfavorable meteorological circumstances such as high wetness. This causes an increase in the protease activity to double, which raises the concentration of ASN in flour formed from grains of sprouted wheat and raises AA levels in loaves prepared with that flour. Therefore, even when combined with other flours, flours made by using sprouted wheat or rye should not be utilized in baked goods. The harvest year influences ASN levels in wheat and rye significantly. The ASN concentration of samples collected in 2004 was much lower content of ASN than that of those collected in 2003. As a result, loaves baked using 2004 flour had AA levels that were down to as low as 62% (Marti et al. 2020).

Storage conditions of the food being processed can influence the content of AA in the final product. The kind of raw material and how long it is stored are important factors in the creation of AA. Companies need to store raw materials for many months to ensure year-round availability. Potatoes are often kept in cold storage to prevent sprouting. Nonetheless, it is known that storing tubers below 8 °C results in cold sweetening, and buildup hexose in the tubers, which raises RS levels (Viklund et al. 2010). Cooking foods high in carbohydrates for a longer period at temperatures above 120°C results in the natural production of AA (Tepe et al. 2019).

4.5.2 Effect of Additives During Food Processing on the Generation AA

Ammonium carbonate: The increase in AA formation is facilitated by the addition of a baking agent, ammonium carbonate, which alters the system's pH and may promote the Schiff base formation. To investigate the possibilities of producing AA with additional amino acids, ammonium carbonate was added to a system that included glucose and an amino acid combination that had serine, threonine, alanine, proline, and phenylalanine. (resulting in ammonia release). The results of these model tests showed that ammonium carbonate had no noticeable impact on the production of AA. Further studies were carried out in matrices made of wheat flour. Ammonium carbonate was added to unwashed whole wheat flour, and this increased the production of AA. However, definitive conclusions cannot be drawn at this stage, and further comprehensive tests are required (Taeymans et al. 2004). Fluckiger and Salih, on the other hand, investigated the impact of cysteine on the development of AA in crisp bread and discovered no such impact.

Lipids and oils: There has been much discussion on the function of lipids in the synthesis of AA, particularly regarding the impact of frying oil. The process by which AA is generated is the breakdown of lipids, such as glycerol or oxidized fatty acids, which produce acrolein (Figure 4.2) (Umano and Shibamoto 1987). Acrolein can oxidize to form acrylic acid or react further to form intermediate acrylic radicals. Under favorable conditions and in the presence of nitrogen, both species may aid in the synthesis of AA (Yasuhara et al. 2003). Research on the frying of potato slices revealed that the type of oil used had a noticeable impact on generation of AA (Becalski et al. 2003). A small number of trials showed that frying in paraffin or olive oil, as compared to frying in maize oil, resulted in slightly greater AA concentrations (Tareke 2003). Additionally, it has been demonstrated that adding oil to potatoes while they were heating up raised the amounts of AA. Fried foods in sesame oil had the highest AA content when compared to corn and olive oils. Another study discovered that glycerol or edible oil added to a dry potato model before heat treatment had no appreciable effect on the development of AA. Furthermore, frying potatoes in deteriorated oil did not change the amounts of AA, indicating that there was no visible relationship between oil quality and AA generation in actual studies (Biedermann et al. 2003).

Impact of Processing Conditions: The reduction of AA in food products primarily focuses on processing conditions, including heating, frying, temperature, time, and blanching. Research studies highlight the correlation between baking temperature and time with AA formation, showing a 50% reduction with optimized conditions. Blanching, which involves leaching reducing sugars before frying, effectively minimizes the development of AA. Frying conditions, particularly temperature and time, impact AA formation, with lower pressures resulting in higher reductions. The pH adjustment through acid treatment reduces AA formation, while water activity influences its formation and removal. Prolonged fermentation aids in bread and potato products with reduced AA content, especially when united with blanching (Fiselier et al. 2006).

The details above on the sources reveal that raw food sources, their storage, additives, and techniques used to produce food have a significant impact on generation/formation of AA. Careful selection of food raw materials and appropriate conditions and techniques can significantly reduce the AA in industrially produced foods.

4.6 PRACTICES FOR REDUCTION OF ACRYLAMIDE IN PROCESSED FOODS

Acrylamide should be absent in both uncooked and cooked carbohydrate-rich foods, as well as in meat, fish, poultry, and baby formula (Food and COVID 2018). Therefore, any strategies meant to lower the production of AA should ensure that the taste and smell of cooked foods are not adversely affected. Recent researchers have focused on simple matrices like fried potatoes, to examine factors such as potato variety (including storage conditions and age), composition (such as levels of reducing sugars and amino acids), and processing parameters (such as frying temperature and time, pH adjustments using organic acids, and additives) in generation of AA (Biedermann et al. 2002). This results in providing initial "solutions" for considerably reducing AA levels in specific foods. However, there is significant concern regarding the high potential for AA formation during thermal processing. Given that fried products are a major source of this chemical compound, various preventative measures can be taken. These include selecting food varieties with lower concentration of reducing sugars, at the storage temperatures above 8°C, utilizing vacuum frying at temperatures below 120°C, favoring deep frying over surface frying, pH levels within the range between 6 to 8, pre-processing or immersing in glycine and lysine-containing amino acids, employing asparaginase and glucose oxidase to reduce AA precursors, soaking in salt solutions of NaCl and $CaCl_2$, applying lactic fermentation before processing, adding vitamins, using antioxidants, and pre-processing or treatment with pectin hydrocolloids and alginic acid (Mousavi et al. 2023).

A significant amount of effort is being made by researchers to develop strategies that reduce the AA content in processed foods. Seal et al. (2008) investigated the growing season of agricultural plants, variations in the dry matter content, sugar-to-starch ratios, and ASN levels take place. Potato tubers' sugar concentration is mostly determined by their maturity and harvest period, with choosing the right maturity stage being crucial to reducing AA during processing. When potatoes are picked before they reach maturity, their dry matter content is usually lower (16%) and their glucose and fructose contents are higher. Furthermore, the ASN levels in raw potatoes remain relatively consistent, the amount of reducing sugars can vary significantly based on factors such as variety, maturity, and storage conditions. Harvesting mature tubers is therefore essential for effectively reducing AA levels (Seal et al. 2008). In contrast to potatoes, ripe bananas exhibited noticeably larger amounts of reducing sugar, or glucose and fructose, than less mature bananas. Daniali et al. found that for two banana types (Musa paradisiaca variety Awak and Abu), the quantities of free amino acids at various maturation steps were reasonably comparable. This information was related to the synthesis of AA. Additionally, they discovered that an increase in both reducing sugars (fructose and glucose) during maturity increased the generation of AA in both banana cultivars. This study indicated the concentration of fructose and glucose had a greater influence on the production of AA in banana fritters than did the ASN level (Daniali et al. 2013).

4.6.1 Use of Additives

Addition of dietary fiber: Schouten et al. (2023), investigated methods to reduce AA formation in cookies by incorporating three different types of dietary fibers into the dough. These fibers consisted of commercial pectin, negatively charged k-carrageenan, and neutrally charged arabinogalactan. Plant polysaccharides include pectin and arabinogalactan. Pectins are straight line segments of pectic polysaccharides, primarily derived from apples and composed of (α1→4)-D-galacturonic acid residues. Arabinogalactan, on the other hand, is a branching biopolymer that was isolated from larch wood. It is mostly made up of α-L-arabinofuranose residues attached to a 1,4-linked β-D-galactopyranose backbone. Red algae's cell walls are the source of k-carrageenan, a linear sulfated mucopolysaccharide (López-Ruiz et al. 2023).

Addition of amino acids and proteins: Studies on the effects of amino acids including glycine, methionine, cysteine, and glutathione on the synthesis and reduction of AA have been undertaken by a number of researchers. These amino acids' main effect was thought to result from their rivalry with ASN during the Maillard process. Adding glycine to dough resulted in a notable 90% drop in AA content through mixing and a 16% reduction through spraying. Potato cakes dipped in a glycine solution showed an estimated 60% reduction in AA levels (Fink et al. 2006) (Bråthen et al. 2005) Glycine was added to the dough before fermentation in studies by Bråthen et al. (2005), prepared bread and flatbreads by industry standards. In breads, the reduction in AA levels was from 202 to 0.25 mg/kg, or around a 90% decrease, whereas in flatbreads, the reduction was as high as 80%. However, there was a noticeable increase in the glycine administered levels. The effects relative to flour weight were probably seen at concentrations of 1.5% and 3.0% (Fink et al. 2006). An only treatment did not have any effect on AA; however, eight repetitions of the spraying process led to a 16% reduction in AA. In contrast, another researcher found that dipping potato cakes in a glycine solution resulted in a nearly 60% drop, which could have been caused by ASN's competition with the substance for sugar reduction (Claus et al. 2008). As an alternative, AA can potentially interact by Michael addition by a nucleophilic amino group of glycine (Stadler 2006). Given that glycine is more effective than other strategies like process technology optimization, applying it to the surface of commercial bread may be one way to lower its AA concentration. In addition to glycine, other sulfur-bearing amino acids and peptides, like glutathione, cysteine, and methionine, have also been effectively employed to reduce AA in baked goods. By the addition of cysteine and methionine to potato dough or cracker (Levine and Smith 2005), the generated of AA was about 50% reduced (Bazin et al. 2017) Moreover, reactive SH-groups are present in both glutathione and cysteine, and the reaction of these groups with AA may be responsible for the production of cysteine-S propionamide, also known as S-(3-amino-3-oxopropyl)-cysteine (Stadler et al. 2002b). Flückiger and Salih investigated the effects of cysteine on AA in crisp bread, based on their previous assumption. Interestingly, adding cysteine directly to the dough had no effect, as concluded from this investigation (Flückiger and Salih 2006). On the contrary, Claus et al. investigated the addition of cysteine to bread dough preceding fermentation observed a roughly 50% decrease in ASN levels. However, because of cysteine's limited mobility in the dry crust, putting a solution of cysteine to the bread's surface had no effect. This amino acid's poor effect may be due to its interaction with pre-formed AA or due to its competition with ASN Claus et al. (2008). Studies using cracker models demonstrated that while utilizing wheat gluten caused a small decrease in AA levels, employing casein lowered them by roughly 50% (Levine and Smith 2005).

Zeng et al. (2010) explored the effects of eight hydrocolloids of the formation of AA in chemical and snack model systems. Pectin, alginic acid, and xanthan gum, at 2% (w/w) in chemical models, notably reduced AA formation by altering Maillard reaction interactions. These hydrocolloids, commonly used as gelling or thickening agents, demonstrated potential in reducing AA levels. For most hydrocolloids to effectively prevent 30% of AA production in a fried snack model, a greater 5% concentration was needed. Alginic acid and pectin emerged as favorable AA inhibitors, with immersion time proving vital in their efficacy against production of AA in fried potato products (Cai et al. 2014).

Addition of divalent cations: Before baking, adding divalent cations like Ca^{2+} or Mg^{2+} to the dough had significant effects on the amount of AA in the final goods (Elder et al. 2004). Furthermore, most of this research is conducted in small-scale or in vitro settings, which may prevent them from providing results that are comparable under commercial or industrial conditions. The addition of mono and divalent cations (Na^+ , Ca^{2+}, or Mg^{2+}) to the dough had a notable impact on the decrease of AA (Elder et al. 2004). Several investigations indicated a considerable effect of NaCl on AA reduction during polymerization and discovered that by administering these ions, AA was decreased by around 20% and as much as 50% in settings with a pH of 5.5 which is common for various meals. Comparable results were observed for bread, where Ca^{2+} led to a 30% decrease in AA. Gokmen (2007) examined how bivalent ions affected the production of AA in potato products and found an almost linearly negative relationship between the amount of Ca^{2+} supplied and the synthesis of AA. Since the interaction between carbohydrates and the ASN a-amino group is the first step, the AA-decreasing impact of bivalent cations was attributed to an inhibiting the production of the Schiff base and consequently AA synthesis. It's interesting to note that while the resultant low AA products were sufficiently brown, the Maillard reaction process did not seem to have any impact on color formation (Gökmen and Şenyuva 2007). Furthermore, polyvalent cations have the ability to decrease the production of AA when heated. The formation of Schiff reduce by the interaction of polyvalent cations with ASN and the subsequent production of AA (Tomoda et al. 2004).

Acrylamide might be reduced by up to 63% by simply soaking and washing potatoes for at least 15 minutes prior to frying. This is explained by the sugars and ASN that are run-off the potato surface (Ismial et al. 2013).

Addition of enzymes: The ASNs' side chain amide group integrates into the amide bond of AA. Evaluating the amount of AA present in food after hydrolyzing the amide bond with an acid or an enzyme would be a sensible way to confirm this process. Using the enzyme asparaginase to promote ASN's breakdown into aspartic acid and ammonia is one way to complete this, especially with ASN (Hendriksen et al. 2009). In their 2003 study, Zyzak and colleagues investigated the efficiency of asparaginase in lowering the production of AA in products made with microwave-heated mashed potatoes. They found that the enzyme led to a decrease of more than 99% in AA levels associated with samples processed without it. This finding confirms that the main route for AA formation in potato products involves ASN. Nonetheless, the practicality of implementing such interventions in actual food manufacturing environments needs thorough assessment, considering their possible effects on the final product's quality and safety (Zyzak et al. 2003). French fries, wheat crackers, and corn-based products have all been tested for the impact of enzymes such asparaginase on AA reduction. According to reports, adding asparaginase enzyme to dough decreased the amount of AA by 70% in wheat crackers and up to 90% in corn-based snacks, all while maintaining the goods' color and flavor (Ciesarová et al. 2010). When slices were blanched before asparaginase immersion, Pedreschi et al. (2011) investigated the impact of these treatments on the decrease of total AA in potato chips. They found that this method might reduce AA content by as much as 90%. (Pedreschi et al. 2009). Enzyme additions to cereal-based foods showed to reduce AA levels up to 85–90% while not influencing the taste or appearance of the final product (Tuta et al. 2010).

Addition of antioxidants: The antioxidants are shown to reduce AA formation. Antioxidants are compounds that hinder the creation of free radicals within food and neutralize free radicals within the body. Numerous research carried out between 2007 and 2013 showed that adding antioxidants to dietary items, such as phenolic compounds, bamboo extract, wild oregano, thyme, cinnamon, bougainvillaea, green tea, piquin pepper, oleo resin, and rosemary, could reduce the concentration of AA (Salazar et al. 2014, Kotsiou et al. 2010). Antioxidants have been shown to trap sugars or the 3-deoxy-2-hexosulose intermediate, which prevents the generation of Maillard products. The effects of common phenolic compounds on formation of AA in an emulsion system were studied by Kotsiou et al. these compounds included trolox, phenolic acids (ferulic, gallic, protocatechuic, and caffeic acid), and secoiridoids derived from virgin olive oil (VOO). Standard phenolic compounds were added, and this resulted in a reduction in AA production that reached up to 70% during the first 20 minutes of heating (Kotsiou et al. 2010). According to research conducted by Mulla et al. (2010), it is possible to decrease the AA levels in pooris, a traditional Indian dish, by incorporating 10 μmol of citric acid per gram of flour and 100 μmol/g of calcium chloride per gram of flour. The reduction in AA content can be accomplished without adversely affecting the sensory evaluation of the end product (Mulla et al. 2011). Shiyi Ou et al. in 2010 conducted research to find the antioxidants effects on acrylamide synthesis in foods processed at high temperatures. Their study explored three probable ways that antioxidants could decrease the production of AA and removal in model reaction systems. They investigated the impact of various antioxidants, including ferulic acid, butylated hydroxy anisole (BHA), butylated hydroxytoluene (BHT), tert-butyl hydroquinone (TBHQ), epigallocatechin gallate (EGCG), and vitamin C, along with oxidation products related to them, on both the inhibition of AA formation and its removal (Ou et al. 2010). To tackle the reduction or mitigation of AA in foods, eliminating AA post-formation during heating is considered. However, this method was not found to be effective (Mousavi et al. 2023). Cai et al. (2014) investigated how chlorogenic acid affected the production of amyloid by inducing the synthesis of HMF and 3-aminopropionamide deamination. The addition of chlorogenic acid markedly boosted the production of AA and prevented its removal in the ASN/glucose Maillard reaction system (pH = 6.8). On the other hand, AA production was reduced by the quinone derivative of chlorogenic acid.

Addition of organic acids: Over the past ten years, several researchers have examined the influence of organic acids and amino acids on the decrease in the amount of AA. The researchers showed that adding organic acid to food matrices was a successful method to dramatically limit the formation of AA by decreasing the pH, leaching away free ASN, and reducing sugars from the top layer. Reducing the pH through the use of food system organic acids may result in the protonation of ASN α-amino group, which prevents it from participating in nucleophilic addition reactions with carbonyl sources (Zhang et al. 2007). Jung et al. (2003) studied the efficacy of soaking potato slices in the solution of citric acid (10 g/L) to decrease AA formation while frying, achieving a notable reduction ranging from 28% to 86% however it may result in flavor sourness and oil rancidity during frying. Furthermore, tartaric acid is commonly incorporated into baking agents consisting of $NaHCO_3$ to improve leavening. Research with additional tartaric acid in dough demonstrated a decrease in AA levels of up to 33% as the tartaric acid content rose from 0.195% to 0.244%, and up to 44% as it increased from

0.244% to 0.293%, this suggested that the addition of tartaric acid reduces the pH, thus influencing the Maillard reaction and diminishing AA formation (Pedreschi et al. 2007). Every research found that the concentration of acid is inversely proportional to AA. This is because aspartic acid is produced at lower pH levels due to the carboxamide group's hydrolysis (Graf et al. 2006).

4.6.2 Processing Condition

Optimal processing conditions, including temperature, time duration, and moisture levels, significantly influence food processing. Research and industry studies emphasize the importance of controlling these factors to minimize formation of AA in fried potato products. Frying at temperatures exceeding 175°C intensifies AA formation, while prolonged frying exacerbates the issue. Lowering frying temperatures to less than 175°C and controlling frying period can mitigate AA levels, with additional reduction possible by lowering temperatures to 150°C to the cooking end. Vacuum frying offers strategic advantages in reducing AA production, with studies showing a 94% reduction in AA levels in potato chips (Granda et al. 2004). Ahrne et al. (2007) revealed a significant link between AA formation, baking conditions, and time. The research explored AA levels across three baking methods: deck oven with lack of the air circulation, steam baking, and falling-temperature baking, at 200°C to 260°C. AA formation initiated at 120–130°C, with traces evident at 150°C crust temperature. Steam injection halved AA formation compared to dry baking, while reducing oven temperature during baking intervals decreased AA levels by 67%, 36%, and 35%, respectively (Ahrné et al. 2007). During coffee roasting, AA forms rapidly, exceeding 7 mg/kg within the initial minutes, but its concentration diminishes with prolonged roasting. Studies by Taeymans et al. (2004) using kinetic models and spiking experiments revealed over 95% reduction of AA during roasting. However, optimizing roasting conditions to reduce AA while preserving the desired aroma and taste remains unachieved. Recent findings indicate instability of AA in commercial coffee stored in original packaging (Taeymans, Wood, Ashby, Blank, Studer, Stadler, Gondé, Eijck, Lalljie, Lingnert, et al. 2004). See Table 4.2.

TABLE 4.2 Good practices for reduction in AA content in food processing*

PROCESS	*RECOMMENDATION*
Raw material selection	1. Raw materials produced under dry metrological conditions and controlled fertilizer use. 2. Foods with low ASN and low reducing sugar content. 3. No sprouted food and no sweetened stored potatoes
Thermal processing	120 °C or preferably vacuum frying
Pre-process steps	Blanching is recommended; 15 minutes of potato soaking in water
Additives	1. Appropriate selection of oils (higher oxidative stability is preferred) 2. Addition of enzymes at industrial processing levels 3. Amino acids, antioxidants, or organic acids

Note: * Average dietary intake is estimated by the WHO to be 0.001 mg/kg/day.

4.7 PREVALENCE AND HEALTH IMPACT STUDIES

To determine the AA concentration in the Slovak Republic, the Regional Republic Health Authorities performed laboratory investigations on 164 food products intended for babies and young children. From 2013 to 2015, the average quantity of AA (midpoint bound) detected in cereal-based goods and baby food meant for babies and young children aged 6 months to 3 years was 7.5 $\mu g.kg^{-1}$. The indicated values outlined in Commission Recommendation 2013/647/EU (European Commission 2013) reports that within the period 2013–2015, none of the individual categories of food intended for infants and young children aged 6 months to 3 years exceeded the recommended levels, which ranged from 4 to 15% of the revealing values established by the Commission Recommendation. The risk evaluation of AA exposure from food for infants and young children was based on samples collected from 2013–2015. In the Slovak Republic, the average daily consumption of AA from food for the overall population from 2006–2014 was 249 $\mu g.kg^{-1}$ body weight per day. Specifically for infants and young children, the average daily consumption of AA from cereal-based foods (based on average consumption) and specified infant foods was 0.6 $\mu g.kg^{-1}$ body weight per day, and for young children, it was 0.2 $\mu g.kg^{-1}$ body weight per day (Bušová et al. 2020).

Sweden's population-based case-control study with 538 healthy controls and cases of kidney, bladder, and large bowel cancer was reanalyzed in one research. High users of 14 different food products with a high range between 300–1,200 µg/kg or moderate range between 30–299 µg/kg AA concentration did not exhibit an elevated risk of colon, bladder, or kidney cancer, nor performed they show any significant pattern in this regard (Mucci et al. 2005). Multiple multigenerational investigations conducted on rodents have also demonstrated the impacts of exposure to AA. In a continuous breeding study, mice subjected to 3, 10, or 30 ppm AA concentrations in their drinking water for 14 weeks exhibited diminished litter sizes in the F1 generation. (Tyl and Friedman 2003). Rats administered varying doses of AA, ranging from 0.5 to 60 mg/kg/d via drinking water, exhibited reproductive consequences across generations F0 through F2, particularly at higher doses. These consequences encompassed a decline in the count of live offspring, diminished survival rates among offspring, and a decrease in mating activity. Additionally, rats treated with ACR displayed decreased levels of testosterone and prolactin hormones. Several studies documented dominant poisonous effects in both rats and mice found in elevated concentrations of AA in their drinking water (Tyl et al. 2000).

4.8 CONCLUSION

This chapter has provided valuable insights into the existence of AA in various food items and the potential health implications of this compound. Acrylamide is categorized as a potential carcinogen to humans, and its neurotoxic effects have been observed in both animal studies and human cases. Collaborative efforts by organizations such as the FAO and WHO have played a pivotal role in identifying knowledge gaps and promoting ongoing research in this area. Accurate establishment of AA in foods is key to ensuring food safety and maintaining quality control. Analytical methods, particularly those utilizing HPLC and GC techniques, have been developed and refined over time, enabling precise and sensitive detection of AA. The AA formation precursors are influenced by various factors, including the selection of raw materials, fertilization techniques, environmental conditions, and storage practices. This highlights the importance of a comprehensive understanding and effective management within the food processing industry. Mitigating AA formation requires a multifaceted approach. Recommendations include the implementation of raw materials with reduced AA concentrations precursors, optimization of processing conditions, and the incorporation of additives that inhibit AA formation. The formation of AA in food products is significantly influenced by factors such as the type of food, storage conditions, and processing parameters, underscoring the importance of meticulous attention and control at every stage of production. To effectively manage AA formation in food products, a comprehensive approach is required, taking into account multiple factors across the entire food production process. This includes ongoing research, collaboration, and the implementation of efficient mitigation strategies. These methods are crucial in order to minimize AA levels in commonly consumed foods, ensuring the safety and well-being of consumers. By embracing these strategies, we can make progress in reducing AA levels and protecting public health.

ACKNOWLEDGMENT

We would like to express our sincere gratitude to the Scientific and Technological Research Council of Turkey (TUBITAK) for their generous support through the 2221-TUBITAK Scholarship program under the number 1059B212200690.

REFERENCES

Ahrné, Lilia, Claes-Göran Andersson, Per Floberg, Johan Rosén, and Hans Lingnert. 2007. "Effect of crust temperature and water content on acrylamide formation during baking of white bread: Steam and falling temperature baking." *LWT – Food Science and Technology* 40:1708–1715. doi: 10.1016/j.lwt.2007.01.010.

Ali, S Fatehyab, Jau-Shyong Hong, William E Wilson, Lynda L Uphouse, and Stephen C Bondy. 1983. "Effect of acrylamide on neurotransmitter metabolism and neuropeptide levels in several brain regions and upon circulating hormones." *Archives of Toxicology* 52:35–43.

Ali Zaber, Mohammad. 2022. "Acrylamide quantitation and quality assessment of locally available breads in Chattogram." Chattogram Veterinary & Animall Sciences University.

Alves, Rita C, C Soares, Susana Casal, JO Fernandes, and M Beatriz Oliveira. . 2010. "Acrylamide in espresso coffee: Influence of species, roast degree and brew length." *Food Chemistry* 119 (3):929–934.

Arisseto, Adriana Pavesi, Maria Cecilia de Figueiredo Toledo, Yasmine Govaert, Joris van Loco, Stéphanie Fraselle, Jean-Marie Degroodt, and Daniela Cristina Rosseto Caroba. 2009. "Contribution of selected foods to acrylamide intake by a population of Brazilian adolescents." *LWT-Food Science and Technology* 42 (1):207–211.

Bazin, Ingrid, Scherrine A Tria, Akhtar Hayat, and Jean-Louis Marty. 2017. "New biorecognition molecules in biosensors for the detection of toxins." *Biosensors and Bioelectronics* 87:285–298.

Becalski, Adam, Benjamin P-Y Lau, David Lewis, and Stephen W Seaman. 2003. "Acrylamide in foods: occurrence, sources, and modeling." *Journal of Agricultural and Food Chemistry* 51 (3):802–808.

Biedermann, Maurus, Sandra Biedermann-Brem, Anja Noti, Koni Grob, Peter Egli, and H Mandli. 2002. "Two GC-MS methods for the analysis of acrylamide in foods." *Mitteilungen aus Lebensmitteluntersuchung und Hygiene* 93 (6):638–652.

Biedermann, Maurus, Sandra Biedermann-Brem, Anja Noti, and Konrad Grob. 2003. "Methods for determining the potential of acrylamide formation and its elimination in raw materials for food preparation, such as potatoes." *Mitteilungen aus Lebensmitteluntersuchung und Hygiene*, 93 (2002), 653–667.

Borrelli, Rose C, and Vincenzo Fogliano. 2005. "Bread crust melanoidins as potential prebiotic ingredients." *Molecular Nutrition & Food Research* 49 (7):673–678.

Bråthen, Erland, Agnieszka Kita, Svein Halvor Knutsen, and Trude Wicklund. 2005. "Addition of glycine reduces the content of acrylamide in cereal and potato products." *Journal of Agricultural and Food Chemistry* 53 (8):3259–3264.

Bušová, Milena, Vladimír Bencko, Katarína Kromerová, Iris Nadjo, and Jana Babjaková. 2020. "Occurrence of acrylamide in selected food products." *Central European Journal of Public Health* 28 (4):320–324.

Cai, Y., Z. Zhang, S. Jiang, M. Yu, C. Huang, R. Qiu, Y. Zou, Q. Zhang, S. Ou, H. Zhou, Y. Wang, W. Bai, and Y. Li. 2014. "Chlorogenic acid increased acrylamide formation through promotion of HMF formation and 3-aminopropionamide deamination." *Journal of Hazardous Materials* 268:1–5. doi: 10.1016/j.jhazmat.2013.12.067.

Capuano, Edoardo, and Vincenzo Fogliano. 2011. "Acrylamide and 5-hydroxymethylfurfural (HMF): A review on metabolism, toxicity, occurrence in food and mitigation strategies." *LWT-Food Science and Technology* 44 (4):793–810.

Chain, EFSA Panel on Contaminants in the Food. 2015. "Scientific opinion on acrylamide in food." *Efsa Journal* 13 (6):4104.

Christova-Bagdassarian, Valentina L, JA Ticshkova, and Terry MVrabcheva. 2012. "Acrylamide in processed food." *Bulgarian Journal Chemical* 1:123–132.

Ciesarová, Zuzana, Kristína Kukurová, and Cyntia Benešová. 2010. "Enzymatic elimination of acrylamide in potato-based thermally treated foods." *Nutrition Benešová and Food Science* 40 (1):55–63.

Claus, Achim, Reinhold Carle, and Andreas Schieber. 2008. "Acrylamide in cereal products: A review." *Journal of Cereal Science* 47 (2):118–133.

Claus, Achim, Melanie Mongili, Georg Weisz, Andreas Schieber, and Reinhold Carle. 2008. "Impact of formulation and technological factors on the acrylamide content of wheat bread and bread rolls." *Journal of Cereal Science* 47 (3):546–554.

Claus, Achim, Pat Schreiter, Albrecht Weber, Simone Graeff, Wilfried Herrmann, Wilhelm Claupein, Andreas Schieber, and Reinhold Carle. 2006. "Influence of agronomic factors and extraction rate on the acrylamide contents in yeast-leavened breads." *Journal of Agricultural and Food Chemistry* 54 (23):8968–8976.

Claus, Achim, Georg M Weisz, Andreas Schieber, and Reinhold Carle. 2006. "Pyrolytic acrylamide formation from purified wheat gluten and gluten-supplemented wheat bread rolls." *Molecular Nutrition and Food Research* 50 (1):87–93.

Curtis, Tanya Y, Nira Muttucumaru, Peter R Shewry, Martin AJ Parry, Stephen J Powers, J Stephen Elmore, Donald S Mottram, Simon Hook, and Nigel G Halford. 2009. "Effects of genotype and environment on free amino acid levels in wheat grain: implications for acrylamide formation during processing." *Journal of Agricultural and Food Chemistry* 57 (3):1013–1021.

Curtis, Tanya Y, Stephen J Powers, Dimitrios Balagiannis, J Stephen Elmore, Donald S Mottram, Martin AJ Parry, Mariann Rakszegi, Zoltan Bedo, Peter R Shewry, and Nigel G Halford. 2010. "Free amino acids and sugars in rye grain: implications for acrylamide formation." *Journal of Agricultural and Food Chemistry* 58 (3):1959–1969.

Daniali, G, Selamat Jinap, NL Hanifah, and P Hajeb. 2013. "The effect of maturity stages of banana on the formation of acrylamide in banana fritters." *Food Control* 32 (2):386–391.

Doerge, Daniel R, Gonçalo Gamboa da Costa, L Patrice McDaniel, Mona I Churchwell, Nathan C Twaddle, and Frederick A Beland. 2005. "DNA adducts derived from administration of acrylamide and glycidamide to mice and rats." *Mutation Research/Genetic Toxicology and Environmental Mutagenesis* 580 (1–2):131–141.

Dybing, Erik. 2007. "Risk assessment of acrylamide." In Erik Dybing (ed), *Thermal processing of food: potential health benefits, and risks,* Chapter 7 Risk assessment of acrylamide, pp. 75–89. Deutsche Forschungsgemeinschaft (DFG), Wiley-VCH Verlag GmbH & Co. KGaA. https://doi.org/10.1002/9783527611492.ch7, ISBN:9783527319091, Online ISBN:9783527611492. https://onlinelibrary.wiley.com/doi/abs/10.1002/9783527611492.ch7

Dybing, Erik, PB Farmer, M Andersen, Timothy R Fennell, SPD Lalljie, DJG Müller, S Olin, BJ Petersen, J Schlatter, and Gabriele Scholz. 2005. "Human exposure and internal dose assessments of acrylamide in food." *Food and Chemical Toxicology* 43 (3):365–410.

Eisenbrand, Gerhard. 2020. "Revisiting the evidence for genotoxicity of acrylamide (AA), key to risk assessment of dietary AA exposure." *Archives of Toxicology* 94 (9):2939–2950.

Elder, VA, JG Fulcher, H Leung, and MG Topor. 2004. "Method for reducing acrylamide in thermally processed foods." Patent US20040058045.

Elmore, J Stephen, Donald S Mottram, Nira Muttucumaru, Andrew T Dodson, Martin AJ Parry, and Nigel G Halford. 2007. "Changes in free amino acids and sugars in potatoes due to sulfate fertilization and the effect on acrylamide formation." *Journal of Agricultural and Food Chemistry* 55 (13):5363–5366.

EFSA Panel on Contaminants in the Food Chain. 2015. "Scientific opinion on acrylamide in food." *EFSA Journal* 13 (6):4104.

European Commission. 2013. "Commission recommendation of 8 November 2013 on investigations into the levels of acrylamide in food." *Official Journal of European Union* 56:15–17. Retrieved from https://eur-lex.europa.eu/legal-content/EN/TXT/?uri=CELEX:32013H0647

Fink, Martin, Roger Andersson, Johan Rosén, and Per Åman. 2006. "Effect of added asparagine and glycine on acrylamide content in yeast-leavened bread." *Cereal Chemistry* 83 (2):218–222.

Fiselier, Katell, Diego Bazzocco, Fabiana Gama-Baumgartner, and Koni Grob. 2006. "Influence of the frying temperature on acrylamide formation in French fries." *European Food Research and Technology* 222:414–419. doi: 10.1007/s00217-005-0046-6.

Flückiger, Rudolf, and Erdjan Salih. 2006. "Method to limit acrylamide in heated foods." Google Patents.

Food COVID. 2018. "Drug Administration (USFDA)." United States Development, and Licensure of Vaccines to Prevent COVID 19.

Gökmen, Vural, and Hamide Z Şenyuva. 2007. "Acrylamide formation is prevented by divalent cations during the Maillard reaction." *Food Chemistry* 103 (1):196–203.

Graf, Maya, Thomas M Amrein, Stephan Graf, Réka Szalay, Felix Escher, and Renato Amadò. 2006. "Reducing the acrylamide content of a semi-finished biscuit on industrial scale." *LWT-Food Science and Technology* 39 (7):724–728.

Granda, C, RG Moreira, and SE Tichy. 2004. "Reduction of acrylamide formation in potato chips by low-temperature vacuum frying." *Journal of Food Science* 69 (8):E405–E411.

Hendriksen, Hanne V, Beate A Kornbrust, Peter R Østergaard, and Mary A Stringer. 2009. "Evaluating the potential for enzymatic acrylamide mitigation in a range of food products using an asparaginase from Aspergillus oryzae." *Journal of Agricultural and Food Chemistry* 57 (10):4168–4176.

Ismial, Samir Abdel-Monem Ahmed, Rehab Farouk Mohammed Ali, Mohsen Askar, and Wafaa Mahmoud Samy. 2013. "Impact of pre-treatments on the acrylamide formation and organoleptic evolution of fried potato chips." *American Journal of Biochemistry and Biotechnology* 9 (2):90.

Joint FAO/WHO Expert Committee on Food Additives. Meeting, and World Health Organization. Safety Evaluation of Certain Contaminants in Food. Vol. 82. World Health Organization, 2006.

Jung, MY, DS Choi, and JW Ju. 2003. "A novel technique for limitation of acrylamide formation in fried and baked corn chips and in French fries." *Journal of Food Science* 68 (4):1287–1290.

Kotsiou, Kali, Maria Tasioula-Margari, Kristína Kukurová, and Zuzana Ciesarová. 2010. "Impact of oregano and virgin olive oil phenolic compounds on acrylamide content in a model system and fresh potatoes." *Food Chemistry* 123 (4):1149–1155.

Krishnakumar, T, and R Visvanathan. 2014a. "Acrylamide in food products: a review." *Journal of Food Processing & Technology* 5 (7):1.

Krishnakumar, T, and R Visvanathan. 2014b. "Acrylamide in food products: A review." *Journal of Food Processing & Technology* 5 (7):344. https://doi.org/10.4172/2157-7110.1000344

Kusnin, N, Mohd Arif Syed, and Siti Aqlima Ahmad. 2015. "Toxicity, pollution and biodegradation of acrylamide â€“a mini review." *Journal of Biochemistry, Microbiology and Biotechnology* 3 (2):6–12.

Lea, Peter J, Ladaslav Sodek, Martin AJ Parry, Peter R Shewry, and Nigel G Halford. 2007. "Asparagine in plants." *Annals of Applied Biology* 150 (1):1–26.

Lee, Maw-Rong, Li-Yo Chang, and Jianpeng Dou. 2007. "Determination of acrylamide in food by solid-phase microextraction coupled to gas chromatography–positive chemical ionization tandem mass spectrometry." *Analytica Chimica Acta* 582 (1):19–23.

Lerner, SE, ML Seghezzo, ER Molfese, NR Ponzio, M Cogliatti, and WJ Rogers. 2006. "N-and S-fertiliser effects on grain composition, industrial quality and end-use in durum wheat." *Journal of Cereal Science* 44 (1):2–11.

Levine, Robert A, and Robert E Smith. 2005. "Sources of variability of acrylamide levels in a cracker model." *Journal of Agricultural and Food Chemistry* 53 (11):4410–4416.

Lineback, David R, James R Coughlin, and Richard H Stadler. 2012. "Acrylamide in foods: a review of the science and future considerations." *Annual Review of Food Science and Technology* 3:15–35.

López-Ruiz, Rosalía, Jesús Marin-Saez, Sara C Cunha, Ana Fernandes, Victor de Freitas, Olga Viegas, and Isabel MPLVO Ferreira. 2023. "Fibre enrichment of cookies to mitigate acrylamide formation and gastrointestinal bioaccessibility." *LWT* 182:114835.

Lyon, F. (2014). IARC monographs on the evaluation of carcinogenic risks to humans. World health organization, International agency for research on cancer. Available at publication@ iarc. fr.

Maan, Abid Aslam, Muhammad Adeel Anjum, Muhammad Kashif Iqbal Khan, Akmal Nazir, Farhan Saeed, Muhammad Afzaal, and Rana Muhammad Aadil. 2022. "Acrylamide formation and different mitigation strategies during food processing–a review." *Food Reviews International* 38 (1):70–87.

Marti, Alessandra, Gaetano Cardone, and M. Pagani. 2020. "Sprouted Cereal Grains and Products." In Milica Pojić, Uma Tiwari, (eds), Innovative Processing Technologies for Healthy Grains, pp. 113–141. John Wiley & Sons Ltd. ISBN:9781119470168, Online ISBN:9781119470182. DOI:10.1002/9781119470182. https://doi.org/10.1002/9781119470182.ch6

Mottram, Donald S, Bronislaw L Wedzicha, and Andrew T Dodson. 2002. "Acrylamide is formed in the Maillard reaction." *Nature* 419 (6906):448–449.

Mousavi, Mir-Michael, Mansour Mahmoudpourb, Mojtaba Yousefi, and Amir M Mortazavian. 2023. "Strategies for reduction of acrylamide in fried potatoes and potato chips: A review." *Middle East Journal of Rehabilitation and Health Studies* (In Press).

Mucci, Lorelei A, Sven Sandin, Katarina Bälter, Hans-Olov Adami, Cecilia Magnusson, and Elisabete Weiderpass. 2005. "Acrylamide intake and breast cancer risk in Swedish women." *Jama* 293 (11):1322–1327.

Mulla, Mehrajfatema Z, Vikas R Bharadwaj, Uday S Annapure, and Rekha S Singhal. 2011. "Effect of formulation and processing parameters on acrylamide formation: A case study on extrusion of blends of potato flour and semolina." *LWT-Food Science and Technology* 44 (7):1643–1648.

Ou, Shiyi, Jianjun Shi, Caihuan Huang, Guangwen Zhang, Jiuwei Teng, Yue Jiang, and Baoru Yang. 2010. "Effect of antioxidants on elimination and formation of acrylamide in model reaction systems." *Journal of Hazardous Materials* 182 (1–3):863–868.

Ouyang, Qin, Yanna Rong, Baoning Wang, Waqas Ahmad, Shuangshuang Liu, and Quansheng Chen. "Transforming on-site detection of N-Nitrosodimethylamine: A novel fluorescent biosensor utilizing Zein Film and upconversion nanoparticles." Available at SSRN 4478145.

Pedreschi, Franco, Jorge Leon, Domingo Mery, Pedro Moyano, Romina Pedreschi, Karl Kaack, and Kit Granby. 2007. "Color development and acrylamide content of pre-dried potato chips." *Journal of Food Engineering* 79 (3):786–793.

Pedreschi, Franco, Ximena Travisany, Carolina Reyes, Elizabeth Troncoso, and Romina Pedreschi. 2009. "Kinetics of extraction of reducing sugar during blanching of potato slices." *Journal of Food Engineering* 91 (3):443–447.

Postles, Jennifer, Stephen J Powers, J Stephen Elmore, Donald S Mottram, and Nigel G Halford. 2013. "Effects of variety and nutrient availability on the acrylamide-forming potential of rye grain." *Journal of Cereal Science* 57 (3):463–470.

Renn, Ortwin. 2003. "Acrylamide: Lessons for risk management and communication." *Journal of Health Communication* 8 (5):435–441.

Rosén, J, and KE Hellenäs. 2002. "Analysis of acrylamide in cooked foods by liquid chromatography tandem mass spectrometry." *Analyst* 127 (7):880–2. doi: 10.1039/b204938d.

Rudén, Christina. 2004. "Acrylamide and cancer risk—expert risk assessments and the public debate." *Food and Chemical Toxicology* 42 (3):335–349.

Sakamoto, Junko, and Kazuo Hashimoto. 1986. "Reproductive toxicity of acrylamide and related compounds in mice—effects on fertility and sperm morphology." *Archives of toxicology* 59:201–205.

Salazar, R, Geronimo Arámbula-Villa, G Luna-Bárcenas, JD Figueroa-Cárdenas, E Azuara, and Pedro Alberto Vazquez-Landaverde. 2014. "Effect of added calcium hydroxide during corn nixtamalization on acrylamide content in tortilla chips." *LWT-Food Science and Technology* 56 (1):87–92.

Sarion, Cristina, Georgiana Gabriela Codină, and Adriana Dabija. 2021. "Acrylamide in bakery products: A review on health risks, legal regulations and strategies to reduce its formation." *International Journal of Environmental Research and Public Health* 18 (8):4332.

Schouten, Maria Alessia, et al. 2023. "Mitigation strategies to reduce acrylamide in cookies: Effect of formulation." Food Reviews International 39 (8):4793–4834

Seal, Chris J, Anika de Mul, Gerhard Eisenbrand, AJ Haverkort, Knut Franke, SPD Lalljie, H Mykkänen, E Reimerdes, Gabriele Scholz, and V Somoza. 2008. "Risk-benefit considerations of mitigation measures on acrylamide content of foods–a case study on potatoes, cereals and coffee." *British Journal of Nutrition* 99 (S2):S1–S46.

Semla, Magdalena, Zofia Goc, Monika Martiniaková, Radoslav Omelka, and Grzegorz Formicki. 2017. "Acrylamide: a common food toxin related to physiological functions and health." *Physiological Research* 66 (2):205.

Sharp, David. 2003. "Acrylamide in food." *The Lancet* 361 (9355):361–362.

Shepherd, LVT, JE Bradshaw, MFB Dale, JW McNicol, SDA Pont, Donald Stewart Mottram, and HV Davies. 2010. "Variation in acrylamide producing potential in potato: segregation of the trait in a breeding population." *Food Chemistry* 123 (3):568–573.

Shojaee-Aliabadi, Saeedeh, Hooshang Nikoopour, Farzad Kobarfard, Mahdi Parsapour, Maryam Moslehishad, Hassan Hassanabadi, Jesus M Frias, Maryam Hashemi, and Ezzat Dahaghin. 2013. "Acrylamide reduction in potato chips by selection of potato variety grown in Iran and processing conditions." *Journal of the Science of Food and Agriculture* 93 (10):2556–2561.

Sirot, Véronique, Frédéric Hommet, Alexandra Tard, and Jean-Charles Leblanc. 2012. "Dietary acrylamide exposure of the French population: Results of the second French Total Diet Study." *Food and Chemical Toxicology* 50 (3–4):889–894.

Slayne, Martin A, and David R Lineback. 2005. "Acrylamide: considerations for risk management." *Journal of AOAC International* 88 (1):227–233.

Soares, Cristina Maria Dias. 2015. "Assessment of the dietary intake of acrylamide in Portugal. Development and evolution of strategies for reduction of acrylamide formation in thermally processed foods." Universidade do Porto (Portugal).

Stadler, Richard H. 2006. "The formation of acrylamide in cereal products and coffee." In . Skog and J. Alexander (eds), *Acrylamide, and other hazardous compounds in heat-treated foods, chapter 2,* Pp. 23–40. Cambridge, UK Woodhead Publishing.

Stadler, Richard H, Imre Blank, Natalia Varga, Fabien Robert, Jörg Hau, Philippe A Guy, Marie-Claude Robert, and Sonja Riediker. 2002a. "Acrylamide from Maillard reaction products." *Nature* 419 (6906):449–450.

Stadler, Richard H, Imre Blank, Natalia Varga, Fabien Robert, Jörg Hau, Philippe A Guy, Marie-Claude Robert, and Sonja Riediker. 2002b. "Acrylamide from Maillard reaction products." *Nature* 419 (6906):449–450.

Svensson, K, L Abramsson, W Becker, A Glynn, K-E Hellenäs, Y Lind, and J Rosen. 2003. "Dietary intake of acrylamide in Sweden." *Food and Chemical Toxicology* 41 (11):1581–1586.

Sweeney, Lisa M, Christopher R Kirman, Michael L Gargas, M Leigh Carson, and Robert G Tardiff. 2010. "Development of a physiologically-based toxicokinetic model of acrylamide and glycidamide in rats and humans." *Food and Chemical Toxicology* 48 (2):668–685.

Taeymans, Dominique, Anders Andersson, Peter Ashby, Imre Blank, Pierre Gondé, Paul van Eijck, Virginie Faivre, Sam PD Lalljie, Hans Lingnert, and Marianne Lindblom. 2005. "Acrylamide: update on selected research activities conducted by the European food and drink industry." *Journal of AOAC International* 88 (1):234–241.

Taeymans, Dominique, John Wood, Peter Ashby, Imre Blank, Alfred Studer, Richard H Stadler, Pierre Gondé, Paul Eijck, SAM Lalljie, and Hans Lingnert. 2004. "A review of acrylamide: an industry perspective on research, analysis, formation, and control." *Critical Reviews in Food Science and Nutrition* 44 (5):323–347.

Taeymans, Dominique, John Wood, Peter Ashby, Imre Blank, Alfred Studer, Richard H Stadler, Pierre Gondé, Paul Eijck, SAM Lalljie, and Hans Lingnert. 2004. "A review of acrylamide: an industry perspective on research, analysis, formation, and control." *Critical Reviews in Food Science and Nutrition* 44 (5):323–347.

Tardiff, Robert G, Michael L Gargas, Christopher R Kirman, M Leigh Carson, and Lisa M Sweeney. 2010. "Estimation of safe dietary intake levels of acrylamide for humans." *Food and Chemical Toxicology* 48 (2):658–667.

Tareke, Eden. 2003. "Identification and origin of potential background carcinogens: Endogenous isoprene and oxiranes, dietary acrylamide." Department of Environmental Chemistry.

Tepe, Yalçın, and Ayşegül Çebi. 2019. "Acrylamide in environmental water: A review on sources, exposure, and public health risks." *Exposure and Health* 11:3–12.

Timmermann, Clara Amalie Gade, Signe Sonne Mølck, Manik Kadawathagedara, Anne Ahrendt Bjerregaard, Margareta Törnqvist, Anne Lise Brantsæter, and Marie Pedersen. 2021. "A review of dietary intake of acrylamide in humans." *Toxics* 9 (7):155.

Tomoda, Y, A Hanaoka, T Yasuda, T Takayama, and A Hiwatashi. 2004. "Preparing food to be cooked by heat treatment and removing polyvalent cations by flushing with water." US Patent Application 20040126469.

Tuta, Sezin, T Koray Palazoğlu, and Vural Gökmen. 2010. "Effect of microwave pre-thawing of frozen potato strips on acrylamide level and quality of French fries." *Journal of Food Engineering* 97 (2):261–266.

Tyl, Rochelle, and Marvin Friedman. 2003. "Effect of acrylamide on rodent reproductive performance." *Reproductive Toxicology (Elmsford, N.Y.)* 17: 1–13. doi: 10.1016/S0890-6238(02) 00078-3.

Tyl, Rochelle, Marvin Friedman, Patricia Losco, Louann Fisher, Keith Johnson, Dale Strother, and Claude Wolf. 2000. "Rat two-generation reproduction and dominant lethal study of acrylamide in drinking water." *Reproductive Toxicology (Elmsford, N.Y.)* 14: 385–401. doi: 10.1016/S0890-6238(00)00097-6.

Umano, Katsumi, and Takayuki Shibamoto. 1987. "Analysis of acrolein from heated cooking oils and beef fat." *Journal of Agricultural and Food Chemistry* 35 (6):909–912.

van Boekel, Martinus, Vincenzo Fogliano, Nicoletta Pellegrini, Catherine Stanton, Gabriele Scholz, Sam Lalljie, Veronika Somoza, Dietrich Knorr, Pratima Rao Jasti, and Gerhard Eisenbrand. 2010. "A review on the beneficial aspects of food processing." *Molecular Nutrition & Food Research* 54 (9):1215–1247.

Viklund, Gunilla ÅI, Kerstin M Olsson, Ingegerd M Sjöholm, and Kerstin I Skog. 2010. "Acrylamide in crisps: Effect of blanching studied on long-term stored potato clones." *Journal of Food Composition and Analysis* 23 (2):194–198.

Wang, Hao, Pan Huang, Tietao Lie, Jian Li, Reinhold J Hutz, Kui Li, and Fangxiong Shi. 2010. "Reproductive toxicity of acrylamide-treated male rats." *Reproductive Toxicology* 29 (2):225–230.

Wei, Quanwei, Jian Li, Xingmei Li, Lei Zhang, and Fangxiong Shi. 2014. "Reproductive toxicity in acrylamide-treated female mice." *Reproductive Toxicology* 46:121–128.

Wenzl, Thomas, M Beatriz De La Calle, and Elke Anklam. 2003. "Analytical methods for the determination of acrylamide in food products: a review." *Food Additives and Contaminants* 20 (10):885–902.

Whittaker, Anne, Ilaria Marotti, Giovanni Dinelli, Luca Calamai, Sigfrido Romagnoli, Marco Manzelli, Enrico Palchetti, Vincenzo Vecchio, and Stefano Benedettelli. 2010. "The influence of tuber mineral element composition as a function of geographical location on acrylamide formation in different Italian potato genotypes." *Journal of the Science of Food and Agriculture* 90 (12):1968–1976.

Xu, Yi, Bo Cui, Ran Ran, Ying Liu, Huaping Chen, Guoyin Kai, and Jianxin Shi. 2014. "Risk assessment, formation, and mitigation of dietary acrylamide: current status and future prospects." *Food and Chemical Toxicology* 69:1–12.

Yasuhara, Akio, Yuuka Tanaka, Matt Hengel, and Takayuki Shibamoto. 2003. "Gas chromatographic investigation of acrylamide formation in browning model systems." *Journal of Agricultural and Food Chemistry* 51 (14):3999–4003.

Zamani, Ehsan, Mohammad Shokrzadeh, Marjan Fallah, and Fatemeh Shaki. 2017. "A review of acrylamide toxicity and its mechanism." *Pharmaceutical and Biomedical Research* 3 (1):1–7.

Zeng, Xiaohui, et al. 2010. "Activities of hydrocolloids as inhibitors of acrylamide formation in model systems and fried potato strips." *Food Chemistry* 121(2):424–428.

Zhang, Yu, Jie Chen, Xiaoling Zhang, Xiaoqin Wu, and Ying Zhang. 2007. "Addition of antioxidant of bamboo leaves (AOB) effectively reduces acrylamide formation in potato crisps and French fries." *Journal of Agricultural and Food Chemistry* 55 (2):523–528.

Zhang, Yu, Yi Dong, Yiping Ren, and Ying Zhang. 2006. "Rapid determination of acrylamide contaminant in conventional fried foods by gas chromatography with electron capture detector." *Journal of Chromatography A* 1116 (1–2):209–216.

Zyzak, David V, Robert A Sanders, Marko Stojanovic, Daniel H Tallmadge, B Loye Eberhart, Deborah K Ewald, David C Gruber, Thomas R Morsch, Melissa A Strothers, George P Rizzi. 2003. "Acrylamide formation mechanism in heated foods." *Journal of Agricultural and Food Chemistry* 51 (16):4782–4787.

Hydroxymethyl-furfural (HMF) and Safety

5

Food Processing and Manufacturing Related Effects

Fatmagul Hamzaoglu and Sena Bakir

5.1 INTRODUCTION

5-hydroxymethylfurfural (HMF) is a heat-processing product in foods which contain carbohydrates. In the presence of proteins or amino acids, reactants reduce hexoses, known as the Maillard reaction. It is produced by heat dehydration of fructose, sucrose, and, to a lesser level, glucose, which is catalyzed by acid. The structure and quantity of reducing sugars, amino acids, moisture content, temperature, pH, water activity, and the presence or absence of catalyst and inhibitor compounds all affect the rate of the HMF. This product is absent in fresh, unprocessed food, but when foods high in sugar are heated to high temperatures, HMF quickly builds up in those foods (Ameur, Mathieu, Lalanne, Trystram, & Birlouez-Aragon, 2007). Besides, HMF is a commonly used indicator of the nutritional value of foods, including coffee and baked diets (Erbersdobler & Somoza, 2007). Moreover, HMF was found in heat-sterilized glucose/fructose solutions intended for parenteral feeding. It is regarded as the primary grading system for molasses (pekmez) and other culinary items (Dogan, Sienkiewicz, & Oral, 2005),

DOI: 10.1201/9781032712291-5

and HMF formation in foods accelerates with a few factors, including the temperature during processing, the kind of sugar, pH, water activity, and the amount of divalent cations in the medium. According to Abraham et al. (2011), the buildup of HMF in thermally processed food is deemed undesirable, and its existence in food raises some possible toxicological concerns. Food composition, ingredient levels, and heat-processing levels affect how much HMF is in products. While there is ongoing debate on the toxicity of HMF to humans, food experts emphasize HMF reduction techniques more (Lee et al., 2019). In this section, the underlying mechanism of HMF production in foods is assessed. That is followed by the general evaluation on the effect of food processing techniques and storage time as well as the storage conditions on HMF formation, as well as the health effects of HMF consumption are presented.

5.2 STRUCTURAL INFORMATION ABOUT HMF AND ITS BY-PRODUCTS PRODUCTION

HMF is a heterocyclic aldehyde (a furan derivative) that contains an alcohol functional group in its structure. The ring structure features a formyl group at position 2 and a hydroxymethyl group at position 5 (Figure 5.1). HMF's alcohol and aldehyde groups can undergo various chemical reactions, including oxidation, esterification, dehydration, reduction, decarbonylation, and halogenation. Formic acid and levulinic acid are produced by the hydrolysis of HMF. 2,5-dimethylfuran (DMF) is obtained by catalytic hydrogenation, 2,5-diformylfuran (DFF) by oxidation with stoichiometric oxidants and electrophilic agents, and 2,5-furandicarboxylic acid (FDA) by aerobic oxidation. These compounds have a broad range of applications in industries, such as perfumery,

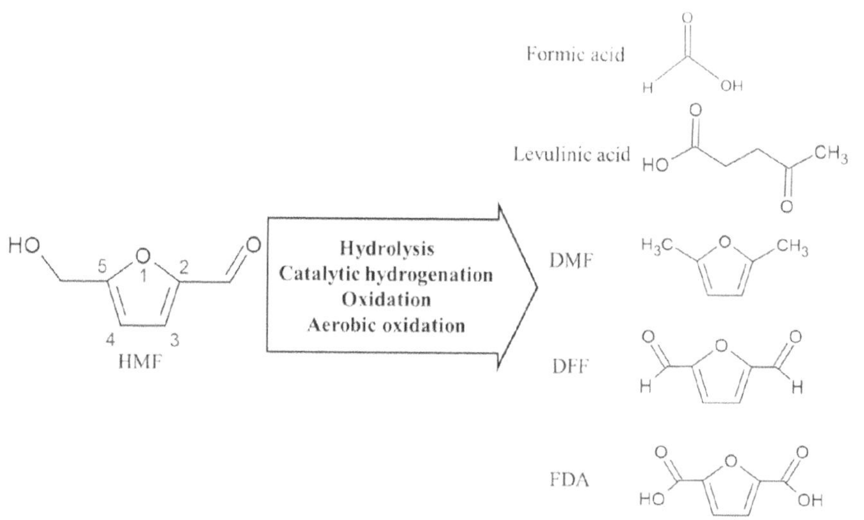

FIGURE 5.1 Structure of HMF and its by-products.

pharmaceuticals, fungicides, bioplastics, and biofuels (Rose et al., 2013; Rosenboom, Langer, & Traverso, 2022; Takei et al., 2012; Ventura, Dibenedetto, & Aresta, 2018).

5.2.1 Synthesis and Chemistry

HMF is a reliable indicator of storage and processing in carbohydrate-rich foods. The Codex Alimentarius legally limits the amount of HMF in honey to 40 mg/kg. However, if honey is obtained in a country or region with a tropical ambient temperature, this value should not exceed 80 mg/kg. It is worth noting that there is no limit for HMF in fruit juices, but the analysis method is specified. HMF in foods is formed through two distinct processes. The first process is the acid-catalyzed thermal dehydration of sugars. HMF formation was discovered in 1875 as an intermediate product of levulinic acid production through H_2SO_4-catalyzed sugar dehydration in an aqueous medium (Grote & Tollens, 1875). Ketoses have a higher HMF production performance than aldoses. Fructose is a superior substrate for HMF production compared to glucose and sucrose. HMF is formed from fructose by losing three water molecules under acid catalysis and heat. For the production of HMF from sucrose under the same conditions, sucrose must be inverted to glucose and fructose, and then isomerization of glucose in the formed inverted sugar to fructose is required (Figure 5.2).

Another way to form HMF is the Maillard reaction. The Maillard reaction was initially described by French chemist Louis Camille Maillard (Maillard, 1912). The detailed mechanism for the reaction was later proposed by Hodge (1953). The Maillard

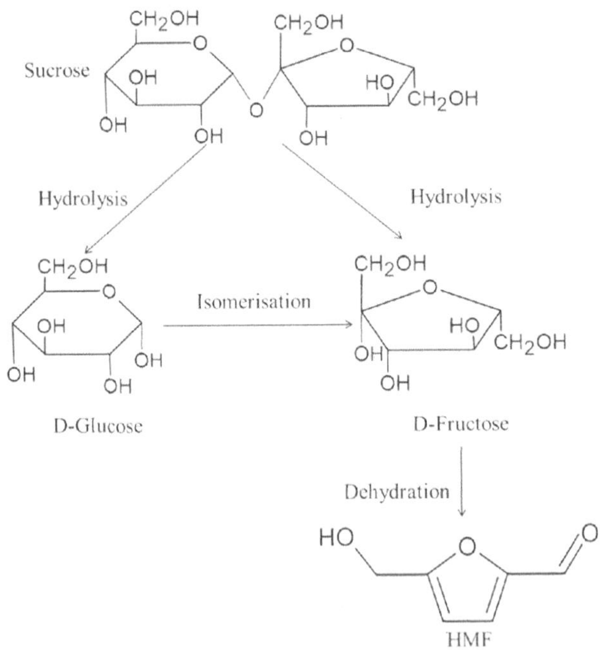

FIGURE 5.2 HMF formation from sucrose.

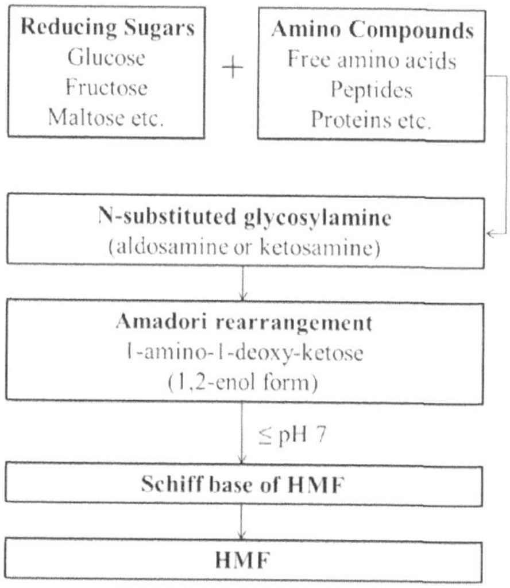

FIGURE 5.3 Formation of HMF by Maillard reaction.

reaction is a non-enzymatic browning reaction between reducing sugars and amino compounds. Reducing sugars have a free anomeric carbon, such as glucose, fructose, and maltose. Amino compounds include free amino acids, peptides, or proteins that contain free amino groups. Essentially, the Maillard reaction first forms glycosylamine (aldosamine or ketosamine) by condensing reducing sugars and amino compounds. Amadori rearrangement products are then formed from the glycosylamine. At pH 7.0 and below, Schiff base and then HMF are formed from the Amadori products (Figure 5.3). The rate of the Maillard reaction is influenced by various factors, including the type of amino compound and sugar, pH, aw, and processing temperature. Maillard reaction occurs more readily in basic amino acids than in acidic amino acids and in aldoses more than ketoses.

5.3 THE EFFECTS OF FOOD PROCESSING TECHNIQUES AND STORAGE ON HMF

Various food processing techniques, such as pasteurization, drying, heating, fermentation, baking, roasting, are applied to food products with different purposes. These purposes could be to increase the volume of the product, inhibit the microbial growth, inactivate the enzymes, extending the product shelf life. Significant changes in the chemical composition of carbohydrates can occur during food product manufacturing and storage. In any event, non-enzymatic browning reactions brought on by the heating operations may compromise the quality of such items, and HMF as a result of subsequent responses. HMF accumulation in several processed food items is listed in Table 5.1.

TABLE 5.1 HMF content in some processed food products

PROCESS	*FOOD PRODUCTS*	*APPLICATION*	*HMF CONTENT (MG/KG)*	*REFERENCE*
Pasteurization	Lactose-Free Milk	20–40 min at 60°C	12.51–14.84 mg/kg	Bi, Wang, Guo, Liao, and Na (2023)
		20–40 min at 65°C	13.32–16.18 mg/kg	
		15 min at 70, 75, 80, 85, and 90°C	13.58, 14.26, 14.48, 14.78 and 16.13 mg/kg	
		30 min at 70, 75, 80, 85, and 90°C	14.72, 14.8, 15.84, 16 and 16.83 mg/kg	
Kilning and germination	Malt	0h, 2h, 4h and 6h at 50°C	n.d.	Nobis et al. (2019)
		7h at 60°C		
		8h at 70°C		
		9h at 80°C		
		12h at 60 and 70°C		
		10h at 90°C	0.40 mg/L (G1: **7min-14°C**)	
			0.25 mg/L (G2: **6min-12°C**)	
			0.18 mg/L (G3: **5min-12°C**)	
		12h at 80°C	0.33 mg/L (G1)	
			0.15 mg/L (G2)	
			0.13 mg/L (G3)	
		12h at 90°C	1.52 mg/L (G1)	
			1.49 mg/L (G2)	
			1.25 mg/L (G3)	
		12h at 100°C	4.15 mg/L (G1)	
			3.02 mg/L (G2)	
			2.56 mg/L (G3)	
Drying	Pomegranate leather (pestil)	Traditional	376.43 mg/kg	Tontul and Topuz (2017)
		Hot air (50, 60 and 70°C)	314.90, 489.38 and 626.11 mg/kg	
		Microwave assisted hot-air drying (90 W) (50, 60 and 70°C)	184.76, 208.60 and 182.65 mg/kg	

		Microwave assisted hot-air drying (180 W) (50, 60 and 70°C)	439.59, 344.12 and 154.17 mg/kg	
		Refractance windows (90, 95 and 98°C)	162.98, 142.80 and 154.58 mg/kg	
	Eggplant	Hot-air convective drying	4.12 mg/kg	Chouaibi, Snoussi, Attouchi, and Ferrari (2021)
		Microwave drying	10.24 mg/kg	
		Ultrasonically assisted convective drying	2.70 mg/kg	
		Freeze-drying	1.23 mg/kg	
		İnfrared drying	7.56 mg/kg	
	Goji berries	Microwave drying	40.3 mg/kg	Tang et al. (2023)
		Freeze-drying	21.8 mg/kg	
		Hot-air drying	24.6 mg/L	
Roasting	Coffee	First 10 min (up to 170°C)	From 0 to 5.6 mg/kg	Hu, Liu, Jiang, Zhang, and Zhang (2021)
		From 170°C to 220°C	From 5.6 to 0.43 mg/kg	
	Cacao	Before roasting	1.21 mg/kg	Sacchetti et al. (2016)
		74 min at 125°C	3.38 mg/kg	
		35 min at 135°C	6.09 mg/kg	
		40 min at 145°C	7.32 mg/kg	
Baking	Bread	Control	0.5–7.3 mg/L	Mesías and Morales (2017)
		0.3 g glucose addition	121.9–180.1 mg/kg	
	Cookie	7 min at 180 °C	n.d.–2.1 mg/kg	Žilić, Aktağ, Dodig, and Gökmen (2021)
		10 min at 180 °C	1.8–21.3 mg/kg	
		13 min at 180 °C	8.1–57.2 mg/kg	

(continued)

TABLE 5.1 (Continued)

PROCESS	FOOD PRODUCTS	APPLICATION	HMF CONTENT (MG/KG)	REFERENCE
	Cake	Wholemeal cake White cake	60–132 µg/kg 45–138 µg/kg	de Borba, Lemos, Rodrigues, Cerqueira, and Badiale–Furlong (2023)
Storage	Beer	Before Storage 40 days at 30°C 40 days at 40°C 40 days at 50°C	3.75 mg/L 4.91 mg/L 9.27 mg/L 17.8 mg/L	Viegas, Prucha, Gökmen, and Ferreira (2018)
	Orange juice	For 180 days at 4°C For 180 days at 25°C	5.07–146.27 µg/kg 5.07–2744.67 µg/kg	Akyildiz, Mertoglu, and Agcam (2021)
	Honey	2 years in the dark at room temperature	From 0.451–4.27 to 9.41–20.2 mg/kg	Czipa, Phillips, and Kovács (2019)
Fermentation	Beer		n.d.	Paszkot, Kawa-Rygielska, and Anioł (2021)
	Black Garlic	10–30 days at 60°C	6.42–8.24 mg/100g	Dursun Capar, Inanir, Cimen, Ekici, and Yalcin (2022)

	Common wheat	Naitive S. cerevisiae and Sourdough	From 6.5 to 4.7 and 64.2 mg/kg	Yıltırak, Kocadağlı, Çelik, Özkaynak Kanmaz, and Gokmen (2021)
		Sprouted *S. cerevisiae* and Sourdough	From 35 to 7.3 and 43 mg/kg	
	Barley	Naitive *S. cerevisiae* and Sourdough	From 13.7 to 5.5 and 15.8 mg/kg	
		Sprouted *S. cerevisiae* and Sourdough	From 47.7 to 3.0 and 10.0 mg/kg	
	Rye	Naitive *S. cerevisiae* and Sourdough	From 7.8 to 8.3 and 115 mg/kg	
		Sprouted *S. cerevisiae* and Sourdough	From 140.3 to 55 and 600 mg/kg	
	Oat	Naitive *S. cerevisiae* and Sourdough	From 10.3 to 3.7 and 170.2 mg/kg	
		Sprouted *S. cerevisiae* and Sourdough	From 21.2 to 5.3 and 551.8 mg/kg	
	Einkorn wheat	Naitive *S. cerevisiae* and Sourdough	From 7.7 to 2.8 and 15.5 mg/kg	
		Sprouted *S. cerevisiae* and Sourdough	From 49 to 6.0 and 13.2 mg/kg	
	Buckwheat	Naitive *S. cerevisiae* and Sourdough	From 29 to 21.1 and 52.5 mg/kg	
		Sprouted *S. cerevisiae* and Sourdough	From 44 to 23 and 62.5 mg/kg	
Mashing	Wort		18.92 mg/L	Paszkot, Kawa-Rygielska, and Anioł (2021)
Hopping	Wort		24.05–26.72 mg/L	
Decrystallization	Honey	Before decrystallization	0.21–0.77 mg/kg	Dżugan, Grabek-Lejko, Sidor, and Tomczyk (2021)
		Conventional (at 40°C)	0.35–0.92 mg/kg (re-crystallization after 30 days)	
		Ultrasound (max 53°C)	0.34–1.02 mg/kg (re-crystallization after 4 months)	

Different food processing applications are presented here to discuss their potential effect on HMF production and stability.

5.3.1 Thermal Processing

Heat treatment is a widely used preservation method to prevent spoilage caused by microorganisms in food. It reduces the microbial load in products such as fruit and vegetable juices, dairy products, beer, wine, milk, and pickles. Additionally, it is used to prevent crystallization in honey. The application of heat treatment renders food both physically and microbially safer. The formation of HMF is affected by the composition of the product and the temperature-time combination used in the heat treatment process. In a comparative study conducted by Katsch, Methner, and Schneider (2021) to investigate the impact of heat treatment stages on various juices, it was demonstrated that high pH and low invert sugar content can slow down HMF formation. Heat treatments consist of two stages: the non-isothermal and isothermal phases. The study investigated HMF changes during two stages of pasteurization. It was determined that the amount of HMF formed in the isothermal stage was twice that of the non-isothermal stage at the end of the same time (200 min). It should be noted that in industrial applications, the isothermal phase is typically shorter than the non-isothermal phase. In another comparative study investigating the effect of heat treatments on HMF formation in milk, high temperature and short time (HTST) pasteurization, ultrahigh pasteurization (UP), and ultrahigh temperature (UHT) sterilization were compared (Xing et al., 2020). The study observed the lowest HMF formation at 70°C for 15 s (7.10×10^{-2} mg/kg). Although the treatment temperature was only 4 seconds for UP and UHT milk, the amount of HMF was significantly higher (ranging from 23.35 to 82.97×10^{-2} mg/kg) compared to HTST-treated milk, which was subjected to a lower treatment temperature (120–140°C).

HMF formation also occurs in new pasteurization technologies, such as microwave and ultrasound, which are still on a pilot scale. The rapid attainment of the target temperature in microwave pasteurization reduces the adverse effects of temperature on food. A study was conducted to investigate the impact of conventional and microwave pasteurization on the microbiological and physicochemical properties of raisins. The study found that microwave pasteurization reached the target temperature 3.8–6.7 times faster than conventional pasteurization. Additionally, the amount of HMF in raisins pasteurized by microwave was 44.8% less than those pasteurized by convection pasteurization (Çağlar, Rayman Ergün, & Baysal, 2021).

In contrast to microwave and traditional methods, the ultrasound method does not heavily hcat thc product. Its pasteurization power is based on the destructive effect of instantaneous and locally high temperature and pressure caused by acoustic cavitation on microorganisms. A study by X. Cao, Cai, Wang, and Zheng (2019) investigated the impact of ultrasound treatment and convection pasteurization on the physicochemical properties of bayberry juice. The HMF content increased by 100% after the conventional pasteurization process for 1 minute, while the 10-minute ultrasound process resulted in an 18% increase. Furthermore, increments at ultrasound intensity amount led to more significant HMF formation. In summary, regardless of the type, increasing the duration and intensity of heat treatment accumulates the formation of HMF.

5.3.2 Drying and Dehydration

Drying and dehydration are standard methods for preserving food. For these processes, food is exposed to a specific temperature for a specified time. Drying method and utilized product are key factors to determine the temperature and duration of exposure. Several factors, including the shape and size of the cut, the presence of sugar and other organic compounds, air temperature, speed, and humidity, can affect the drying rate of food. The traditional sun drying, hot-air drying under atmospheric pressure, hot surface drying, vacuum drying, and freeze drying are commonly used drying methods in the food industry. The drying temperature can prolong to 300°C under atmospheric pressure, but in vacuum and freeze-drying methods, the drying application is completed in a moderate environment below 100°C temperature. The amount of HMF accumulated depends on the drying temperature and duration. Kayacan et al. (2020) conducted a study involving ultrasound-assisted vacuum drying and hot-air drying at a drying temperature of 55°C, while infrared drying at an infrared power of 88W to screen the quality characteristics of persimmon under different drying applications. Ultrasound-assisted vacuum drying and hot-air drying were conducted at a drying temperature of 55°C, while infrared drying was conducted at an infrared power of 88W. Drying continued until the dry product's moisture content reached 20%. The drying process took 255, 310, and 742 minutes for infrared, ultrasound-assisted vacuum, and hot-air drying, respectively. Freeze drying was completed using the standard mode of the freeze dryer. The highest amount of HMF was found in the persimmon dried by hot-air drying, followed by infrared and ultrasound-assisted-vacuum drying. The amount of HMF was not determined in the freeze-dried persimmon.

The drying method is as effective as the drying temperature and time in HMF formation. In a comparative study by Kanar and Mazı (2019), different drying methods and different temperatures and microwave powers in these methods were compared. Although the drying process took a considerable amount of time (4.5–257 hours), there was no statistically significant difference in the amount of HMF between fresh bee pollen and bee pollen that was freeze-dried, hot air-dried, or vacuum-dried ($p<0.05$). During microwave drying, there was no significant change in the amount of HMF at 300W and 900W power. However, at 450W and 600W power, there was a significant increase in the amount of HMF despite the short drying sequence of 8 to 12 minutes. It was stated that this was due to the rapid temperature increase in microwave application and the bee pollen was exposed to high temperatures for a longer time at 450 and 600W power compared to other microwave powers until reaching the target moisture (<8%). It was also observed that vacuum application slowed down the formation of HMF when microwave drying and microwave assisted vacuum drying were compared in fresh bee pollen.

5.3.3 Roasting

Roasting is a process widely used in the food industry for various purposes. Roasting is often used as part of the process for purposes such as revealing the characteristic color, aroma, and flavor of foods, easily separating the shell of nuts, facilitating processing and providing a crisper texture. For these reasons, roasting is applied to many foods, such as coffee, cocoa, nuts, seeds, grains and malt. In addition to the food processing benefits, it also increases the digestibility and antioxidant activity of foods.

Tahini is a condiment commonly used in West Asian, Eastern Mediterranean, South Caucasian, and North African cuisines. It is made by peeling, roasting, and grinding sesame seeds. There are two varieties of tahini, white tahini and double-roasted tahini, which differ in their roasting time. The longer the roasting time, the darker, more aromatic, and easier to digest the product becomes. However, the high temperature that sesame is exposed to during roasting inevitably leads to the formation of HMF. Berk, Hamzalıoğlu, and Gökmen (2019) conducted a study on the impact of roasting sesame seeds at different temperatures (150–220°C) on the formation of Maillard reaction products. The study found that high temperatures led to increased formation of HMF.

The amount of HMF that may occur during roasting is affected not only by temperature and time, but also by the content of the roasted product. During the first production stage of halva, flour and oil are roasted together. Akpınar, Karakaş, Akça, & Certel, 2011 examined the impact of roasting flour:oil mixtures at various ratios and temperatures (120–200°C) on the formation of HMF. The results showed that higher temperatures led to increased HMF formation. Additionally, HMF formation increased with higher fat ratios in the mixture, which was attributed to increased heat transfer.

Roasting is also used in the process of malt production. Green malt or kilned malt is roasted both to halt germination and to impart a characteristic taste and flavor to the malt, depending on the type of beer to be brewed. Light-colored beers are roasted at lower temperatures (70–80°C), while dark-colored beers are roasted at higher temperatures. For instance, black (patent) malt is roasted at approximately 220°C. Herdt et al. (2021) conducted a study to compare the effect of different roasting times and temperatures on the formation of HMF at malts. The study found no systematic change in the amount of HMF with varying temperatures and time.

As is known, HMF is an intermediate product of the Maillard reaction. While its amount increases with temperature and time, it transforms into other compounds in the later stages of the Maillard reaction. Zhu et al. (2022) examined the change in HMF during roasting at different temperatures (190, 215, and 240°C) in a study investigating thermal contaminants formed by roasting coffee beans. At all three temperatures, HMF levels initially increased. After reaching their peak, HMF levels decreased. As the roasting temperature increased, the maximum level of HMF also increased, while the time to reach this peak decreased. The decrease at HMF level after reaching the peak point may be due to its reaction with the free amino acids or its transformation into other compounds.

5.3.4 Baking

Baking is traditional cooking with cereals that has a significant place in the modern food processing industry. Due to their organoleptic impressions of flavor, color, texture, and scent, consumers prefer eating baked goods. Traditionally, baking takes place in an enclosed stone oven heated with firewood. A hot-air oven with convectional heat transfer is the most popular baking method. Because high temperatures have a more significant effect on generating free radicals and a weaker effect on antioxidant defense, they can speed up the oxidative process in lipids and proteins (Devu et al., 2022).

The oven type had a significant impact on moisture and HMF concentration but had no effect on the F content. During standard baking, there was a noticeable drop in moisture and an increase in HMF. Microwave-baked cakes showed a notable rise in HMF (up to 16.84 mg/kg dry weight at 2.5 min). The cakes from steam ovens had the maximum moisture content and showed no discernible variations in furfural and HMF. The type of oven also influenced cake volatility. The findings show that using a steam oven to bake a cake may lower its HMF content without sacrificing its moisture content or volatile profile (Petisca, Henriques, Pérez-Palacios, Pinho, & Ferreira, 2013).

The effects of batter composition and baking parameters on the development of HMF and F were investigated in sponge cakes. The findings demonstrated that HMF and F formation in sponge cakes was significantly influenced by sugar type, baking temperature, and baking time. The citric acid content in sponge cake batter positively correlated with the circumstances for creating HMF and F improve. In sponge cakes, glucose and fructose can produce higher HMF than sucrose, lactose, and maltose. The pH of the sponge cake dropped and the concentrations of HMF and F rose as the baking temperature and duration increased (Zhang et al., 2012).

First fermentation for 30 minutes, second fermentation for 60 minutes, and baking for 20 minutes all work together to reduce the creation of acrylamide (AA) and HMF (Lemos, de Borba, de Medeiros Burkert, Scaglioni, & Badiale-Furlong, 2023).

According to Ameur, Trystram, and Birlouez-Aragon (2006), the conditions are more conducive to the formation of HMF when the water activity of the cookies diminishes during baking. Lowering the temperature below 200°C stops sucrose from breaking down too much and, thus, from forming too much HMF. Sodium bicarbonate can be substituted for ammonium bicarbonate to keep the pH of cookies between 9.0 and 10.0 while they bake, limiting the amount of sucrose that breaks down. Reactive sugar content can be reduced to reduce the amount of HMF that forms in cookies (Gökmen, Açar, Serpen, & Morales, 2008).

In biscuits, fructose – not glucose – was primarily responsible for producing AA. It was proposed that the caramelization of the reducing sugars, fructose but not glucose, was the primary cause of HMF production in biscuits. The current study's findings imply that in order to lower the concentrations of AA and HMF during biscuit production, wheat flour's glucose, fructose, and asparagine concentrations should be taken into account (Nguyen, van der Fels-Klerx, & Van Boekel, 2017).

By examining the variance in flavonoids, this study attempted to assess the nutritional changes that occurred in the blueberry fillings of pastries under thermal processing techniques. Stir-frying, which is a pleasant cooking method that involves heating and combining blueberries with other culinary ingredients like butter, syrup, sugar, and starch, dramatically increased the creation of HMF (which is four times higher than that of cookies) and significantly reduced the amount of flavonoids in the blueberry filling. In contrast to stir-frying, baking had a negligible effect on these chemicals' concentrations (Yang et al., 2019).

According to the research findings, AA synthesis can be virtually entirely inhibited, while HMF creation can be restricted when sulfur dioxide is added to the baking atmosphere.

The current study raised some worries because it looked at potential allergies and the adverse sensory effects of sulfur dioxide. Because carbon dioxide has a low AA and HMF concentration and good sensory properties, it may be recommended as a potential substitute for sulfur dioxide in baking environments (Gülcan, Uslu, Mutlu, Arslan-Tontul, & Erbaş, 2020).

Higher quantities of HMF were seen at all tested temperatures, while AA concentrations were decreased at 180 and 190°C but not at 200°C due to the presence of NaCl in the biscuit formulation. Even though a sizable amount of AA had been produced before the start of sucrose hydrolysis, this step was crucial in synthesizing AA and HMF while baking biscuits. Before sucrose hydrolysis, there was a lag phase, which may have been caused by the crystalline sucrose melting at 180°C (Van Der Fels-Klerx et al., 2014).

5.3.5 Storage

According to reports, as time passed, the amount of HMF and brown pigments in grape juice and the simulated Maillard reaction model grew noticeably (Bozkurt, Gogus, & Eren, 1999).

Within eight weeks of storage, the concentration change of MRPs was measured at 37°C, a temperature where sauces were frequently stored in the kitchen around the cooker, creating a warmer environment than room temperature. It appears that the HMF level was significantly increased in vinegar and light soy sauce but significantly decreased in dark soy sauce and oyster sauce at two-week intervals (Liu, He, Xiao, Zhou, & Wang, 2021).

The content of 5-HMF and the rate of browning both climbed in tandem with the rise in acetic acid. The degree to which food samples are browning is directly correlated with acetic acid, ionization, and hydrolysis content. Iron ions, particularly Fe^{2+}, tended to accelerate the browning rate and the formation of 5-HMF. On the other hand, adding Na_2SO_3 suppressed them – there was even no detectable 5-HMF (Chen, Lin, Li, Lin, & Chen, 2019).

Vinegar containing acetic acid or other acids lowered pH levels, accelerating the HMF production (Kowalski, Lukasiewicz, Duda-Chodak, & Zięć, 2013). As time passed, the amount of HMF and brown pigments in grape juice and the associated Maillard reaction model grew dramatically (Bozkurt et al., 1999). It was also determined how important phenolics and 5-HMF (5-hydroxymethylfurfural) are to the indicator of sparkling wines. A noteworthy linear increase in browning was noted for all types of aging in all four Cava series. There were no apparent distinctions in browning between the two aging processes (commercial vs biological). On the other hand, the 5-HMF concentration climbed linearly with time, even though the kind of aging (biological or storage) affected how quickly it formed. Trans-coumaric and p-coumaric, two phenolic compounds, grew dramatically over time and strongly correlated with browning. 5-HMF can be used as a quality marker and was found to be connected with browning in wines (Serra-Cayuela, Aguilera-Curiel, Riu-Aumatell, Buxaderas, & López-Tamames, 2013).

In China, a special fermentation method is used to make msalais a type of wine produced by fermenting boiled grape juice. During fermentation, the browning index and absorbance at 420 nm of the sample decreased, but increased slightly during wine storage (Zhang, Zhang, Pu, & Zhu, 2022).

The concentration changes of MRPs at 37°C after eight weeks of storage were investigated and according to the results, at two-week intervals, the HMF level was significantly lower in oyster sauce and dark soy sauce and significantly higher in vinegar and light soy sauce. When dark soy sauce's HMF content decreases and its MF content increases with time, it is possible that HMF was demethylated into MF during storage (Liu et al., 2021).

The HMF contents rose with increasing temperature (25, 35, and 45°C) and time (0, 45, and 90 days). Furthermore, it was observed that the molasses samples' HMF level could be accurately anticipated based on storage temperature and duration, which may be utilized to estimate the samples' shelf life or rate of HMF creation (Toker, Dogan, Ersöz, & Yilmaz, 2013).

In place of pectin, the peel of a dried orange (Citrus x sinensis) was used in an orange marmalade, which has the potential to be utilized in the production of food preservatives and nutraceuticals, and also it can additionally be a helpful component in confections like marmalades. Because HMF is formed when the peel desiccates, freshly prepared marmalade has higher HMF values than marmalade made with dried peel. The enriched samples were kept at 20°C after storage and beat those kept at 35°C regarding analytical parameter maintenance (Sicari, Pellicanò, Laganà, & Poiana, 2018).

The HMF content of rosehip nectar in the glass bottle was found to be lower (26.41 mg/L), whereas it was higher in the cardboard box at the start of storage. The eight months of storage at 25, 35, and 45°C caused increments at HMF levels in cardboard-packaged samples increased by 60.66%, 62.93%, and 267.4%, respectively. The findings suggest that high-temperature rosehip nectars in glass bottles are the ideal conditions for HMF to arise (Duru, Karadeniz, & Erge, 2012).

Due to its high carbohydrate content, HMF is a crucial ingredient in jams and this led researchers to investigate the effects of time (0, 2, 4, and 6 months) and storage temperature (10, 20, and 37◦C) on the HMF and color values in strawberries, cherries, and apricot jams.

In every jam examined, minimal HMF production was noted at 10 and 20°C during storage. Nonetheless, storage at 37°C resulted in a significant increase in HMF. In all jams, HMF reached its peak concentrations at 37°C after six months (Aslanova, Bakkalbasi, & Artik, 2010).

Ninety samples of commercially produced plant-based food items were included in the test material. These samples included: cereal-based baby foods, such as ten jarred baby meals that were ready to eat, ten instant baby foods that were in powder form, and ten infant biscuits. A greater temperature rater (25°C) than a lower one (4°C) showed a significant increase in HMF during storage, indicating a significant dependence of HMF creation on storage duration and temperature. Furthermore, it was discovered that HMF creation is favorably impacted by an aw of almost 0.4. HMF and first aw had a statistically significant negative connection. It was observed that over the entire study period, HMF generation was restricted in products with very high and low product water activity (Michalak, Gujska, Czarnowska, Klepacka, & Nowak, 2016).

The methodical investigation of catechin and catechin-iron, founded on advanced glycation end-product (AGEs) inhibition processes, will offer a fresh perspective on inhibitors and enhance vinegar quality. For this aim, the effects of catechin on the development of AGEs after two and five days of incubation at 50°C, with or without iron, were investigated. The inhibitory effect was similar in the presence of catechin with the addition of iron, and it remained unaltered as the number of incubation days increased. This investigation showed that HMF production can be prevented by adding catechin to vinegar and the same catechin could prevent HMF creation by interacting with iron at further (Wu et al., 2019).

The storage at various temperatures (4, 20, and 35°C) and how it affected the hue and HMF in different honey samples were also examined. These characteristics were not significantly affected by storage at 4 or 20°C. However, HMF increased after storage at 35°C, and the results are mostly above the permitted limit (40 mg/Kg). The Maillard reaction also simultaneously highlights the samples' color (Mouhoubi-Tafinine, Ouchemoukh, Bey, Louaileche, & Tamendjari, 2018).

5.3.6 Others

HMF formation is a predictable consequence of food processing stages where temperature and time are applied. In addition, other process stages applied to reach the targeted product also affect HMF formation. In a study by Li et al. (2019) the effects of industrial-scale apple juice production stages (mashing, pressing, preconcentration, enzymatic clarification, ultrafiltration, resin decolorization, quintuple effect concentration, and pasteurization) on HMF formation were investigated. No HMF formation was observed in the mashing and pressing stages. HMF formation commenced with increased in temperature and solids content in the preconcentration stage. Even though apple juice was not subjected to high temperatures, HMF formation proceeded pending the enzymatic clarification and ultrafiltration stages. However, its amount decreased in the resin decolorization stage, which is probably due to resin adsorption. The prolonged temperature applications at concentration and pasteurization stages are increased the amount of HMF increased.

The colloids causing turbidity in fruit juices are removed by the ultrafiltration process in producing clear juices. Prominent molecular weight turbidity factors, including fruit tissue, live and dead microorganisms, phenolics, and proteins, which caused turbidity, are removed from depectinized fruit juices. In a study performed by Makhlouf-Gafsi et al. (2018), the effect of ultrafiltration on HMF formation in male and female date palm sap syrups was investigated in male and female date palm sap syrups with and without ultrafiltration, it was observed that the amount of HMF was less with the ultrafiltration process in both syrup samples after concentration processes. The removal of sugars and proteins by ultrafiltration could have caused less amount of HMF.

An alternative process to ultrafiltration to produce clear juice is using a clarifying agent in depectinized juice. A clarifying agent allows the colloids that cause turbidity in the juice to settle as flocs. Naknean and Meenune (2015) pointed out that, the effect of different evaporation techniques (open pan, vacuum evaporator) and different clarifying agents (chitosan, gelatin, bentonite, and PVPP) could cause various HMF formation in palm sap syrups. The syrups concentrated by open pan exhibited elevated HMF levels relative to those obtained by vacuum evaporation due to prolonged exposure to elevated temperatures. The syrups concentrated without a clarifying agent exhibited the highest HMF levels. In contrast, the syrups concentrated by vacuum evaporation with a clarifying agent exhibited the lowest HMF contents, potentially as a consequence of the removal of proteins that can cause both turbidity and HMF formation through the Maillard reaction by clarifying agents. The results of this study indicate that the use of clarifying agents results in a more apparent final product and a reduction in HMF formation.

Sulfur dioxide (SO_2) is one of the most commonly used additives in the food industry due to its antimicrobial and antioxidant properties. SO_2 is crucial in preserving the color and maintaining microbial stability in the dried fruit production. SO_2 is generally used in drying light-colored fruits. With the sulfurization process, SO_2 reacts with the water in the fruit and it hydrates to sulfurous acid (H_2SO_3). Sulfurous acid dissociates to form hydrogen sulfide (HSO_3^-) and sulfite (SO_2^-) ions. During this reaction, sulfurous acid releases hydrogen ions into the medium. The pH value of the food decreases due to the increased concentration of hydrogen ions. As the pH of the medium decreases, the food becomes more prone to forming hydroxymethylfurfural (HMF). In a study conducted by Hamzaoğlu, Türkyılmaz, and Özkan (2018), the changes in HMF in dried apricots containing different levels of SO_2 (0, 451, 832, 1594, 2112 and 3241 mg/kg) were investigated during storage. While no HMF was found in dried apricots at the beginning of storage, HMF formation increased with storage. Furthermore, the rate of HMF formation increased as the SO_2 concentration increased. In the study, the sugar distribution of dried apricots was also analyzed. As the amount of SO_2 increased, the sucrose content of dried apricots dropped while the glucose and fructose content increased. The decrease in pH and the increase in reducing sugars, which are substrates of the Maillard reaction, were found to promote HMF formation. SO_2 is an additive also used in wine production. Wines are sulphurized to prevent both the growth of harmful microorganisms in the wine and the browning of the wine due to the bleaching effect of sulfur. The European Wine Regulation limits on the maximum concentration of SO_2 in the final product according to the wine color and sugar content. The limit for dry red wines is 150 mg/L, while for dry white and rosé wines it is 200 mg/L. The above-mentioned limit is higher for wines with a sugar content exceeding 5% (200 mg/L for red wine, 250 mg/L for white and rosé wines). For other exceptional wines, the limit is 300 mg/L (Reg. (EU) 2019/934). Before alcohol fermentation, wines are sulphurized, and the sulphurized product undergoes a lengthy process. In a study by Filimon et al. (2016), the levels of HMF in wines with different sugar contents (10, 30 and 50 g/L) were analyzed after heat treatment at temperatures of 45, 75 and 100°C. The wines with the highest sugar content reached the highest HMF levels, regardless of the must in which the sugar content was adjusted. The highest HMF increase was observed in wines with sugar content adjusted to 50 mg/L with oversulfited must (21.2 times) after 120 minutes at 45°C. This increase was 10.3 times in wines with sugar content adjusted with concentrated must.

Another pretreatment utilized in the food industry is blanching. Blanching is employed in the freezing, drying, and canning processes of fruits and vegetables to inactivate enzymes, modifying texture, and removing the peel. Briefly, this process is carried out by placing the fruits and vegetables in boiling water for a few minutes and then rapidly cooling them to prevent further cooking. In a study by Huang et al. (2022), the effect of different pretreatments (ultrasound, blanching, sodium alginate) and combinations of these pretreatments on HMF formation in fried potatoes was investigated. All pretreatments resulted in a decrease in HMF formation in fried potatoes. The most effective pretreatment in preventing HMF formation was the blanching+sodium alginate combination (40.9%). It was stated that this reduction in HMF formation was due to removing reducing sugars and amino acids with water during blanching. Blanching pretreatment is also used in frozen peach production. In a study by Al-Jeddawi, Northcutt, and Dawson (2023), the effect of blanching and dipping in an ascorbic acid-citric acid mixture on HMF formation in peach slices was investigated. The formation of HMF during storage was higher in blanching-treated samples than in untreated samples. It was stated that this was due to the peach being rich in reducing sugars and the high temperature to which it was briefly exposed during blanching. As evidenced by these studies, in addition to temperature and storage, other processing steps that affect the amount of Maillard reaction substrates during processing also indirectly affect the formation of HMF in processed foods.

5.4 THE EFFECTS OF HMF ON HUMAN HEALTH

The presence of HMF in a number of food products prompted an overview of dietary consumption and potential health implications. Certain foods lose their nutritional content when heated to high temperatures, while heating treatment makes others more nutritious. The Codex Alimentarius reported the Acceptable Daily Intake (ADI) for furfural to be 0.5 mg/kg body weight (Kamalabadi, Ghaemi, Mohammadi, & Alizadeh, 2015).

HMF is quickly absorbed from the gastrointestinal tract, as evidenced by the effects of HMF in tests assessing the metabolism of HMF and the oral gavage of radioactive carbon-14 (14C) to rats and mice with HMF or HMF at various levels (5–500 mg/kg) (Godfrey, Chen, Griffin, Lebetkin, & Burka, 1999). Furthermore, enteric bacterial strains have been proposed to convert HMF, which makes it through the digestive process in the gastrointestinal tract, into furfuryl alcohol (Boopathy, Bokang, & Daniels, 1993).

It is evident from several studies employing (14C) HMF that dietary HMF is quickly digested and mainly eliminated through the urine. Nonetheless, it has been noted that the kidneys, bladder, and liver may also experience some covalent binding (Germond, Philippossian, Richli, Bracco, & Arnaud, 1987). During its metabolism, HMF is changed into 5-hydroxymethyl-2-furoic acid (HMFA), which is eliminated by the urine of mammals (Husoy et al., 2008). However, during the digestion stage, some health-related beneficial and adverse effects of HMF occur.

5.4.1 Positive Effects of HMF on Human Health

Reactive oxygen species (ROS) comprise oxygen ions, peroxides, and oxidized free radicals. They are a consequence of biological anaerobic metabolism. Reports on HMF's antioxidant qualities – namely, its ability to remove ROS – are growing. The advantageous impacts of MRPs' antioxidant qualities have been identified, and HMF was discovered to be crucial to honey's antioxidant capacity (Gheldof, Wang, & Engeseth, 2002).

Electron spin resonance spectroscopy-based in vitro tests (Y. X. Li, Li, Qian, Kim, & Kim, 2009) and 1,1,2-diphenyl-2-picrylhydrazide (DPPH), 2,2′-azobis(2-amidinopropane) dihydrochloride (AAPH), and 2,2′-azinobis-3-ethylbenzothiazoline-6-sulfonic acid (ABTS) assays (Zhao et al., 2013) showed that HMF could eliminate intracellular ROS and scavenge free radicals.

In a cellular experiment, HMF dose-dependently reduced calcium ion inflow and β-hexosaminidase degranulation (S. F. Wu et al., 2011). Additionally, as the dosage was reduced, the experimental group's levels of IL-4 and IFN-γ also fell. Combining 5-HMF and other related substances can prevent or treat type I hypersensitivity allergies by having an anti-allergic effect (Yamada, Nemoto, Shigemori, Yokota, & Isoda, 2011). OVA-immunized rats were researched and, after treatment with HMF, the animals' serum IgE and OVA-specific IgE significantly dropped compared to controls (Hamzalioglu & Gökmen, 2017).

The melanoidins extracted from seven amino acid-glucose model systems were all shown to produce inhibition of ACE, which is a regulatory enzyme responsible for the elevation of blood pressure, as indicated by Rufian-Henares and Morales (2007). The Maillard reaction can successfully increase the ACE inhibitory activity of casein hydrolysate when the right circumstances are met (Hong, Meng, & Lu, 2015).

HMF can be used to treat diabetes because, as discovered by Cao, Cai, Cai, and Tu (2013), it can quickly block induced IL-8 activity and down-regulate P-Akt, protecting human endothelium oxidative damage caused by elevated glucose levels.

Zhang et al. (2015) discovered that 5-HMF blocked the decrease of APE/Ref-1 protein levels in PC12 cells, preventing cerebral ischemia or oxidative damage brought on by hydrogen peroxide. Moreover, Yalcin and Cabrales (2012) found that the administration of 20 mg/kg 5-HMF to golden hamsters resulted in an increase in hemoglobin affinity and a decrease in tissue hypoxia. Additionally, Bharate and Bharate (2014) noted that 5-HMF has strong anti-ischemic properties.

In a study, the hepatoprotective properties of 5-HMF produced from processed dried fruits were examined, and it found that 5-HMF improved LO2 cell survival and, in a dose-dependent way, prevented LO2 cell apoptosis after injuring them with H2O2 (Ding, Wang, Yao, Li, & Cai, 2010).

5.4.2 Negative Effects of HMF on Human Health

The mutagenicity of 5-HMF and four isoforms of furfuryl alcohol (FFA) compounds was investigated against strains of S. typhimurium TA100 and TA100-derived bacteria expressing human and rodent SULTs by Glatt, Schneider, Murkovic, Monien, and

Meinl (2012). The findings indicated that S. typhimurium TA100 expressing human SULT1C2 was mutagenic to 2,5-bishydroxymethylfuran (a metabolite of HMF), FFA, and 5-methyl-FFA, yet not to the original strain. The reasons for variances between positive and negative outcomes of in vitro tests could stem from various factors, such as the expression of sulphotransferase (SULT) in the cell lines employed and the existence or lack of 3'-phosphoadenosine-5'-phosphosulphate in the metabolizing system (Abraham et al., 2011).

After injecting 500 mg/kg of 5-HMF and 25 mg/kg of 5-SMF into 3- to 6-day-old mice, Svendsen, Husoy, Glatt, Paulsen, and Alexander (2009) observed that, after 12 weeks, the mice treated with 5-HMF or 5-SMF were experiencing intestinal adenoma and flatness in the large intestine. According to Zhao et al. (2014) 5-HMF can cause G0/G1 phase arrest and apoptosis through the ROS-mediated signaling pathway. This finding suggested that 5-HMF may be a powerful anti-proliferative and anti-carcinogen component.

It has been discovered that certain HMF compounds, including sulfoxymethyl and chloromethyl, have a significant propensity to initiate skin tumors (Surh, Liem, Miller, & Tannenbaum, 1994).

Hepatocytes are toxically affected when furfural is oxidized in the liver and converted to pyromucic acid. Fortunately, the liver also produces glycine, which aids in the detoxification of furfural by conjugating it with pyromucic acid and excreting it in the urine (Maruyama et al., 1992).

In particular, HMF reduced glutathione levels, an essential antioxidant that guards against reactive oxygen species damaging cellular components (Janzowski, Glaab, Samimi, Schlatter, & Eisen Brand, 2000).

5.5 CONCLUSION

A growing body of published data has highlighted the detrimental impact of MRPs on human health, in tandem with the increasing awareness of food safety in recent years. One popular food preservation technique that protects food from spoiling due to microorganisms is heat treatment. Nevertheless, even if it lowers the microbial load in products, there is an HMF formation, which is affected by the product's composition and the temperature-time combination that is employed during the heat treatment procedure. The bakery sector must take processing circumstances into account to minimize AA and HMF generation while preserving the nutritional and sensory qualities of the baking products. Additionally, high temperatures cause some foods to lose their nutritional value, whereas heating processes increase the nutritional value of other foods. Mammals excrete HMFA, which is produced during the metabolism of HMF. However, it should be kept in mind that there are certain beneficial and detrimental health effects of HMF that happen throughout the digestive stage.

ABBREVIATIONS

DFF 2,5-diformylfuran
DMF 2,5-dimethylfuran
F Furfural
FDA 2,5-furandicarboxylic acid
HMF 5-Hydroxy methyl furfural
HMFA 5-hydroxymethyl-2-furoic acid
HTST High temperature and short time
UHT Ultrahigh temperature
UP Ultrahigh pasteurization

REFERENCES

Abraham, K., Gürtler, R., Berg, K., Heinemeyer, G., Lampen, A., & Appel, K. E. (2011). Toxicology and risk assessment of 5-Hydroxymethylfurfural in food. *Molecular Nutrition & Food Research, 55*(5), 667–678.

Akpınar, K., Karakaş, B., Akça, H., & Certel, M. (2011). Determination of HMF in roasted flour/oil mixtures and effect of solvent used in the extraction procedure. *Food Chemistry, 128*(3), 790–794.

Akyildiz, A., Mertoglu, T. S., & Agcam, E. (2021). Kinetic study for ascorbic acid degradation, hydroxymethylfurfural and furfural formations in Orange juice. *Journal of Food Composition and Analysis, 102*, 103996.

Al-Jeddawi, W., Northcutt, J. K., & Dawson, P. (2023). Effect of pre-treatment and freezing on the polyphenol oxidase activity and color stability of sliced peaches. *Journal of the Science of Food and Agriculture, 103*(7), 3376–3389.

Ameur, L. A., Mathieu, O., Lalanne, V., Trystram, G., & Birlouez-Aragon, I. (2007). Comparison of the effects of sucrose and hexose on furfural formation and browning in cookies baked at different temperatures. *Food Chemistry, 101*, 1407–1416.

Ameur, L. A., Trystram, G., & Birlouez-Aragon, I. (2006). Accumulation of 5-hydroxymethyl-2-furfural in cookies during the backing process: Validation of an extraction method. *Food Chemistry, 98*(4), 790–796.

Aslanova, D., Bakkalbasi, E., & Artik, N. (2010). Effect of storage on 5-hydroxymethylfurfural (HMF) formation and color change in jams. *International Journal of Food Properties, 13*(4), 904–912.

Berk, E., Hamzalıoğlu, A., & Gökmen, V. (2019). Investigations on the Maillard reaction in sesame (Sesamum indicum L.) seeds induced by roasting. *Journal of Agricultural and Food Chemistry, 67*(17), 4923–4930.

Bharate, S. S., & Bharate, S. B. (2014). Non-enzymatic browning in citrus juice: Chemical markers, their detection and ways to improve product quality. *Journal of Food Science & Technology, 51*, 2271–2288.

Bi, H., Wang, Y., Guo, Y., Liao, Z., & Na, Z. (2023). Influence of Pasteurization on Maillard reaction in Lactose-Free milk. *Molecules, 28*(20), 7105.

Boopathy, R., Bokang, H., & Daniels, L. (1993). Biotransformation of furfural and 5-hydroxymethyl furfural by enteric bacteria. *Journal of Industrial Microbiology and Biotechnology, 11*(3), 147–150.

Bozkurt, H., Gogus, F., & Eren, S. (1999). Nonenzymic browning reactions in boiled grape juice and its models during storage. *Food Chemistry, 64*(1), 89–93.

Çağlar, D., Rayman Ergün, A., & Baysal, T. (2021). Effects of microwave pasteurization on the thermal resistance of Zygosaccharomyces rouxii and on the physicochemical properties of organic intermediate moisture raisin. *Journal of Food Processing and Preservation, 45*(4), e15382.

Cao, G., Cai, H., Cai, B., & Tu, S. (2013). Effect of 5-hydroxymethylfurfural derived from processed Cornus officinalis on the prevention of high glucose-induced oxidative stress in human umbilical vein endothelial cells and its mechanism. *Food Chemistry, 140*, 273–279.

Cao, X., Cai, C., Wang, Y., & Zheng, X. (2019). Effects of ultrasound processing on physicochemical parameters, antioxidants, and color quality of bayberry juice. *Journal of Food Quality, 2019*, 7917419.

Chen, Y., Lin, H., Li, Y., Lin, M., & Chen, J. (2019). Non-enzymatic browning and the kinetic model of 5-hydroxymethylfurfural formation in residual solution of vinegar soaked-soybean. *Industrial Crops and Products, 135*(146–152).

Chouaibi, M., Snoussi, A., Attouchi, S., & Ferrari, G. (2021). Influence of drying processes on bioactive compounds profiles, hydroxymethylfurfural, color parameters, and antioxidant activities of Tunisian eggplant (Solanum melongena L.). *Journal of Food Processing and Preservation, 45*(6), e15460.

Czipa, N., Phillips, C. J., & Kovács, B. (2019). Composition of acacia honeys following processing, storage and adulteration. *Journal of Food Science and Technology, 56*, 1245–1255.

de Borba, V. S., Lemos, A. C., Rodrigues, M. H. P., Cerqueira, M. B. R., & Badiale–Furlong, E. (2023). Acrylamide and hydroxymethylfurfural in cakes: An approach to reduce the formation of processing contaminants in sweet bakery products. *Food Research International, 165*, 112518.

Devu, S. S., Dileepmon, R., Kothakota, A., Venkatesh, T., Pandiselvam, R., Garg, R., … Khaneghah, A. M. (2022). Recent advancements in baking technologies to mitigate formation of toxic compounds: A comprehensive review. *Food Control, 135*, 108707.

Ding, X., Wang, M. Y., Yao, Y. X., Li, G. Y., & Cai, B. C. (2010). Protective effect of 5-hydroxymethylfurfural derived from processed Fructus Corni on human hepatocyte LO2 injured by hydrogen peroxide and its mechanism. *Journal of Ethnopharmacology, 128*(2), 373–376.

Dogan, M., Sienkiewicz, T., & Oral, R. A. (2005). Hydroxymethylfurfural content of some commercial whey protein concentrates. *Milk Science International Milchwissenschaft, 60*(3), 309–311.

Dursun Capar, T., Inanir, C., Cimen, F., Ekici, L., & Yalcin, H. (2022). Black garlic fermentation with green tea extract reduced HMF and improved bioactive properties: Optimization study with response surface methodology. *Journal of Food Measurement and Characterization*, 1340–1353.

Duru, N., Karadeniz, F., & Erge, H. S. (2012). Changes in bioactive compounds, antioxidant activity and HMF formation in rosehip nectars during storage. *Food and Bioprocess Technology, 5*, 2899–2907.

Dżugan, M., Grabek-Lejko, D., Sidor, E., & Tomczyk, M. (2021). The impact of ultrasound decrystallization on enzymatic, antioxidant and antibacterial properties of honey. *Innovative Food Science & Emerging Technologies, 71*,102709.

Erbersdobler, H. F., & Somoza, V. (2007). Forty years of furosine–Forty years of using Maillard reaction products as indicators of the nutritional quality of foods. *Molecular Nutrition & Food Research, 51*, 423–430. https://doi.org/10.1002/mnfr.200600154

Filimon, V. R., Băetu, M., Nechita, A., Filimon, R., Niculaua, M., & Rotaru, L. (2016). Assessment of 5-hydroxymethylfurfural content in dry and sweetened white wines. *Lucrări Ştiinţifice, Seria Horticultură, 59*(2), 149–154.

Germond, J. E., Philippossian, G., Richli, U., Bracco, I., & Arnaud, M. J. (1987). Rapid and complete urinary elimination of [14C]-5-(hydroxymethyl)-2-furaldehyde. *Journal of Toxicology and Environmental Health, 22*(1), 79–89.

Gheldof, N., Wang, X., & Engeseth, N. (2002). Identification and quantification of antioxidant components of honeys from various floral sources. *Journal of Agricultural and Food Chemistry, 50*, 5870–5877. https://doi.org/10.1021/jf0256135

Glatt, H., Schneider, H., Murkovic, M., Monien, B. H., & Meinl, W. (2012). Hydroxymethyl-substituted furans: Mutagenicity in Salmonella typhimurium strains engineered for expression of various human and rodent sulphotransferases. *Mutagenesis, 27*, 41–48.

Godfrey, V. B., Chen, L. J., Griffin, R. J., Lebetkin, E. H., & Burka, L. T. (1999). Distribution and metabolism of (5-hydroxymethyl)furfural in male F344 rats and B6C3F1 mice after oral administration. *Journal of Toxicology and Environmental Health Part A, 57* 199–210.

Gökmen, V., Açar, Ö. Ç., Serpen, A., & Morales, F. J. (2008). Effect of leavening agents and sugars on the formation of hydroxymethylfurfural in cookies during baking. *European Food Research and Technology, 226*, 1031–1037.

Grote, A. F. V., & Tollens, B. (1875). Untersuchungen über kohlenhydrate. I. Ueber die bei Einwirkung von Schwefelsäure auf Zucker entstehende Säure (Levulinsäure). *Justus Liebigs Annalen der Chemie, 175*(1–2), 181–204.

Gülcan, Ü., Uslu, C. C., Mutlu, C., Arslan-Tontul, S., & Erbaş, M. (2020). Impact of inert and inhibitor baking atmosphere on HMF and acrylamide formation in bread. *Food Chemistry, 332*, 127434.

Hamzalioglu, A., & Gökmen, V. (2017). Formation and elimination reactions of 5-hydroxymethylfurfural during in vitro digestion of biscuits. *Food Research International, 99*, 308–314.

Hamzaoğlu, F., Türkyılmaz, M., & Özkan, M. (2018). Effect of SO2 on sugars, indicators of Maillard reaction, and browning in dried apricots during storage. *Journal of the Science of Food and Agriculture, 98*(13), 4988–4999.

Herdt, D., Teumer, T., Balan, S. N., Kamarulzman, N. A., Kunz, T., Kühnemuth, S., … Rädle, M. (2021). Malt roasting quality control by mid-infrared spectroscopy. *Journal of Brewing and Distilling, 10*(1), 1–16.

Hodge, J. E. (1953). Dehydrated foods, chemistry of browning reactions in model systems. *Journal of Agricultural and Food Chemistry, 1*(15), 928–943.

Hong, X., Meng, J., & Lu, R. R. (2015). Improvement of ACE inhibitory activity of casein hydrolysate by Maillard reaction with xylose. *Journal of the Science of Food and Agriculture, 95*, 66–71.

Hu, H., Liu, X., Jiang, L., Zhang, Q., & Zhang, H. (2021). The relationship between acrylamide and various components during coffee roasting and effect of amino acids on acrylamide formation. *Journal of Food Processing and Preservation, 45*(5), e15421.

Huang, Y., Li, M., Lu, J., Hu, H., Wang, Y., Li, C., … Xie, M. (2022). nhibitory effect of hydrocolloids and ultrasound treatments on acrylamide and 5-hydroxymethylfurfural formation in French fries. *Food Hydrocolloids, 133*, 107839.

Husoy, T., Haugen, M., Murkovic, M., Joebstl, D., Stolen, L. H., Bjellaas, T., … Alexander, J. (2008). Dietary exposure to 5-hydroxymethylfurfural from Norwegian food and correlations with urine metabolites of short-term exposure. *Food and Chemical Toxicology*, 46. https://doi.org/10.1016/j.fct.2008.09.048;)

Janzowski, C., Glaab, V., Samimi, E., Schlatter, J., & Eisen Brand, G. (2000). 5-Hydroxymethylfurfural: assessment of mutagenicity, DNA-damaging potential and reactivity towards cellular glutathione. *Food and Chemistry Toxicology, 38*, 801–809.

Kamalabadi, M., Ghaemi, E., Mohammadi, A., & Alizadeh, N. (2015). Determination of Furfural and Hydroxymethylfurfural from Baby Formula Using Headspace Solid Phase Microextraction Based on Nanostructured Polypyrrole Fiber Coupled with Ion Mobility Spectrometry. *Food Chemistry, 181*, 72–77. DOI: 10.1016/j.foodchem.2015.02.069

Kanar, Y., & Mazı, B. G. (2019). HMF formation, diastase activity and proline content changes in bee pollen dried by different drying methods. *LWT – Food Science and Technology, 113*(108273).

Katsch, L., Methner, F. J., & Schneider, J. (2021). Kinetic studies of 5-(hydroxymethyl)-furfural formation and change of the absorption at 420 nm in fruit juices for the improvement of pasteurization plants. *International Journal of Food Engineering, 17*(9), 703–713.

Kayacan, S., Karasu, S., Akman, P. K., Goktas, H., Doymaz, I., & Sagdic, O. (2020). Effect of different drying methods on total bioactive compounds, phenolic profile, in vitro bioaccessibility of phenolic and HMF formation of persimmon. *LWT – Food Science and Technology, 118*, 108830.

Kowalski, S., Lukasiewicz, M., Duda-Chodak, A., & Zięć, G. (2013). 5-Hydroxymethyl-2-Furfural (HMF)-Heat-induced formation, occurrence in food and biotransformation-a Review. *Polish Journal of Food and Nutrition Sciences, 63*(4), 207–225.

Lee, C.-H., Chen, K.-T., Lin, J.-A., Chen, Y.-T., Chen, Y.-A., Wu, J.-T., & Hsieh, C.-W. (2019). Recent advances in processing technology to reduce 5-hydroxymethylfurfural in foods. *Trends in Food Science & Technology, 93*, 271–280.

Lemos, A. C., de Borba, V. S., de Medeiros Burkert, J. F., Scaglioni, P. T., & Badiale-Furlong, E. (2023). Role of white bread matrix components and processing parameters on 5-hydroxymethylfurfural (HMF) and acrylamide formation. *Food Control, 145*, 109407.

Li, Y. X., Li, Y., Qian, Z. J., Kim, M. M., & Kim, S. K. (2009). In vitro antioxidant activity of 5-HMF isolated from marine red alga Laurencia undulata in free-radical-mediated oxidative systems. *Journal of Microbiology and Biotechnology, 19*, 1319–1327.

Li, Z., Yuan, Y., Yao, Y., Wei, X., Yue, T., & Meng, J. (2019). Formation of 5-hydroxymethylfurfural in industrial-scale apple juice concentrate processing. *Food Control, 102*, 56–68.

Liu, D., He, Y., Xiao, J., Zhou, Q., & Wang, M. (2021). The occurrence and stability of Maillard reaction products in various traditional Chinese sauces. *Food Chemistry, 342*, 128319.

Maillard, L. C. (1912). Action des acides amines sur les sucres; formation des melanoidies par voie methodique. *CR Academic Science, 154*, 66–68.

Makhlouf-Gafsi, I., Krichen, F., Mansour, R. B., Mokni, A., Sila, A., Bougatef, A., … Besbes, S. (2018). Ultrafiltration and thermal processing effects on Maillard reaction products and biological properties of date palm sap syrups (Phoenix dactylifera L.). *Food Chemistry, 256*, 397–404.

Maruyama, H., Amanuma, T., Takashima, Y., Yoshiji, H., Nakae, D., Tsutsumi, M., … Konishi, Y. (1992). Possible enhancing effect of the immunosuppressive agent, 6-mercaptopurine(6-MP) on focal lesion development in cirrhotic liver induced by carbon tetrachloride but not furfural in F344 rats. *Carcinogenesis, 13*, 1365–1369.

Mesías, M., & Morales, F. J. (2017). Effect of different flours on the formation of hydroxymethylfurfural, furfural, and dicarbonyl compounds in heated glucose/flour systems. *Foods, 6*(2), 14.

Michalak, J., Gujska, E., Czarnowska, M., Klepacka, J., & Nowak, F. (2016). Effect of storage on acrylamide and 5-hydroxymethylfurfural contents in selected processed plant products with long shelf-life. *Plant Foods for Human Nutrition, 71*, 115–122.

Mouhoubi-Tafinine, Z., Ouchemoukh, S., Bey, B. M., Louaileche, H., & Tamendjari, A. (2018). Effect of storage on hydroxymethylfurfural (HMF) and color of some Algerian honey. *International Food Research Journal, 25*(3), 1044–1050.

Naknean, P., & Meenune, M. (2015). Impact of clarification of palm sap and processing method on the quality of palm sugar syrup (Borassus flabellifer Linn.). *Sugar Technology, 17*, 195–203.

Nguyen, H. T., van der Fels-Klerx, H. J., & Van Boekel, M. A. J. S. (2017). Acrylamide and 5-hydroxymethylfurfural formation during biscuit baking. Part II: Effect of the ratio of reducing sugars and asparagine. *Food Chemistry,* 230 , 14–23).

Nobis, A., Röhrig, A., Hellwig, M., Henle, T., Becker, T., & Gastl, M. (2019). Formation of 3-deoxyglucosone in the malting process. *Food Chemistry, 290*, 187–195.

Paszkot, J., Kawa-Rygielska, J., & Anioł, M. (2021). Properties of dry hopped dark beers with high xanthohumol content. *Antioxidants, 10*(5), 763.

Petisca, C., Henriques, A. R., Pérez-Palacios, T., Pinho, O., & Ferreira, I. M. (2013). Study of hydroxymethylfurfural and furfural formation in cakes during baking in different ovens, using a validated multiple-stage extraction-based analytical method. *Food Chemistry, 141*(4), 3349–3356.

Rose, M., Weber, D., Lotsch, B. V., Kremer, R. K., Goddard, R., & Palkovits, R. (2013). Biogenic metal–organic frameworks: 2, 5-Furandicarboxylic acid as versatile building block. *Microporous and Mesoporous Materials, 181*, 217–221.

Rosenboom, J. G., Langer, R., & Traverso, G. (2022). Bioplastics for a circular economy. *Nature Reviews Materials, 7*(2), 117–137.

Rufian-Henares, J. A., & Morales, F. J. (2007). Functional properties of melanoidins: in vitro antioxidant, antimicrobial and antihypertensive activities. *Food Research International, 40*(8), 995–1002.

Sacchetti, G., Ioannone, F., De Gregorio, M., Di Mattia, C., Serafini, M., & Mastrocola, D. (2016). Non enzymatic browning during cocoa roasting as affected by processing time and temperature. *Journal of Food Engineering, 169*, 44–52.

Serra-Cayuela, A., Aguilera-Curiel, M. A., Riu-Aumatell, M., Buxaderas, S., & López-Tamames, E. (2013). Browning during biological aging and commercial storage of Cava sparkling wine and the use of 5-HMF as a quality marker. *Food Research International, 53*(1), 226–231.

Sicari, V., Pellicanò, T. M., Laganà, V., & Poiana, M. (2018). Use of orange by-products (dry peel) as an alternative gelling agent for marmalade production: Evaluation of antioxidant activity and inhibition of HMF formation during different storage temperature. *Journal of Food Processing and Preservation, 42*(2), e13429.

Surh, Y.-J., Liem, A., Miller, J. A., & Tannenbaum, S. R. (1994). 5-Sulfooxymethylfurfural as a possible ultimate mutagenic and carcinogenic metabolite of the Maillard reaction product, 5-hydroxymethylfurfural. *Carcinogenesis, 15*, 2375–2377.

Svendsen, C., Husoy, T., Glatt, H., Paulsen, J. E., & Alexander, J. (2009). 5-Hydroxymethylfurfural and 5-sulfooxymethylfurfural increase adenoma and flat ACF number in the intestine of Min/+ mice. *Anticancer Research, 29*, 1921–1926.

Takei, T., Akita, T., Nakamura, I., Fujitani, T., Okumura, M., Okazaki, K., … Haruta, M. (2012). Heterogeneous catalysis by gold. In Bruce C. Gates and Friederike C. Jentoft (Eds.) *Advances in catalysis* (Vol. 55, pp. 1–126). Academic Press. www.sciencedirect.com/bookseries/advances-in-catalysis/vol/55/suppl/C

Tang, X., Zhang, Y., Li, F., Zhang, N., Yin, X., Zhang, B., … Fan, J. (2023). Effects of traditional and advanced drying techniques on the physicochemical properties of Lycium barbarum L. polysaccharides and the formation of Maillard reaction products in its dried berries. *Food Chemistry, 409*, 135268).

Toker, O. S., Dogan, M., Ersöz, N. B., & Yilmaz, M. T. (2013). Optimization of the content of 5-hydroxymethylfurfural (HMF) formed in some molasses types: HPLC-DAD analysis to determine effect of different storage time and temperature levels. *Industrial Crops and Products, 50*, 137–144.

Tontul, I., & Topuz, A. (2017). Effects of different drying methods on the physicochemical properties of pomegranate leather (pestil). *LWT – Food Science and Technology, 80*, 294–303.

Van Der Fels-Klerx, H. J., Capuano, E., Nguyen, H. T., Mogol, B. A., Kocadağlı, T., Taş, N. G., … Gökmen, V. (2014). Acrylamide and 5-hydroxymethylfurfural formation during baking of biscuits: NaCl and temperature–time profile effects and kinetics. *Food Research International, 57*, 210–217.

Ventura, M., Dibenedetto, A., & Aresta, M. (2018). Heterogeneous catalysts for the selective aerobic oxidation of 5-hydroxymethylfurfural to added value products in water. *Inorganica Chimica Acta, 470*, 11–21.

Viegas, O., Prucha, M., Gökmen, V., & Ferreira, I. M. (2018). Parameters affecting 5-hydroxymethylfurfural exposure from beer. *Food Additives & Contaminants: Part A, 35*(8), 1464–1471.

Wu, Q., Luo, Q., Xiao, J., Tang, S., Chen, Y., Shen, Y., … Feng, N. (2019). Catechin-iron as a new inhibitor to control advanced glycation end-products formation during vinegar storage. *LWT – Food Science and Technology, 112*, 108245.

Wu, S. F., Chen, Y. H., Lin, C. L., Yang, J. C., Du, Y. C., Lee, J. J., … Chang. F. (2011). Qualitative and quantitative analyses of the anti-allergic constituent of commercial prunus mume products in Taiwan. *Journal of Food and Drug Analysis, 19*(1), 1.

Xing, Q., Ma, Y., Fu, X., Cao, Q., Zhang, Y., & You, C. (2020). Effects of heat treatment, homogenization pressure, and overprocessing on the content of furfural compounds in liquid milk. *Journal of the Science of Food and Agriculture, 100*(14), 5276–5282.

Yalcin, O., & Cabrales, P. (2012). Increased hemoglobin O2 affinity protects during acute hypoxia. *American Journal of Physiology – Heart and Circulatory Physiology, 303*, H271–H281.

Yamada, P., Nemoto, M., Shigemori, H., Yokota, S., & Isoda, H. (2011). Isolation of 5-(hydroxymethyl)furfural from Lycium chinense and its inhibitory effect on the chemical mediator release by basophilic cells. *Planta Medica, 77*, 434–440.

Yang, N., Qiu, R., Yang, S., Zhou, K., Wang, C., Ou, S., & Zheng, J. (2019). Influences of stir-frying and baking on flavonoid profile, antioxidant property, and hydroxymethylfurfural formation during preparation of blueberry-filled pastries. *Food Chemistry, 287*, 167–175.

Yıltırak, S., Kocadağlı, T., Çelik, E. E., Özkaynak Kanmaz, E., & Gokmen, V. (2021). Effects of sprouting and fermentation on free asparagine and reducing sugars in wheat, einkorn, oat, rye, barley, and buckwheat and on acrylamide and 5-hydroxymethylfurfural formation during heating. *Journal of Agricultural and Food Chemistry, 69*(32), 9419–9433.

Zhang, J. H., Di, Y., Wu, L. Y., He, Y. L., Zhao, T., Huang, X., … Zhu, L. L. (2015). 5-HMF prevents against oxidative injury via APE/Ref-1. *Free Radical Research, 49*, 86–94.

Zhang, R. L., Zhang, M. M., Pu, Y. F., & Zhu, L. X. (2022). Evolution of nonenzymatic browning during the simulated Msalais-production process in models of grape juice. *Food Science & Nutrition, 10*(7), 2132–2140.

Zhang, Y. Y., Song, Y., Hu, X. S., Liao, X. J., Ni, Y. Y., & Li, Q. H. (2012). Effects of sugars in batter formula and baking conditions on 5-hydroxymethylfurfural and furfural formation in sponge cake models. *Food Research International, 49*(1), 439–445.

Zhao, L., Chen, J., Su, J., Li, L., Hu, S., Li, B., … Chen, T. (2013). In vitro antioxidant and antiproliferative activities of 5-hydroxymethylfurfural. *Journal of Agricultural and Food Chemistry, 61*, 10604–10611.

Zhao, L., Su, J., Li, L., Chen, J., Hu, S., Zhang, X., & Chen, T. (2014). Mechanistic elucidation of apoptosis and cell cycle arrest induced by 5-hydroxymethylfurfural, the important role of ROS-mediated signaling pathways. *Food Research International, 66*, 186–196.

Zhu, M., Long, Y., Ma, Y., Huang, Y., Wan, Y., Yu, Q., … Chen, Y. (2022). Investigation of thermal contaminants in coffee beans induced by roasting: A kinetic modeling approach. *Food Chemistry, 378*, 132063.

Žilić, S., Aktağ, I. G., Dodig, D., & Gökmen, V. (2021). Investigations on the formation of Maillard reaction products in sweet cookies made of different cereals. *Food Research International, 144*, 110352.

ABBREVIATIONS

DFF	2,5-diformylfuran
DMF	2,5-dimethylfuran
F	Furfural
FDA	2,5-furandicarboxylic acid
HMF	5-Hydroxy methyl furfural
HMFA	5-hydroxymethyl-2-furoic acid
HTST	High temperature and short time
UHT	Ultrahigh temperature
UP	Ultrahigh pasteurization

6 Polycyclic Aromatic Hydrocarbons and Safety

Food Processing and Manufacturing-Related Effects

Júlia Vianna de Pinho, Ludmila Rosa Bergsten-Torralba, Júlia Scarpa-de-Souza, Antonio Eugenio Castro Cardoso de Almeida, Rachel Ann Hauser-Davis, and Carlos Adam Conte-Junior

6.1 INTRODUCTION

Polycyclic aromatic hydrocarbons (PAH) are a group of ubiquitous persistent organic pollutants (POPs) consisting of two or more benzene rings that may present different structural configurations (Haritash and Kaushik, 2009). These compounds can be categorised as light PAH, which are highly volatile compounds consisting of up to four aromatic rings, and heavy PAH, which contain more than four rings and, thus, more stable, lipophilic and with greater adsorption potential to particulate matter (Lawal, 2017; Singh et al., 2016).

DOI: 10.1201/9781032712291-6

The main natural PAH sources encompass incomplete organic matter combustion or pyrolysis in forest fires and volcanic eruptions. Human activities, however, comprise the main current sources of these compounds, including the burning of coal, wood and agricultural residues (Ravindra et al., 2008), vehicle emissions (Ravindra et al., 2008), tobacco smoke (Abdel-Shafy and Mansour, 2016) and wastewater, industrial and oil residue discharges (Kuppusamy et al., 2020; Moon et al., 2010; Neff et al., 2005; Sun et al., 2019; Zhao et al., 2019).

Following their environmental input, PAHs can either be absorbed by plants or animals from surrounding air, water bodies or soil or volatilize from contaminated surfaces into the atmosphere, influenced by temperature, wind speed, and the physical and chemical properties of the PAH compounds themselves. These dynamics depend on species, environmental PAH concentrations and physicochemical PAH properties, such as lipophilicity and hydrophobicity. For example, soil PAH assimilation is usually more frequent for high molecular weight (HMW) PAH compared to low molecular weight (LMW) compounds (Seo et al., 2020), as LMW PAH are more likely to be found in the atmosphere due to their high volatility and the fact that they undergo photochemical degradation during atmospheric transport, which prevents them from accumulating at high concentrations in soil. On the other hand, HMW PAH exhibit a greater tendency to adhere to suspended particles and can be deposited in the soil through dry and wet deposition processes (Hussain et al., 2018). The main environmental PAH dynamics are depicted in Figure 6.1.

Once absorbed, these compounds can be assimilated by a vast array of living organisms, as they are highly volatile, facilitating their dispersion between environmental compartments, and transferred via diverse food chains (WHO, 1998), posing

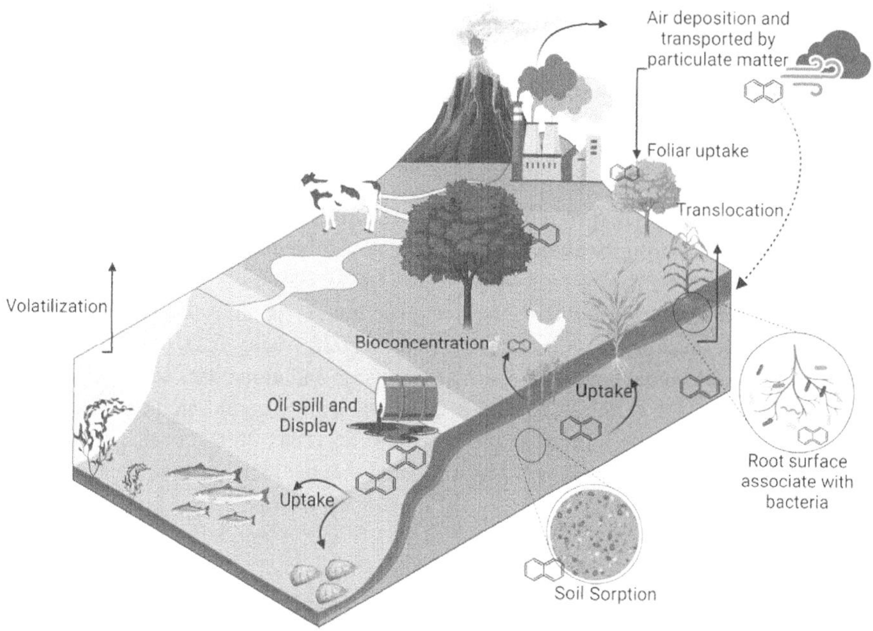

FIGURE 6.1 Overview of environmental pathways for PAHs, highlighting areas with significant pollution sources and accumulation pathways. (Created by the authors.)

environmental and human health risks (Jafarabadi et al., 2020). Naturally contaminated products include agricultural crops and seafood and shellfish from both marine and freshwater environments. Food products may also become contaminated by different food processing techniques that involve high temperatures or smoking procedures, among others (Stołyhwo and Sikorski, 2005).

6.2 HUMAN POLYCYCLIC AROMATIC HYDROCARBONS HEALTH EFFECTS

The International Agency for Research on Cancer (IARC), as well as the Joint Food and Agriculture Organization/World Health Organization (FAO/WHO), the Scientific Committee on Food (SCF) and the European Food Safety Authority (EFSA) have, therefore, classified and evaluated the carcinogenicity of several PAHs (Table 6.1), with several PAHs categorised as carcinogens, highlighting the need to understand and mitigate their health impacts. A total of 16 PAH have been set as priority compounds (Cavalcante et al., 2009; Harner et al., 2018). Benzo(a)pyrene is highlighted as the most toxic PAH, categorised by the IARC as a Group 1 agent, due to its strong carcinogenic effects and environmental prevalence, making it an indicator for the entire PAH group (Ailijiang et al., 2022; Polak-Śliwińska et al., 2022; Rosińska and Dąbrowska, 2018). It is also the most studied PAH to date (Eldaly et al., 2016). The Chronic Daily Intake (CDI) for PAH, however, depends on exposure levels, behaviour patterns and individual physiological characteristics (Wu et al., 2016).

In general, in addition to the formation of diol epoxides followed by the generation of DNA adducts, the main mechanisms of toxicity and carcinogenicity of PAHs in human cells are through the formation of free radicals and formation of orthoquinone (Ewa and Danuta, 2016). In this sense, PAHs cause significant changes at the cellular and molecular level and promote cancer in different organs, including lung cancer: Exposure to PAHs through inhalation of tobacco smoke, air pollution or occupational exposure is strongly associated with the development of lung, skin, bladder and liver cancer (Figure 6.2)

Studies have demonstrated that PAH can pose human health risks both through both short- and long-term exposures (Leizando & Ashraf, 2022; Ma et al., 2019). Adverse health effects of PAHs in humans include eye irritation, skin inflammation, celiac disease, and kidney stones (Derafsh et al., 2022; Shiue, 2016), as well as carcinogenic and mutagenic effects (Kelly et al., 2003; Kim et al., 2013; Naumova et al., 2002). In this sense, recent studies in the field of epidemiology have suggested that the increased incidence of cancer in humans and animals may be influenced by dietary exposure to polycyclic aromatic hydrocarbons, in addition to other multiple factors (Luper Tsenum and Yarkwan, 2020).

Occupational exposure to PAHs results in vomiting, diarrhoea, nausea, eye irritation and confusion, while chronic health effects involve jaundice, kidney damage, asthma-like symptoms, cataracts, impaired immunity, respiratory problems, redness and skin inflammation. Degradation of red blood cells (RBCs) has also been reported. Animal studies have shown that prolonged exposure to PAHs can lead to the development of

TABLE 6.1 International Agency for Research on Cancer polycyclic aromatic hydrocarbon categorization

POLYCYCLIC AROMATIC HYDROCARBON	*ACRONYM*	*IARC GROUP*	*DEFINITION*
Benzo[a]pyrene	BaP	1	Carcinogenic to humans
Dibenzo[a,h]anthracene	DBahA		
7,12-Dimethylbenz[a]anthracene	BMBA		
2-Acetylaminofluorene	-		
4-Aminodiphenyl	-		
4-Nitroquinoline-N-oxide	-		
2-Aminofluorene	-		
2-Aminoanthracene	-		
2-Aminonaphthalene	-		
4-Aminobiphenyl			
Benzo[g,h,i]perylene	BghiP	2A	Probably carcinogenic to humans
Benzo[e]pyrene	BeP		
Indeno[1,2,3-cd]pyrene	IcdP		
5-Methylcrystallene	-		
6-Nitrobenzo[a]pyrene	-		
Benzo[a]anthracene	BaA	2B	Probably carcinogenic to humans
Benzo[b]fluoranthene	BbF		
Benzo[j]fluoranthene	BjF		
Benzo[k]fluoranthene	BkF		
Benzo[c]fluorene	BcF		
Benzo[c]fluorene	BcF	3	Not classifiable regarding carcinogenicity to humans
Benzo[e]fluorene	BeF		
Benzo[g,h,i]fluoranthene	BghiF		
Benzo[k]fluoranthene	BkF		
Crystallene	Crys	4	Probably not carcinogenic to humans
Fluorene	Flu		
Phenanthrene	Phe		
Anthracene	Ant		
Naphthalene	Nap		

stomach, skin and lung cancers, depending on the exposure route (Rengarajan et al., 2015). Furthermore, PAHs have been associated with a series of adverse embryo effects, including premature birth, decreased body weight in offspring, cardiac malformations, reduced intelligence quotient, and behavioural problems in infants. Dietary PAH exposure represents between 88% and 98% of total PAH burdens, with an estimated PAH dose of about 3 μg d^{-1} obtained through this route. Regular cigarette smoking can add 2 to 5 μg d^{-1} to total exposure, although both smokers and passive smokers exhibit a higher dietary exposure to PAHs (Yu et al., 2015).

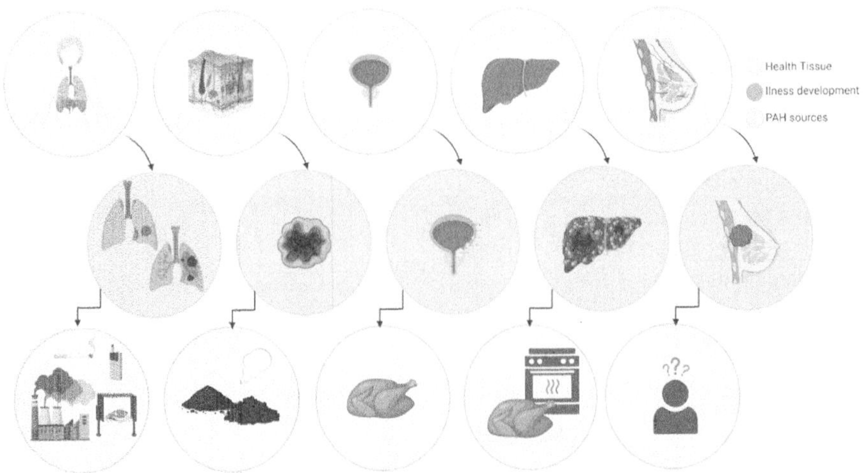

FIGURE 6.2 Diagram depicting the biological mechanisms through which exposure to PAHs leads to tissue damage and the subsequent development of cancer. This illustration focuses on the cellular and molecular pathways activated by PAHs, highlighting areas of potential intervention for prevention and treatment. (Created by the authors.)

Once metabolised in mammals, these hydrophobic compounds transform into diol epoxides, which are highly reactive compounds with the potential to interact with cellular macromolecules, such as DNA, causing mutations and errors during nucleic acid replication and triggering carcinogenic effects (Chen and Chen, 2005). In addition to the diol epoxide route, PAH toxicity and carcinogenicity mechanisms in human cells take place through the formation of free radicals and of orthoquinone (Ewa and Danuta, 2016), when PAHs are metabolised by enzymes of the cytochrome P450 system, resulting in the formation of reactive metabolites that can bind to DNA and form DNA adducts (Stowers and Anderson, 1985). Another toxicological pathway is the induction of oxidative stress through phase 1 of PAH detoxification, in addition to the activation of cell signalling pathways that trigger uncontrolled cell proliferation. These compounds can also inhibit intercellular communication through GAP junctions (Upham and Troski, 2009), which allow the transfer of ions and metabolites between cells, and whose inhibition can lead to changes in cell growth and differentiation, or even stimulate the activity of Mitogen-Activated Protein Kinase (MAPK), a family of proteins that enable transduction of intracellular signals, regulating cell growth and consequent proliferation (Bozman et al., 2020).

6.3 FOODSTUFF POLYCYCLIC AROMATIC HYDROCARBON LEVEL PROCESSING EFFECTS

Food processing has been carried out since the dawn of human society, and became crucial to the establishment of food conservation methods and food diversification. The emergence of ultra-processed foods, especially after the Second World War, however, led to a significant food culture transformation, with the prevalence of ready-to-eat products and consequent decreases in fresh food consumption (Levy et al., 2024).

Processed foods undergo changes in their physical form and/or chemical composition before being consumed. These changes may be minimal, such as washing, cutting and packaging, or more complex, such as the addition of other ingredients, such as sugars, oil and fats, preservatives, seasoning and food colouring, with the aim of increasing food durability and/or palatability. Processed foods may also undergo further treatment steps such as grinding, fermentation, cooking and drying (Figure 6.3).

The consumption of processed foods is a significant contributor to dietary PAH intake, as PAH levels can be significantly influenced by food processing and preparation methods (Duedahl-Olesen, 2013; Wang et al., 2022). Different PAH are differently affected by thermal, photooxidative and other processes (Jia et al., 2018), such as the fuel used in processing, processing duration and method, applied temperature, distance between the product and the heat source, process flow, cleaning and maintenance (Iko Afé et al., 2020; Iwegbue et al., 2020; Kim, Cho, and Jang, 2021; Mahugija and Njale, 2018; Onopiuk et al., 2021; Wang et al., 2019a, 2019b). Total PAH content also depends on the types of food materials, chemical composition, fat content, casings and presence of antioxidant agents and food additives, all of which are further influenced by fat, protein and carbohydrate content (Wang et al., 2022).

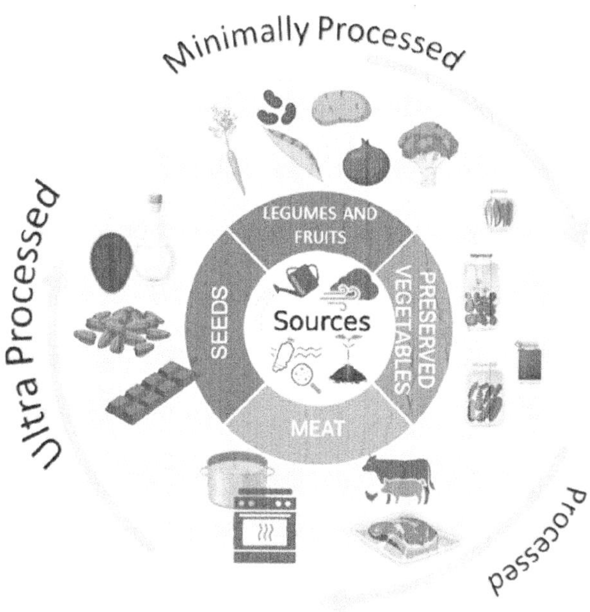

FIGURE 6.3 Classification system for food items based on the extent of processing. This figure categorizes various food types from minimally processed to ultra-processed, illustrating the typical changes in nutrient composition and potential health impacts. (Created by the authors.)

High-temperature food processing operations such as smoking, frying, drying, and grilling are significant PAH sources, despite being necessary for food treatment and enhancement (Onopiuk et al., 2021). In these cases, PAH can be generated either by the incomplete combustion of organic matter, resulting in the dripping of oils or fats onto the heat sources used in the cooking process, or by pyrolysis. In pyrolysis, HMW compounds can be formed from LMW precursors, while in pyrogenesis reactions PAH formation occurs from PAH-forming precursors and radicals (Palade et al., 2023). Smoking techniques, in particular, are an important method of food preservation, as they inhibit microbial activities and increase food shelf lives (Allen et al., 2012). These methods can also be employed to alter food sensory attributes, such as odour, flavour, colour and texture (Sigurgisladottir et al., 2000), in order to add product value to marketed items. Different processes have been evaluated in this regard concerning PAH foodstuff contamination. Boiling, for example, has been reported as leading to the formation of 5.36 μg kgV of PAH in fish (hake) and 3.15 μg/kg of PAH in potatoes (Perelló et al., 2009) while direct cooking led to 1.22 μg/kg of total PAH in cooked ham and 1.09 μg/kg in cooked beef samples (Martorell et al., 2010). Coroian et al. (2020), for example, assessed pork, ham, shoulder and bacon samples, reporting the highest phenanthrene value (35.76 μg/kg) in cured ham and the lowest (29. 25 μg/kg) in boiled-only samples. The same trend was observed for pork, ham and bacon. In bacon, for example, benzo(a)pyrene was present at the highest value after curing, followed by roasting (0.53 μg/kg) and boiling (0.51 μg/kg). The highest values of other PAHs (naphthalene, acenaphthene, fluorene, anthracene, fluoranthene, pyrene, benzo(k)fluoranthene) were also observed in cured samples followed by roasting and boiling processes.

In one study, Martins et al. (2000) suggested that the formation of PAH may be related to the Maillard reaction. This non-enzymatic chemical reaction occurs between amino acids and reducing carbohydrates (such as glucose and fructose) in foods during heating, forming a variety of products that contribute to the sensory characteristics of foods, such as color, aroma, flavor and texture (Neves et al., 2021).

Artificial drying and heating steps are also routinely applied for the processing of grains and raw materials, such as vegetable seeds, as climatic conditions (*i.e.,* rain and inadequate storage) can lead to mould growth in grains and cereals (Beyoz et al., 2021). Seed and grain drying treatments are, in fact, considered one of the main PAH contamination sources of edible oils derived from these foodstuffs (Marszałkiewicz et al., 2020).

Furthermore, PAHs can be transferred to food products through films or plasticizing materials that compose packaging materials. Materials such as low-density polyethylene (LDPE), for example, are recognized as potential sources of PAH contamination in grilled or smoked foods (Bansal and Kim, 2015), The sorption of these compounds in the films is affected mainly by van der Waals forces (Simko and Brunckova, 1993). Furthermore, the smaller the thickness of the film used, the more likely it is that PAHs will remelt from the film into the food (Simko et al., 1995).

Chromophores and auxochromes present in packaging pigments also affect PAH levels, as they directly affect the transfer kinetics between chemical agents and packaging. Li et al. (2017) reported that even water used as a food simulant alters PAH migration profiles in foodstuffs, which is not observed in simulated models. The authors also highlight that migration dynamics can also be affected by time-temperature combinations, which, in turn, alter the common pathways of by-product formation, with the potential to generate more toxic metabolites. In this sense, any adhered impurities or external materials can migrate based on the plastic structure and food ingredients, also potentially influencing PAH migration.

6.4 POLYCYCLIC AROMATIC HYDROCARBON CONTAMINATION IN DIFFERENT FOODSTUFFS

6.4.1 Grains and Cereals

Food contamination from environmental sources stands out mainly in agricultural activities, due to the transfer of PAH into the soil in contact with the roots, as well as through irrigation with contaminated water and atmospheric absorption during photosynthesis (Tarigholizadeh et al., 2023). Vegetables and grains grown in contaminated environments may potentially reach considerable PAH contamination levels (Cheng et al., 2019).

In addition to their nutritional importance for different crops, grains and seeds constitute one of the main economic forces in several countries (Zingale et al., 2023) and their regular consumption is associated with health benefits, such as reducing the risk of chronic diseases, such as cardiovascular disease, type 2 diabetes and certain types of cancer. Furthermore, oils of vegetable origin are widely used in cooking, either directly, as a fuel in frying or as an ingredient, and can be used in a variety of food products, such as margarine, mayonnaise, sauces and bakery products (Ciecierska and Obiedzinski, 2013). They may, however, comprise an important source of PAH.

Grains such as wheat, oats, rye, and barley usually contain low PAH levels, making them safe for consumption and export (Einolghozati et al., 2022), while rice and corn tend to have higher PAH levels due to drying processes used in their production (Camargo et al., 2003). The use of industrial ovens and furnaces directly influences grain quality, as combustion is followed by pyrolysis and pyro-synthesis, which results in both the formation of low-grade PAH and complex PAH, which vary depending on the fuel used in the furnaces. For example, B(a)P in corn samples has been reported as significantly higher in kernels dried in light oil-based ovens than those dried in natural gas-based ovens (Kiš et al., 2009). However, another study reported that drying grains with eucalyptus leaves did not compromise the quality of soybeans and maintained PAH levels below the maximum values allowed by EU legislation (Quequeto et al., 2021; Quequeto et al., 2022; Wen et al., 2017). In another study, Nieva-Cano et al. (2001) verified about 7.38–18.0 μg/kg of PAH in toasted bread.

6.4.2 Oils and Fats

The process of extracting and refining vegetable oils aims to produce high-quality items with desirable sensory characteristics and long shelf life, but due to the lipophilic nature of PAHs, they are particularly susceptible to PAH contamination, which are commonly found at unsafe levels in these food products (Sánchez-Arévalo et al., 2020). The presence of PAH in vegetable oils can be attributed to both oilseed drying processes and contamination during their harvesting, as well as of their packaging materials, soil burning and raw materials (Bansal and Kim, 2015).

Oils with higher levels of monounsaturated fatty acids, such as canola oil for example, may contain higher total PAH levels (Onopiuk et al., 2022). In one study, PAH levels ranging from 6.40 to 76.08 μg $^{-1}$ were reported in corn oil (Dost and İdeli, 2012). In another assessment, Alomirah et al. (2010) investigated 115 samples of olive oils, cooking oils (corn oil, sunflower oil, sesame oil, palm olein oil, soybean oil, canola oil, mustard oil, peanut oil and mixed vegetable oil) and fats (butter and margarine) and estimated that populations in countries such as Kuwait are exposed to 196 ng day−1 of Sum PAH8 equivalents (benz[a]anthracene, chrysene, benzo[b]fluoranthene, benzo[k]fluoranthene, benzo [a]pyrene, indeno[1,2,3-cd]pyrene, dibenz[a,h]anthracene and benzo[ghi]perylene) by consuming these products alone.

The use of vegetable oils as fat substitutes is the simplest strategy to inhibit the formation of PAH in meat products (Lü et al., 2017). In one study, Siddique et al. (2021) evaluated the influence of the type of oil and preparation conditions for rabbit meat, noting the importance of evaluating application conditions such as the volume and temperature of the oil used to estimate the impact of these ingredients. Mocek and Ciemniak (2015) reported that the PAH content in sunflower oil gradually decreased along with increasing temperatures (100 °C) from 0.5 to 2 4h, with the most prominent reduction noted for BkFA (51.99%) and CHR (36.20%). The same study observed that when subjected to direct UV radiation, PAH values decreased from 354.44 μg/L to 234.36 μg/L.

6.4.3 Leaves, Fruits and Seeds

Polycyclic aromatic hydrocarbon contamination in vegetables during food processing can occur at different stages, from cultivation to final preparation (Sampaio et al., 2021). During cultivation, PAH contamination in plants can occur both directly (air-leaf) and indirectly (air-soil-root) (Desalme et al., 2013), varying between leaves, branches, roots and fruits (Qiu et al., 2018). In vegetables, high PAH levels can lead to chlorophyll degradation and fluctuations in the activities of antioxidant enzymes, and further damage to photosynthetic capacity impacts the growth of vegetables and can trigger significant economic losses (Marwood et al., 2001; Song et al., 2012).

Concerning processing, one study verified that roasting led to increased PAH levels from < LOQ up to 13.8 μg/kg in raw and roasted pumpkin seeds, respectively (Hoed et al., 2017). Polycyclic aromatic hydrocarbons in leaves may also be transferred into tea infusions (Gao et al., 2017). However, tea contamination levels vary widely. Values have been reported as ranging from 14 up to 2,662 μg/kg for various types of tea (Li et al., 2011), with the highest levels observed in green tea and black tea (497 and 1,162 μg/kg, respectively) (Lin and Zhu, 2004), and from 1.4 to 2.662 μg/kg in herbal and fruit tea, black tea, green tea, white tea and mate teas (Sadowska-Rociek et al., 2013), while other studies indicate that the transfer of PAH from the leaves to infusions is very small. Furthermore, people only consume small amounts of tea (Jiao et al., 2023). However, the variability of these compounds in tea is high, due to varying leaf/water ratios, the number of water additions, tea washing before preparation and infusion covering (Oranuba et al., 2018).

Fruits, vegetables and plants may absorb PAH either through the soil or through autogenous biosynthesis (Koszucka et al., 2019), affecting both leafy vegetables (Guillén et al., 1997), due to larger, waxy areas on the surface of leaf cuticles, facilitate PAH deposition and assimilation, and non-leafy vegetables that translocate these compounds from roots to aerial portions. The contact surface is also a relevant factor for roots. Tuberous roots such as potatoes and carrots generally contain higher total PAH contents due to high PAH soil adsorption (13 μg/kg) compared to other vegetables such as turnips (10.9 μg/kg) (Ashraf and Salam, 2012; Ashraf et al., 2013; Taqvi, 2018).

In the case of leafy vegetables, the absorption and transport of PAH by the roots plays a significant role in contamination and transport to other plant parts. Therefore, PAH are less frequent in fruits (Ashraf et al., 2013), with absorption and translocation reported as ranging from 0.01 to 5 μg/kg throughout the vegetable body (Deng et al., 2018). However, several fruits have been noted as contaminated with high PAH levels, such as fluorene, anthracene, fluoranthene, benzo(a)pyrene, chrysene, benzo[a]anthracene, benzo[k]fluoranthene, dibenzo(a,h)anthracene, pyrene, indo[123-cd]pyrene, ranging from 1.9–6.0 and 4.1–15 μg/kg in strawberries and apples (Wennrich et al., 2002).

The lipophilic profile of PAH tends to cause PAH plant translocation and accumulation mainly in seeds. Chocolate produced from cocoa seeds, stands out as a significant human PAH exposure route, (Żyżelewicz et al., 2016). The average daily consumption of chocolate worldwide has been estimated as 28 a day, consumed pure or used as an ingredient in sweets, bars, milkshakes and cookies (Vanderschueren et al., 2023). Some commercial chocolate samples have been reported as containing BaP levels above European Union limits (07–0.63 μg/kg) and ≤0.07 μg/kg of Dibenzo[a]pyrene origined from roasting. Other PAH levels have also been detected, such as Naphthalene, Fluorene, Phenanthrene, Anthracene, Fluoranthene, Pyrene, Benzo[a]pyrene, Benzo[a]anthracene, Benzo[b] fluoranthene, Benzo[k]fluoranthene, and Dibenzo[a,h]anthracene, ranging from 2.7 to 235 μg/kg. In one study conducted in Nigeria, although total PAH levels were below EU limits, the concentrations of Σ16 PAH in 29 commercial brands of sweets and chocolates, ranged from 1.09 to 149.46 μg/kg and from 0.95 to 222.62 μg/kg, respectively, with a predominance of LMW PAH and Sum4PAH (benzo(a)anthracene + benzo(b)fluoranthene + benzo(a)pyrene + chrysene) above the limit of 35 μg/kg (Iwegbue et al., 2015).

6.4.4 Beverages

In tropical and subtropical countries, sugarcane production plays an important role in the local economy and culture, being used in the production of ethanol, obtained through the fermentation of sugar from juice and distillate to produce different types of alcoholic beverages. However, high PAH amounts are released due to the burning of sugarcane fields, which have been reported as present in higher levels in products derived from burned sugarcane compared to unburned sugarcane (Galinaro et al., 2007; Shitandayi et al., 2019; Verma et al., 2019), also affecting refined products (Verma et al., 2019). Around 50% of sugarcane juice may contain PAH due to the use of crop residues, as farmers evaporate water from harvested sugar cane to obtain higher sugar concentrations (Bansal and Kim, 2015). In cachaça, for example, the typical Brazilian alcoholic drink, the prevalence of carcinogenic PAHs, such as BaP, is high (Galinaro et al., 2007; Shitandayi et al., 2019).

High PAH levels have also been noted in whole milk, semi-skimmed milk, ultra-treated milk and pasteurized milk (Naccari et al., 2011). Naccari et al. (2011) reported an increasing order in PAH levels in different heat-treated milk samples, which may have been influenced by environmental sources and pasture feeding (raw milk) and the levels of fats and triglycerides (whole and semi-skimmed milk). For this reason, milk and its derivatives have been widely investigated after undergoing heat treatment and processing (Amirdivani et al., 2019).

Concerning juices, B(a)P levels have been reported as significantly lower in fruit juices, such as pineapple and grape (1.94 μg/L), apple (0.483 μg/L) and orange (2.707 μg/L) (Rascón et al., 2019).

6.4.5 Animal Protein

Meat is an important protein source for humans and it is estimated that food products of animal origin represented 21% of total global food production, providing 40% of the total protein supply and 44.2% of the total fat supply. The process of grilling meat contributes to around 21% of the daily intake of BaP in the USA (Kazerouni et al., 2001). Farhadian et al. (2010) reported BaP concentration of 12.5 μg/kg in beef products such as beef kebabs, chicken kebabs and grilled chicken.

On the other hand, fishing and aquaculture provides around 109.3 million tons of products to humans, including bony fish, crustaceans and molluscs (FAO, 2022). Seafood is preferred in many regions due to its flavour and lower cost than other sources of animal protein, maintaining an adequate content of fatty acids. However, fish products are usually highly contaminated by PAH as, despite their efficiency in metabolising these compounds, some species such as sardines and salmon are preserved by smoking and many are prepared by frying and grilling (Bansal and Kim, 2015), all of which may increase PAH levels. In an analysis of fish sold in Korea, BaP levels between non-detectable and 2.87 μg/kg were reported, with salmon containing the highest PAH levels (Cho and Shin, 2012). Other seafood, such as crustaceans and bivalves, have been reported as containing fluoranthene, phenanthrene and anthracene, among other PAH (Zhang et al., 2010).

6.5 POLYCYCLIC AROMATIC HYDROCARBON REGULATIONS CONCERNING FOODSTUFFS

As these compounds are present in several food and beverages in the form of mixtures containing different PAH ratios, the European Commission has identified four main PAHs (ΣPAH4) in foodstuffs, including benz[a]anthracene (BaA), chrysene (Chr), benzo[b]fluoranthene (BbF) and benzo[a]pyrene (BaP) and since 2005 has monitored these agents in different groups such as oils, fruits, and seafood. Another important measure was the creation of the Scientific Committee on Food, whose data collected by each Member State is used to prepare prompt responses to food contamination events.

In 2003, this committee evaluated a total of 33 PAHs, 16 of which exhibit genotoxic and carcinogenic properties, which are currently included in the 16 priority PAHs.

European regulations are considered the most stringent in relation to PAH levels in foods, using benzo[a]pyrene (BaP) as a general marker, despite it representing only 1–20% of the total concentration of carcinogenic PAHs. Therefore, it is also recommended to monitor other compounds for long-term exposure assessments, checking whether the use of BaP is justified. However, regarding contaminant transfer during storage periods, Regulation (EU) No. 10/20111 on plastic materials in contact with food does not establish Specific Migration Limits (LME) for many substances.

The EFSA plays a key role in assessing and managing risks related to food safety, including the presence of PAH in food, developing regulations and guidelines that establish maximum limits for these compounds in various food products, using a total of eight PAHs as indicators of PAH toxicity in food (Martena et al., 2011). The PAHs listed as priority are determined by standardised analytical methods common to both EFSA and the United States Environmental Protection Agency (Huang et al., 2018). Typically PAH are observed at levels below the maximum permissible limit by EFSA (1.0 μg/kg) for cereals and cereal-based products (Mastanjević et al., 2019; Odika et al., 2022). In terms of international legislation, the European Community Commission (2006) has established maximum levels at 5.0 μg/kg for benzo(a)pyrene and 10.0 μg/kg for the sum of benzo(a) anthracene, benzo (a)pyrene, benzo(b)fluoranthene and chrysene in food items.

In the United States, the Food and Drug Administration (FDA) and the Environmental Protection Agency (EPA) are responsible for regulating the presence of PAH in foods. The FDA focuses primarily on food products of plant and animal origin and has also proposed monitoring the 16 priority PAHs due to toxicity. The FDA has also established tolerable limits for PAH in foods and beverages and implemented monitoring strategies to ensure compliance with these limits (Sampaio et al., 2021). The EPA is responsible for monitoring environment PAH levels, including in drinking water.

However, despite regulatory efforts, there are still flaws in monitoring, which is not very comprehensive in terms of types of food and, in many countries, is inefficient or non-existent. Therefore, efforts to reduce exposure to PAHs in food are still necessary and essential to protect public health. As they can be formed at different food production stages, one of the best alternatives is to invest in the development and implementation of methods effective to eliminate them and minimise human exposure to AH (Huang et al., 2018).

6.6 POLYCYCLIC AROMATIC HYDROCARBON FOOD MITIGATING STRATEGIES

Concerns about the adverse effects of PAHs on food and the environment, and therefore on human health, are reflected in the development of studies on removal methodologies that, if employed efficiently, could reduce the health risks related to PAHs (Singh et al., 2016; Singh et al., 2023). Table 6.2 indicates several foodstuff preparation approaches for different types of foodstuffs and their effects on PAH contamination.

TABLE 6.2 Foodstuff preparation approaches for different types of foodstuffs and their effects on polycyclic aromatic hydrocarbon contamination

TYPE OF FOODSTUFF	*PREPARATION METHOD*	*POLYCYCLIC AROMATIC HYDROCARBON EFFECT*
Oilseeds and oils	Drying processes at lower temperatures	↓
	Direct toasting	↑
	Refrain direct combustion contact	↓
Cereals	Flour type	↑
	Dehusking and polishing	↓
	Baking at low distances and high temperature	↑
Fruits, leaves and vegetables	Drying	↑
	Maintaining crops far from highways and industrial areas	↓
	Cultivation against the wind direction in areas displaying potential for fire and vapour emissions	↓
	Washing raw vegetables	↓
Seafood	Cultured or obtained in areas affected by industrial and oil effluents	↑
	Collected after oil spills	↑
	Use of ecological wood and sustainable tanks for mussel cultivation	↓
Meat, chicken and fish	Choosing lean meat and fish	↓
	Adding fat to grilling food	↑
	Avoiding direct contact between food and flames	↓
	Cooking at lower temperatures	↓
	Use of traditional or creosate-treated firewood	↑
	Choosing electric or gas heat sources over charcoal as fuel	↓
	Applying acid marination prior to grilling	↓

Several factors influence PAH transfer to foods, including PAH diffusion coefficients, migration properties, surface affinity based on polarity, liquid medium viscosity, molar mass and compound size, packaging exposure to UV radiation, the nature and existing contamination of the storage container or packaging, and the presence of auxochromes and chromophores, among others (Chen and Chen, 2005; Li et al., 2017; Singh et al., 2022).

Reducing pH levels, processing temperatures and cooking times have also been suggested as general food PAH mitigation strategies, minimising the formation of unwanted by-products, although the effectiveness of each option varies depending on the food matrix and PAH. Low-temperature cooking, in fact, comprises one of the main approaches to reducing PAHs as it prevents charring and can be controlled by reducing direct contact with heat sources (grills and surfaces) (Farhadian et al., 2011; Yebra-Pimentel et al., 2013), minimising fat dripping onto the heat source and opting for leaner cuts of meat (Duedahl-Olesen & Ionas, 2021; Duedahl-Olesen et al., 2015).

Using efficient ventilation systems or indirect drying methods as well as adequate temperature controls aiming at avoiding high temperature variations throughout drying processes, have also been noted as efficient in reducing PAH contamination (Yi et al., 2017).

Modifying preparation procedures, such as using marinades rich in antioxidants (Rey-Salgueiro et al., 2008), changing ingredients, adjusting pre-treatment conditions (Wang et al., 2022; Zhang et al., 2021), as well as selecting low-fat meats (Arab et al., 2023), are also common approaches. During cooking, optimising the smoking processes for dry-cured meats, such as limiting cold smoking to up to 20 h has also been verified as an adequate PAH-reducing approach (Duedahl-Olesen & Ionas, 2021; Fraqueza et al., 2020). Silva et al. also highlight that the use of different materials to smoke fish, such as sawdust, charcoal and firewood, also affect PAH contamination (Uzomah et al., 2021).

Ozonation and UV irradiation, the use of lactic acid bacteria and fuel catalysts in food also demonstrate high potential in reducing PAH foodstuff contamination, although not yet widespread (Wang et al., 2022; Yousefi et al., 2019). Khalil et al. (2016), for example, observed that, depending on the amount of applied radiation, PAH content may be reduced, while Kiralan et al. (2017) emphasise that the prior introduction of microwaves still requires the implementation of modelling studies to ensure the effectiveness of these processes. The most significant reductions in this sense have been observed for fluoranthene in steam and microwave preheating treatments (up to 46% and 37% for beef), as well as when using banana leaf wrappings (39% for beef and 72% for chicken meat) (Farhadian et al., 2011).

Fruits have also been used as PAH-reducing agents. For example, the use of lemon juice in marination, besides pleasing the general public's palate, has been noted as reducing PAH levels by up to 72% compared to oil ingredients and the same is noted for pork loin smoking (Orecchio et al., 2009). The use of cranberries as a filling has also been reported as capable of inhibiting 58% of BaP formation. The same potential has also been described for functional foods (herbs, red fruit extracts, ginger, garlic, etc.) which are a source of bioactive phenolic compounds and remove PAH and free radicals involved in the transformation and formation of these substances (Hazra et al., 2010).

The use of natural and edible casings, such as carrageenan packaging, instead of synthetic packaging, can both improve food quality and preserve sensory properties such as aroma and appearance, while also preventing contamination (Savitri et al., 2022).

Using recycled plastics in packaging production can also reduce PAH transfer from packaging materials to non-polar foods such as oils, bacon and dairy fats. Semanová et al. (2016) discusses the potential of polyethylene packaging to reduce PAH levels in smoked meat sausages. In that study, sausages were packaged in a laminated polyamide/low-density polyethylene film and the content of four PAHs in the packaged samples was reduced from

30.1 to 5.7 μg/kg (benzo[a]anthracene decreased from 11.5 to 2.1 μg/kg, chrysene from 9.4 to 1.9 μg/kg, benzo[b]fluoranthene from 5.3 to 0.6 μg/kg and benzo[a]pyrene from 3.9 to 1.1 μg/kg), indicating the migration of the ∑4PAH from sausages to the low-density polyethylene packaging. Despite this, some studies indicate that recycling may not have an impact on the PAHs present in recycled packaging materials as the temperature at which PAHs boil is higher than that of the recycling process (Simko 2002).

Another interesting approach comprises the use of physical barriers, *i.e.*, filters while the foodstuffs are still in the oven, especially for fish rich in fat, as these barriers prevent fat from dripping and burning. Furthermore, when combined with the use of activated carbon, PAH adsorption onto these filters has been noted as significantly higher, reducing PAH foodstuff contamination (Essumang, 2014). For fish products, another suggested approach is to remove the skin from the fish prior to preparation (Bomfeh, 2021).

Absorbents are commonly used in external food processing stages to reduce contamination. Graphene-based materials and carbon nanomaterials, for example, exhibit potential for removing PAH from water and heavy naphthenic oil (Cortés-Arriagada, 2017; Luna et al., 2016; Yang et al., 2006). Activated charcoal has also been widely used due to its ability to efficiently remove impurities, odours, unwanted flavours, toxins and contaminants from water used in food processing (Chaemsai et al., 2016), where it is incorporated into water filtration systems and used in subsequent production stages, in beverage production (Singh et al., 2023), and even in decolorizing sugar and removing residual odours from vegetable oils, contributing to improving the sensorial quality of food products. Activated charcoal is also employed in food detoxification, where it is added to certain processes to adsorb toxins and harmful chemicals. However, although it offers a number of benefits in food processing, excessive or inappropriate use of activated charcoal can negatively affect sensory food characteristics, such as colour, flavour and aroma. Data on the specific adsorption capacity of PAH on activated carbons of different commercial grades, however, are still incipient (Yebra-Pimentel et al., 2014). It is important to note that surface properties, pore size distribution, porosity and the presence of functional groups may differ between activated carbons, influencing PAH removal efficiency (Shin et al., 2015).

Concerning teas, the process of rolling tea leaves leads to loss of moisture and cell breakdown and release of smell (Olesen, Navaratnam, Jewula, & Jensen, 2015). This change in the surface area of the leaves during drying can help reduce the levels 3- to 211-fold compared to whole leaves (Rociek, Surma, and Cieślik, 2014).

In the case of fruits and vegetables, recommendations include the preservation of cultivation areas and the use soils far from industrial sources or areas influenced by forest fires and volcanic activity (Singh et al., 2023). For areas managed by small producers, it is simple and efficient to monitor winds and establish cultivation in areas that blow in the opposite direction (Rey-Salgueiro et al., 2008).

Regarding home care with fruits and vegetables, washing, removing dirt, can remove up to 50% of PAH deposited on the surface (SCF, 2002). The use of tap water combined with soaps or hydrogen peroxide (H_2O_2) and potassium permanganate ($KMnO_4$) has been noted as contributed significantly to the removal of PAH from 65.5–97.8% and 79.4–99.9%, respectively (Zohair et al., 2006).

In the case of aquatic ecosystems, the PAH contamination load is multifactorial. Oil spills and industrial effluent discharges, especially near coastal areas, require sustainable strategies such as bioremediation (Bamforth and Singleton, 2005). Microbial degradation has attracted much attention, as it also employs benzene rings as carbon and energy sources (Juhasz et al., 2000). Therefore, by modifying some variables such as pH or humidity, microorganisms can degrade the structure of multiple PAH rings simultaneously (Samanta et al., 2002).

REFERENCES

Abdel-Shafy, H. I., and Mansour, M. S. (2016). A review on polycyclic aromatic hydrocarbons: Source, environmental impact, effect on human health and remediation. *Egyptian Journal of Petroleum, 25*(1), 107–123. https://doi.org/10.1016/j.ejpe.2015.03.011

Aikin, T. J., Peterson, A. F., Pokrass, M. J., Clark, H. R., and Regot, S. (2020). MAPK activity dynamics regulate non-cell autonomous effects of oncogene expression. *eLife*, 9, e60541. https://doi.org/10.7554/eLife.60541

Ailijiang, N., Li, R., Sun, H., Yao, Y., Zhang, L., Chen, Y., and Zhang, D. (2022). Source identification and health risk assessment of polycyclic aromatic hydrocarbons (PAHs) in urban areas. *Scientific Reports, 12*, 2326. https://doi.org/10.1038/s41598-022-06437-8

Allen, K., Cornforth, D., and Whittier, D. (2012). Functionality of liquid smoke as an all-natural antimicrobial in food preservation. *Journal of Food Science, 77*(3), R106–R111.

Alomirah, H., Al-Zenki, S., Husain, A., Sawaya, W. N., Ahmed, N., Gevao, B., … and Kannan, K. (2010). Benzo[a]pyrene and total polycyclic aromatic hydrocarbons (PAHs) levels in vegetable oils and fats do not reflect the occurrence of the eight genotoxic PAHs. *Food Additives and Contaminants: Part A, 27*(6), 869–878. https://doi.org/10.1080/19440040903493793

Amirdivani, S., Khorshidian, N., Ghobadi Dana, M., Mohammadi, R., Mortazavian, A. M., Quiterio de Souza, S. L., ... and Raices, R. (2019). Polycyclic aromatic hydrocarbons in milk and dairy products. *International Journal of Dairy Technology, 72*(1), 120–131.

Arab, W. S., Edris, A. M., Ahmed, E. M., and Mansour, A. H. (2023). The existence of polycyclic aromatic hydrocarbons in commercially grilled beef kofta and chicken meat in El-Gharbia governorate markets. *Benha Veterinary Medical Journal, 44*(1), 5–8.

Ashraf, M. W., and Salam, A. (2012). Polycyclic aromatic hydrocarbons (PAHs) in vegetables and fruits produced in Saudi Arabia. *Bulletin of Environmental Contamination and Toxicology, 88*, 543–547. https://doi.org/10.1007/s00128-012-0528-8

Ashraf, M. W., Taqvi, S. I. H., Solangi, A. R., and Qureshi, U. A. (2013). Distribution and risk assessment of polycyclic aromatic hydrocarbons in vegetables grown in Pakistan. *Journal of Chemistry*, 2013, Article ID 873959. https://doi.org/10.1155/2013/873959

Bamforth, S. M., and Singleton, I. (2005). Bioremediation of polycyclic aromatic hydrocarbons: Current knowledge and future directions. *Journal of Chemical Technology and Biotechnology: International Research in Process, Environmental and Clean Technology, 80*(7), 723–736.

Bansal, V., and Kim, K. (2015). Review of PAH contamination in food products and their health hazards. *Environment International, 84*, 26–38. https://doi.org/10.1016/j.envint.2015.06.016

Beyoz, T., Yildirim, I., and Karabulut, O. (2021). Effects of artificial drying on mold growth in grains and cereals. *Journal of Agricultural Sciences*, 27(3), 345–352.

Bomfeh, K. (2021). The shift from a traditional to an improved fish smoking oven in Ghana: Implications for food safety and public health. *Afrika Focus, 34*(1), 155–166.

Camargo, M., and Toledo, M. (2003). Polycyclic aromatic hydrocarbons in Brazilian vegetables and fruits. *Food Control, 14*, 49–53. https://doi.org/10.1016/S0956-7135(02)00052-X

Cavalcante, R. M., Medeiros, A. C., dos Santos, M. R., Silva, F. V., and Nascimento, R. F. (2009). Distribution and sources of polycyclic aromatic hydrocarbons in sediments of a tropical estuary in Northeastern Brazil. *Journal of Environmental Science and Health, Part A, 44*(13), 1435–1440.

Chaemsai, S., Kunanopparat, T., Srichumpuang, J., Nopharatana, M., Tangduangdee, C., and Siriwattanayotin, S. (2016). Reduction of the polycyclic aromatic hydrocarbon (PAH) content of charcoal smoke during grilling by charcoal preparation using high carbonisation and a preheating step. *Food Additives and Contaminants: Part A, 33*(3), 385–390.

Chen, J., and Chen, S. (2005). Removal of polycyclic aromatic hydrocarbons by low density polyethylene from liquid model and roasted meat. *Food Chemistry, 90*(3), 461–469.

Cheng, J., Zhang, X., Ma, Y., Zhao, J., and Tang, Z. (2019). Concentrations and distributions of polycyclic aromatic hydrocarbon in vegetables and animal-based foods before and after grilling: Implication for human exposure. *Science of the Total Environment, 690*, 965–972.

Cho, H.-K., and Shin, H.-S. (2012). Analysis of benzo[a]pyrene content from smoked food products in Korea. *Food Science and Biotechnology, 21*(4), 1095–1100. https://doi.org/10.1007/s10068-012-0142-x

Ciecierska, M., and Obiedziński, M. (2013). Polycyclic aromatic hydrocarbons in vegetable oils from unconventional sources. *Food Control, 30*(2), 556–562. https://doi.org/10.1016/j.foodcont.2012.07.046

Coroian, A., Vioara, M., Ovidiu, C. C., Răducu, C., Cocan, D., Andronie, L., and Marchiş, Z. (2020). The level of polycyclic aromatic hydrocarbons (PAHs) from pork meat depending on the heat treatment applied. *Romanian Biotechnological Letters, 25*(3), 1601–1606. https://doi.org/10.25083/rbl/25.3/1601.1606

Cortés-Arriagada, D. (2021). High stability and properties of adsorbed polycyclic aromatic hydrocarbons (PAHs) onto phosphorene: An atomistic DFT study. *Journal of Molecular Liquids, 341*, 117465. https://doi.org/10.1016/j.molliq.2021.117465

Deng, S., Ke, T., Wu, Y., Zhang, C., Hu, Z., Yin, H., ... and Zhang, D. (2018). Heavy metal exposure alters the uptake behavior of 16 priority polycyclic aromatic hydrocarbons (PAHs) by Pak Choi (Brassica chinensis L.). *Environmental Science and Technology, 52*(22), 13457–13468.

Derafsh, E., Ghaderpoori, M., Ashrafi, S. D., Yousefi, M., Ghanizadeh, G., Sahebi, S., and Sobhi, H. R. (2022). Polycyclic aromatic hydrocarbons in drinking water: A systematic review and health risk assessment. *Science of the Total Environment, 802*, 149828.

Desalme, D., Binet, P., and Chiapusio, G. (2013). Challenges in tracing the fate and effects of atmospheric polycyclic aromatic hydrocarbon deposition in vascular plants. *Environmental Science and Technology, 47*(9), 3967–3981. https://doi.org/10.1021/es304964b

Dost, K., and İdeli, C. (2012). Determination of polycyclic aromatic hydrocarbons in edible oils and barbecued food by HPLC/UV-Vis detection. *Food Chemistry, 133*(1), 193–199. https://doi.org/10.1016/j.foodchem.2012.01.001

Duedahl-Olesen, L. (2013). Polycyclic aromatic hydrocarbons (PAHs) in foods. In *Persistent Organic Pollutants and Toxic Metals in Foods* (pp. 308–333). https://doi.org/10.1533/9780857098917.2.308

Duedahl-Olesen, L., and Ionas, A. C. (2021). Formation and mitigation of PAHs in barbecued meat - A review. *Critical Reviews in Food Science and Nutrition, 62*(13), 3553–3568. https://doi.org/10.1080/10408398.2020.1867056

Duedahl-Olesen, L., Navaratnam, M. A., Jewula, J., and Jensen, A. H. (2015). Polycyclic aromatic hydrocarbons (PAHs) in some brands of tea and coffee. *Polycyclic Aromatic Compounds, 35*(1), 74–90.

Einolghozati, M., Talebi-Ghane, E., and Amirsadeghi, S. (2022). Evaluation of polycyclic aromatic hydrocarbons (PAHs) in processed cereals: A meta-analysis study, systematic review, and health risk assessment. *Heliyon, 8*(12), e12297. *(Verificar número do artigo para completar caso necessário)*

Eldaly, E. A., Hussein, M. A., El-Gaml, A. A., and Mishref, M. A. (2016). Polycyclic aromatic hydrocarbons (PAHs) in charcoal grilled meat (kebab) and kofta and the effect of marinating on their existence. *Zagazig Veterinary Journal*, 44(1), 40–47. https://doi.org/10.21608/zvjz.2016.7830

Essumang, D. K. (2014). Polycyclic aromatic hydrocarbon (PAH) in cooked (charred) rice: Levels, risk assessment and source characterization. *Journal of Basic and Applied Sciences, 1*(2), 63–83.

European Community Commission. (2006). Commission Regulation (EC) No 1881/2006 of 19 December 2006 setting maximum levels for certain contaminants in foodstuffs. *Official Journal of the European Union*, L364, 5–24.

Ewa, B., and Danuta, M.-Š. (2016). Polycyclic aromatic hydrocarbons and PAH-related DNA adducts. *Journal of Applied Genetics, 58*(3), 321–330. https://doi.org/10.1007/s13353-016-0380-3

Farhadian, A., Jinap, S., Abas, F., and Sakar, Z. I. (2010). Determination of polycyclic aromatic hydrocarbons in grilled meat. *Food Control, 21*(5), 606–610. https://doi.org/10.1016/j.foodcont.2009.09.002

Farhadian, A., Jinap, S., Hanifah, H. N., and Zaidul, I. S. (2011). Effects of meat preheating and wrapping on the levels of polycyclic aromatic hydrocarbons in charcoal-grilled meat. *Food Chemistry, 124*(1), 141–146.

Food and Agriculture Organization (FAO). (2022). *The State of World Fisheries and Aquaculture 2022: Towards Blue Transformation*. Rome: FAO.

Fraqueza, M. J., Laranjo, M., Alves, S., Fernandes, M. H., Agulheiro-Santos, A. C., Fernandes, M. J., ... and Elias, M. (2020). Dry-cured meat products according to the smoking regime: Process optimization to control polycyclic aromatic hydrocarbons. *Foods, 9*(1), 91.

Galinaro, C. A., Mendonça, M. C., Nogueira, F. N., de Souza, R. M., and Valente Soares, L. M. (2007). Determination of polycyclic aromatic hydrocarbons in cachaça by gas chromatography-mass spectrometry. *Ciência e Agrotecnologia, 31*(5), 1452–1458. https://doi.org/10.1590/S1413-70542007000500033

Gao, G., Chen, H., Liu, P., Hao, Z., Ma, G., Chai, Y., Wang, C., and Lu, C. (2017). Residue pattern of polycyclic aromatic hydrocarbons during green tea manufacturing and their transfer rates during tea brewing. *Food Additives and Contaminants: Part A, 34*(6), 990–999. https://doi.org/10.1080/19440049.2017.1316873

Guillén, M. D., Sopelana, P., and Partearroyo, M. A. (1997). Food as a source of polycyclic aromatic carcinogens. *Reviews on Environmental Health*, 12(3), 133–146.

Haritash, A. K., and Kaushik, C. P. (2009). Biodegradation aspects of polycyclic aromatic hydrocarbons (PAHs): A review. *Journal of Hazardous Materials, 169*(1–3), 1–15.

Harner, T., Su, K., Genualdi, S., Karpowicz, J., Jantunen, L., Ahrens, L., and Lane, D. (2018). Review of polycyclic aromatic hydrocarbons (PAHs) in the Athabasca Oil Sands Region. *Environmental Reviews, 26*(4), 456-479. https://doi.org/10.1139/er-2018-0039

Hazra, B., Ghosh, M., and Sil, S. (2010). Role of bioactive phenolic compounds from functional foods in the removal of polycyclic aromatic hydrocarbons and free radicals involved in their formation. *Food and Chemical Toxicology, 48*(5), 1378–1385. https://doi.org/10.1016/j.fct.2010.02.038

Hoed, V., Sampaio, K., Felkner, B., Bavec, F., Scippo, M., Brose, F., ... and Verhé, R. (2017). Tocopherols and polyphenols in pumpkin seed oil are moderately affected by industrially relevant roasting conditions. *European Journal of Lipid Science and Technology, 119*(12). https://doi.org/10.1002/ejlt.201700110

Huang, S., Dai, C., Zhou, Y., Peng, H., Yi, K., Qin, P., ... and Zhang, X. (2018). Comparisons of three plant species in accumulating polycyclic aromatic hydrocarbons (PAHs) from the atmosphere: A review. *Environmental Science and Pollution Research, 25*, 16548–16566.

Hussain, K., Hoque, R. R., Balachandran, S., Medhi, S., Idris, M. G., Rahman, M., and Hussain, F. L. (2018). Monitoring and risk analysis of PAHs in the environment. In C. M. Hussain (Ed.), *Handbook of Environmental Materials Management* (pp. 1–35). Springer.

Iko Afé, O. H., Saegerman, C., Kpoclou, Y. E., Anihouvi, V. B., Douny, C., Igout, A., Mahillon, J., Hounhouigan, D. J., and Scippo, M.-L. (2020). Polycyclic aromatic hydrocarbons contamination of traditionally grilled pork marketed in South Benin and health risk assessment for the Beninese consumer. *Food Additives and Contaminants: Part A, 37*(5), 742–752. https://doi.org/10.1080/19440049.2020.1726502

Iwegbue, C. M. A., Obi, G., Tesi, G. O., Bassey, F. I., Overah, L. C., and Martincigh, B. S. (2015). Concentrations of polycyclic aromatic hydrocarbons in some commercial brands of candies and chocolates in Nigeria. *Quality Assurance and Safety of Crops and Foods, 7*(5), 661–676. https://doi.org/10.3920/qas2014.0466

Iwegbue, C. M. A., Osijaye, K. O., Igbuku, U. A., Egobueze, F. E., Tesi, G. O., Bassey, F. I., and Martincigh, B. S. (2020). Effect of the number of frying cycles on the composition, concentrations and risk of polycyclic aromatic hydrocarbons (PAHs) in vegetable oils and fried fish. *Journal of Food Composition and Analysis, 94*, 103633. https://doi.org/10.1016/j.jfca.2020.103633

Jafarabadi, A. R., Mashjoor, S., Bakhtiari, A. R., and Jadot, C. (2020). Dietary intake of polycyclic aromatic hydrocarbons (PAHs) from coral reef fish in the Persian Gulf—Human health risk assessment. *Food Chemistry, 329*, 127035.

Jia, H., Zhao, S., Shi, Y., Zhu, L., Wang, C., and Sharma, V. K. (2018). Transformation of polycyclic aromatic hydrocarbons and formation of environmentally persistent free radicals on modified montmorillonite: The role of surface metal ions and polycyclic aromatic hydrocarbon molecular properties. *Environmental Science and Technology, 52*(10), 5725–5733.

Jiao, W., Zhu, L., Li, Q. X., Shi, T., Zhang, Z., Wu, X., Yang, T., Hua, R., and Cao, H. (2023). Pyrrolizidine alkaloids in tea (*Camellia sinensis* L.) from weeds through weed-soil-tea transfer and risk assessment of tea intake. *Journal of Agricultural and Food Chemistry, 71*(48), 19045–19053. https://doi.org/10.1021/acs.jafc.3c04339

Kazerouni, N., Sinha, R., Hsu, C., Greenberg, A., and Rothman, N. (2001). Analysis of 200 food items for benzo[a]pyrene and estimation of its intake in an epidemiologic study. *Food and Chemical Toxicology, 39*(5), 423–436. https://doi.org/10.1016/S0278-6915(00)00158-7

Khalil, A., Albachir, M., and Kiš, D. (2016). A comparative analysis of corn drying in Stabil 3000 dry kiln using different energy sources. *Agriculturae Conspectus Scientificus, 74*(3), 205–207.

Kim, H., Cho, J., and Jang, A. (2021). Effect of charcoal type on the formation of polycyclic aromatic hydrocarbons in grilled meats. *Food Chemistry, 343*, 128453. https://doi.org/10.1016/j.foodchem.2020.128453

Kim, K.-H., Jahan, S. A., Kabir, E., and Brown, R. J. C. (2013). A review of airborne polycyclic aromatic hydrocarbons (PAHs) and their human health effects. *Environment International, 60*, 71–80. https://doi.org/10.1016/j.envint.2013.07.019

Kiralan, S. S., Erdogdu, F., and Tekin, A. (2017). Reducing polycyclic aromatic hydrocarbons (PAHs) formation in olive pomace oil using microwave pre-heating of olive pomace. *European Journal of Lipid Science and Technology, 119*(1), 1600241.

Koszucka, A., and Nowak, A. (2019). Thermal processing food-related toxicants: A review. *Critical Reviews in Food Science and Nutrition, 59*(22), 3579–3596.

Kuppusamy, S., Maddela, N. R., Megharaj, M., Venkateswarlu, K., ... and Venkateswarlu, K. (2020). Fate of total petroleum hydrocarbons in the environment. In *Total Petroleum Hydrocarbons: Environmental Fate, Toxicity, and Remediation* (pp. 57–77). Springer.

Lawal, A. T. (2017). Polycyclic aromatic hydrocarbons: A review. *Cogent Environmental Science, 3*(1), 1339841.

Leizando, L. N., and Ashraf, M. W. (2022). Dietary exposure to polycyclic aromatic hydrocarbons and health risk assessment: An emerging global concern. *Food Chemistry: X, 13*, 100232.

Levy, R. B., Barata, M. F., Leite, M. A., and Andrade, G. C. (2024). How and why ultra-processed foods harm human health. *Proceedings of the Nutrition Society, 83*(1), 1–8.

Li, J., Li, F., and Liu, Q. (2017). PAHs behavior in surface water and groundwater of the Yellow River estuary: Evidence from isotopes and hydrochemistry. *Chemosphere, 178*, 143–153.

Li, X., Li, N., Luo, H., Lin, L., Zou, Z., Jia, Y., ... and Li, Y. (2011). A novel synchronous fluorescence spectroscopic approach for the rapid determination of three polycyclic aromatic hydrocarbons in tea with simple microwave-assisted pretreatment of sample. *Journal of Agricultural and Food Chemistry, 59*(11), 5899–5905. https://doi.org/10.1021/jf104873g

Lin, D., and Zhu, L. (2004). Polycyclic aromatic hydrocarbons: Pollution and source analysis of a black tea. *Journal of Agricultural and Food Chemistry, 52*(26), 8268–8271.

Luna, F. M. T., Oliveira Filho, A. N., Araújo, C. C., Azevedo, D. C., and Cavalcante Jr, C. L. (2016). Adsorption of polycyclic aromatic hydrocarbons from heavy naphthenic oil using commercial activated carbons. 1. Fluid-particle studies. *Industrial and Engineering Chemistry Research, 55*(29), 8176–8183.

Luper Tsenum, J., and Yarkwan, B. (2020). Polycyclic aromatic hydrocarbons (PAHs) food contamination: A contributing factor to rising carcinogenicity. *ScienceOpen Preprints. (Verificar se já foi publicado em periódico revisado por pares caso necessário)*

Ma, Y., Liu, Y., and Wang, L. (2019). Exposure to polycyclic aromatic hydrocarbons and risk of lung cancer: A meta-analysis. *Environmental Research*, 170, 33–42. https://doi.org/10.1016/j.envres.2018.12.006

Mahugija, J. A., and Njale, E. (2018). Effects of washing on the polycyclic aromatic hydrocarbons (PAHs) contents in smoked fish. *Food Control, 93*, 139–143. https://doi.org/10.1016/j.foodcont.2018.05.050

Marszałkiewicz, S., Siger, A., Gawrysiak-Witulska, M., Kmiecik, D., and Rudzińska, M. (2020). The effect of drying temperature of milk thistle seeds on quality and bioactive compounds in the lipid fraction. *Journal of Food Science and Technology, 57*, 4003–4013.

Martena, M., Grutters, M., Groot, H., Konings, E., and Rietjens, I. (2011). Monitoring of polycyclic aromatic hydrocarbons (PAH) in food supplements containing botanicals and other ingredients on the Dutch market. *Food Additives and Contaminants: Part A, 28*(7), 925–942. https://doi.org/10.1080/19440049.2011.569573

Martins, S. I., Jongen, W. M., and Van Boekel, M. A. (2000). A review of Maillard reaction in food and implications to kinetic modelling. *Trends in Food Science and Technology, 11*(9–10), 364–373. https://doi.org/10.1016/S0924-2244(01)00022-X

Martorell, I., Perelló, G., Martí-Cid, R., Castell, V., Llobet, J. M., and Domingo, J. L. (2010). Polycyclic aromatic hydrocarbons (PAH) in foods and estimated PAH intake by the population of Catalonia, Spain: Temporal trend. *Environment International, 36*(5), 424–432.

Marwood, C., Solomon, K., and Greenberg, B. (2001). Chlorophyll fluorescence as a bioindicator of effects on growth in aquatic macrophytes from mixtures of polycyclic aromatic hydrocarbons. *Environmental Toxicology and Chemistry, 20*(4), 890–898. https://doi.org/10.1002/etc.5620200425

Mastanjević, K., Kartalović, B., Petrović, J., Novakov, N., Puljić, L., Kovačević, D., ... and Mastanjević, K. (2019). Polycyclic aromatic hydrocarbons in the traditional smoked sausage Slavonska kobasica. *Journal of Food Composition and Analysis, 83*, 103282.

Mocek, K., and Ciemniak, A. (2015). Influence of physical factors on polycyclic aromatic hydrocarbons (PAHs) content in vegetable oils. *Journal of Environmental Science and Health, Part B, 51*(2), 96–102. https://doi.org/10.1080/03601234.2015.1092820

Moon, H. B., Kim, H. S., Choi, M., and Choi, H. G. (2010). Intake and potential health risk of polycyclic aromatic hydrocarbons associated with seafood consumption in Korea from 2005 to 2007. *Archives of Environmental Contamination and Toxicology, 58*, 214–221.

Naccari, C., Cristani, M., Giofrè, F., Ferrante, M., Siracusa, L., and Trombetta, D. (2011). PAHs concentration in heat-treated milk samples. *Food Research International, 44*(3), 716–724. https://doi.org/10.1016/j.foodres.2010.12.029

Naumova, Y. Y., Eisenreich, S. J., Turpin, B. J., Weisel, C. P., Morandi, M. T., Colome, S. D., and Stock, T. H. (2002). Polycyclic aromatic hydrocarbons in the indoor and outdoor air of three cities in the U.S. *Environmental Science & Technology*, *36*(12), 2552–2559.

Neff, J. M., Stout, S. A., and Gunster, D. G. (2005). Ecological risk assessment of polycyclic aromatic hydrocarbons in sediments: Identifying sources and ecological hazard. *Integrated Environmental Assessment and Management, 1*(1), 22–33.

Neves, T. D. M., da Cunha, D. T., de Rosso, V. V., and Domene, S. M. Á. (2021). Effects of seasoning on the formation of heterocyclic amines and polycyclic aromatic hydrocarbons in meats: A meta-analysis. *Comprehensive Reviews in Food Science and Food Safety, 20*(1), 526–541.

Nieva-Cano, M. J., Rubio-Barroso, S., and Santos-Delgado, M. J. (2001). Determination of PAH in food samples by HPLC with fluorimetric detection following sonication extraction without sample clean-up. *The Analyst, 126*(8), 1326–1331. https://doi.org/10.1039/b102546p

Odika, I. M., Okoye, C. B., Nwanisobi, G. C., Odionyenma, O. M., and Okpala, O. (2022). Risk assessment of polycyclic aromatic hydrocarbons in foreign and local rice consumed in South East Nigeria. *(Fonte do periódico não informada — incluir nome do periódico se disponível).*

Onopiuk, A., Kołodziejczak, K., Szpicer, A., Stelmasiak, A., and Poltorak, A. (2021). The effect of partial substitution of beef tallow on selected physicochemical properties, fatty acid profile and PAH content of grilled beef burgers. *Foods, 11*(13), 1986. https://doi.org/10.3390/foods11131986

Onopiuk, A., Kołodziejczak, K., Szpicer, A., Stelmasiak, A., and Poltorak, A. (2022). The effect of partial substitution of beef tallow on selected physicochemical properties, fatty acid profile and PAH content of grilled beef burgers. *Foods, 11*(13), 1986. https://doi.org/10.3390/foods11131986

Onopiuk, A., Kołodziejczak, K., Szpicer, A., Wojtasik-Kalinowska, I., Wierzbicka, A., and Półtorak, A. (2021). Analysis of factors that influence the PAH profile and amount in meat products subjected to thermal processing. *Trends in Food Science & Technology*, 115, 366–379. https://doi.org/10.1016/j.tifs.2021.06.007

Oranuba, E., Deng, H., Peng, J., Dawsey, S. M., and Kamangar, F. (2018). Polycyclic aromatic hydrocarbons as a potential source of carcinogenicity of mate. *Journal of Environmental Science and Health, Part C, 37*(1), 26–41.

Orecchio, S., Ciotti, V. P., and Culotta, L. (2009). Polycyclic aromatic hydrocarbons (PAHs) in coffee brew samples: Analytical method by GC-MS, profile, levels and sources. *Food and Chemical Toxicology, 47*(4), 819–826.

Palade, L. M., Negoiță, M., Adascălului, A. C., and Mihai, A. L. (2023). Polycyclic aromatic hydrocarbon occurrence and formation in processed meat, edible oils, and cereal-derived products: A review. *Applied Sciences, 13*(13), 7877. https://doi.org/10.3390/app13137877

Perelló, G., Martí-Cid, R., Castell, V., Llobet, J. M., and Domingo, J. L. (2009). Concentrations of polybrominated diphenyl ethers, hexachlorobenzene and polycyclic aromatic hydrocarbons in various foodstuffs before and after cooking. *Food and Chemical Toxicology, 47*(4), 709–715. https://doi.org/10.1016/j.fct.2008.12.030

Polak-Śliwińska, M., Paszczyk, B., and Śliwiński, M. (2022). Evaluation of polycyclic aromatic hydrocarbons in smoked cheeses made in Poland by HPLC method. *Molecules*, 27(20), 6909. https://doi.org/10.3390/molecules27206909

Qiu, Y., Qiu, H., Li, J., and Zhang, G. (2018). Bioaccumulation and cycling of polycyclic aromatic hydrocarbons (PAHs) in typical mangrove wetlands of Hainan Island, South China. *Archives of Environmental Contamination and Toxicology, 75*(3), 464–475. https://doi.org/10.1007/s00244-018-0548-4

Quequeto, W. D., Resende, O., Tfouni, S. A. V., Gomes, F. M. L., Borges, A. X., Almeida, A. B. de, Cabral, J. C. O., Costa, L. M., and Oliveira, L. P. de. (2022). Corn grains drying using direct-fired furnace with wood chips: Performance, quality and polycyclic aromatic hydrocarbons. *Food Science and Technology, 42*, e27122. https://doi.org/10.1590/fst.27122

Quequeto, W. D., Resende, O., Tfouni, S. A. V., Leme Gomes, F. M., Borges, A. X., Santos, M. R. B., Costa, E. R., Ferreira Junior, W. N., Glasenapp, M., Quirino, J. R., and Rosa, E. S. (2021). Drying of soybean grains with direct-fired furnace using wood chips: Performance, quality and polycyclic aromatic hydrocarbons. *Drying Technology*, 1–11. https://doi.org/10.1080/07373937.2021.1929293

Rascón, A. J., Azzouz, A., and Ballesteros, E. (2019). Use of semi-automated continuous solid-phase extraction and gas chromatography–mass spectrometry for the determination of polycyclic aromatic hydrocarbons in alcoholic and non-alcoholic drinks from Andalucía (Spain). *Journal of the Science of Food and Agriculture*, 99(3), 1117–1125. https://doi.org/10.1002/jsfa.9279

Ravindra, K., Sokhi, R., and Van Grieken, R. (2008). Atmospheric polycyclic aromatic hydrocarbons: Source attribution, emission factors and regulation. *Atmospheric Environment, 42*(13), 2895–2921.

Rengarajan, T., Rajendran, P., Nandakumar, N., Lokeshkumar, B., Rajendran, P., and Nishigaki, I. (2015). Exposure to polycyclic aromatic hydrocarbons with special focus on cancer. *Asian Pacific Journal of Tropical Biomedicine, 5*(3), 182–189.

Rey-Salgueiro, L., García-Falcón, M. S., Martínez-Carballo, E., and Simal-Gándara, J. (2008). Effects of toasting procedures on the levels of polycyclic aromatic hydrocarbons in toasted bread. *Food Chemistry, 108*(2), 607–615.

Rey-Salgueiro, L., Martínez-Carballo, E., García-Falcón, M. S., and Simal-Gándara, J. (2008). Effects of a chemical company fire on the occurrence of polycyclic aromatic hydrocarbons in plant foods. *Food Chemistry, 108*(1), 347–353.

Rosińska, A., and Dąbrowska, D. (2018). Polycyclic aromatic hydrocarbons in the environment: Their occurrence and health risk. *Environmental Monitoring and Assessment, 190*, 253. https://doi.org/10.1007/s10661-018-6590-1

Sadowska-Rociek, A., Surma, M., and Cieślik, E. (2013). Comparison of different modifications on QuEChERS sample preparation method for PAHs determination in black, green, red and white tea. *Environmental Science and Pollution Research, 21*(2), 1326–1338. https://doi.org/10.1007/s11356-013-2022-1

Samanta, S. K., Singh, O. V., and Jain, R. K. (2002). Polycyclic aromatic hydrocarbons: Environmental pollution and bioremediation. *Trends in Biotechnology, 20*(6), 243–248.

Sampaio, G. R., Guizellini, G. M., da Silva, S. A., de Almeida, A. P., Pinaffi-Langley, A. C. C., Rogero, M. M., ... and Torres, E. A. (2021). Polycyclic aromatic hydrocarbons in foods: Biological effects, legislation, occurrence, analytical methods, and strategies to reduce their formation. *International Journal of Molecular Sciences, 22*(11), 6010.

Savitri, L. A., Hermayanti, Y., and Isnansetyo, A. (2022). Carrageenan-based edible film as food packaging: Potential to improve quality and prevent contamination. *International Journal of Food Science, 2022*, 1–10. https://doi.org/10.1155/2022/9465073

Scientific Committee on Food (SCF). (2002). *Opinion of the Scientific Committee on Food on the risks to human health of polycyclic aromatic hydrocarbons in food.* European Commission, Health & Consumer Protection Directorate-General. https://ec.europa.eu/food/fs/sc/scf/out153_en.pdf

Semanová, J., Skláršová, B., Šimon, P., and Šimko, P. (2016). Elimination of polycyclic aromatic hydrocarbons from smoked sausages by migration into polyethylene packaging. *Food Chemistry, 201*, 1–6.

Seo, S., Kwon, H., Park, M., Lee, I., and Choi, S. (2020). Contamination characteristics of polycyclic aromatic hydrocarbons in river and coastal sediments collected from the multi-industrial city of Ulsan, South Korea. *Marine Pollution Bulletin, 160*, 111666.

Sheehan, J., O'Connor, T., Sheehy, P., Buckley, D., and FitzGerald, R. (1998). Stability of astaxanthin and canthaxanthin in raw and smoked salmon during frozen storage. *Journal of Food Science, 63*(5), 730–734.

Shin, B. R., Yang, S. O., Song, H. W., Chung, M. S., and Kim, Y. S. (2015). Effects of adsorbents on benzo[a]pyrene, sesamol, and sesamolin contents and volatile component profiles in sesame oil. *Food Science and Biotechnology, 24*, 2017–2022.

Shitandayi, E., Serem, J. C., Ngule, C. M., and Mugera, G. W. (2019). Assessment of polycyclic aromatic hydrocarbons (PAHs) in sugarcane juice from burnt and unburnt canes in Kenya. *African Journal of Food Science, 13*(6), 119–126. https://doi.org/10.5897/AJFS2019.1791

Shiue, I. (2016). Urinary polycyclic aromatic hydrocarbons are associated with adult celiac disease and kidney stones in the USA, 2003–2006. *Environmental Science and Pollution Research, 23*(2), 1627–1631.

Siddique, R., Zahoor, A. F., Ahmad, H., Zahid, F. M., and Karrar, E. (2021). Impact of different cooking methods on polycyclic aromatic hydrocarbons in rabbit meat. *Food Science and Nutrition, 9*(6), 3219-3227. https://doi.org/10.1002/fsn3.2284

Sigurgisladottir, S., Torrissen, O., and Vallet, J. (2000). Relation of smoking parameters to the yield, colour and sensory quality of smoked Atlantic salmon (*Salmo salar*). *Food Research International, 33*(6), 847–855.

Simko, P. (2002). Determination of polycyclic aromatic hydrocarbons in smoked meat products and smoke flavouring food additives. *Journal of Chromatography B, 770*(1–2), 3–18.

Simko, P., and Brunckova, B. (1993). Lowering of polycyclic aromatic hydrocarbons concentration in a liquid smoke flavour by sorption into polyethylene packaging. *Food Additives and Contaminants, 10*(2), 257–263.

Singh, L., Agarwal, T., and Simal-Gandara, J. (2022). Summarizing minimization of polycyclic aromatic hydrocarbons in thermally processed foods by different strategies. *Food Control*, 138, 109017. https://doi.org/10.1016/j.foodcont.2022.109017

Singh, L., Agarwal, T., and Simal-Gandara, J. (2023). Summarizing minimization of polycyclic aromatic hydrocarbons in thermally processed foods by different strategies. *Food Control, 146*, 109514.

Singh, L., Varshney, J., and Agarwal, T. (2016). Polycyclic aromatic hydrocarbons' formation and occurrence in processed food. *Food Chemistry, 199*, 768–781. https://doi.org/10.1016/j.foodchem.2015.12.074

Song, H., Wang, Y., Sun, C., Wang, Y., Yang, P., and Cheng, H. (2012). Effects of pyrene on antioxidant systems and lipid peroxidation level in mangrove plants, *Bruguiera gymnorrhiza. Ecotoxicology, 21*(6), 1625–1632. https://doi.org/10.1007/s10646-012-0945-9

Stading, R., Gastelum, G., Chu, C., Jiang, W., and Moorthy, B. (2021). Molecular mechanisms of pulmonary carcinogenesis by polycyclic aromatic hydrocarbons (PAHs): Implications for human lung cancer. *Seminars in Cancer Biology*, 76, 3–16. https://doi.org/10.1016/j.semcancer.2021.07.001

Stołyhwo, A., and Sikorski, Z. E. (2005). Polycyclic aromatic hydrocarbons in smoked fish - A critical review. *Food Chemistry, 91*(2), 303–311.

Stowers, J., and Anderson, M. (1985). Formation and persistence of benzo[a]pyrene metabolite-DNA adducts. *Environmental Health Perspectives, 62*, 31–39. https://doi.org/10.1289/EHP.856231

Sun, Y., Wu, S., and Gong, G. (2019). Trends of research on polycyclic aromatic hydrocarbons in food: A 20-year perspective from 1997 to 2017. *Trends in Food Science and Technology, 83*, 86–98.

Taqvi, S. I. H. (2018). Potential risk assessment of selected polycyclic aromatic hydrocarbons in vegetables grown in polluted and non-polluted areas of Sindh, Pakistan. *Pakistan Journal of Analytical and Environmental Chemistry, 19*(2), 128–134.

Tarigholizadeh, S., Sushkova, S., Rajput, V. D., Ranjan, A., Arora, J., Dudnikova, T., ... and Wong, M. H. (2023). Transfer and degradation of PAHs in the soil-plant system: A review. *Journal of Agricultural and Food Chemistry, 72*(1), 46–64.

Upham, B. L., and Trosko, J. E. (2009). Oxidative-dependent integration of signal transduction with intercellular gap junctional communication in the control of gene expression. *Antioxidants & Redox Signaling*, 11(2), 297–307. https://doi.org/10.1089/ars.2008.2146

Uzomah, A., Lundebye, A. K., Kjellevold, M., Chuku, F. A., and Stephen, O. A. (2021). A review of chemical contaminants in marine and freshwater fish in Nigeria. *Foods, 10*(9), 2013.

Vanderschueren, R., Doevenspeck, J., Goethals, L., Andjelkovic, M., Waegeneers, N., and Smolders, E. (2023). The contribution of cacao consumption to the bioaccessible dietary cadmium exposure in the Belgian population. *Food and Chemical Toxicology*, 181, 113599. https://doi.org/10.1016/j.fct.2023.113599

Verma, N., Bharti, A., and Jaiswal, R. K. (2019). Impact of sugarcane field burning on sugar quality and polycyclic aromatic hydrocarbon contamination. *Sugar Tech, 21*(5), 783–789. https://doi.org/10.1007/s12355-019-00721-7

Wang, C., Xie, Y., Wang, H., Bai, Y., Dai, C., Li, C., Xu, X., and Zhou, G. (2019a). Phenolic compounds in beer inhibit formation of polycyclic aromatic hydrocarbons from charcoal-grilled chicken wings. *Food Chemistry, 294*, 578–586. https://doi.org/10.1016/j.foodchem.2019.05.094

Wang, W., Wang, P., Wang, S., Duan, X., Wang, T., Feng, X., Li, L., Zhang, Y., Li, G., Zhao, J., Li, L., Wang, Y., Yan, Z., Feng, F., Zhou, X., Yao, W., Zhang, Y., and Yang, Y. (2019b). Benchmark dose assessment for coke oven emissions-induced telomere length effects in occupationally exposed workers in China. *Ecotoxicology and Environmental Safety, 182*, 109453. https://doi.org/10.1016/j.ecoenv.2019.109453

Wang, Z., Ng, K., Warner, R. D., Stockmann, R., and Fang, Z. (2022). Reduction strategies for polycyclic aromatic hydrocarbons in processed foods. *Comprehensive Reviews in Food Science and Food Safety, 21*(2), 1598–1626. https://doi.org/10.1111/1541-4337.12905

Wen, Y.-Q., Liu, Y.-L., Xu, L.-L., Yu, W., and Ma, Y. (2017). Occurrence of polycyclic aromatic hydrocarbons in various types of raw oilseeds from different regions of China. *Food Additives and Contaminants: Part B - Surveillance*, 1–9. https://doi.org/10.1080/19393210.2017.1345993

Wennrich, L., Popp, P., and Zeibig, M. (2002). Polycyclic aromatic hydrocarbon burden in fruit and vegetable species cultivated in allotments in an industrial area. International *Journal of Environmental Analytical Chemistry,* 82(9), 667–690. https://doi.org/10.1080/0306731021000075401

World Health Organization (WHO). (1998). *Environmental health criteria 202: Polycyclic aromatic hydrocarbons (PAHs).* Geneva: World Health Organization.

Wu, C., Zhu, H., Luo, Y., and Wang, J. (2016). Concentrations and potential health hazards of polycyclic aromatic hydrocarbons in shallow groundwater of a metal smelting area in Southeastern China. *Science of the Total Environment, 569*, 1561–1569.

Yang, K., Zhu, L., and Xing, B. (2006). Adsorption of polycyclic aromatic hydrocarbons by carbon nanomaterials. *Environmental Science and Technology, 40*(6), 1855–1861.

Yebra-Pimentel, I., Fernández-González, R., Martínez-Carballo, E., and Simal-Gándara, J. (2013). Optimization of purification processes to remove polycyclic aromatic hydrocarbons (PAHs) in polluted raw fish oils. *Science of the Total Environment, 470*, 917-924.

Yebra-Pimentel, I., Martínez-Carballo, E., Regueiro, J., and Simal-Gándara, J. (2014). Adsorption of polycyclic aromatic hydrocarbons onto activated carbon: Effect of physical and chemical properties. *Environmental Science and Pollution Research, 21*(14), 8365–8373.

Yi, J. Y., Kim, H. J., and Chung, M. S. (2017). Manufacture of low-benzo[a]pyrene sesame seed (*Sesamum indicum* L.) oil using a self-designed apparatus. *PLOS ONE, 12*(3), e0173585.

Yousefi, M., Shariatifar, N., Ebrahimi, M., Mortazavian, A. M., Mohammadi, A., Khorshidian, N., … and Hosseini, H. (2019). *In vitro* removal of polycyclic aromatic hydrocarbons by lactic acid bacteria. *Journal of Applied Microbiology, 126*(3), 954–964. https://doi.org/10.1111/jam.14163

Yu, Y., Li, Q., Wang, H., Wang, B., Wang, X., Ren, A., and Tao, S. (2015). Risk of human exposure to polycyclic aromatic hydrocarbons: A case study in Beijing, China. *Environmental Pollution, 205*, 70–77. https://doi.org/10.1016/j.envpol.2015.05.022

Zhang, H., Xue, M., and Dai, Z. (2010). Determination of polycyclic aromatic hydrocarbons in aquatic products by HPLC-fluorescence. *Journal of Food Composition and Analysis, 23*(5), 469–474. https://doi.org/10.1016/j.jfca.2009.12.016

Zhang, Y., Chen, X., and Zhang, Y. (2021). Analytical chemistry, formation, mitigation, and risk assessment of polycyclic aromatic hydrocarbons: From food processing to *in vivo* metabolic transformation. *Comprehensive Reviews in Food Science and Food Safety, 20*(2), 1422–1456. https://doi.org/10.1111/1541-4337.12705

Zhao, L., Deng, J., Hou, H., Li, J., and Yang, Y. (2019). Investigation of PAH and oil degradation along with electricity generation in soil using an enhanced plant-microbial fuel cell. *Journal of Cleaner Production, 221*, 678–683.

Zingale, S., Spina, A., Ingrao, C., Fallico, B., Timpanaro, G., Anastasi, U., and Guarnaccia, P. (2023). Factors affecting the nutritional, health, and technological quality of durum wheat for pasta-making: A systematic literature review. *Plants, 12*(3), 530.

Zohair, A., Salim, A. B., Soyibo, A. A., and Beck, A. J. (2006). Residues of polycyclic aromatic hydrocarbons (PAHs), polychlorinated biphenyls (PCBs) and organochlorine pesticides in organically-farmed vegetables. *Chemosphere, 63*(4), 541–553.

Żyżelewicz, D., Oracz, J., Krysiak, W., Budryn, G., and Nebesny, E. (2016). Effects of various roasting conditions on acrylamide, acrolein, and polycyclic aromatic hydrocarbons content in cocoa bean and the derived chocolates. *Drying Technology, 35*(3), 363–374. https://doi.org/10.1080/07373937.2016.1175470

7 Heterocyclic Amines and Safety

Food Processing and Manufacturing Related Effects

Yong Li and Maomao Zeng

7.1 INTRODUCTION

Heterocyclic amines (HAs) are a broad class of dangers formed by protein-rich dietary raw materials (such as meat and fish) during thermal processing. Numerous studies have revealed that these chemicals can covalently bond to DNA to induce carcinogenic and mutagenic damage (Pezdirc, Zegura, & Filipic, 2013; Turesky et al., 2015). The HAs are absorbed in the small intestine and enter the liver, where they undergo activation through N-oxidation by cytochrome enzymes. Subsequently, it was activated to form N-acetoxyamine, which can potentially induce DNA damage, including chain hydrogen bond breaking, mutations in the site, and deletions and insertions in the DNA (Turesky et al., 2015). Multiple studies have established a strong correlation between consuming red meat and heat-processed meat and an elevated risk of developing cancer, namely colon cancer (Gibis, 2016). Human case-control studies have demonstrated a significant association between the occurrence of colon cancer and the high consumption of 2-amino-3,8-dimethylimidazo [4,5-f]quinoxaline (MeIQx), 2-amino-1-methyl-6-phenyl-imidazo[4,5-b]pyridine (PhIP), and DiMeIQx (Barbir et al., 2012). In 1993, the International Agency for Research on Cancer categorized 2-amino-3-methylimidazo[4,5-f] quinoline (IQ) as a class 2A carcinogen, and 2-amino-3,4-dimethylimidazo[4,5-f]

DOI: 10.1201/9781032712291-7

quinolone (MeIQ), 4,8-MeIQx, PhIP, AαC, MeAαC, Trp-P-1, Trp-P-2, and Glu-P-1 as class 2B carcinogens (IARC, 1993). Epidemiological studies have demonstrated that heterocyclic amine intake elevates the risk of esophagus, stomach, colon, and pancreas (Augustsson, Skog, Jägerstad, Dickman, & Steineck, 1999) (Gibis, 2016). The study discovered that MeIQx was the only HAs linked to the development of lung cancer. Additionally, it was shown that the risk of colon adenoma rose by 5% for every 10 ng increase in MeIQx intake (Sinha, Kulldorff, Chow, Denobile, & Rothman, 2001). Moreover, experimental trials have verified the carcinogenic nature of HAs. Consequently, considerable attention has been devoted to the formation, influencing elements, and reduction and management strategies for HAs in thermal-processed foods.

The precursors of HAs are creatine (creatinine), hexose, amino acids, and some dipeptides compounds in muscle tissue. Elucidating the formation pathways of HAs and the influencing factors in processing are conducive to controlling the formation of HAs in food. However, a wide range of HAs exists, and their production process is quite intricate, and identical HAs can be generated via distinct pathways. For instance, IQ, MeIQ, MeIQx, and 4,8-DiMeIQx are formed via free radical and Maillard reactions. PhIP, however, can arise from phenylacetaldehyde, a pyrolytic product of phenylalanine, via complex reactions with creatinine, or from the reaction of leucine, isoleucine, or tyrosine with creatinine (Meurillon & Engel, 2016). Bound-HAs have been found in processed foods. They form when free HAs combine with proteins or meat proteins directly react with other precursors (J. Chen, He, Qin, Chen, & Zeng, 2017; Q. C. Chen, Xue, He, et al., 2021).

Heat treatment is an essential technique in meat processing and a sterilizing procedure. The thermal processing methods often used for meat products include baking, grilling, frying, smoking, low-temperature vacuum cooking, etc., which involve high temperatures, promoting the synthesis of HAs (Shabbir, Raza, Anjum, Khan, & Suleria, 2015). Therefore, thermal processing methods significantly impact the content and types of HAs present in food products. Multiple factors influence HA levels in heat-treated meat and meat products, also including meat type, cooking conditions, pH, water content, carbohydrates, amino acids, creatinine, heat and mass transfer, lipid oxidation, and antioxidants. Some additives inhibit HA formation, but their mechanism of action is poorly understood. Their inhibition is mainly based on targeting specific generation pathways, such as targeting free radicals, and targeting phenylacetaldehyde intermediates. However, these pathways are not the only ones, and the same additive may have multiple inhibition mechanisms. Understanding the formation, chemistry, and occurrence of HAs in food is essential for developing practical risk assessment and management strategies to protect human health. This chapter introduces the types of HAs formed in food processing, the formation mechanism, the effects of processing, and mitigation strategies.

7.2 CHEMICAL STRUCTURES AND CLASSIFICATION OF HAS

Currently, close to 30 different types of HAs have been discovered in heat-processed foods and model systems. These can be categorized into aminoimidazoazarenes (AIAs) and amino-carbolines (ACs) based on their chemical structures, as illustrated in Figure 7.1. The former can be classified into imidazoquinolines, imidazoquinoxalines, and imidazopyridine HAs. Pyridines include PhIP, 2-amino-1,6-dimethylimidazo[4,5-b]pyridine (DMIP), etc.; imidazoquinolines include IQ, MeIQ, IQ[4,5-b], etc.; and imidazoquinoxalines include MeIQx, 2-amino-3,4,8-trimethylimidazo[4,5-f]quinoxaline (4,8-DiMeIQx), 2-amino-3,7,8-trimethylimidazo[4,5-f]quinoxaline (7,8-TriMeIQx), etc.; ACs include α-carbolines (2-amino-9H-pyrido[2,3-b]indole (AαC), 2-amino-3-methyl-9H-pyrido[2,3-b] indole (MeAαC)), β-carbolines (1-methyl-9H-pyrido[4,3-b]indole (Harman), 9H-pyrido[4,3-b] indole(Norharman)) and γ-carbolines (3-amino-1,4-b-indole(4,3-b)indole (3-amino-1, 4- dimethyl-5H-pyrido[4,3-b]indole (Trp-P-1), 3-amino-1-methyl-5H-pyridino[4,3-b] indole (Trp-P-2), 2-amino-6-methyldipyridino[1,2-a:3',2'-d]imidazole (2- amino-6-methyldipyrido[1,2-a:3',2'-d] imidazole (Glu-P-1), etc.). AIAs or IQ-type Has, which involves the reaction of amino acids and hexose at temperatures ranging from 100°C to 300°C; whereas, ACs or non IQ-type HAs are formed by pyrolytic reactions, which refers to the degradation of proteins or amino acids at temperatures above 300°C. AIAs, compared to ACs, have much higher mutagenicity and are more prone to being generated during routine cooking processes, garnering more attention.

7.3 THE FORMATION PATHWAYS OF HAS

7.3.1 Imidazoquino(x)alines

The imidazoquino(x)alines such as IQ、MeIQ、MeIQx、4,8-DiMeIQx, and 7,8-DiMeIQx share similar chemical compositions, each containing an imidazole ring along with a pyrazine or pyridine ring. These compounds are formed through analogous reaction pathways, quickly produced at regular cooking temperatures (e.g., frying, grilling, and roasting) below 300°C. There are two primary pathways through which imidazoquino(x)alines compounds are commonly formed, namely the Maillard pathway and the free radical pathway (Zamora & Hidalgo, 2020). (1) the Maillard reaction pathway: Reducing sugar (e.g., hexose) degrades to form dicarbonyl compounds, which and amino acids form pyridines and pyrazines via Strecker degradation reactions in the Maillard reaction as building blocks for quinolines and quinoxalines (Figure 7.2 A). When heated, creatine undergoes cyclization to form creatinine, which then reacts with vinyl-pyridine or vinyl-pyrazine generated in the Strecker degradation, and through a series of cyclization, dehydration, and desaturation steps to generate imidazoquinoline and imidazoquinoxaline, respectively. Aldehydes and creat(in)ine are crucial in synthesizing the imidazole ring found in IQ-type HAs (E. Oz et al., 2023). Additionally, aldehydes and creatinine can be directly combined and connected to Schiff's base. Reactive carbonyls, generated from carbohydrates, lipids, and amino acids, are essential in these processes. These carbonyls can proceed with cyclization processes to generate different heterocyclic structures, which in turn helps to promote the production of HAs (Hidalgo & Zamora, 2022; Zamora, Lavado-Tena, & Hidalgo, 2020a, 2020b). IQx, MeIQx, and 4,8-DiMeIQx formation process have been verified using 14C-labeled glucose in model systems (Skog & Jagerstad, 1993). (2) the free radical pathway: Alkylpyridine radicals and dialkylpyrazine radicals are involved,

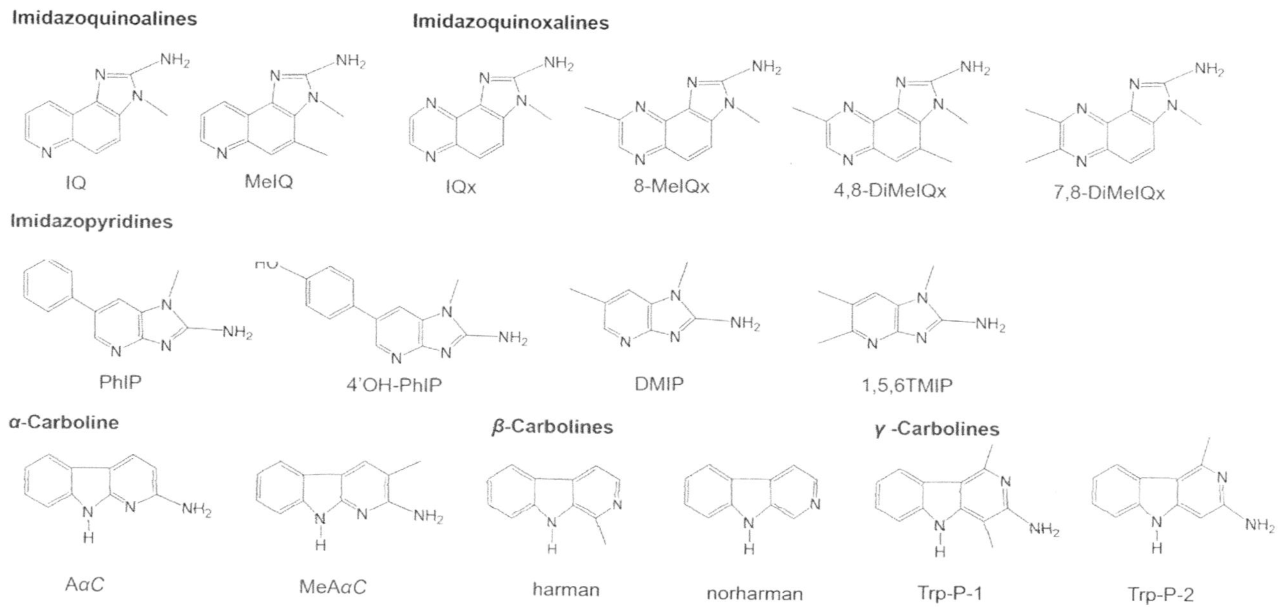

FIGURE 7.1 The chemical structure and abbreviation of HAs commonly found in food.

but this mechanism is still controversial (Pearson, Chen, Gray, & Aust, 1992; Zamora & Hidalgo, 2020). As shown in Figure 7.2 B, sugar reacts with amino groups to produce ethanolaldehyde alkyl amines, which subsequently react with active carbonyl groups to form glyoxal-monoalkyl amines and other 2-C fragments, respectively. The former undergoes a sequence of reactions to produce alkyl pyridine radicals, which combine with creatinine to produce quinoline. The latter generates dialkyl pyrazine radicals, which react with creatinine to produce quinoxaline (Meurillon & Engel, 2016).

7.3.2 Imidazopyridine (PhIP)

Phenylalanine and creatinine serve as precursors in the synthesis of PhIP. The experiments conducted on the 13C-labeled phenylalanine model confirmed that the C-3, amino nitrogen, and benzene ring of phenylalanine were fully transferred to the PhIP molecule (Murkovic, Weber, Geiszler, Fröhlich, & Pfannhauser, 1999). The methyl carbon, 1-N, and amino nitrogen from creatinine are incorporated into the PhIP molecule. Therefore, the carbon atom in phenylalanine constitutes the pyridine component of PhIP, whereas creatinine contributes to the formation of imidazole. Phenylacetaldehyde is a crucial active carbonyl group involved in PhIP production (Gibis, 2016). The specific reaction pathway is illustrated in Figure 7.2 C. Initially, phenylalanine undergoes degradation through the Strecker reaction, forming an active carbonyl group known as phenylacetaldehyde. Subsequently, the reaction between phenylacetaldehyde and creatinine resulted in a carbonyl-creatinine adduct. Specifically, the C-5 of creatinine and phenylacetaldehyde underwent a nucleophilic addition to produce 2-amino-1-methyl-5-(1-hydroxy-2-phenyletyl)-imidazol-4-one, which was then followed by dehydration and formed 2-amino-1-methyl-5-(2-propenylphenyl)-imidazol-4-one. Finally, the imidazole rings of PhIP were formed through the participation of ammonia and formaldehyde (Zamora, Alcón, & Hidalgo, 2013, 2014; Zamora & Hidalgo, 2015).

7.3.3 Carbolines

Limited research has been conducted on the formation pathway of amino-carbolines. β-carbolines, including Norharman and Harman, can be synthesized at lower temperatures (Figure 7.2 D illustrates the formation of Norharman) (Geng et al., 2023; Gibis, 2016). The initial step involves the conversion of tryptophan into tryptophan-Amadori rearrangement products by the Maillard reaction in the presence of glucose. Tetrahydroβ-carboline is then formed by cyclization (Pictet-Spengler reaction) with active carbonyl compounds, followed by oxidation to form β-carboline (Geng et al., 2023). The tryptophan-Amadori rearrangement products can undergo a dehydration elimination reaction to produce conjugated oxygen ions. These intermediates then participate in intermolecular substitution events, subsequently intermolecularly substituting to form β-carbolines. Norharman and Harman lacked free amino groups outside the ring and did not show mutagenicity in Ames tests. Compared to other carbolines (pyrolytic HAs), the Norharman and Harman formation occurred at lower temperatures, primarily in meat and processed meat products (Gibis, 2016).

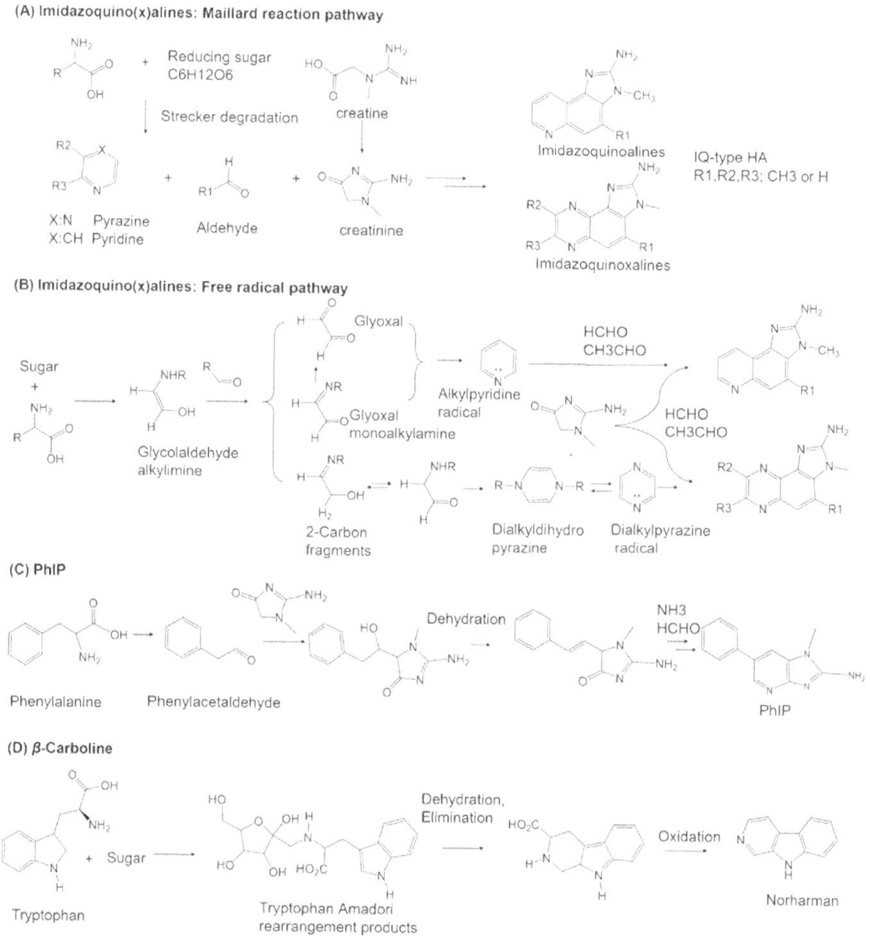

FIGURE 7.2 The formation pathways of imidazoquino(xa)lines(A, B), imidazopyridine (C), and carbolines (D) (Hidalgo & Zamora, 2022; Pearson et al., 1992; Zamora & Hidalgo, 2020).

7.4 ENDOGENOUS FACTORS OF PROCESSED FOOD

7.4.1 The Animal Species

Meats and meat products contain varying levels of HAs at nanogram per gram. However, the types and contents of HAs in meat products under different thermal processing conditions differ significantly. The presence and amount of specific HAs, such as PhIP, MeIQx, 4,8-DiMeIQx, MeIQ, harman, and norharman commonly detected in meat products, differed widely based on types of meat, such as fish, poultry, pork, and beef and the cooking conditions. For instance, studies have found that PhIP concentrations in chicken breasts barbecued at 177–260°C reached 480 ng/g (Sinha et al., 1995), grilled chicken breast had 270 ng/g of PhIP, while 4,8-DiMeIQx generally under one ng/g, and 7,8-DiMeIQx under 0.1 ng/g. In some instances, IQ levels were not detectable. PhIP is more likely to be generated in chicken than in beef, pork, or fish when subjected to cooking. Conversely, the concentration of MeIQx is often lower in cooked chicken than in cooked beef and pork (E. Oz et al., 2023). MeIQx and PhIP are relatively abundant HAs found in meat products. Pure meat generally has higher HAs than combined meat products like meatballs and sausage. The soup contains negligible HAs that cannot even be detected. The concentration of total HAs in pan-fried duck breast was measured to be 53.3 ng/g, which was about twice as high as the concentration found in pan-fried chicken breast (27.4 ng/g) (Liao, Wang, Xu, & Zhou, 2010). The concentration of HAs in dried products from Chinese yellow cattle exceeded that of other chosen breeds, including Aberdeen Angus cattle and Chinese yak. The Sichuan beef jerky contained more HAs than the beef jerky from Inner Mongolia (Hou et al., 2018). Only 7,8-DiMeIQx was detected in beef bouillons, while 7,8-DiMeIQx, and IQx were detected in chicken bouillons. The low quantities of HA in bouillons might be linked to meat (precursors for HA formation) and the manufacturing temperature (Elbir & Oz, 2021).

Bound-HAs, namely Trp-P-2, MeIQx, Harman, AαC, Norharman, and so on, were discovered in the dry heated SPI (90–220°C). When heated for 10 min, the rapid rise inflection points are at 150°C (Trp-P-2, MeIQx, Harman, Total HAs), 170°C (AαC, Norharman), and 210°C (DMIP, IQ), respectively (Xi & Chen, 2021). Deng et al. (Deng et al., 2022) examined the HA content of commercially available raw materials to the final products of plant-based hamburger meats. They revealed that the total free HA levels ranged from 4.74 to 6.63 ng/g in raw plant proteins, while textured vegetable proteins exhibited a higher range of 5.81 to 20.23 ng/g following extrusion.

It is vital to recognize the complexities of the relationship between the type and content of HAs and animal species. Although many studies have investigated the effect of meat on the content of HAs in various samples with different processing methods and cooking doneness, it is challenging to draw definitive conclusions about the relationship between HAs and meat type based solely on the existing literature. This is mainly due to differences in processing conditions between studies and the complexity of other influencing factors, such as differences in cooking methods and other ingredients (marinades, spices, fats, HAs precursors, etc.). Therefore, to further clarify the relationship between meat type and the formation of HAs, standard procedures for cooking meat (such as meat cutting and cooking techniques) must be taken into account (E. Oz et al., 2023). This will facilitate a more comprehensive comprehension of the correlation between the formation of HAs and the type of meat.

7.4.2 Categories of Commercial Products

The composition and concentrations of HAs varied among different meat products. Consumers are greatly concerned about the levels of HAs in meat products sold commercially. Assessing the risk of HAs from the perspective of human cancer development requires determining the level of exposure to a specific amount of HAs in food. Consequently, research has investigated the HA levels across various commercially consumed meat products, such as chicken breast, sausages, deep-fried meat, chargrilled chicken, bacon, burgers, hot dogs, crispy chicken drumsticks, traditional Chinese nuggets, and ready-to-eat (RTE) meat, etc. However, HA content varies significantly among different types of commercial meat products. Harman and norharman were the predominant HAs found at levels ranging from 0.052 to 34.98 ng/g in various commercially available meat samples, including deep-fried chicken breast, deep-fried pork ribs, smoked chicken, pork sausages, charcoal-grilled Pacific sauries, and marinated pork (Chang et al., 2018). (Khan et al., 2019) investigated the content of HA in some fast-food meat products available in the Chinese market. The total HA levels varied among different types of products. Specifically, crispy chicken drumsticks had the highest levels at 24.18 ng/g, followed by crispy fried chicken burgers at 19.99 ng/g, traditional Chinese nuggets at 19.17 ng/g, and shrimp cake burgers at 13.17 ng/g (Khan et al., 2019). Upon analysis of various ready-to-eat (TRE) meat products available in the UK market, it was observed that chargrilled chicken breast slices exhibited the highest concentration of IQ (up to 22.68 ng/g). Conversely, roasted bacon, ham, meatballs, and sausage did not yield any detectable levels of IQ. However, MeIQ, MeIQx, and 4,8-DiMeIQx were present in most of the RTE meat products. The most abundant HA in these RTE samples were 4,8-DiMeIQx (nd–12.61 ng/g) and IQ (Lu, Kuhnle, & Cheng, 2017). In another study conducted by Puangsombat et al. (Puangsombat, Gadgil, Houser, Hunt, & Smith, 2011), they found that PhIP (nd–7.9 ng/g) and MeIQx (nd–3.6 ng/g) were the major HAs in RTE products available in the United States market. The HAs detected in the selected RTE products ranged from 0.05 to 13.07 ng/g. The products can be ranked according to their HA content in the following order: pepperoni (0.05 ng/g), hot dogs and deli meats (0.5 ng/g) < fully cooked bacon (1.1 ng/g) < rotisserie chicken flesh (1.9 ng/g) < rotisserie chicken skin (163 ng/g) (Puangsombat, Gadgil, et al., 2011). As for RTE meat products, it was observed that pork and bacon items contained relatively low levels of HAs, which may be attributed to their specific processing methods. Commercially available ham and smoked ham commonly undergo cooking through immersion in hot water (80–90°C) or steam until reaching an internal temperature of 72°C. This method results in high moisture content and lower amounts of HAs. They were smoking at temperatures ranging from 15–85°C, resulting in the generation of a relatively low amount of HA in smoked ham and crispy bacon.

Additional representative meat products from China, such as cured meat, sauced meat, roasted meat, dried meat, fried meat, sausage, and ham, were also analyzed (H. Pan, Wang, Guo, Ni, & Zhang, 2014). Dry meat and sauced meat displayed higher levels of total HAs at 36.60 ng/g and 32.15 ng/g, respectively, followed by cured (18.93 ng/g) and roasted meats (18.10 ng/g). In contrast, fried meat exhibited the lowest concentration of HAs at 10.26 ng/g. Harman and norharman in samples at levels of 1.09–63.97 and 1.19–62.30 ng/g, respectively, while concentrations of MeIQx, PhIP, 4,8-DiMeIQx, and AαC were lower than 10 ng/g (H. Pan et al., 2014). The concentration of total HAs in the commercial braised chicken ranged from 6.20 to 27.7 ng/g. IQ was detected in all samples at levels ranging from 2.72 to 13.5 ng/g), while PhIP was present at lower levels (0.14–0.41 ng/g). Norharman and Harman were identified as the sample's main carboline components (Y. Q. Cheng et al., 2019). Haskaraca et al. (Haskaraca, Soncu, Kolsarici, Öz, & Juneja, 2017) discovered that the total HAs in chicken nuggets (CNs) and chicken burgers (CBs) from fast-food restaurants ranged from nd to 22.01 ng/g and not quantified to 9.72 ng/g, respectively. Commercial ultra-high temperature processed (UHT) milk has been found to contain HAs, such as IQx, IQ, MeIQx, MeIQ, and PhIP (F. Oz et al., 2021).

7.4.3 Moisture Content of Products

The proximate composition of meat, including its moisture content, has been found to impact the formation of HAs significantly. Water is a carrier for chemical reactions and heat transportation while cooking meat and meat products (Gibis, 2016). Water facilitates the transportation of precursor components, such as creatine/creatinine, amino acids, glucose, and others to the product's surface. These substances are exposed to elevated temperatures at the surface, thereby facilitating the formation of HAs. Meat products with higher moisture content seemed to have lower levels of HAs. The presence of excessive moisture in the product may result in the reduction of the concentration of HAs produced during the processing. Furthermore, water can potentially hinder the migration of HAs precursors to the food surface, reducing the concentration of carcinogens (Gibis, 2016). More HAs were found in roast chicken with skin. This might be due to the exposure of skin to higher temperatures, as well as the loose skin tissue compared to the interior of the meat, which allows the transportation of precursors easier to skin, resulting in increased HAs in the chicken skin (Yao, Zhou, Li, & Ma, 2023).

The addition of substances with water-holding capacities, such as tripolyphosphate, sodium chloride (Persson, Sjöholm, & Skog, 2003), polysaccharides, and fiber to minced meat products, water was bound, leading to a reduction in HAs formation (Mehr, Hosseini, Ardebili, & Khaneghah, 2020; Persson, Sjöholm, Nyman, & Skog, 2004). Vangnai et al. (Vangnai, Houser, Hunt, & Smith, 2014) discovered that injecting 12% salt/phosphate into pork loin resulted in an increase in moisture content and a decrease in the level of total HAs and individual HAs (PhIP by 42.7%, MeIQx by 79.0%, and DiMeIQx by 75.0%) when the meat was pan-frying at 200°C for 16 min. However, adding higher phosphates and sodium chloride and roasting at higher temperatures can easily lead to a hard outer crust formation. While this may increase the overall water content of the roasted meat, the lack of moisture on the surface causes an increase in surface temperature, leading to higher levels of HA formation (Li et al., 2020).

7.4.4 Fat/Oils and Lipid Oxidation

The impact of fat on the formation of HAs in meat involves a nuanced integration of both chemical and physical factors. During heating, fats undergo oxidation and generate free radicals, which can affect the Maillard reaction by either enhancing the reaction or interacting with specific intermediates. Additionally, carbonyl compounds intermediates formed during lipid oxidation can also participate in HA formation, further promoting HA formation. Additionally, at elevated temperatures, the melting of lipids serves as essential precursors and heat transfer media, ultimately impacting the production of HAs. High-fat meat products can also reduce the amount of HAs produced by diluting the precursor content in the meat. However, it is essential to note that high-fat meat products also lead to an increase in transfer during thermal processing, which in turn leads to a rise in HAs content compared to leaner meats processed for the same duration (Persson, Oroszvári, Tornberg, Sjöholm, & Skog, 2008).

Vegetable oil is an essential ingredient in cooking meat products. It is rich in polyphenols, which also provide meat flavor, and can be used as a heat transfer medium. Fat content, lipid oxidation, and plant phenolic compounds all affect the formation of HAs (Zamora, 2021). Monounsaturated and polyunsaturated fatty acids in fats and oils undergo oxidation and decomposition when exposed to high temperatures during cooking (Shahidi, Wanasundara, & Wanasundara, 1997). Linoleic acid in oils and fats contributes to the generation of aldehydes implicated in the heterocyclic amine synthesis pathway. Zamora (Zamora, 2021) proposed that lipid oxidation promotes the formation of HAs in meat products, possibly through the Strecker degradation reaction of oxidation products and amino acids, as well as the promotion of free radicals during the oxidation of unsaturated fatty acids.

Vegetable oils, on the other hand, contain a range of antioxidants, including polyphenols, which scavenge free radicals or form adducts with reactive carbonyl compounds, potentially inhibiting the production of HAs. The lower MeIQx and PhIP contents of beef patties fried in virgin olive oil instead of refined olive oil may be attributable to the higher phenolic content of virgin olive oil. Palm oil-treated meat loaf exhibited elevated Norharman, Harman, AαC, and MeAαC concentrations compared to the less saturated fatty acid-rich soybean oil (T. Pan, Wang, Chen, Hui, & Zhang, 2020). The PhIP content in roasted beef patties was substantially inhibited by grape seed, sunflower, and walnut oil. Conversely, MeIQx, 4,8-DiMeIQx, and IQx were significantly promoted by corn oil, soybean oil, peanut oil, and rapeseed oil at varying concentrations (Li et al., 2021). In addition, Hoffmann et al. (Hoffmann et al., 2018) reported that the fatty acid compositions of vegetable oils impact the properties (density, thermal conductivity, and specific heat) of oils and fats at a given temperature. This, in turn, can lead to variations in convective heat transfer coefficients, eventually influencing the formation of HAs in the cooked meat.

7.4.5 Precursors and Other Factors

The balance and the relative quantities of creatinine and glucose affect the formation of HAs. Various types of meat contain differing levels of HAs' precursors. Compared to other meats, chicken has lower glucose content but exhibits higher levels of PhIP. Excessive amounts of reducing sugars can alter the formation pathway of PhIP and generate various secondary products. A higher level of glucose hinders PhIP production while promoting the creation of harman and norharman in grilled chicken (Gibis & Loeffler, 2019). The levels of glucose, fructose, and sucrose had a significant impact on the formation of MeIQx and DiMeIQx (128°C). The addition of 9% and 14% sugar increased HA levels in fried fish, but at the concentration of 19% sugar, both total and individual HAs levels, including Norharman, Harman, and AαC, decreased (Tai, Lee, & Chen, 2001). The results indicate that adding more sugar during frying could reduce the formation of harmful HAs by generating Maillard products with lower levels.

Heating causes the conversion of creatine into creatinine, which is an essential component in the formation of the imidazole ring. In the absence of creatinine, the formation of imidazoquino(x)aline does not occur. In contrast, creatinine does not influence the formation of carbolines HAs (eg. harman or norharman). A further study indicated that beef with a high creatine level (1.5 mg/g) had more carcinogenic activity than beef with a low creatine (Jackson & Hargraves, 1994). The formation of HAs requires the availability of essential free amino acids. A clear correlation exists between increased concentrations of free amino acids and total creatine and the synthesis of larger quantities of HAs. The specific HAs generated in the model system were contingent upon the type of amino acid present, in addition to creatinine and sugar (Gibis, 2016; Meurillon & Engel, 2016). Ile, Phe, Tyr are the precursors of PhIP, while Ala, Arg, Asn, Gln, Gly, His, Ile, Leu, Lys, Met, etc. are associated with the formation of MeIQx, 7,8-DIMeIQx, and IQx (Meurillon & Engel, 2016). Tyr increased the synthesis of IQx molecules, and under wet conditions, large concentrations of Ala, Phe, and Gly promoted the development of PhIP and MeIQx (Skog, 2002).

7.5 EFFECTS OF FOOD PROCESSING METHODS ON THE FORMATION OF HAS

7.5.1 Thermal Processing Time and Temperature

The amount of HAs generated by thermal processing is highly related to the duration and temperature of cooking. Increasing the temperature and time of cooking resulted in a greater content of HAs (Gibis, 2016). MeIQx, PhIP, norharman, and harman were detected in pan-fried bacon slices (150 to 220°C) at concentrations ranging from 1.5 to 5.6 ng/g, 0.1 to 2.6 ng/g, 5.0 to 19.9 ng/g, and 0.3 to 1.7 ng/g, respectively. The bacon exhibited significantly elevated levels of HAs when subjected to prolonged frying durations and temperatures ranging from 200–220°C and 150–170°C (Gibis, Kruwinnus, & Weiss, 2015). While it is often believed that HAs are produced at high temperatures, it is worth noting that some HAs may also accumulate at temperatures below 100°C. For example, trout fillets cooked sous-vide at 65°C, 75°C, and 85°C showed variable amounts of IQx (up to 0.16 ng/g) and MeIQx (up to 5.66 ng/g) (F. Oz & Seyyar, 2016). Low-temperature sausages included AαC, harman, norharman, Phe-P-1, PhIP, and DMIP, which increased with heat exposure (Yang et al., 2019; 2020). The quantity of MeIQx present in pan-fried steak varied from 1.3 ng/g of meat for rare samples to 8.2 ng/g for extremely well-done samples (Sinha et al., 1998). The content of HAs in beef increased at higher grilling temperatures. Raising the grill temperature from 180°C to

220°C doubled the content of 4,8-DiMeIQx and increased the level of 7,8-DiMeIQx by nine times. It is advisable to subject beef to heat treatment at temperatures lower than 180°C (Bula, Przybylski, Jaworska, & Kajak-Siemaszko, 2019). Recent studies have employed a backpropagation-artificial neural network to predict the formation of HAs in roasted chicken based on heating temperature and time (Yao et al., 2023). This helps to correctly understand the essential variables to control the generation of HA and guide the optimization and improvement of the cooking process.

7.5.2 Different Thermal Processing Methods

The cooking methods significantly affect both the formation and the types of HAs generated during thermal processing. The three types of thermal processing procedures with the highest creation of HAs in meat products are grilling, frying, and roasting. The content of HAs in meat products, such as beef patties and steaks, is significantly higher in fried samples than in roasted ones, which strongly correlated with heat transfer during processing (Gibis, 2016). Charcoal grilling and pan-frying are two commonly used techniques for preparing poultry that might result in greater levels of HAs compared to other cooking methods. After grilling the chicken breast at 200°C for 10 min on each side, the total HAs content of this sample was 112 ng/g, with norharman having the highest amount (32.2 ng/g), followed by Harman (32 ng/g) and PhIP (31.1 ng/g) (Liao et al., 2010). On the other hand, the total HAs content in the pan-fried chicken breast was 27.4 ng/g (PhIP > Harman > IQ= MeIQx > Norharman > 4,8-DiMeIQx > AαC > MeAαC). Deep-fried chicken breast had a lower total HAs content (21.3 ng/g) than pan-fried samples, with Harman having the highest content (12.3 ng/g), followed by Norharman (5.4 ng/g), PhIP (2.2 ng/g), MeIQx (1 ng/g), 4, 8-DiMeIQx (0.4 ng/g), AαC (0.3 ng/g), and MeAαC (0.02 ng/g) (Liao et al., 2010). Nevertheless, roasting does not appear to lead to the production of a high amount of HAs. Norharman exhibited the greatest concentration (5 ng/g) in roasted chicken breast (200°C), followed by Harman (1 ng/g). In contrast, the levels of other HAs were below 0.1 ng/g (Liao et al., 2010). Pan-fried lamb chops did not contain IQx, MeIQx, MeIQ, 7,8-DiMeIQx, 4,8-DiMeIQx, PhIP, AαC, and MeAαC. However, varied amounts of IQx, IQ, and MeIQx compounds were detected in these samples. In grilled lamb chops, 4,8-DiMeIQx was the most abundant HA, with no other HAs detected in most samples (F. Oz, Kaban, & Kaya, 2010). The concentrations of HAs in roasted, fried, pan-fried, and stewed lamb patties ranged from 4.39 to 123.15 ng/g, 3.59 to 43.24 ng/g, 0.71 to 10.05 ng/g, and 51.07 to 120.32 ng/g, respectively. PhIP was the most prevalent HAs in deep-roasted and fried samples. In contrast, Harman and Norharman were the primary HAs in pan-fried and stewed samples (Guo et al., 2014). The content of HAs in Chinese dried meat products is decreasing according to the following processing methods: roasting > non-deep-frying > natural air-drying > deep-frying (Hou et al., 2018).

Cooking methods involving dry heat, such as pan-frying, have been found to produce considerably higher levels of mutagens than those that employ moist heat, like oven roasting. The disparity in mutagen production can be attributed to several factors, including the cooking temperature, duration of heat exposure, and the mode of heat transfer involved in the cooking process. Dry-heating techniques typically rely on conduction, where heat is transferred directly through contact with hot surfaces, or radiation, where heat is emitted from a source to cook food. Oven roasting often involves convection, where heat is circulated through the food, allowing for a more even cooking temperature. Moreover, the medium surrounding the food – be it metal in pan-frying or air in oven-roasting –plays a crucial role in determining the heat transfer rate, leading to varying heat transfer coefficients. pan-frying resulted in significantly higher mutagenic activity and HAs concentrations compared to oven roasting of hamburgers. When hamburgers were prepared using a deep-fat fryer, a convection oven, or a contact grill, notable variances in the levels of HAs were observed (Persson, Sjöholm, & Skog, 2002). Specifically, the MeIQx levels were approximately ten times higher when the meat was cooked on a contact grill versus the other cooking methods. Deep-fat frying and convection oven cooking resulted in lower HAs formation, which can be attributed to the differences in heat transfer. Direct contact methods like grilling or pan-frying facilitate a more efficient heat transfer, leading to higher HAs levels. In contrast, oven roasting involves an indirect application of heat through convection, and microwave cooking relies on radiation to transfer energy, which results in lower HAs production compared to direct heat methods (Gibis, 2016).

7.5.2.1 Deep-frying

Deep-frying is highly appreciated due to its distinct color and flavor, and it has become one of the most popular meat cooking techniques. The frying process involves immersing the food in hot oil (150–190°C), so in this process, the transfer of heat and mass occurs simultaneously between the oil and the food (Hosseini, Ghorbani, Meshginfar, & Mahoonak, 2016). When oils and fats are employed in the frying process, thermal and oxidative degradation occur, resulting in the development of volatile and non-volatile degradation products, which can affect HA formation. Frying commonly uses a variety of vegetable oils, including corn oil, soybean oil, sunflower oil, palm oil, and canola oil.

The type of oil also significantly impacted the production of HAs, with peanut and rapeseed oils resulting in higher HA levels in deep-fat fried chicken than palm, lard, soybean, and sunflower seed oil. Furthermore, the oxidation of frying oil increased HAs in deep-fat fried chicken. The repeated use of frying oil causes lipid oxidation, which promotes the development of HAs (C. X. Zhang, Xi, Wang, Ma, & Wang, 2022). There was an observed increase in the variety of HAs present in the grass carp fish patties as the number of frying cycles increased. Following the 25th frying cycle, seven distinct categories of HAs were identified. The formation of IQ, PhIP, MeIQx, and AαC was facilitated by repeated frying (Y. Wang et al., 2015).

The unsaturation of oils significantly influenced the formation of HAs. The levels of HAs in fried pork balls cooked in peanut oil (1.89 μg/kg) and the colza oil group (1.71 μg/kg) were considerably more significant than those in the palm oil, soybean oil, and corn oil groups (Sun et al., 2023). Meat floss treated with lard and palm oil exhibited greater concentrations of Norharman, Harman, AαC, and MeAαC than that treated with soybean oil, which contained fewer saturated fatty acids. Significantly increased concentrations of total HAs were found in meat floss treated with palm oil (150°C and 180°C) compared to that treated with soybean oil and lard (T. Pan et al., 2020). MeIQx can be generated during the deep-fat frying of meatballs using several frying oils. The meatballs fried in hazelnut oil had the lowest MeIQx, whereas those fried in commercially mixed oil had the highest MeIQx content (Ekiz & Oz, 2019). The oxidative decomposition of unsaturated fatty acids in frying oil at high temperatures generates reactive oxygen species, which can trigger the breakdown of Maillard reactants into HAs intermediates, including 1- and 3-deoxyketone. This process ultimately impacts the formation of HAs (Sun et al., 2023). In fried beef patties, PhIP and MeIQx were the predominant HAs. The greatest amount of total HA was formed by frying beef patties with sunflower oil, followed by canola, peanut, maize, and soybean oils (Fan et al., 2022). Soybean oil produced fewer total HAs than palm oil in fried pork spareribs marinated and coated in sweet potato powder. When frying crispy pork spareribs, Harman and Norharman were the predominant HA in these samples (Lai, Inbaraj, Chen, & Smith, 2024).

7.5.2.2 Pan-frying

When frying meat, a Teflon-coated pan with minimal oil or grease is utilized as the base to prepare the meat quickly. The primary difference between pan-frying and deep-frying is the amount of oil used, which affects heat transfer during heating. The type of cooking oil also affects the content of HAs in pan-frying. The utilization of frying oils with lower saturated fatty acids has been beneficial in mitigating HA formation of HAs in meat floss, a traditional delicacy commonly made from shredded, simmered, and fried pork (T. Pan et al., 2020). Among the several vegetable oils (peanut, maize, rapeseed, sunflower seed, and soybean), bacon fried in corn oil had the greatest levels of HAs. Peanut, rapeseed, and sunflower seed oil had the least promoting effect on total HAs in pan-fried bacon. In contrast, rapeseed and soybean oils exhibited the lowest total HAs (Du et al., 2022). When aiming to decrease the levels of HAs in fried bacon, it is crucial to consider the kind of vegetable oil used. The concentration of total HAs in pork fried in sesame oil was higher than that fried in lard (Lai, Inbaraj, & Chen, 2023).

7.5.2.3 Roasting and grilling

Roasted meat enjoys global popularity as a delectable meat product commonly prepared through oven roasting. Yao et al. (Yao et al., 2023) examined the influences of roasting conditions (The time and temperature ranged from 30–130 min and 175–250°C, respectively) on HAs and precursor profiles, recording a significant temperature- and time-dependent increase in HAs formation. Similarly, the grilling of beef samples on an electric grill (180°C and 280°C) formed several HAs (4,8-DiMeIQx, 4.50–7.55 ng/g; MeIQx,1.10–7.20 ng/g; MeIQ,1.70–5.30 ng/g; and IQ, 0.38–2.00 n ng/g) (Szterk, 2015). Yan (Yan et al., 2021) revealed that times have a more pronounced effect on the formation of HAs at higher roasting temperatures (250°C) in roasted pork, and the variations in the HAs profile were notably attributed to PhIP, 4'-OH-PhIP, IQ[4,5-b], and MeIQ. Three different processing stages: marinating, low-temperature roasting (180°C, 30 min), and high-temperature roasting (240°C, 10min) on the accumulation of HAs in roasted chicken drumsticks were conducted, recording both low-temperature and high-temperature roasting were remarkedly induced the formation of PhIP, Harman, and Norharman (L. Zhang et al., 2022). Therefore, to reduce the formation of HAs while cooking meat products, it is advisable to use lower roasting/grilling temperatures (< 240°C) (Polak, Demsar, Zahija, & Polak, 2020).

7.5.2.4 Other moderate cooking methods

Various studies have shown that moderate cooking methods, such as smoking, stewing, steaming, and boiling at temperatures below 120°C, typically result in negligible or minimal HAs (Skog & Solyakov, 2002). Smoking not only helps preserve meat products but also enhances their visual appeal, provides a distinctive flavor, and contributes to their exceptional taste. HA formations are influenced by various smoking parameters, including the type of wood, methods of smoke generation, and smoking times and temperatures (Y. Q. Cheng, Leible, Weiss, & Gibis, 2023). The contents of HAs in smoked meat products were generally regarded as being quite low (Y. Q. Cheng et al., 2023). Additionally, both free and bound forms of the HAs were detected in smoked sausages (Yang et al., 2019; Yang et al., 2020). A limited number of free HAs with modest concentrations (ranging from 3.05 to 22.9 ng/g DW) were detected in Frankfurter-type sausages. However, more bound-HAs with much higher levels (ranging from 10.8 to 396 ng/g DW) were observed. Shen et al. (Shen et al., 2023) reported that the traditional smoked pork patty contains both free and protein-bound HAs. The smoking process considerably increased the amount of HAs, remarkably free Norharman, found at a concentration of 3.26 ng/g in control samples and 82.24 ng/g in meat smoked with cypress (Shen et al., 2023). Yin (Yin, Du, Xu, Chen, & Kong, 2021) discovered that traditional Harbin red sausages in the northern Chinese market had greater HAs levels (average: 173.20 ± 33.40 ng/g) compared to conventional sausages (average: 44.14 ± 2.93 ng/g). Norharman and Harman were the predominant HAs in all the samples, despite Harbin red sausages having a maximum processing temperature of 85°C.

The presence of carbonyl compounds, furan derivatives, aldehydes, or ketones in smoke is primarily responsible for the formation of HAs in smoked meat products. These chemicals may speed up the Maillard reactions, which not only give cooked food its taste and appearance but can also produce harmful HAs (Gibis, 2016; Zamora et al., 2020b). Traditional sausage smoking introduces all volatile chemicals and carbon particles straight to the meat products. However, in modern smokehouses, the smoke that reaches the surface of the sausages via pipes may include fewer particles associated with HAs. Furthermore, traditional smoking processes require longer smoking time than modern ones, leading to an elevated level of HAs in traditionally smoked sausages. In the production of low-temperature sausages, HAs tend to accumulate at four processing stages (raw, dried, roasted, and steamed). Harman, norharman, Glu-P-1, and PhIP were present in raw sausages, with their concentrations notably increasing after the steaming. Conversely, AαC, MeAαC, DMIP, and 1,5,6-TMIP are generated during the drying stage. MeIQ is the only HA detected to form exclusively during the steaming stage (Yang et al., 2020).

The sous-vide cooking is characterized by vacuumizing the piece of meat and cooking in a water bath with low-temperature (<100°C) long-time cooking, thereby improving the juiciness of meat and retaining small water-soluble nutrients of interest and flavor. Therefore, the sous-vide method of cooking enhances the flavor and texture of ready-to-eat foods and significantly elevates their safety profile. (Supaphon et al.,

2021) (Khalid et al., 2023). OZ and Zikirov (F. Oz & Zikirov, 2015) have compared sous-vide cooking, boiling, and pan-frying on the formation of HA in beef chops. As expected, total HAs amounts in pan-fried samples were higher than in sous-vide cooked and boiled samples. 4,8-DiMeIOx was the only HAs detected in all sous-vide cooking treatments at 85–95°C for 120–240 min. In another report, Only IQx and MeIQx were found in sous-vide-cooked trout fillets. However, IQ MeIQ, 7,8-DiMeIQx, 4,8 DiMeIQx, PhIP, AαC, and MeAαC could not be detected. The concentration of MeIQ in the total HAs was relatively high (F. Oz & Seyyar, 2016). However, the utilization of superheated steam in cooking slices of Longissimus thoracic (LT) muscle has been observed to significantly increase the production of HAs. Notably, when LT muscle slices are subjected to superheated steam at elevated temperatures (170°C or 200°C), they significantly increase HAs content. Comparative analysis revealed that while the concentrations of MelQx and PhIP remain relatively consistent between LT slices and the gravy, lQx and 4,8-DiMeIQx are markedly lower in the LT slices, with quantities three to four times less than those detected in gravy (Kondjoyan et al., 2010).

7.6 MITIGATION OF HAS: STRATEGIES AND CHEMICAL MECHANISMS

The key elements that influence the formation of HAs have been discussed above, including the compositions and amounts of precursors, processing parameters (such as time, temperature, heating technique, etc.) and the added food ingredients. Thus, these factors can be the foundation for HAs regulation techniques. Multiple aspects involved in the processing methods, including temperature, duration, moisture content, and heat treatment techniques, can be further enhanced through optimization. In addition, several antioxidant strategies have been proposed to reduce the formation of HAs and lower the possible health concerns associated with the processed meats, the most practical and broadly applicable of which is the use of natural antioxidants (Xiong, 2017). Phenolic compounds, which include a variety of phenolic acids, flavonoids, and other secondary metabolites, are abundant in plant material and have gained much attention. Other natural substances, such as organic sulfur derivatives found in garlic and onions (Janoszka, 2010), as well as vitamins, play a vital role in suppressing HAs. However, the mechanisms behind the inhibition process vary. As a result, many mechanisms involved may be exploited to explore methods for reducing HAs.

7.6.1 Selecting Suitable Processing Methods

Heat processing methods have a significant impact on HAs formation. Heat transfer to the surface of the meat product and the precursors available on the skin surface affect the formation of HAs (Meurillon & Engel, 2016). Microwave heating is a gentle method that transforms electromagnetic energy into thermal energy, making it suitable for processing various meats, including bacon, meatballs, and meatloaf. Due to internal heat generation, the surface temperature of the meat is lower than the temperature of the remainder of the product throughout the heating process. Therefore, microwave treatment can effectively reduce the HAs content in the product. Frying pans produced higher amounts of HAs in sliced bacon than microwaves. Microwave heating had a lower influence on the synthesis of carcinogenic chemicals in bacon, with negligible effects on sensory qualities (Soladoye et al., 2017). For example, pre-treatment of meat with microwaves (0–3 min) before frying decreased water, fat, and HAs precursors (such as amino acids, glucose, creatine, and creatinine) by around 30%, leading to a 95% reduction in HAs mutagenic activity. After frying, samples with microwave pre-treatment had a 3–9 times reduction in total HAs (IQ, MeIQx, DiMeIQx, and PhIP) concentration compared to untreated samples (Felton, Fultz, Dolbeare, & Knize, 1994). Microwave cooking mostly yields carboline HAs (norharman, harman, Trp-P-1, AαC, etc.), whereas frying results in both imidazoquinoline and carbolines. Thus, a reduction in the formation of imidazoquinolines could be achieved by substituting microwave cooking for frying (Chiu, Yang, & Chen, 1998).

Cold-cut beef and pre-cooked meat products (hams and sausages) have no detectable HAs or deficient amounts of HAs owing to the low temperatures during processing. Generally, the three high-temperature cooking methods, frying, grilling, and oven roasting, produce higher HA levels, especially PhIP. Frying and grilling produce more 8-MeIQx and PhIP, probably because this processing takes less time for the meat to reach the same internal temperature (Zimmerli, Rhyn, Zoller, & Schlatter, 2001); roasting, a cooking method that utilizes the transmission of heat via the air to food, generates a lower level of HAs in comparison to frying (Skog, Augustsson, Steineck, Stenberg, & Jagerstad, 1997). Using new processing technology is another option. For example, pulsed electric field treatment of food raw materials may remove amino acids and reduce sugars from the food matrix, limiting the Maillard reaction and perhaps inhibiting the formation of HAs. Ohmic heating converts electrical energy into thermal energy via the food matrix's resistance, enabling short-time high-temperature heat treatment. The food is cooked uniformly, and local overheating is prevented, allowing control or reduction of the Maillard reaction. The use of cold atmospheric plasma during the curing of superheated steam beef enhanced its taste and did not increase HAs in samples (Hui, Fang, Ma, Hamid, & Li, 2023).

7.6.2 Addition of Functional Ingredients

Polyphenols from various plant sources have garnered much attention as Maillard reaction inhibitors in food processing because they are natural substances and have greater acceptance than synthetic chemicals. In addition to phenolic phytochemicals, other naturally occurring non-phenolic phytochemicals, such as carotenoids, terpenoids, vitamins, and amino acids, are extensively found in nature. These compounds have also been discovered to inhibit HAs by targeting active sites, intermediates, or products. To enhance the taste and reduce hazardous components, spices, herbs, fruits, seeds, essential oils, and various plant extracts are also extensively employed in the further processed prepared meat products (Figure 7.3).

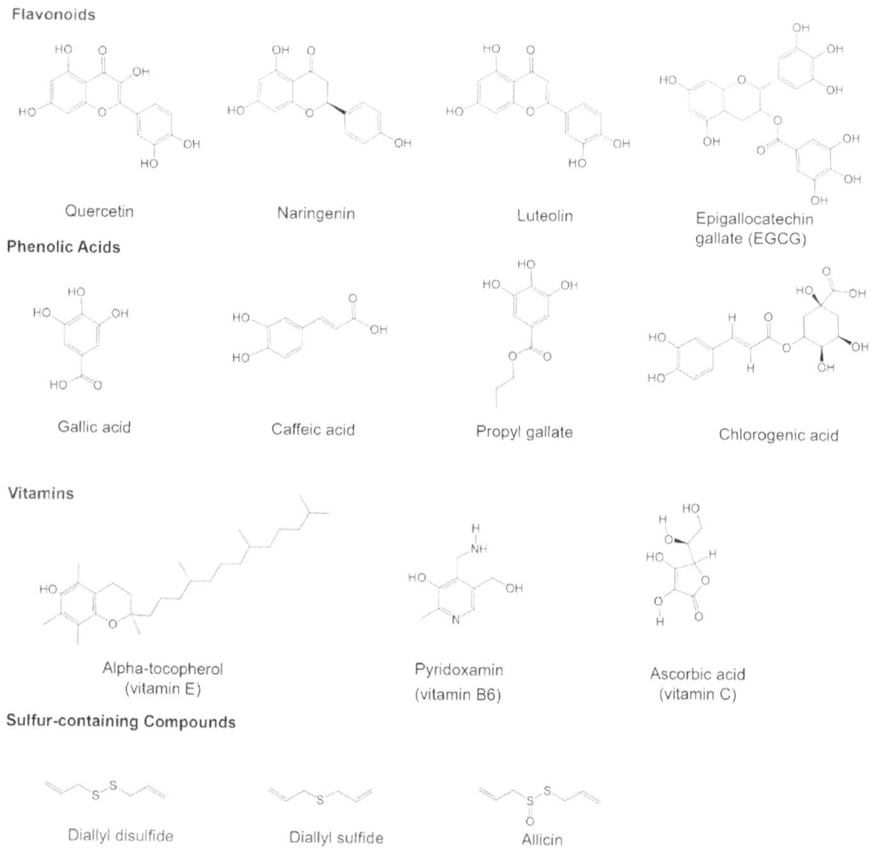

FIGURE 7.3 Chemical structures of common plant components inhibiting the formation of HA.

7.6.2.1 Scavenging free radicals by antioxidants

The free radical reaction is a critical step in the synthesis of HAs, particularly quinoline and quinoxaline (Zamora & Hidalgo, 2020). Antioxidants most likely prevent the production of HAs by scavenging free radicals involved in their formation route (E. Oz et al., 2023). In addition to neutralizing free radicals, antioxidants also employ various inhibitory mechanisms. These mechanisms include scavenging reactive carbonyls (Hidalgo & Zamora, 2023), impeding the production of free radicals that is caused by lipid and protein oxidation (Xiong, 2017; Zamora, Alcón, & Hidalgo, 2012), competitively capturing precursors of HAs (Xue et al., 2020), and restricting the migration of water-soluble HAs precursors to the surface of the meat (Khan, et al., 2022). A variety of natural antioxidants, including polyphenols, flavonoids, citrus flavonoids, ellagic acid, vitamins E and C, propyl gallate, plant or fruit extracts and various spices, have an impact on the development of HAs in meat products. Polyphenols are believed to hinder the synthesis of HAs by impeding the formation of free radicals generated by the Maillard reaction (Figure 7.4 A and Table 7.1).

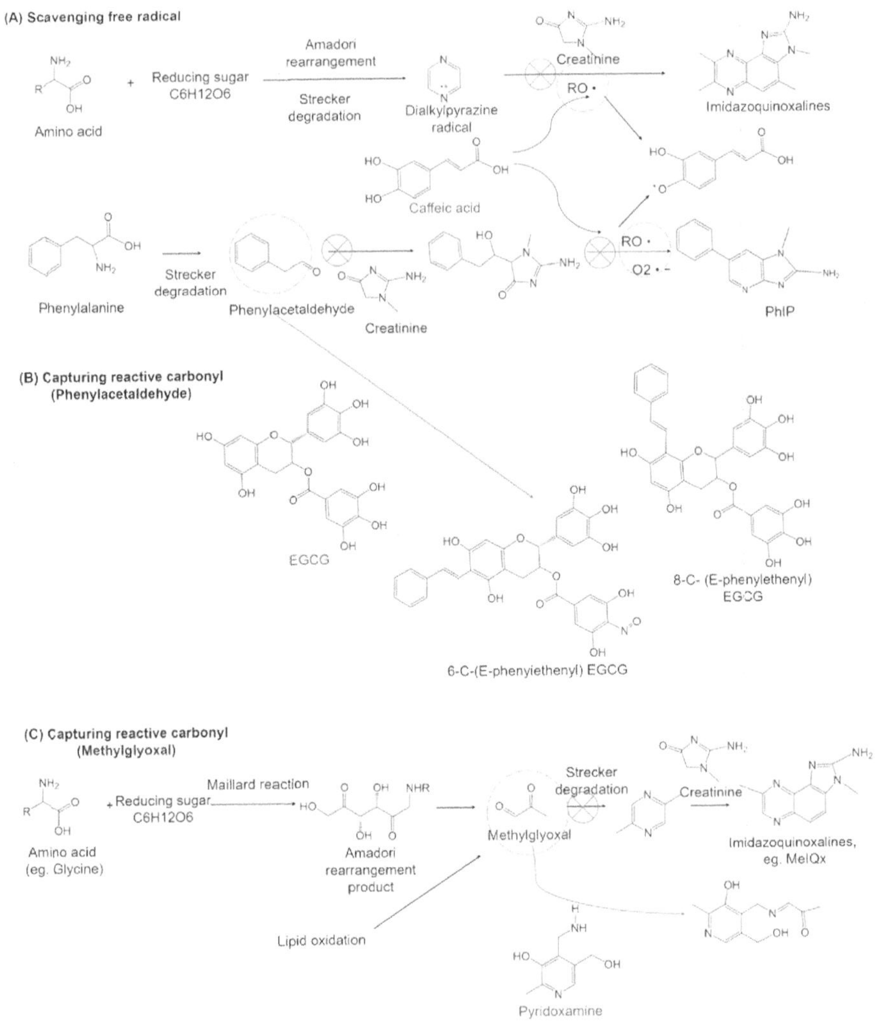

FIGURE 7.4 Chemical pathways for the inhibition of HA by radical-scavenging (A) and the capture of reactive carbonyl intermediates phenylacetaldehyde (B) and methylglyoxal (C) (Bao et al., 2020; B. Li et al., 2023; Xiong, 2017; Zamora et al., 2012).

Nevertheless, contradictory findings have been reported between the antioxidant (free radical scavenging) activity of some compounds and their inhibition rates of HAs; antioxidants may show prooxidant or antioxidant activity depending on their concentration, which could potentially facilitate the formation of HAs (Zeng et al., 2017). Furthermore, phenolic compounds with meta-hydroxyl groups exhibited the most remarkable ability to capture carbonyl compounds. In contrast, phenolic compounds with substitution patterns that promote electron delocalization showed a lesser capacity for carbonyl trapping but a better capacity for scavenging free radicals. Hence, various phenols have distinct inhibitory effects in the production of HAs (Salazar,

TABLE 7.1 Plant and spice extracts, and phenolic compounds already used to inhibit HA formation in processed meats

	PLANT MATERIAL	*DOSAGE*	*PROCESSED MEATS*	*HAS*	*INHIBITION %*	*MECHANISM OF INHIBITION*	*REFERENCE*
Plant and spices extracts or powder	Rosemary extract	1.5%	Fried beef patties	MeIQx,PhIP	50–90	Free radical scavenging	(Gibis & Weiss, 2012)
		0.3%	Fried beef patties	MeIQx,PhIP	48–77	Free radical scavenging	(Tsen, Ameri, & Smith, 2006)
	Grape seed extract	0.8%	Fried beef patties	MeIQx,PhIP	57–90	Free radical scavenging	(Gibis & Weiss, 2012)
		0.1%	Fried beef patties	MeIQx,DiMeIQx,PhIP	66–72	Free radical scavenging	(K. W. Cheng, Wu, et al., 2007)
	Apple extract	0.1%	Fried beef patties	MeIQx,DiMeIQx,PhIP	59–69	Free radical scavenging	(K. W. Cheng, Wu, et al., 2007)
	Apple skin extract	5%	Grilled beef patties	MeIQx,PhIP	76–82	Free radical scavenging	(Rounds, Havens, Feinstein, Friedman, & Ravishankar, 2012)
	Green tea extract	5%	Grilled beef patties	MeIQx,PhIP	31–86	Free radical scavenging	(Rounds et al., 2012)
	Grape seed extract	5%	Grilled beef patties	MeIQx,PhIP	50–79	Free radical scavenging	(Rounds et al., 2012)
	Pomegranate seed extract	1%	Fried meatballs	IQ,MeIQx,PhIP, harman, norharman	18–68	Free radical scavenging	(Keskekoglu & Üren, 2014)
	Rosemary powder	0.2%	Fried beef patties	MeIQx,PhIP	44	Free radical scavenging	(Puangsombat, Jirapakkul, & Smith, 2011)

(continued)

TABLE 7.1 (Continued)

PLANT MATERIAL	DOSAGE	PROCESSED MEATS	HAS	INHIBITION %	MECHANISM OF INHIBITION	REFERENCE
Turmeric powder	0.2%	Fried beef patties	MelQx,PhIP	39	Free radical scavenging	(Puangsombat, Jirapakkul, et al., 2011)
Fingerroot powder	0.2%	Fried beef patties	MelQx,PhIP	34	Free radical scavenging	(Puangsombat, Jirapakkul, et al., 2011)
Green tea infusion	0.8%	Pan-fried beef	PhIP	74	Free radical scavenging	(Quelhas et al., 2010)
Sichuan pepper	0.5%	Roasted Beef patties	PhIP, IQx, MelQx, and 4,8-DiMelQx	28–82	Free radical scavenging	(Zeng et al., 2018)
Black pepper	0.01%	Fried meatballs	IQ, MelQx,4,8-DiMelQx, PhIP, total HA	12–100	Free radical scavenging	(F. Oz & Kaya, 2011a)
Red pepper	0.01%	Fried beef chops	MelQx, IQ, MelQ, PhIP, 4,8-DiMelQx	75–100	Free radical scavenging	(F. Oz & Kaya, 2011b)
Ginger powder	1%, 1.5%	roast beef patties	MelQx, MelQ, PhIP, 7,8-DiMelQx,harman, norharman	28–100	Eliminating alkyl free radicals and singlet oxygen	(Xue et al., 2022)
Chilli pepper	0.5–1.5%	roasted Beef patties	PhIP, DMIP, TMIP, IQx,MelQx, 4,8-DiMelQx, Harman	16–100	Free radical scavenging	(Zeng et al., 2017)
Onion powder	5%	Grilled beef patties	MelQx,PhIP	78–94	Reaction with glucose	(Rounds et al., 2012)
Turmeric powder	5%	Grilled beef patties	MelQx,PhIP	49–74	Free radical scavenging	(Rounds et al., 2012)
Cumin powder	5%	Grilled beef patties	MelQx,PhIP	24–52	Reaction with glucose	(Rounds et al., 2012)
Garlic powder	5%	Grilled beef patties	MelQx,PhIP	66–85	Reaction with glucose	(Rounds et al., 2012)

Phenolic compounds	Carnosic acid	0.001%	pan-frying beef	MeIQx,DiMeIQx,PhIP,total HA	18–33	Free radical scavenging	(K. W. Cheng, Chen, et al., 2007)
	Rosmarinic acid	0.001%	pan-frying beef	MeIQx,DiMeIQx,PhIP,total HA	33–48	Free radical scavenging	(K. W. Cheng, Chen, et al., 2007)
	Naringenin	0.001%	pan-frying beef	MeIQx,DiMeIQx,PhIP,total HA	63–74	Free radical scavenging	(K. W. Cheng, Chen, et al., 2007)
	EGCG	0.2 mol/kg	Roasted beef patties	MeIQx,DiMeIQx,PhIP	69–86	Free radical scavenging Trapping of phenylacetaldehyde	(Zhu, Zhang, Wang, Chen, & Zheng, 2016)
	Epicatechin	0.025, 0.125, and 0.625 mmol/100g	Charcoal roasted lamb	IQx, 8-MeIQx, Norharman, Harman and PhIP	42–100		(Ding, Zhang, Liu, Wang, & Hui, 2022)
Vitamins	Vitamins E	0.1%	Pork floss	Norharman, Harman PhIP,AαC,MeAαC, total HA	31–78	Free radical scavenging	(Liao et al., 2009)
		1%	Beef patties	IQ, MeIQ, MeIQx, DiMeIQx, PhIP, and total Has	45–76		(Balogh et al., 2000)
	Pyridoxamine (Vitamin B6)	0.03 mmol/g	Deep-fried tilapia cakes	MeIQx	78.54%	Free radical scavenging Trapping of phenylacetaldehyde	(Balogh et al., 2000)
		0.0067mmol/g beef	Fried beef patties	PhIP, 4,8-DiMeIQx, and MeIQx, total HA	38–43%		(Wong, Cheng, & Wang, 2012)
	Ascorbic acid	0.0067mmol/g beef	Fried beef patties	PhIP, 4,8-DiMeIQx, and MeIQx, total HA	14–19%	Free radical scavenging;	(Wong et al., 2012)
	Niacin	0.0067mmol/g beef	Fried beef patties	PhIP, 4,8-DiMeIQx, and MeIQx, total HA	15–19%	Free radical scavenging;	(Wong et al., 2012)

Arámbula-Villa, Hidalgo, & Zamora, 2014). Thus, the structure of phenols determines the inhibitory characteristics of each for the formation of HAs.

Numerous research has investigated the inhibitory effects of plant extracts or antioxidants on HAs (Geng et al., 2023; H. J. Wang et al., 2023). Mulberry leaf extract and its phenolic compounds were also tested in deep-fried braised muscovy duck (180°C) (Xu, Huang, Huang, & Huang, 2024). The addition of the marinades (0.3 mg/L) of chlorogenic acid, quercetin, kaempferol, luteolin, tetrahydroxyisoflavone, and the mixture of these compounds to muscovy duck before frying (180°C) lowered the content of free HAs by 33.98%, 25.23%, 49.78%, 55.28%, 64.43% and 52.85%, respectively. The mixture of these compounds (equimolar ratio) also effectively prevented the formation of bound-HAs (43.44%). Zhao (Zhao, Zhang, Ma, & Wang, 2023) simultaneously applied Adinandra nitida leaf extract (ANE) (200 mg/100g) to chicken and tea polyphenols, sesamol, or rosemary extract (40 mg/100g) to soybean oil, recording a significant reduction in total HAs contents by 33.98%–61.46%, 15.86%–44.94%, and 25.22%–48.01%, respectively, during 40 frying cycles. The addition of ANE (200 mg/100 g) to chicken alone also lowered the total HAs (11.98%–38.00%). Gumus et al. (Gumus & Kizil, 2022) found that blueberry extract (0.25%, 0.5%, 1%) significantly inhibited the production of IQ, IQx, MeIQx, MeIQ, 7,8-DiMeIQx, Harman, Trp-P-2, AαC, and total HAs by 100%, 100%, 77.1–100%, 63.4–100%, 77–100%, 100%, 100%, 100%, and 91.9–100%, respectively, in oil-free pan-fried beef (200°C). After marinating beef with both grape seed extract (0.8g/100g) and rosemary extract (1.5g/100g), reduced MeIQx and PhIP formation by up to 57% and 90%, respectively, in fried patties (230°C). The antioxidant capabilities of grape seed and rosemary extract have been verified to be strongly associated with their inhibition of HAs (Gibis & Weiss, 2012). Zhang et al. (Y. Zhang, Luo, Shao, Yu, & Wang, 2014) investigated bamboo leaves nine flavonoids on the inhibition of PhIP formation in the solution of creatinine and phenylalanine. The order of inhibitory ability on PhIP formation was homoorientin > orientin > luteolin > vitexin > vitexin > apigenin > isorhamnetin > fisetin > hesperetin. The inhibitory ability was strongly dependent on the concentration of added flavonoids. Red pepper contains quercetin, luteolin, other flavonoids, and terpenoids. The addition of 1% red pepper to beef patties before roasting (225°C) reduced the total HAs (IQ, MeIQ, 4, 8-DiMeIQx, and PhIP) by 94% (Zeng et al., 2017). The capacity of grape seed and rosemary extracts (Gibis & Weiss, 2012), bamboo leaf flavonoids (Y. Zhang et al., 2014) and certain other flavonoids (C. D. Yu, Shao, Liu, Zhang, & Wang, 2016) to inhibit HAs is directly related to their ability to scavenge ABTS free radicals, superoxide anion radicals and alkoxy free radicals, respectively. In addition, the suppressive impacts of beer (Viegas, Moreira, & Ferreira, 2015), tea extract (Jamali et al., 2016), pomegranate seed extract (Keskekoglu & Üren, 2014), sugarcane molasses extract (D. Yu, Chen, & Yu, 2016), and vegetable oils (Erdogan & Özdestan-Ocak, 2022; Li et al., 2021) on HAs were correlated with their antioxidant ability.

Phenolic acids and flavonoids are the primary antioxidants found in plant extracts. Spices and condiments used in processed meats are a subclass of plant-derived food additives that are high in phenolics, which may include gallic acid, caffeic acid, catechins, quercetin, apigenin, kaempferol, naringenin, epigallocatechin gallate (EGCG), eugenol, thymol, and menthol, depending on their specific source. Hence, the use of spices, Chinese herbs, and their extracts in meat products can serve as antioxidants, effectively eliminating free radicals, preventing the degradation products of oxidized lipids, and eventually minimizing the production of HAs (Table 7.1) (Khan, Luo, et al., 2022; Sepahpour et al., 2018; Xiong, 2017). Ginger and curcumin exhibited variable degrees of inhibition on both free and bound-HAs in roast beef patties by eliminating alkyl free radicals and singlet oxygen ($^1O_2$). The highest levels of inhibition observed were 59.97% for free HAs (with 0.015% curcumin) and 27.42% for bound-HAs (with 1.5% ginger) (Xue et al., 2022). The addition of 0.5% Sichuan pepper (Zanthoxylum bungeanum) on ground beef before grilling (225 °C) inhibited the formation of PhIP, IQx, MeIQx, and 4,8-DiMeIQx by 82%, 61%, 28%, and 79%, respectively. Similarly, at the concentrations of 0.005%, sanshoamide extract reduced the formation of PhIP, IQx, MeIQx, and 4,8-DiMeIQx by 27%, 45%, 44%, and 26%, respectively (Zeng et al., 2018). All these individual HA concentrations in beef patties were reduced by over 70% with Sichuan Pepper (1.0%) and sanshoamide extract (0.010%) treatments, respectively. Star anise, alpinia galanga, and black pepper (0.5%, 1.0%, and 1.5%, w/w) were also tested on HA formation in the roast barbecued pork. At the concentrations of 1.5%, all three spices lowered the content of Norharman with 66%, 66%, and 63%, respectively, whereas promoted the formation of PhIP (Y. J. Zhou et al., 2023).

The water-soluble vitamins, like pyridoxamine, could directly react with reactive carbonyl compounds, such as MGO and phenylacetaldehyde, to inhibit HAs, while vitamin E as an effective inhibitor of HAs formation in ground beef patties is related to its free radical scavenging ability (Table 7.1). At the concentrations of 1% (fat basis), vitamin E and oleoresin rosemary added to the fried ground beef patties (225°C) reduced the MeIQ content by 78.6% and 87%, respectively. They also effectively prevented the formation of IQ (85.7% and 72.4%), MelQx (48% and 31%), DiMelQx (79.2% and 77%), and PhIP (69% and 44%). The addition of 1% (based on fat content) vitamin E to the surface of ground beef patties reduced IQ, MeIQ, MeIQx, DiMeIQx, PhIP, and total HAs production by 61%, 75.6%, 45.1%, 74.6%, 73%, and 71%, respectively (Balogh, Gray, Gomaa, & Booren, 2000). Similarly, the addition of vitamin E (0.1%) before preparing the pork floss significantly reduced the total HAs, PhIP, Norharman, MeAαC, and AαC contents in pork floss when stir-fried at 150°C (Liao, Xu, & Zhou, 2009).

The antioxidants described above have inhibitory effects on some HAs, and the inhibition rate can even reach more than 90% or 100% in some cases. Still, there are also many HAs with lower inhibition rate. These compounds exhibit efficacy against limited HAs, such as PhIP, MeIQx, 4, 8-DiMeIQx, IQ, MeIQ, Harman, Norharman, and Trp-P-1. Certain antioxidants can suppress specific HAs while being ineffective against others and may even increase the formation of some. For instance, grape seed extract can hinder aminoimidazoquinolines but does not affect pyrolytic HAs like Harman and Norharman. The resistance to some pyrolysis HAs might be due to spatial hindrance at the reaction sites, complicating flavonoids' ability to neutralize free radicals. The reduction of PhIP and Trp-P-1 in charcoal-grilled pork marinated with pilsner beer can reach 71% and >99%, respectively; however, it can raise the levels of MeAαC by 28% (Viegas et al., 2015). Cheng et al. (K. W. Cheng, Chen, & Wang, 2007) revealed that quercetin and its glycoside variant significantly reduced PhIP synthesis by over half. In contrast, rutin was ineffective, and rosmarinic acid heightened PhIP formation by over 50%. The primary phenolic antioxidants in Sichuan peppercorns, star anise, cumin, and chili pepper, including rutin, protocatechuic acid, ferulic acid, and p-coumaric acid, notably enhanced the formation of pyridines, quinolines, and quinoxalines HAs (Zeng, Li, He, Qin, & Chen, 2016). These results show that the free radical reaction is not the only or common pathway for the formation of various HAs; non-free radical pathways also generate a large number of HAs. The mechanism by which antioxidants promote the production of HAs from non-free radical pathways also needs to be investigated in the future.

7.6.2.2 Capturing reactive carbonyl intermediates

Reactive carbonyl compounds serve as crucial intermediates in the Maillard reaction and play an essential role in the formation of HAs (Hidalgo, Lavado-Tena, & Zamora, 2021; Hidalgo & Zamora, 2022; Zamora et al., 2020b). Regulating reactive carbonyls may prevent the production of HAs. Carbohydrates, lipids, and amino acids are all sources of the reactive carbonyl intermediates required to generate HAs (Hidalgo & Zamora, 2022; Zamora et al., 2012). The most abundant active dicarbonyls produced by carbohydrates in food are 3-deoxyglucosanone, 3-deoxygalactosone, and glucosanone, and C2 and C3 active dicarbonyl compounds, including glyoxal (GO) and methylglyoxal (MGO) (Zamora & Hidalgo, 2020). Active carbonyl compounds are also involved in the Strecker degradation of amino acids, which contributes to the formation of pyrazines and other carbonyl groups. In addition to participating in the carbonyl amine reaction, active carbonyl compounds can also react with phenolic compounds. Consequently, certain phenolic compounds can capture the active carbonyl group necessary for HAs formation, thus effectively inhibiting the production of such harmful substances (Hidalgo & Zamora, 2023). Therefore, the active carbonyl compounds present in processed foods undergo a continuous transformation, and circumstances that favor the formation of carbonyl compounds may also facilitate the formation of HAs and vice versa. Strategies that restrict lipid oxidation and the Maillard reaction can be used to limit the generation of reactive carbonyl compound and prevent the formation of HAs.

The water-soluble vitamins on the formation of HAs were conducted in model systems and meat products. Niacin and ascorbic acid (VC) (0.2mmol/30g beef) inhibited the formation of PhIP, 4,8-DiMeIQx, and MeIQx around 14–19% in fried beef (200°C), whereas pyridoxamine reduced the levels of all three HAs by approximately 38–43% (Table 7.1). Pyridoxamine directly reacts with phenylacetaldehyde, a vital intermediate in PhIP formation, to form an adduct, therefore significantly reducing the level of PhIP (Wong et al., 2012). Another investigation has shown that pyridoxamine inhibited MeIQx more effectively, with inhibition rates of 82.72% and 78.54% in the glycine-glucose-creatinine model system and deep-fried tilapia cakes (180°C), respectively (Bao, Miao, Huang, & Lai, 2020). MGO plays a crucial role in the formation of MeIQx. The ability of pyridoxamine to capture MGO and inhibit MeIQx was shown by quantum chemistry calculations and experimental testing (Bao et al., 2020).

Any compound with nucleophilic properties that combines with the reactive carbonyl group involved in HAs or competes with the carbonyl-creatinine reaction can affect HAs (Hidalgo & Zamora, 2022). The condensation reaction between phenylacetaldehyde, a Strecker degradation product of phenylalanine, and creatinine is one of the main pathways for PhIP. EGCG inhibits the formation of PhIP by scavenging phenylacetaldehyde, a key intermediate in the PhIP formation pathways (Figure 7.4 B). Cheng et al. (K. W. Cheng, Chen, et al., 2007) conducted experiments in a model system and found that EGCG inhibited the formation of nearly 90% phenylacetaldehyde, and adduct was formed between EGCG and phenylacetaldehyde by LC-MS. Therefore, it inhibited the formation of PhIP mainly by forming adduct with the intermediates. Consequently, several research have focused on inhibiting PhIP by targeting this route and using certain chemicals to react with phenylacetaldehyde, resulting in the formation of adducts (Figure 7.4 B). For instance, pyridoxamine (Wong et al., 2012) and dihydromyricetin (B. Zhou et al., 2018) have been shown to prevent the formation of PhIP by capturing phenylacetaldehyde in beef patties (Table 7.1). The inhibition rates of VB1, VB3, pyridoxine, VB7, and VC on PhIP varied from 60% to 80%; these water-soluble vitamins similarly inhibited MeIQx in the model systems, but with the exception of pyridoxine, which exhibited an inhibition rate close to 80%, all other inhibition rates were less than 60%. Dietary flavonoids such as phlorizin, EGCG, and quercetin prevent the formation of total HA (55–70%) and PhIP (60–80%) in roasted beef patties. However, the antioxidant activity of these phenolics and HA inhibition did not show a statistically significant relationship. The binding of phenolic compounds with HA intermediates (such as phenylacetaldehyde) was a more credible explanation for preventing conversion to PhIP or other products. (Zhu et al., 2016). In addition, several amino acids prevented the synthesis of PhIP by attaching their amino groups to the carbonyl group of phenylglyoxal in the glucose/creatinine/phenylalanine system heated at 180°C for 1h. The sequence of their inhibitory effects, ranked from highest to lowest, was as follows: tryptophan=lysine > proline > leucine > methionine > valine > isoleucine > threonine > phenylalanine > aspartic acid (Linghu, Karim, & Smith, 2017).

The application of ultraviolet C-treated gallic acid (UVC-GA) and ultraviolet B-treated gallic acid (UVB-GA) resulted in substantial inhibition of MeIQx formation and carbonyl valence formation rates in oil-fried squid. The UVB-GA suppressed the production of formaldehyde, and UVC-GA notably decreased the levels of formaldehyde, acetaldehyde, and 2,5-dimethyl pyrazine. UV-GA decreased the carbonyl generated by lipid oxidation, weakening its catalysis and causing the MeIQx precursor to degrade into intermediates during Strecker degradation (Ma et al., 2023). Quercetin, when added, and the implementation of an oil-water separation system effectively reduced the formation of Harman (by 27.3% and 28.2%, respectively) and Norharman (by 28.7% and 64.1%, respectively) during the frying of braised chicken drumsticks. This is due to lesser consumption of tryptophan, precursors of β-carboline, and reduced formation of intermediate carbonyl compounds and 1,2,3,4-Tetrahydro-8-carboline-3-carboxylic acid (X. F. Li et al., 2023). Many studies have used α-dicarbonyl trapping agents, such as glyoxal (GO) and MGO, to control the Maillard reaction (Figure 7.4 C). It employs phenolic compounds that bind with sugar remnants and reactive carbonyl species, forming adducts primarily through electrophilic aromatic substitution. This action helps hinder the formation of harmful products. The effectiveness of trapping depends dramatically on the structure of the phenolic compounds, specifically the arrangement and number of hydroxyl groups (Rannou, Laroque, Renault, Prost, & Sérot, 2016).

7.6.2.3 Eliminating the final harmful products

The primary HAs found in the food system, including pyridine, quinoline, and quinoxaline, have an exocyclic free amino group. The HAs' exocyclic free amino groups sequentially undergo hydroxylation by cytochrome P450 enzymes, esterification by acetyltransferases or sulfotransferases, and cleavage into electrophilic nitrogen ion intermediates. These intermediates subsequently react with DNA to form carcinogenic and mutagenic adducts. Therefore, the produced HAs and their toxicity may thus be removed by reacting with certain compounds via their free amino groups to form stable admixtures. Some non-heterocyclic amine precursors' amino acids have active carboxyl groups that may react with HAs' exocyclic free amine groups to form stable amides. Kataoka et al. conducted an experiment where PhIP and glycine were heated at 200°C for 5 min, resulting in the creation of PhIP-glycine adduct, which may be formed by dehydration condensation of PhIP's amino group and glycine's carboxyl group (Kataoka, Miyake, Saito, & Mitani, 2012). PhIP and proline might condense to form the adduct by direct heating, as shown in the PhIP-producing model system and beef patties containing proline (Q. C. Chen, Xue, Chen, et al., 2021). PhIP can form adducts with several other amino acids, including glycine, histidine, and phenylalanine. Proline and histidine can also react with other HAs like IQ, MeIQx, and Trp-P-1 to form adducts (Kataoka et al., 2012). This reaction may potentially help eliminate HAs. Nevertheless, serine, cysteine, and lysine exhibit infrequent binding with HAs. The occurrence of heterocyclic amine-amino acid adducts has been established in the model system, indicating their potential significance in evaluating the carcinogenic risk associated with HAs. Additionally, these adducts provide a method for the elimination of HAs from processed foods.

7.6.2.4 Competitive inhibition of the Maillard reaction

Garlic and Onions are rich in sulfur-containing compounds such as allicin, diallyl disulfide, diallyl sulfide, and dipropyldisulfide, which have been shown to inhibit the formation of HAs (Table 7.1) (Shin, Strasburg, & Gray, 2002). Mince garlic or marinades containing garlic, onion, and lemon juice can also inhibit HAs formation. This effect is caused by the fact that allicin, diallyl disulfide and other thiosulfinates may interact directly with glucose to suppress the Maillard reaction (Gibis, 2007). Diallyl disulfide was reported to block 73% PhIP and 82% MeIQx in the model system after heating at 180°C for 30 min. Several sulfur compounds, including acetylcysteine, cysteine, diallyl disulfide, and glutathione, have been shown to suppress MeIQx production in meat model systems (Murkovic, 2004). Other sulfur compounds that reduce non-enzymatic browning include glutathione, L-cysteine, and L-cysteine, which all prevent the generation of HAs in model systems and fried beef patties. $NaHSO_3$ is the most extensively investigated sulfur-containing chemical for preventing the formation of HAs. The addition of $NaHSO_3$ may considerably reduce the development of HAs in canned food. Non-precursor amino acids, such as histidine, leucine, proline, and methionine, can effectively prevent the formation of beta-carbolines in beef patties by both alkyl radical scavenging (up to 82.59%) and competitive inhibition (up to 65.60%) (Xue et al., 2020).

Hasnol et al. (2014) reported a decrease in the levels of MeIQ, PhIP, DiMeIQx, IQ, IQx, and norharman in grilled chicken that was marinated with honey prior to cooking (Hasnol, Jinap, & Sanny, 2014). The glucose and fructose in honey reduced the development of total HAs (i.e., MeIQx, DiMeIQx, and PhIP) in fried beef by 41%. Jinap et al. (Jinap, Hasnol, Sanny, & Jahurul, 2018) found that using brines with four organic acids, including Tamarind, Lemon, Lime, and Calamansi, can notably decrease the formation of HAs in grilled chicken. This could be attributed to the decrease in pH, which increased the overall acidity of the marinades. Low pH values resulted in a higher concentration of free protons (H^+), which subsequently reacted with the negatively charged areas of amino acids (COO^-) to create COOH. The COOH functional group is chemically stable and does not undergo reactions with the carbonyl groups of reducing sugars in the Maillard reaction, which leads to the formation of HAs. Variables such as time, temperature, processing conditions, pH, water activity, reactant concentration, metal ions and most importantly reactant availability can be changed to prevent the Maillard reaction and subsequent quality alterations. This objective can be accomplished via thermal and non-thermal processing methods (Kathuria, Hamid, Gautam, & Thakur, 2023). Manipulating factors such as pH, temperature, and heating time during the carbonyl-ammonia reaction might effectively slow the interaction between carbonyl and creatinine (Zamora et al., 2020b).

7.7 CONCLUSIONS

Diverse types of HAs are produced during the cooking and manufacturing of meat products, primarily through the Maillard reaction, among these, PhIP and specific imidazoquinoxalines are prevalent. Variables such as the type of meat, fat content, vegetable oils, concentrations of amino acids and creatinine in raw materials, and the moisture content all affected the formation of HAs. Furthermore, different cooking methods have a significant impact on the formation of HAs, with pan-frying and roasting being more prone to their formation compared to smoking, stewing, steaming, and boiling. Controlling the HAs in food is thus a crucial subject that can be accomplished via scavenging free radicals, entrapping reactive carbonyl compounds, and inhibiting the Maillard reaction, among other methods.

The use of phenolic antioxidants derived from plants and spices has received substantial attention due to their capacity to alleviate chemical toxicity in meat products. However, it is worth noting that present research has mostly concentrated on limiting the generation of toxic HAs, with little attention paid to other hazardous substances and flavor formation simultaneously. As meat is a complex system in which these harmful substances and flavor precursors coexist, their formation pathways often involve the Maillard reaction. Therefore, future food safety considerations regarding HAs should encompass the simultaneous generation of HAs, other harmful substances and flavors during food thermal processing, as well as the simultaneous impact mechanisms of specific interventions on HAs, other harmful substances and flavors. Furthermore, it is critical to evaluate the safety and health implications of the degradation products of these plant-derived active components, the adducts of these components that capture reactive carbonyl compounds, and other newly produced products during food processing and manufacturing.

REFERENCES

Augustsson, K., Skog, K., Jägerstad, M., Dickman, P. W., & Steineck, G. (1999). Dietary heterocyclic amines and cancer of the colon, rectum, bladder, and kidney: A population-based study. *Lancet*, 353(9154), 703–707. doi:10.1016/S0140-6736(98)06099-1

Balogh, Z., Gray, J. I., Gomaa, E. A., & Booren, A. M. (2000). Formation and inhibition of heterocyclic aromatic amines in fried ground beef patties. *Food and Chemical Toxicology*, 38(5), 395–401. doi:10.1016/S0278-6915(00)00010-7

Bao, X. X., Miao, J. J., Huang, Y. Q., & Lai, K. Q. (2020). Revealing a key inhibitory mechanism of 2-amino-3,8-dimethylimidazo[4,5-f] quinoxaline via trapping of methylglyoxal. *Journal of Food Science*, 85(7), 2090–2097. doi:10.1111/1750-3841.15305

Barbir, A., Linseisen, J., Hermann, S., Kaaks, R., Teucher, B., Eichholzer, M., & Rohrmann, S. (2012). Effects of phenotypes in heterocyclic aromatic amine (HCA) metabolism-related genes on the association of HCA intake with the risk of colorectal adenomas. *Cancer Causes & Control*, 23(9), 1429–1442. doi:10.1007/s10552-012-0017-8

Bula, M., Przybylski, W., Jaworska, D., & Kajak-Siemaszko, K. (2019). Formation of heterocyclic aromatic amines in relation to pork quality and heat treatment parameters. *Food Chemistry*, 276, 511–519. doi:10.1016/j.foodchem.2018.10.073

Chang, C. C., Kao, T. H., Zhang, D. Q., Wang, Z. Y., Inbaraj, B. S., Hsu, K. Y., & Chen, B. H. (2018). Application of QuEChERS coupled with HPLC-DAD-ESI-MS/MS for determination of heterocyclic amines in commercial meat products. *Food Analytical Methods*, 11(11), 3243–3256. doi:10.1007/s12161-018-1302-2

Chen, J., He, Z. Y., Qin, F., Chen, J., & Zeng, M. M. (2017). Formation of free and protein-bound heterocyclic amines in roast beef patties assessed by UPLC-MS/MS. *Journal of Agricultural and Food Chemistry*, 65(22), 4493–4499. doi:10.1021/acs.jafc.7b01828

Chen, Q. C., Xue, C. Y., Chen, J., He, Z. Y., Wang, Z. J., Qin, F., … & Zeng, M. M. (2021). Simultaneous determination of the PhIP-proline adduct and related precursors by UPLC-MS/MS for confirmation of direct elimination of PhIP by proline. *Food Chemistry*, 365, 130484. doi:10.1016/j.foodchem.2021.130484

Chen, Q. C., Xue, C. Y., He, Z. Y., Wang, Z. J., Qin, F., Chen, J., & Zeng, M. M. (2021). Generation of sarcoplasmic and myofibrillar protein-bound heterocyclic amines in chemical model systems under different heating temperatures and durations. *Journal of Agricultural and Food Chemistry*, 69(10), 3232–3246. doi:10.1021/acs.jafc.0c06885

Cheng, K. W., Chen, F., & Wang, M. F. (2007). Inhibitory activities of dietary phenolic compounds on heterocyclic amine formation in both chemical model system and beef patties. *Molecular Nutrition & Food Research*, 51(8), 969–976. doi:10.1002/mnfr.200700032

Cheng, K. W., Wu, Q., Zheng, Z. P., Peng, X., Simon, J. E., Chen, F., & Wang, M. (2007). Inhibitory effect of fruit extracts on the formation of heterocyclic amines. *Journal of Agricultural and Food Chemistry*, 55(25), 10359–10365. doi:10.1021/jf071820z

Cheng, Y. Q., Leible, M., Weiss, J., & Gibis, M. (2023). The impact of temperature-controlled smoldering smoking on polycyclic aromatic hydrocarbons and heterocyclic amines contents in Frankfurter-type sausages. *Food Chemistry*, 423, 136258. doi:10.1016/j.foodchem.2023.136258

Cheng, Y. Q., Yao, M. J., Zhu, Z. S., Dong, X. L., Khan, I. A., Huang, J. C., ... & Zhou, G. H. (2019). Content, causes and analysis of heterocyclic amines in Chinese traditional braised chicken. *Food Additives and Contaminants Part A-Chemistry Analysis Control Exposure & Risk Assessment*, 36(7), 1032–1041. doi:10.1080/19440049.2019.1615136

Chiu, C. P., Yang, D. Y., & Chen, B. H. (1998). Formation of heterocyclic amines in cooked chicken legs. *Journal of Food Protection*, 61(6), 712–719. doi:10.4315/0362-028X-61.6.712

Deng, P., Chen, Y., Xie, S. Y., Xue, C. Y., He, Z. Y., Chen, Q. M., ... & Zeng, M. M. (2022). Accumulation of heterocyclic amines and advanced glycation end products in various processing stages of plant-based burgers by UHPLC-MS/MS. *Journal of Agricultural and Food Chemistry*, 70(46), 14771–14783. doi:10.1021/acs.jafc.2c06393

Ding, X. Q., Zhang, D. Q., Liu, H., Wang, Z. Y., & Hui, T. (2022). Chlorogenic acid and Epicatechin: An efficient inhibitor of heterocyclic amines in charcoal roasted lamb meats. *Food Chemistry*, 368, 130865. doi:10.1016/j.foodchem.2021.130865

Du, H. Z., Wang, Z. Y., Li, Y. X., Liu, Q., Chen, Q., & Kong, B. H. (2022). Understanding the development of heterocyclic aromatic amines in fried bacon and in the remaining oil after pan-frying in five different vegetable oils. *Foods*, 11(21), 3491. doi:10.3390/foods11213491

Ekiz, E., & Oz, F. (2019). The effects of different frying oils on the formation of heterocyclic aromatic amines in meatballs and the changes in fatty acid compositions of meatballs and frying oils. *Journal of the Science of Food and Agriculture*, 99(4), 1509–1518. doi:10.1002/jsfa.9325

Elbir, Z., & Oz, F. (2021). The assessment of commercial beef and chicken bouillons in terms of heterocyclic aromatic amines and some of their precursors. *International Journal of Food Science and Technology*, 56(1), 504–513. doi:10.1111/ijfs.14666

Erdogan, B., & Özdestan-Ocak, Ö. (2022). Determination the effects of pumpkin and rosehip seed oils on heterocyclic aromatic amine formation in meatballs by high-performance liquid chromatography. *Journal of Food Processing and Preservation*, 46(2), e16299. doi:10.1111/jfpp.16299

Fan, H. Y., Hu, H. Y., Li, C., Xie, J. H., Chen, J., Zeng, M. M., ... & Xie, M. Y. (2022). Effects of cooking factors on the formation of heterocyclic aromatic amines in fried beef patties. *Journal of Food Processing and Preservation*, 46(2), e16288. doi:10.1111/jfpp.16288

Felton, J. S., Fultz, E., Dolbeare, F. A., & Knize, M. G. (1994). Effect of microwave pretreatment on heterocyclic aromatic amine mutagens/carcinogens in fried beef patties. *Food and Chemcial Toxicology*, 32(10), 897–903. doi:10.1016/0278-6915(94)90087-6

Geng, Y. Q., Xie, Y. F., Li, W., Ji, J. F., Chen, F., Liao, X. J., ... & Ma, L. J. (2023). Heterocyclic amines in meat and meat products: Occurrence, formation, mitigation, health risks and intervention. *Food Reviews International*, 40(5), 1503–1519. doi:10.1080/87559129.2023.2221346

Gibis, M. (2007). Effect of oil marinades with garlic, onion, and lemon juice on the formation of heterocyclic aromatic amines in fried beef patties. *Journal of Agricultural and Food Chemistry*, 55(25), 10240–10247. doi:10.1021/jf071720t

Gibis, M. (2016). Heterocyclic aromatic amines in cooked meat products: Causes, formation, occurrence, and risk assessment. *Comprehensive Reviews in Food Science and Food Safety*, 15(2), 269–302. doi:10.1111/1541-4337.12186

Gibis, M., Kruwinnus, M., & Weiss, J. (2015). Impact of different pan-frying conditions on the formation of heterocyclic aromatic amines and sensory quality in fried bacon. *Food Chemistry*, 168, 383–389. doi:10.1016/j.foodchem.2014.07.074

Gibis, M., & Loeffler, M. (2019). Effect of creatine and glucose on formation of heterocyclic amines in grilled chicken breasts. *Foods*, 8(12), 616. doi:10.3390/foods8120616

Gibis, M., & Weiss, J. (2012). Antioxidant capacity and inhibitory effect of grape seed and rosemary extract in marinades on the formation of heterocyclic amines in fried beef patties. *Food Chemistry*, 134(2), 766–774. doi:10.1016/j.foodchem.2012.02.179

Gumus, D., & Kizil, M. (2022). Comparison of the reducing effects of blueberry and propolis extracts on heterocyclic aromatic amines formation in pan fried beef. *Meat Science*, 186, 108746. doi:10.1016/j.meatsci.2022.108746

Guo, H. T., Wang, Z. Y., Pan, H., Li, X., Chen, L., Rao, W. L., … & Zhang, D. Q. (2014). Effects of traditional Chinese cooking methods on formation of heterocyclic aromatic amines in lamb patties. *Food Science and Biotechnology*, 23(3), 747–753. doi:10.1007/s10068-014-0101-9

Haskaraca, G., Soncu, E. D., Kolsarici, N., Öz, F., & Juneja, V. K. (2017). Heterocyclic aromatic amines content in chicken burgers and chicken nuggetssold in fast food restaurants and effects of green tea extract and microwave thawing on their formation. *Journal of Food Processing and Preservation*, 41, e13240. doi:10.1111/jfpp.13240

Hasnol, N. D. S., Jinap, S., & Sanny, M. (2014). Effect of different types of sugars in a marinating formulation on the formation of heterocyclic amines in grilled chicken. *Food Chemistry*, 145, 514–521. doi:10.1016/j.foodchem.2013.08.086

Hidalgo, F. J., Lavado-Tena, C. M., & Zamora, R. (2021). Identification of acrolein as the reactive carbonyl responsible for the formation of 2-amino-3,8-dimethylimidazo[4,5-*f*]quinoxaline (MeIQx). *Food Chemistry*, 343, 128478. doi:10.1016/j.foodchem.2020.128478

Hidalgo, F. J., & Zamora, R. (2022). Carbonyl chemistry and the formation of heterocyclic aromatic amines with the structure of aminoimidazoazaarene. *Journal of Agricultural and Food Chemistry*, 70(1), 79–86. doi:10.1021/acs.jafc.1c06842

Hidalgo, F. J., & Zamora, R. (2023). Carbonyl-trapping by phenolics and the inhibition of the formation of carcinogenic heterocyclic aromatic amines with the structure of aminoimidazoazaarene in beef patties. *Food Chemistry*, 425, 136505. doi:10.1016/j.foodchem.2023.136505

Hoffmann, J. F., Vaitilingom, G., Henry, J. F., Chirtoc, M., Olivès, R., Goetz, V., & Py, X. (2018). Temperature dependence of thermophysical and rheological properties of seven vegetable oils in view of their use as heat transfer fluids in concentrated solar plants. *Solar Energy Materials and Solar Cells*, 178, 129–138.

Hosseini, H., Ghorbani, M., Meshginfar, N., & Mahoonak, A. S. (2016). A review on frying: Procedure, fat, deterioration progress and health hazards. *Journal of the Amercian Oil Chemists Society*, 93(4), 445–466. doi:10.1007/s11746-016-2791-z

Hou, C. L., Kang, N., Schlosser, C., Zhao, M. Y., Wang, Z. Y., & Zhang, D. Q. (2018). Heterocyclic aromatic amines in commercial Chinese dried meat products. *Journal of Food and Nutrition Research*, 57(2), 151–160.

Hui, T., Fang, Z. F., Ma, Q. L., Hamid, N., & Li, Y. L. (2023). Effect of cold atmospheric plasma-assisted curing process on the color, odor, volatile composition, and heterocyclic amines in beef meat roasted by charcoal and superheated steam. *Meat Science*, 196, 109046. doi:10.1016/j.meatsci.2022.109046

IARC. (1993). Some naturally occurring substances: Food items and constituents, heterocyclic aromatic amines and mycotoxins. In: *IARC Monographs on the Evaluation of Carcinogenic Risks to Human* (Vol. 56). Lyon: World Health Organization; International Agency for Research on Cancer.

Jackson, L. S., & Hargraves, W. A. (1994). The effect of processing time and temperature on the formation of meiqx and dimeiqx in a model system. *Abstracts of Papers of the Amercian Chemical Society,* 207, 11-AGFD.

Jamali, M. A., Zhang, Y. W., Teng, H., Li, S., Wang, F. L., & Peng, Z. Q. (2016). Inhibitory effect of *Rosa rugosa* tea extract on the formation of heterocyclic amines in meat patties at different temperatures. *Molecules,* 21(2), 173. doi:10.3390/molecules21020173

Janoszka, B. (2010). Heterocyclic amines and azaarenes in pan-fried meat and its gravy fried without additives and in the presence of onion and garlic. *Food Chemistry*, 120(2), 463–473. doi:10.1016/j.foodchem.2009.10.039

Jinap, S., Hasnol, N. D. S., Sanny, M., & Jahurul, M. H. A. (2018). Effect of organic acid ingredients in marinades containing different types of sugar on the formation of heterocyclic amines in grilled chicken. *Food Control,* 84, 478–484. doi:10.1016/j.foodcont.2017.08.025

Kataoka, H., Miyake, M., Saito, K., & Mitani, K. (2012). Formation of heterocyclic amine-amino acid adducts by heating in a model system. *Food Chemistry*, 130(3), 725–729. doi:10.1016/j.foodchem.2011.07.094

Kathuria, D., Hamid, Gautam, S., & Thakur, A. (2023). Maillard reaction in different food products: Effect on product quality, human health and mitigation strategies. *Food Control*, 153, 109911. doi:10.1016/j.foodcont.2023.109911

Keskekoglu, H., & Üren, A. (2014). Inhibitory effects of pomegranate seed extract on the formation of heterocyclic aromatic amines in beef and chicken meatballs after cooking by four different methods. *Meat Science*, 96(4), 1446–1451. doi:10.1016/j.meatsci.2013.12.004

Khalid, W., Maggiolino, A., Kour, J., Arshad, M. S., Aslam, N., Afzal, M. F., … & Korma, S. A. (2023). Dynamic alterations in protein, sensory, chemical, and oxidative properties occurring in meat during thermal and non-thermal processing techniques: A comprehensive review. *Frontiers in Nutrition,* 9, 1057457. doi:10.3389/fnut.2022.1057457

Khan, I. A., Cheng, Y. Q., Zhu, Z. S., Ijaz, M. U., Brohi, S. A., Ahmad, M. I., … & Ming, H. (2019). Occurrence of heterocyclic amines in commercial fast-food meat products available on the Chinese market and assessment of human exposure to these compounds. *Journal of Food Science*, 84(1), 192–200. doi:10.1111/1750-3841.14418

Khan, I. A., Khan, A., Zou, Y., Zhu, Z. S., Xu, W. M., Wang, D. Y., & Huang, M. (2022). Heterocyclic amines in cooked meat products, shortcomings during evaluation, factors influencing formation, risk assessment and mitigation strategies. *Meat Science*, 184. doi:10.1016/j.meatsci.2021.108693

Khan, I. A., Luo, J., Shi, H. B., Zou, Y., Khan, A., Zhu, Z. S., … & Huang, M. (2022). Mitigation of heterocyclic amines by phenolic compounds in allspice and perilla frutescens seed extract: The correlation between antioxidant capacities and mitigating activities. *Food Chemistry*, 368, 130845. doi:10.1016/j.foodchem.2021.130845

Kondjoyan, A., Chevolleau, S., Grève, E., Gatellier, P., Santé-Lhoutellier, V., Bruel, S., … & Debrauwer, L. (2010). Formation of heterocyclic amines in slices of Longissimus thoracis beef muscle subjected to jets of superheated steam. *Food Chemistry*, 119(1), 19–26. doi:10.1016/j.foodchem.2009.02.081

Lai, Y. W., Inbaraj, B. S., & Chen, B. H. (2023). Effects of oil and processing conditions on formation of heterocyclic amines and polycyclic aromatic hydrocarbons in Pork fiber. *Foods*, 12(18), 3504. doi:10.3390/foods12183504

Lai, Y. W., Inbaraj, B. S., Chen, B. H., & Smith, J. S. (2024). Analysis of polycyclic aromatic hydrocarbons via GC-MS/MS and heterocyclic amines via UPLC-MS/MS in crispy pork Spareribs for studying their formation during frying. *Foods*, 13(2), 185. doi:10.3390/foods13020185

Li, B., Wang, J. X., Cheng, Z., Song, B. G., Shu, C., Chen, Y., … & Tian, J. L. (2023). Flavonoids mitigation of typical food thermal processing contaminants: Potential mechanisms and analytical strategies. *Food Chemistry*, 416, 135793. doi:10.1016/j.foodchem.2023.135793

Li, X. F., Yang, Z. L., Deng, J. Y., Chen, C. G., Xu, B. C., & Li, P. J. (2023). Effect of quercetin and oil water separation system on formation of?-carboline heterocyclic amines during frying process of braised chicken drumsticks. *Current Research in Food Science*, 6. doi:10.1016/j.crfs.2022.100406

Li, Y., He, J. L., Quan, W., He, Z. Y., Qin, F., Tao, G. J., … & Chen, J. (2020). Effects of polyphosphates and sodium chloride on heterocyclic amines in roasted beef patties as revealed by UPLC-MS/MS. *Food Chemistry*, 326, 127016. doi:10.1016/j.foodchem.2020.127016

Li, Y., Quan, W., Wang, J. H., He, Z. Y., Qin, F., Wang, Z. J., … & Chen, J. (2021). Effects of ten vegetable oils on heterocyclic amine profiles in roasted beef patties using UPLC-MS/MS combined with principal component analysis. *Food Chemistry*, 347, 128996. doi:10.1016/j.foodchem.2020.128996

Liao, G. Z., Wang, G. Y., Xu, X. L., & Zhou, G. H. (2010). Effect of cooking methods on the formation of heterocyclic aromatic amines in chicken and duck breast. *Meat Science*, 85(1), 149–154. doi:10.1016/j.meatsci.2009.12.018

Liao, G. Z., Xu, X. L., & Zhou, G. H. (2009). Effects of cooked temperatures and addition of antioxidants on formation of heterocyclic aromatic amines in pork floss. *Journal of Food Processing and Preservation*, 33(2), 159–175. doi:10.1111/j.1745-4549.2008.00239.x

Linghu, Z. Y., Karim, F., & Smith, J. S. (2017). Amino acids inhibitory effects and mechanism on 2-amino-1-methyl-6-phenylimidazo [4,5-b]Pyridine (PhIP) formationin the maillard reaction model systems. *Journal of Food Science*, 82(12), 3037–3045. doi:10.1111/1750-3841.13959

Lu, F., Kuhnle, G. K., & Cheng, Q. F. (2017). Heterocyclic amines and polycyclic aromatic hydrocarbons in commercial ready-to-eat meat products on UK market. *Food Control*, 73, 306–315. doi:10.1016/j.foodcont.2016.08.021

Ma, H. W., Chen, W. J., Lv, M., Qi, X. B., Ruan, Q. F., Pan, C. Y., & Guo, A. L. (2023). The inhibitory mechanism of 2-amino-3,8-dimethylimidazo [4,5-f] quinoxaline (MeIQx) formation by ultraviolet-gallic acid (UV-GA) during the oil-frying process of squid. *Food Chemistry*, 418, 135957. doi:10.1016/j.foodchem.2023.135957

Mehr, A. E., Hosseini, S. E., Ardebili, S. M. S., & Khaneghah, A. M. (2020). The incorporation of polysaccharides in grilled beef patties: influence on the levels of polycyclic aromatic hydrocarbons and heterocyclic aromatic amines. *Journal of Food Measurement and Characterization*, 14(5), 2393–2401. doi:10.1007/s11694-020-00486-1

Meurillon, M., & Engel, E. (2016). Mitigation strategies to reduce the impact of heterocyclic aromatic amines in proteinaceous foods. *Trends in Food Science & Technology*, 50, 70–84. doi:10.1016/j.tifs.2016.01.007

Murkovic, M. (2004). Formation of heterocyclic aromatic amines in model systems. *Journal of Chromatography B-Analytical Technologies in the Biomedical and Life Sciences*, 802(1), 3–10. doi:10.1016/j.jchromb.2003.09.026

Murkovic, M., Weber, H. J., Geiszler, S., Fröhlich, K., & Pfannhauser, W. (1999). Formation of the food associated carcinogen 2-amino-1-methyl-6-phenylimidazo[4,5-b]pyridine (PhIP) in model systems. *Food Chemistry*, 65(2), 233–237. doi:10.1016/S0308-8146(98)00214-3

Oz, E., Aoudeh, E., Murkovic, M., Toldra, F., Gomez-Zavaglia, A., Brennan, C., … & Oz, F. (2023). Heterocyclic aromatic amines in meat: Formation mechanisms, toxicological implications, occurrence, risk evaluation, and analytical methods. *Meat Science*, 205, 109312. doi:10.1016/j.meatsci.2023.109312

Oz, F., Kaban, G., & Kaya, M. (2010). Heterocyclic aromatic amine contents of beef and lamb chops cooked by different methods to varying levels. *Journal of Animal and Veterinary Advances*, 9(10), 1436–1440.

Oz, F., & Kaya, M. (2011a). The inhibitory effect of black pepper on formation of heterocyclic aromatic amines in high-fat meatball. *Food Control*, 22(3–4), 596–600. doi:10.1016/j.foodcont.2010.10.010

Oz, F., & Kaya, M. (2011b). The inhibitory effect of red pepper on heterocyclic aromatic amines in fried beef longissimus dorsi muscle. *Journal of Food Processing and Preservation*, 35(6), 806–812. doi:10.1111/j.1745-4549.2011.00532.x

Oz, F., Oz, E., Aoudeh, E., Abd El-Aty, A. M., Zeng, M. M., & Varzakas, T. (2021). Is ultra-high temperature processed milk safe in terms of heterocyclic aromatic amines? *Foods*, 10(6), 1247. doi:10.3390/foods10061247

Oz, F., & Seyyar, E. (2016). Formation of heterocyclic aromatic amines and migration level of bisphenol-A in Sous-Vide-cooked trout fillets at different cooking temperatures and cooking levels. *Journal of Agricultural and Food Chemistry*, 64(15), 3070–3082. doi:10.1021/acs.jafc.5b05716

Oz, F., & Zikirov, E. (2015). The effects of sous-vide cooking method on the formation of heterocyclic aromatic amines in beef chops. *LWT-Food Science and Technology*, 64(1), 120–125. doi:10.1016/j.lwt.2015.05.050

Pan, H., Wang, Z. Y., Guo, H. T., Ni, N., & Zhang, D. Q. (2014). Heterocyclic aromatic amines in meat products consumed in China. *Food Science and Biotechnology*, 23(6), 2089–2095. doi:10.1007/s10068-014-0284-0

Pan, T., Wang, Z. Y., Chen, B. H., Hui, T., & Zhang, D. Q. (2020). Frying oils with lower levels of saturated fatty acids induce less heterocyclic amine formation in meat floss (boiled, shredded and fried pork). *International Journal of Food Science and Technology*, 55(2), 823–832. doi:10.1111/ijfs.14368

Pearson, A. M., Chen, C. H., Gray, J. I., & Aust, S. D. (1992). Mechanism(s) involved in meat mutagen formation and inhibition. *Free Radical Biology and Medicine*, 13(2), 161–167. doi:10.1016/0891-5849(92)90078-U

Persson, E., Oroszvári, B. K., Tornberg, E., Sjöholm, I., & Skog, K. (2008). Heterocyclic amine formation during frying of frozen beef burgers. *International Journal of Food Science and Technology*, 43(1), 62–68. doi:10.1111/j.1365-2621.2006.01390.x

Persson, E., Sjöholm, I., Nyman, M., & Skog, K. (2004). Addition of various carbohydrates to beef burgers affects the formation of heterocyclic amines during frying. *Journal of Agricultural and Food Chemistry*, 52(25), 7561–7566. doi:10.1021/jf0493831

Persson, E., Sjöholm, I., & Skog, K. (2002). Heat and mass transfer in chicken breasts: Effect on PhIP formation. *European Food Research and Technology*, 214(6), 455–459. doi:10.1007/s00217-001-0462-1

Persson, E., Sjöholm, I., & Skog, K. (2003). Effect of high water-holding capacity on the formation of heterocyclic amines in fried beefburgers. *Journal of Agricultural and Food Chemistry*, 51(15), 4472–4477. doi:10.1021/jf021089q

Pezdirc, M., Zegura, B., & Filipic, M. (2013). Genotoxicity and induction of DNA damage responsive genes by food-borne heterocyclic aromatic amines in human hepatoma HepG2 cells. *Food and Chemical Toxicology,* 59, 386–394. doi:10.1016/j.fct.2013.06.030

Polak, M. L., Demsar, L., Zahija, I., & Polak, T. (2020). Influence of temperature on the formation of heterocyclic aromatic amines in pork steaks. *Czech Journal of Food Sciences*, 38(4), 248–254. doi:10.17221/144/2019-CJFS

Puangsombat, K., Gadgil, P., Houser, T. A., Hunt, M. C., & Smith, J. S. (2011). Heterocyclic amine content in commercial ready to eat meat products. *Meat Science*, 88(2), 227–233. doi:10.1016/j.meatsci.2010.12.025

Puangsombat, K., Jirapakkul, W., & Smith, J. S. (2011). Inhibitory activity of Asian spices on heterocyclic amines formation in cooked beef patties. *Journal of Food Science*, 76(8), T174–T180. doi:10.1111/j.1750-3841.2011.02338.x

Quelhas, I., Petisca, C., Viegas, O., Melo, A., Pinho, O., & Ferreira, Implvo. (2010). Effect of green tea marinades on the formation of heterocyclic aromatic amines and sensory quality of pan-fried beef. *Food Chemistry*, 122(1), 98–104. doi:10.1016/j.foodchem.2010.02.022

Rannou, C., Laroque, D., Renault, E., Prost, C., & Sérot, T. (2016). Mitigation strategies of acrylamide, furans, heterocyclic amines and browning during the Maillard reaction in foods. *Food Research International*, 90, 154–176. doi:10.1016/j.foodres.2016.10.037

Rounds, L., Havens, C. M., Feinstein, Y., Friedman, M., & Ravishankar, S. (2012). Plant extracts, spices, and essential oils inactivate Escherichia coli O157:H7 and reduce formation of potentially carcinogenic heterocyclic amines in cooked beef patties. *Journal of Agricultural and Food Chemistry*, 60(14), 3792–3799. doi:10.1021/jf204062p

Salazar, R., Arámbula-Villa, G., Hidalgo, F. J., & Zamora, R. (2014). Structural characteristics that determine the inhibitory role of phenolic compounds on 2-amino-1-methyl-6-phenylimidazo [4, 5-b] pyridine (PhIP) formation. *Food Chemistry,* 151, 480–486. doi:10.1016/j.foodchem.2013.11.105

Sepahpour, S., Selamat, J., Khatib, A., Manap, M. Y. A., Razis, A. F. A., & Hajeb, P. (2018). Inhibitory effect of mixture herbs/spices on formation of heterocyclic amines and mutagenic activity of grilled beef. *Food Additives and Contaminants Part A-Chemistry Analysis Control Exposure & Risk Assessment*, 35(10), 1911–1927. doi:10.1080/19440049.2018.1488085

Shabbir, M. A., Raza, A., Anjum, F. M., Khan, M. R., & Suleria, H. A. R. (2015). Effect of thermal treatment on meat proteins with special reference to Heterocyclic Aromatic Amines (HAAs). *Critical Reviews in Food Science and Nutrition*, 55(1), 82–93. doi:10.1080/10408398.2011.647122

Shahidi, F., Wanasundara, P. K. J. P. D., & Wanasundara, U. N. (1997). Changes in edible fats and oils during processing. *Journal of Food Lipids*, 4(3), 199–231.

Shen, X., Chen, Y., Omedi, J. O., Zeng, M. M., Xiao, C. W., Zhou, Y. J., & Chen, J. (2023). Effects of volatile organic compounds of smoke from different woods on the heterocyclic amine formation and quality changes in pork patty. *Food Research International*, 173(Pt 1), 113262. doi:10.1016/j.foodres.2023.113262

Shin, H. S., Strasburg, G. M., & Gray, J. I. (2002). A model system study of the inhibition of heterocyclic aromatic amine formation by organosulfur compounds. *Journal of Agricultural and Food Chemistry*, 50(26), 7684–7690. doi:10.1021/jf025707e

Sinha, R., Kulldorff, M., Chow, W. H., Denobile, J., & Rothman, N. (2001). Dietary intake of heterocyclic amines, meat-derived mutagenic activity, and risk of colorectal adenomas. *Cancer Epedemiology Biomarkers & Prevention*, 10(5), 559–562.

Sinha, R., Rothman, N., Brown, E. D., Salmon, C. P., Knize, M. G., Swanson, C. A., ... & Felton, J. S. (1995). High concentrations of the carcinogen 2-amino-1-methyl-6-phenylimidazo-[4, 5-b] pyridine (PhIP) occur in chicken but are dependent on the cooking method. *Cancer Research,* 55(20), 4516–4519.

Sinha, R., Rothman, N., Salmon, C. P., Knize, M. G., Brown, E. D., Swanson, C. A., ... & Levander, O. A. (1998). Heterocyclic amine content in beef cooked by different methods to varying degrees of doneness and gravy made from meat drippings. *Food and Chemical Toxicology,* 36(4), 279–287. doi:10.1016/S0278-6915(97)00162-2

Skog, K. (2002). Problems associated with the determination of heterocyclic amines in cooked foods and human exposure. *Food and Chemical Toxicology*, 40(8), 1197–1203. doi:10.1016/S0278-6915(02)00052-2

Skog, K., Augustsson, K., Steineck, G., Stenberg, M., & Jagerstad, M. (1997). Polar and non-polar heterocyclic amines in cooked fish and meat products and their corresponding pan residues. *Food and Chemical Toxicology*, 35(6), 555–565. doi:10.1016/S0278-6915(97)00021-5

Skog, K., & Jagerstad, M. (1993). Incorporation of carbon atoms from glucose into the food mutagens MeIQx and 4, 8-DiMeIQx using 14C-labelled glucose in a model system. *Carcinogenesis*, 14(10), 2027–2031. doi:10.1093/carcin/14.10.2027

Skog, K., & Solyakov, A. (2002). Heterocyclic amines in poultry products: A literature review. *Food and Chemical Toxicology*, 40(8), 1213–1221. doi:10.1016/S0278-6915(02)00062-5

Soladoye, O. P., Shand, P., Dugan, M. E. R., Gariépy, C., Aalhus, J. L., Estévez, M., & Juárez, M. (2017). Influence of cooking methods and storage time on lipid and protein oxidation and heterocyclic aromatic amines production in bacon. *Food Research International,* 99, 660–669. doi:10.1016/j.foodres.2017.06.029

Sun, M. Y., Wang, J., Dong, J., Lu, Y. S., Zhang, Y., Dong, L., & Wang, S. (2023). Effects of different frying oils composed of various fatty acids on the formation of multiple hazards in fried Pork balls. *Foods*, 12(22), 4182. doi:10.3390/foods12224182

Supaphon, P., Kerdpiboon, S., Vénien, A., Loison, O., Sicard, J., Rouel, J., & Astruc, T. (2021). Structural changes in local Thai beef during sous-vide cooking. *Meat Ssience*, 175, 108442. doi:10.1016/j.meatsci.2021.108442

Szterk, A. (2015). Heterocyclic aromatic amines in grilled beef: The influence of free amino acids, nitrogenous bases, nucleosides, protein and glucose on HAAs content. *Journal of Food Composition and Aanalysis,* 40, 39–46. doi:10.1016/j.jfca.2014.12.011

Tai, C. Y., Lee, K. H., & Chen, B. H. (2001). Effects of various additives on the formation of heterocyclic amines in fried fish fibre. *Food Chemistry*, 75(3), 309–316. doi:10.1016/S0308-8146(01)00200-X

Tsen, S. Y., Ameri, F., & Smith, J. S. (2006). Effects of rosemary extracts on the reduction of heterocyclic amines in beef patties. *Journal of Food Science*, 71(8), C469–C473. doi:10.1111/j.1750-3841.2006.00149.x

Turesky, R. J., Konorev, D., Fan, X. Y., Tang, Y. J., Yao, L. H., Ding, X. X., … & Zhang, Q. Y. (2015). Effect of cytochrome P450 reductase deficiency on 2-amino-9H-pyrido[2,3-b]indole metabolism and DNA adduct formation in liver and extrahepatic tissues of mice. *Chemical Research in Toxicology*, 28(12), 2400–2410. doi:10.1021/acs.chemrestox.5b00405

Vangnai, K., Houser, T. A., Hunt, M. C., & Smith, J. S. (2014). Effect of enhancement on the formation of heterocyclic amines in cooked pork loins: Preliminary studies. *Meat Science*, 98(2), 88–93. doi:10.1016/j.meatsci.2014.05.007

Viegas, O., Moreira, P. S., & Ferreira, Implvo. (2015). Influence of beer marinades on the reduction of carcinogenic heterocyclic aromatic amines in charcoal-grilled pork meat. *Food Additives and Contaminants Part A-Chemistry Analysis Control Exposure & Risk Assessment*, 32(3), 315–323. doi:10.1080/19440049.2015.1010607

Wang, H. J., Chu, X. R., Du, P. F., He, H. J., He, F., Liu, Y. B., … & Abd El-Aty, A. M. (2023). Unveiling heterocyclic aromatic amines (HAAs) in thermally processed meat products: Formation, toxicity, and strategies for reduction–A comprehensive review. *Food Chemistry-X*, 19, 100833. doi:10.1016/j.fochx.2023.100833

Wang, Y., Hui, T., Zhang, Y. W., Liu, B., Wang, F. L., Li, J. K., … & Peng, Z. Q. (2015). Effects of frying conditions on the formation of heterocyclic amines and trans fatty acids in grass carp (Ctenopharyngodon idellus). *Food Chemistry*, 167, 251–257. doi:10.1016/j.foodchem.2014.06.109

Wong, D., Cheng, K. W., & Wang, M. F. (2012). Inhibition of heterocyclic amine formation by water-soluble vitamins in Maillard reaction model systems and beef patties. *Food Chemistry*, 133(3), 760–766. doi:10.1016/j.foodchem.2012.01.089

Xi, J., & Chen, Y. (2021). Analysis of the relationship between heterocyclic amines and the oxidation and thermal decomposition of protein using the dry heated soy protein isolate system. *LWT-Food Science and Technology*, 148, 111738. doi:10.1016/j.lwt.2021.111738

Xiong, Y. L. L. (2017). Inhibition of hazardous compound formation in muscle foods by antioxidative phytophenols. *Annals of the New York Academy of Sciences*, 1398(1), 37–46. doi:10.1111/nyas.13368

Xu, Y., Huang, T. R., Huang, Y. Q., & Huang, M. (2024). Effect of mulberry leaf (Morus alba L.) extract on the quality and formation of heterocyclic amines in braised muscovy duck. *Food Control*, 156, 110137. doi:10.1016/j.foodcont.2023.110137

Xue, C. Y., Chen, Q. C., He, Z. Y., Wang, Z. J., Qin, F., Yang, T. Y., … & Zeng, M. M. (2020). Non-precursors amino acids can inhibit β-carbolines through free radical scavenging pathways and competitive inhibition in roast beef patties and model food systems. *Meat Science*, 169, 108203. doi:10.1016/j.meatsci.2020.108203

Xue, C. Y., Deng, P., Quan, W., Li, Y., He, Z. Y., Qin, F., … & Zeng, M. M. (2022). Ginger and curcumin can inhibit heterocyclic amines and advanced glycation end products in roast beef patties by quenching free radicals as revealed by electron paramagnetic resonance. *Food Control*, 138, 109038. doi:10.1016/j.foodcont.2022.109038

Yan, Y., You, F. H., Zeng, M. M., Chen, J., Huang, J. J., & Jiang, J. (2021). Evaluating the effects of temperature and time on heterocyclic aromatic amine profiles in roasted pork using combined UHPLC-MS/MS and multivariate analysis. *Food Research International*, 141, 110134. doi:10.1016/j.foodres.2021.110134

Yang, D. D., He, Z. Y., Gao, D. M., Qin, F., Deng, S. L., Wang, P., … & Zeng, M. M. (2019). Effects of smoking or baking procedures during sausage processing on the formation of heterocyclic amines measured using UPLC-MS/MS. *Food Chemistry,* 276, 195–201. doi:10.1016/j.foodchem.2018.09.160

Yang, D. D., Jing, J., Zhang, Z. G., He, Z. Y., Qin, F., Chen, J., & Zeng, M. M. (2020). Accumulation of heterocyclic amines across low-temperature sausage processing stages as revealed by UPLC-MS/MS. *Food Research International*, 137, 109668. doi:10.1016/j.foodres.2020.109668

Yao, G. M., Zhou, Y. J., Li, Z. P., & Ma, Q. S. (2023). Heterocyclic aromatic amines in roasted chicken: Formation and prediction based on heating temperature and time. *Food Chemistry*, 405, 134822. doi:10.1016/j.foodchem.2022.134822

Yin, X. Y., Du, H. Z., Xu, M., Chen, Q., & Kong, B. H. (2021). Heterocyclic aromatic amine level and quality characteristics of selected Harbin red sausages in the northern Chinese market. *Meat Science*, 172, 108360. doi:10.1016/j.meatsci.2020.108360

Yu, C. D., Shao, Z. P., Liu, B., Zhang, Y., & Wang, S. (2016). Inhibition of 2-amino-1-methyl-6-phenylimidazo [4,5-b]pyridine (PhIP) formation by alkoxy radical scavenging of flavonoids and their quantitative structure-activity relationship in a model system. *Journal of Food Science*, 81(8), C1908–C1913. doi:10.1111/1750-3841.13381

Yu, D., Chen, M. S., & Yu, S. J. (2016). Effect of sugarcane molasses extract on the formation of 2-amino-1-methyl-6-phenylimidazo[4,5-b]pyridine (PhIP) in a model system. *Food Chemistry*, 197, 924–929. doi:10.1016/j.foodchem.2015.11.052

Zamora, R. (2021). Lipid oxidation and the formation of processing-induced toxicants in foods: Acrylamide and heterocyclic aromatic amines. *Journal of the Amercian Oil Chemists Society,* 98, 97–97.

Zamora, R., Alcón, E., & Hidalgo, F. J. (2012). Effect of lipid oxidation products on the formation of 2-amino-1-methyl-6-phenylimidazo [4, 5-b] pyridine (PhIP) in model systems. *Food Chemistry*, 135(4), 2569–2574. doi:10.1016/j.foodchem.2012.06.062

Zamora, R., Alcón, E., & Hidalgo, F. J. (2013). Comparative formation of 2-amino-1-methyl-6-phenylimidazo[4,5-*b*]pyridine (PhIP) in creatinine/phenylalanine and creatinine/phenylalanine/4-oxo-2-nonenal reaction mixtures. *Food Chemistry*, 138(1), 180–185. doi:10.1016/j.foodchem.2012.09.116

Zamora, R., Alcón, E., & Hidalgo, F. J. (2014). Ammonia and formaldehyde participate in the formation of 2-amino-1-methyl-6-phenylimidazo[4,5-*b*]pyridine (PhIP) in addition to creati(ni)ne and phenylacetaldehyde. *Food Chemistry*, 155, 74–80. doi:10.1016/j.foodchem.2014.01.039

Zamora, R., & Hidalgo, F. J. (2015). 2-Amino-1-methyl-6-phenylimidazo[4,5-b]pyridine (PhIP) formation and fate: An example of the coordinate contribution of lipid oxidation and Maillard reaction to the production and elimination of processing-related food toxicants. *RSC Advances*, 5(13), 9709–9721. doi:10.1039/c4ra15371e

Zamora, R., & Hidalgo, F. J. (2020). Formation of heterocyclic aromatic amines with the structure of aminoimidazoazarenes in food products. *Food Chemistry*, 313, 126128. doi:10.1016/j.foodchem.2019.126128

Zamora, R., Lavado-Tena, C. M., & Hidalgo, F. J. (2020a). Identification of precursors and formation pathway for the heterocyclic aromatic amine 2-amino-3-methylimidazo(4,5-f)quinoline (IQ). *Journal of Agricultural and Food Chemistry*, 68(28), 7474–7481. doi:10.1021/acs.jafc.0c02869

Zamora, R., Lavado-Tena, C. M., & Hidalgo, F. J. (2020b). Reactive carbonyls and the formation of the heterocyclic aromatic amine 2-amino-3, 4-dimethylimidazo (4, 5-f) quinoline (MeIQ). *Food Chemistry*, 324, 126898. doi:10.1016/j.foodchem.2020.126898

Zeng, M. M., Li, Y., He, Z. Y., Qin, F., & Chen, J. (2016). Effect of phenolic compounds from spices consumed in China on heterocyclic amine profiles in roast beef patties by UPLC-MS/MS and multivariate analysis. *Meat Science*, 116, 50–57. doi:10.1016/j.meatsci.2016.02.002

Zeng, M. M., Wang, J. H., Zhang, M. R., Chen, J., He, Z. Y., Qin, F., ... & Chen, J. (2018). Inhibitory effects of Sichuan pepper (Zanthoxylum bungeanum) and sanshoamide extract on heterocyclic amine formation in grilled ground beef patties. *Food Chemistry*, 239, 111–118. doi:10.1016/j.foodchem.2017.06.097

Zeng, M. M., Zhang, M. R., He, Z. Y., Qin, F., Tao, G. J., Zhang, S., ... & Chen, J. (2017). Inhibitory profiles of chilli pepper and capsaicin on heterocyclic amine formation in roast beef patties. *Food Chemistry*, 221, 404–411. doi:10.1016/j.foodchem.2016.10.061

Zhang, C. X., Xi, J., Wang, S. T., Ma, Y. X., & Wang, X. D. (2022). Effects of deep fat frying conditions on the formation of heterocyclic aromatic amines in chicken meat. *Food Science and Technology*, 42, e67321. doi:10.1590/fst.67321

Zhang, L., Badar, I. H., Chen, Q., Xia, X. F., Liu, Q., & Kong, B. H. (2022). Changes in flavor, heterocyclic aromatic amines, and quality characteristics of roasted chicken drumsticks at different processing stages. *Food Control*, 139, 109104. doi:10.1016/j.foodcont.2022.109104

Zhang, Y., Luo, Z., Shao, Z. P., Yu, C. D., & Wang, S. (2014). Effects of antioxidants of bamboo leaves and flavonoids on 2-amino-1-methyl-6-phenylimidazo[4,5-b]pyridine (PhIP) formation in chemical model systems. *Journal of Agricultural and Food Chemistry*, 62(20), 4798–4802. doi:10.1021/jf500483y

Zhao, T. P., Zhang, C. X., Ma, Y. X., & Wang, X. D. (2023). Effect of simultaneous application of antioxidants in oil and in chicken on heterocyclic aromatic amine formation in deep-fat fried chicken breast. *LWT-Food Science and Technology*, 187, 115287. doi:10.1016/j.lwt.2023.115287

Zhou, B., Zhao, Y. L., Wang, X. C., Fan, D. M., Cheng, K. W., & Wang, M. F. (2018). Unraveling the inhibitory effect of dihydromyricetin on heterocyclic aromatic amines formation. *Journal of the Science of Food and Agriculture*, 98(5), 1988–1994. doi:10.1002/jsfa.8682

Zhou, Y. J., Zhang, M. M., Ma, Z. Y., Li, Z. P., Ma, Q. S., & Wang, L. (2023). Effect of spices on the formation and inhibition of heterocyclic amines in barbecued pork. *Journal of Food Measurement and Characterization*, 883–893. doi:10.1007/s11694-023-02207-w

Zhu, Q., Zhang, S., Wang, M. F., Chen, J., & Zheng, Z. P. (2016). Inhibitory effects of selected dietary flavonoids on the formation of total heterocyclic amines and 2-amino-1-methyl-6-phenylimidazo[4,5-b]pyridine (PhIP) in roast beef patties and in chemical models. *Food & Function*, 7(2), 1057–1066. doi:10.1039/c5fo01055a

Zimmerli, B., Rhyn, P., Zoller, O., & Schlatter, J. (2001). Occurrence of heterocyclic aromatic amines in the Swiss diet: Analytical method, exposure estimation and risk assessment. *Food Additives and Contaminants Part A-Chemistry Analysis Control Exposure & Risk Assessment*, 18(6), 533–551. doi:10.1080/02652030110035084

8 Dioxins or Dioxin-Like Compounds and Safety

Food Processing and Manufacturing Related Effects

Didem Peren Aykas Cinkilic

8.1 INTRODUCTION

Food contaminants can be introduced in various ways and stages of the course of production and manufacturing; hence, they also pose a significant risk to food safety and ultimately to public health. These contaminants include, among others, chemicals that could be described as persistent, bioaccumulative, and toxic chemicals. Possible ways such substances can enter food chains are through environmental exposure, including water, soil, or even air contamination. Human exposure has been mainly through diet, since these substances bioaccumulate over time in food products (Fiedler, 2003b). The persistence of these chemicals in the environment, together with the chemical's ability to bioaccumulate in living organisms, enhances the potential for adverse effects of these substances, including endocrine disruption and carcinogenicity; this requires strict

DOI: 10.1201/9781032712291-8

control measures and monitoring throughout the processing and manufacturing steps of food (X. Li et al., 2024; Y. F. Li et al., 2024; Wang et al., 2024).

The rapid rate of industrial growth throughout the world has improved both the quantity and variety of products we use daily. This, however, carries an equal increase in the volume and types of unwanted by-products from the very same production processes. Most of these by-products are harmful and will thus, unless treated appropriately, be very long-lasting impacts on the environment (Esder & Arkun, 2018). One of the more notorious chemical groups of contaminants includes so-called persistent organic pollutants or POPs. They are notorious for their stability and resistance against natural breakdown processes. Due to their semi-volatile nature, POPs can distribute over long distances by air or on particles carriers, becoming widespread in ecosystems and organisms, including humans. They have, incredibly, even been found in very remote areas such as deserts, oceans, and the polar regions, far away from any anthropogenic or industrial activity, which shows their potential for wide dispersal. POPs include polychlorinated biphenyls (PCBs), polychlorinated dibenzo-p-dioxins and furans (PCDD/Fs), polybrominated diphenyl ethers (PBDEs), along with certain pesticides and fluorinated chemicals. These are mostly lipophilic in nature and tend to bioaccumulate in the fatty tissues of all living organisms, while the concentration increases in species higher up the food chain (Wikoff et al., 2012).

Among the most persistent and toxic contaminants are dioxins and dioxin-like compounds, which pose particular risks due to their ability to bioaccumulate in living organisms and persist in the environment for extremely long periods. Therefore, for the processes related to food and manufacture, it plays a very important role in controlling the presence of dioxins and its related compounds in the interest of ensuring food safety and public health protection (Esder & Arkun, 2018). These compounds, commonly found in soil, water sediments, and air (Institute of Medicine, 2003), are characterized by their high melting point, low vapor pressure, and strong solubility in a non-polar environment. Due to fat solubility and due to strong resistance to metabolic breakdown, they tend to accumulate and biomagnify in the food chain. The presence of dioxins has been found worldwide in a wide variety of environmental samples (Van Overmeire et al., 2001). These chemicals, although present at very low concentrations, enter the human body mainly through the food chain, and this cause serious public health concern (Institute of Medicine, 2003).

Despite concerted efforts to reduce the levels of these compounds in the environment and food over the past few decades, persistence and wide distribution continue to pose a threat. Indeed, the chemical stability allows persistence in the environment and bioaccumulation in the food chain, thus posing a risk of long-term exposure to these compounds (Institute of Medicine, 2003). The issue is most amplified in the sensitive groups, such as infants and young children, as they have a more fragile system; hence, they are easily susceptible to the toxic results of these chemicals. An example would be an infant with breastfeeding that may get higher doses of dioxins and dioxin-like compounds because of the fat-soluble nature of these contaminants, which can accumulate in breast milk. Similarly, the populations ingesting fish, wild game, or other animal products as staple foods are also at greater risk for higher exposures. Exposure during pregnancy and lactation causes a deficiency in neurological, cognitive, and motor development in children. The long-term effects of exposure are incompletely defined and one of the reasons the public remains concerned, and these concerns justify stringent food safety and environmental controls (Bergkvist et al., 2008; Institute of Medicine, 2003; Patandin et al., 1999).

8.2 CHEMISTRY AND TOXICITY OF DIOXINS AND DIOXIN-LIKE COMPOUNDS

The term "dioxin" is a collective term meant to include a group of structurally and chemically related halogenated aromatic hydrocarbons which include polychlorinated dibenzo-*p*-dioxins (PCDDs), polychlorinated dibenzofurans (PCDFs), and "dioxin-like" PCBs (Srogi, 2008). These are persistent environmental pollutants, present in trace levels throughout nearly all compartments of the global ecosystem. Contrary to other chemicals of environmental concern, such as PCBs, polychlorinated naphthalenes (PCNs), and polychlorinated pesticides like dichloro-diphenyl-trichloroethane (DDT) or pentachlorophenol (PCP), are produced on purpose, it never happened in the case of PCDDs and PCDFs (Fiedler, 2003a). Most are formed as unwanted by-products from industrial processes, like incineration, chlorine bleaching of paper and pulp, and production of particular pesticides, herbicides, and fungicides, and hence they are persistent unwanted environmental contaminants, although they are sometimes synthesized in small amounts for scientific research (Schecter et al., 2006).

Dioxins and dioxin-like compounds belong to a broad class of structurally related chemical substances showing environmental and biological persistence. They elicit similar effects owing to their sharing a common mechanism of action (Schecter et al., 2006). These chemicals are extremely toxic and remain effective in the environment for a longer period because of their type of chemical structure. The United Nations Environment Programme (UNEP) has placed dioxins and furans into the "Dirty Dozen" list. Greenpeace describes these compounds as "among the most dangerous chemicals on the planet" (Srogi, 2008). Dioxins and furans are colorless and odorless organic compounds consisting of carbon, hydrogen, oxygen, and chlorine atoms (European Commission, 2001). The toxicity of different dioxins is determined by the number of chlorine atoms and their positions within the dioxin molecule. PCDDs and PCDFs are aromatic hydrocarbons characterized by a three-ring structure, specifically the PCDDs comprising two benzene rings connected by two adjacent oxygen bridges (C-O-C bonds), with varying levels of chlorine substitution on each benzene ring. In contrast, PCDFs have a similar structure but are joined by one C-O-C bond and one direct C-C bond, lacking a bridging atom (Food and Agriculture Organization of the United Nations, 2024) (Figure 8.1). Variations in their structural configuration of these compounds, particularly in the positions where chlorine atoms are attached, create various forms and congeners, with each distinct form of PCDD/PCDF referred to as a congener (Dopico & Gómez, 2015).

There are theoretically 75 distinct PCDD congeners and 135 PCDF congeners, each with physical and chemical properties determined by the number of chlorine atoms and their specific positions on the molecular core (Dopico & Gómez, 2015).

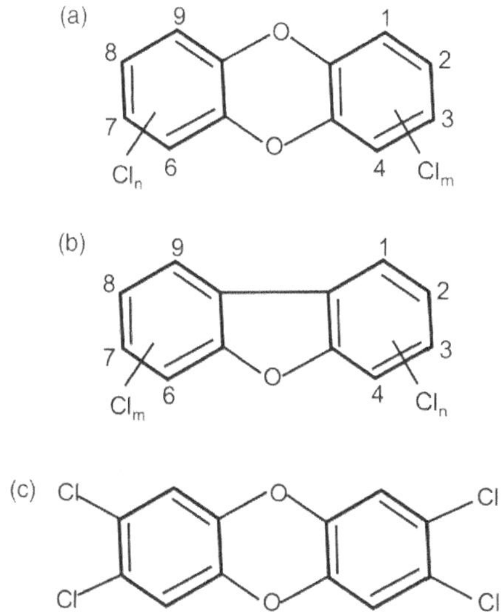

FIGURE 8.1 Chemical structures of a) polychlorinated dibenzo-p-dioxins (PCDDs), b) polychlorinated dibenzofurans (PCDFs), and c) 2,3,7,8-tetrachlorodibenzo-p-dioxin (TCDD).

These compounds belong to two groups of planar, tricyclic ethers, with up to eight chlorine atoms attached to carbon atoms at positions 1 through 4 and 6 through 9. Of the 210 theoretically possible congeners, the 17 that have chlorine atoms it the 2,3,7,8-positions – comprising seven of the PCDDs and ten of the PCDFs – are recognized as toxic to exposed laboratory animals, resistant to chemical, biological, and physical degradation, and therefore tend to accumulate in the environment and living organisms, including humans (Arslan-Alaton & Olmez-Hanci, 2013; Fiedler, 2003a). 2,3,7,8-tetrachlorodibenzo para dioxin (TCDD) is regarded as the most toxic compound in this group, known for its mutagenic, carcinogenic, and teratogenic effects (Food and Agriculture Organization of the United Nations, 2024; World Health Organization (WHO, 2023). The toxicity level decreases as more chlorine atoms are added, with a sharp decline occurring once the number of chlorine atoms exceeds five (Food and Agriculture Organization of the United Nations, 2024; Inserm, 2000).

Evaluating the risk of exposure to a single chemical is difficult due to uncertainties in toxicity estimates, which often depend on data from animal studies being applied to humans. Exposure assessment also involves complicating issues, such as considering different routes of exposure, such as ingestion, skin contact, and inhalation, each with its own rate and pattern. With over 100,000 chemicals in use, both humans and wildlife are exposed to complex mixtures of chemicals that further complicate risk assessment, a concept known as cumulative risk (DeVito, 2012). Cumulative impacts can be described as the potential adverse effects on human health as a result of parallel exposure to various environmental and social stressors. Cumulative risk seeks to measure as accurately as possible the total risks from these combined exposures (Huang et al., 2018). The most accurate estimation of risk from chemical mixtures would depend on data specific for the mixture itself. Testing every possible combination of chemicals is not practical.

Therefore, methods have been developed to evaluate mixtures based on the toxicity of individual chemicals, though these methods involve assumptions about chemical interactions. Chemicals may interact in various ways: additively, synergistically (exceeding expectations), or antagonistically (falling short of expectations). Despite this variability, risk assessors typically assume additivity, which involves either summing the effects of the chemicals (response addition) or summing their doses (dose addition) (DeVito, 2012).

Toxic Equivalency Factors (TEFs) represent the relative on in vivo and in vitro potencies of compounds with dioxin-like activity. Measurements are reported versus a reference chemical, 2,3,7,8-tetrachlorodibenzo-p-dioxin (TCDD). Because of its high toxicity, TCDD has been assigned a TEF value of 1. Other dioxin-like compounds are assigned TEFs relative to the activity of TCDD. TEF values are periodically updated as new experimental data is reported (Huwe, 2002). The TEF method allows for the evaluation of health risks based on the comparative toxicity of other dioxin-like chemicals to that of TCDD. Since toxicological data are available only for a limited number of dioxins, the TEF method is used to express their toxicity in terms of relative potency factors from that of TCDD. For example, just six of the 29 compounds listed in the TEF technique have been studied for carcinogenicity. The TEF methodology represents only a temporary approach and aims to estimate the health risks of dioxin mixtures, with ongoing efforts to refine the methodology and address its limitations (DeVito, 2012).

Dioxins and related compounds have been widely released into the environment, mainly from industrial activities. However, according to regulatory measure, it has been reported that estimated intake levels in the UK have considerably reduced overtime, with daily exposures falling from 7.2 pg TEQ/kg for adults and 18 pg TEQ/kg for toddlers in the year 1982 to 1.8 and 4.6 pg TEQ/kg, respectively, by 1997 (Committee on Toxicity, 2001). In 1990, WHO established a tolerable daily intake (TDI) of 10 pg/kg body weight for TCDD, the most toxic dioxin, but new evidence regarding neurodevelopmental, reproductive, and endocrine effects led to a reassessment (Van Leeuwen et al., 2000; WHO European Centre for Environment and Health International Programme on Chemical Safety, 1998). In 1998, WHO European Center for Environment and Health (WHO- ECEH) and International Programme on Chemical Safety (IPCS) held a consultation to reassess the health risks of dioxins and review the TDI. The TDI reflects a safe level for lifelong exposure, with occasional short-term increases posing minimal risk as long as long-term averages are maintained. Although mild health effects have been noted at current exposure levels (2–6 pg TEQ/kg per day), they are not considered harmful. Experts recommend keeping intake below 4 pg TEQ/kg and ultimately aiming to reduce it to less than 1 pg TEQ/kg body weight per day (WHO European Centre for Environment and Health International Programme on Chemical Safety, 1998).

8.3 HEALTH HAZARDS ASSOCIATED WITH DIOXIN EXPOSURE

While many health risks have been linked to heavy dioxin exposure from unintended releases, the impact of low-level exposure through food on human health remains unclear due to insufficient data for risk assessment and the lack of unexposed populations for comparison. Detecting dioxins is particularly challenging due to their unique chemical structure and extremely low concentrations in food, making testing both time-consuming and costly (Yaktine et al., 2006). Despite these difficulties, evidence suggests that dioxin exposure above background levels poses increased risks, especially for sensitive groups such as developing fetuses and infants, who may experience neurodevelopmental and neurobehavioral effects from dioxins, particularly through breast milk (Boersma, 2001; Huisman et al., 1995; Marinković et al., 2010; Tai et al., 2013). Short-term high-level exposure can damage liver functions and can also result in chloracne, a chronic skin condition characterized by cysts and dark-colored acne. It generally occurs on the face. The shouldering, back, chest, and abdomen are affected during severe poisoning (Marinković et al., 2010).

Long-term exposure to a certain type of dioxin, TCDD, is linked to nervous, immune, reproductive, and endocrine disorders. Due to the persistence of TCDD in the body, this may also lead to the development of atherosclerosis, hypertension, diabetes, and neuropathy. Extensive epidemiological and experimental studies carried out on humans and animals have resulted in the International Agency for Research on Cancer classifying TCDD as a "known human carcinogen." Similarly, the U.S. Environmental Protection Agency (EPA) identifies TCDD as a non-genotoxic carcinogen and a potent tumor promoter, although other congeners are also likely to be carcinogenic (Marinković et al., 2010; Pelclovä et al., 2006; Yaktine et al., 2006).

8.4 SOURCES OF DIOXINS AND NOTABLE CONTAMINATION INCIDENTS

Dioxins result to a large extent from the unintended by-products of various types of human activities, such as manufacturing and combustion processes in industry, improper incineration of waste, and fuels of vehicles. There could be significant releases resulting from accidents in chemical factories, causing contamination in the environment surrounding such factories. Domestic heating systems, burning by agriculture, and backyard disposal of wastes also add the sources of dioxins. Similarly, natural events like volcanic eruptions and forest fires are known to generate these compounds. Waste incinerators also rank among the highest contributors to environmental dioxin release, particularly those which burn solid and hospital waste, because of incomplete combustion processes and large volumes treated (Food and Agriculture Organization of The United Nations (FAO, 2008). Furthermore, dioxins and related toxic compounds are emitted from the open burning of household, municipal, and medical waste, as well as from fires in landfills, agricultural lands, and forests (Dyke et al., 1997; Kulkarni et al., 2008). Additionally, partially burned organic compounds create a favorable environment for carbon to remain in the ash, which promotes the formation of dioxin like chemical compounds. Increased organic content and incomplete combustion lead to a greater concentration of dioxin compounds (Esder & Arkun, 2018). Furthermore, industrial processes, including motor vehicle production, steel manufacturing, and other metal industries, metal tempering, casting, and recycling processes also release significant amounts of these hazardous substances into the environment (Tysklind et al., 1993).

PCBs were commercially produced for around five decades, beginning in the 1920s, through the direct chlorination of biphenyl. These compounds were manufactured as mixtures rather than as individual congeners. The various commercial PCB mixtures are identified by their chlorine content and were marketed under different brand names, including "Aroclor" in the USA, "Clophen" in Germany, "Phenoclor" in France, "Fenclor" in Italy, and "Kanechlor" in Japan. These mixtures found a wide range of applications, including in coatings, inks, flame retardants, and paints, with major uses in electronic devices, heat-transfer systems, and hydraulic fluids. However, because of the persistent nature of PCBs in the environment, many countries implemented bans on their use in open applications during the 1970s. Even so, despite these restrictions, PCBs continue to be used in closed systems such as capacitors and transformers, though their use is supposed to decrease over time. The primary pathway for the release of PCBs into the environment is waste disposal from households and industrial activities (Srogi, 2008).

The paper production industry is one of the main sources of environmental dioxin pollution, largely due to chlorophenol containing chemicals used in the production process (Esder & Arkun, 2018; Gilpin et al., 2003; Thacker et al., 2007). Since the emissions of dioxins and especially PCDDs, PCDFs, and TCDD are significant from pulp and paper processing, strict regulations have been applied especially in Europe. As a result, Europe, especially Scandinavia, USA, and Canada have been implementing chlorine-free bleaching methods, such as elemental chlorine free and total chlorine free, to reduce these emissions. However, conventional chlorine bleaching, particularly in pulp processing, remains common in China, India, Southeast Asia, where PCDD, PCDF, TCDD emissions persist (Esder B & Arkun G, 2018; Gilpin et al., 2003; Xiao et al., 2017).

Dioxins and related compounds are also byproducts of the production of insecticides, fungicides, and bactericides. Their formation is enhanced by the presence of heat (150–600°C), catalysts, such as copper, iron, and aluminum salts, radicals, alkaline conditions, and ultraviolet (UV) light. As the factors promoting dioxin formation are known, their formation can be prevented from taking place during production of pesticides by resorting to other alternative materials and modification in the production conditions, thereby controlling their formation (Holt et al., 2012). Dioxin and related compounds have also been detected in consumer products like coffee filter papers, diapers, napkins plastic plates, PET bottles, and milk and juice cartons (Esder & Arkun, 2018; Hashimoto et al., 1992; Sun et al., 2024). High-temperature exposure, like conditions from direct sunlight or hot liquid contact, increase the risk of dioxin contamination coming from these materials (Esder & Arkun, 2018).

8.4.1 The Seveso Accident

In the latter half of the 20th century, apart from nuclear incidents, the 1976 chemical disaster in Seveso, Italy, was one of the most significant events to raise public awareness about the dangers industrial accidents pose to both people and the environment. The incident underscored the close connection between workplace activities and residential safety, demonstrating how industrial processes can impact human health and the environment. This disaster prompted a stronger focus on occupational and environmental health risks, leading to significant regulatory changes, most notably the European Union's 1984 Seveso Directive, which aimed to prevent and control the effects of similar industrial accidents (Pesatori & Bertazzi, 2012).

The disaster itself took place on July 10, 1976, when a chemical reactor exploded in a subsidiary of the Givaudan firm, an ICMESA plant in Meda, near Seveso, about 25 km north of Milan in Italy. The plant produced 2,4,5-trichlorophenol, a chemical used in cosmetics and pharmaceuticals (Eskenazi et al., 2018; Homberger et al., 1979). Although chemists and engineers were aware of previous accidents in plants producing this chemical, the installation was designed with great care in order to avoid uncontrollable exothermic reactions and dangerous formations of TCDD. In particular, great care was taken in choosing solvents, heating, and cooling systems in order to avoid processes under pressure. But nevertheless, an accident still occurred, triggered by an unexpected exothermic reaction of heating up, gradual decomposition of reaction mass, gas formation, and a rise in pressure (Homberger et al., 1979).

The accident resulted in the release of a pressurized aerosol cloud of the products formed during the runaway reaction that contain harmful substances, including sodium hydroxide, ethylene glycol, sodium trichlorophenate, and an estimated 15 to 30 kg of TCDD, which came into dissemination over an area of 18 km^2. Thus, this accident allowed the highest ever known exposure of a residential population to TCDD (Eskenazi et al., 2018; Homberger et al., 1979). The first symptoms of poisoning came out on the sixth day, when people of region exposed to the cloud complained about nausea, headaches, and irritation in the eye, while 19 children were hospitalized due to skin lesions as a contact result of the skin with caustic chemicals (Eskenazi et al., 2018; Pesatori et al., 2003). In the following weeks, there was also widespread mortality of animals and plants. About 200 instances of chloracne prevail among children and some others (Eskenazi et al., 2018). On a Longer-term scale, excess deaths due to cardiovascular and respiratory diseases have been reported considering the uncontaminated population living nearby. These mortality rates are likely linked with the psychosocial impact of the accident along with the chemical exposure. There was also an increase in diabetes, cancer incidence, especially gastrointestinal, lymphatic, and hematopoietic tissues (Bertazzi et al., 1998; Eskenazi et al., 2018). Accordingly, this disaster showed great danger industrial chemicals are and brought new safety measures all over Europe (Eskenazi et al., 2018).

8.4.2 Yusho and Yu-Cheng Incidents

The Yusho incident is a major Japanese food poisoning disaster in 1968 caused by the consumption of rice bran oil contaminated with PCBs and other toxic by-products such as PCDDs and polychlorinated dibenzofurans (PCDFs). This contamination resulted from a leak in a heating tube through which PCB flowed, causing chloracne-like lesions among the residents of the northern part of Kyushu, eventually resulting in registration of 1,860 cases and 539 confirmed deaths.

A poisoning incident, referred to as Yu-cheng, which also occurred in 1979, when cooking oil contamination with a mixture of PCBs, PCDFs, polychlorinated quaterphenyls, and a small amount of PCDDs caused over 2,000 people from central Taiwan to fall victims (Aoki, 2001; Lida et al., 1999; Mitoma et al., 2015; Ryan et al., 1993; Woolf, 2022). These two poisoning examples underscore the crucial need for strict good manufacturing practices (GMPs) and regulatory oversight to prevent toxic substances from entering food products and causing severe health crises (Woolf, 2022).

8.4.3 Agent Orange in Vietnam, 1961–1971

Between 1961 and 1971, the United States Armed Forces used nearly 19.5 million gallons of herbicides for tactical defoliation and crop destruction across Southeast Asia. The primary mixtures used included Agent Orange (composed of 2,4-dichlorophenoxyacetic acid and 2,4,5-trichlorophenoxyacetic acid), Agent White (composed of triisopropanolamine salts of 2,4-D and picloram), and Agent Blue (which contains cacodylic acid), which each name refers to the identification stripes painted on storage drums. At least 12.6 million gallons of the 2,4,5-T herbicides were contaminated with dioxin (2,3,7,8-tetrachlorodibenzo-p-dioxin – TCDD), leading to the largest recorded dioxin contamination (Schecter et al., 1995; Stellman & Stellman, 2004).

Agent Orange, a defoliant, was containing levels between 0 to 100 ppm of the toxic dioxin TCDD were sprayed over jungle forests, riverbanks, and agricultural areas across southern and central Vietnam (Olson & Morton, 2019; Stellman & Stellman, 2004). Having been the most widespread herbicide used during the Vietnam War, the effects of Agent Orange on the health and environment continue to persist. The persistence presence of TCDD in human tissue and the environment continues to raise significant health concerns for the Vietnamese population, who have been exposed to its effects since the spraying began (Olson & Morton, 2019; Schecter et al., 1995; Stellman & Stellman, 2004).

This chemical defoliation strategy purposed the jungle vegetation to be cleared out to expose enemy positions, to secure military bases, and to destroy food sources for opposing forces. The strategy brought both intended and unintended consequences, affecting the region's soils and sediments, disrupting agriculture and forests, and affecting the civilian population of South Vietnamese, US, and North Vietnamese military personnel (Olson & Morton, 2019; Stellman & Stellman, 2004). Due to its longer half-life, TCDD accumulates in animal and human fat, becoming up to dozen times more concentrated through biomagnification along the food chain. Studies show consistently high levels of dioxins in Vietnam's soil and wildlife, as well as in human samples, including serum, fat tissue, and breast milk (Olson & Morton, 2019; Stellman & Stellman, 2004).

8.4.4 Other Dioxin Contamination Accidents

Besides the rice oil poisoning incidents in Japan and Taiwan, the Seveso accident in Italy, and the use of Agent Orange in Vietnam, several other notable dioxin contamination incidents have occurred around the World, including Belgian dioxin contamination in animal feed in 1999 (Casey et al., 2010), Irish dioxin contamination in animal feed in 2008 (Casey et al., 2010; Heres et al., 2010), both of these crises affected poultry, eggs, pork, and beef, also leads to widespread recalls, export bans, and heightened concerns about food safety regulations across Europe (Casey et al., 2010). In 1982, there was another case of dioxin contamination in Times Beach, Missouri, USA, due to improper waste disposal with dioxins. The highly increased levels of dioxin contamination led to the necessary evacuation and abandonment of the town. This incident brought environmental pollution and waste management problems into the attention of the public across the US (Sun, 1983; United States Environmental Protection Agency (EPA, 2024a). Another case of contamination with chemical wastes involved the Love Canal site in New York, US, during the 1970s with dioxins among other toxic compounds. This led the US President Jimmy Carter to declare the nation's first federal emergency for a man-made environmental disaster (Hoffman, 1995).

Another notable incidents occurred in Saginaw River and Tittabawassee River in Michigan, USA. The Dow Chemical Company facility in Midland, Michigan, which was producing over 1,000 chemicals, including chlorine-based products that generated dioxin and furan by-products since 1897. Past waste disposal practices led to elevated dioxin levels in the Tittabawassee River and over 50 miles downstream into Saginaw Bay, with cleanup activities still ongoing (United States Environmental Protection Agency) (EPA, 2024b).

8.5 ENVIRONMENTAL DISTRIBUTION AND PATHWAYS OF DIOXINS AND RELATED COMPOUNDS INTO THE FOOD SUPPLY CHAIN

The primary environmental concern with dioxins and related compounds arises from their poor solubility in water and high affinity for fat, which causes them to bind with sediment and organic matter. This property leads to their accumulation in the fatty tissues of animals and humans, where they persist due to their non-biodegradable nature. As a result, dioxins bioaccumulate within the food chain, progressively building up in fat tissues once they enter the environment, whether through air or water (European Commission, 2001). Routes of exposure to dioxins and dioxin-like compounds via the food supply chain are given in Figure 8.2. In aquatic environments, dioxins readily attach to mineral and organic particles, leading to their accumulation along the aquatic food chain. Oceans, lakes, and rivers are particularly vulnerable to aerial deposition of these compounds, as well as contamination from wastewater or effluents released by industries such as paper and pulp processing or metallurgical activities (Food and Agriculture Organization of the United Nations) (FAO, 2008). Furthermore, dioxins can enter soil through various pathways, such as deposition from airborne dioxins, the application of sewage sludge contaminated with these compounds to agricultural land, the flooding of pastures with tainted sludge, and the historical use of pesticides and

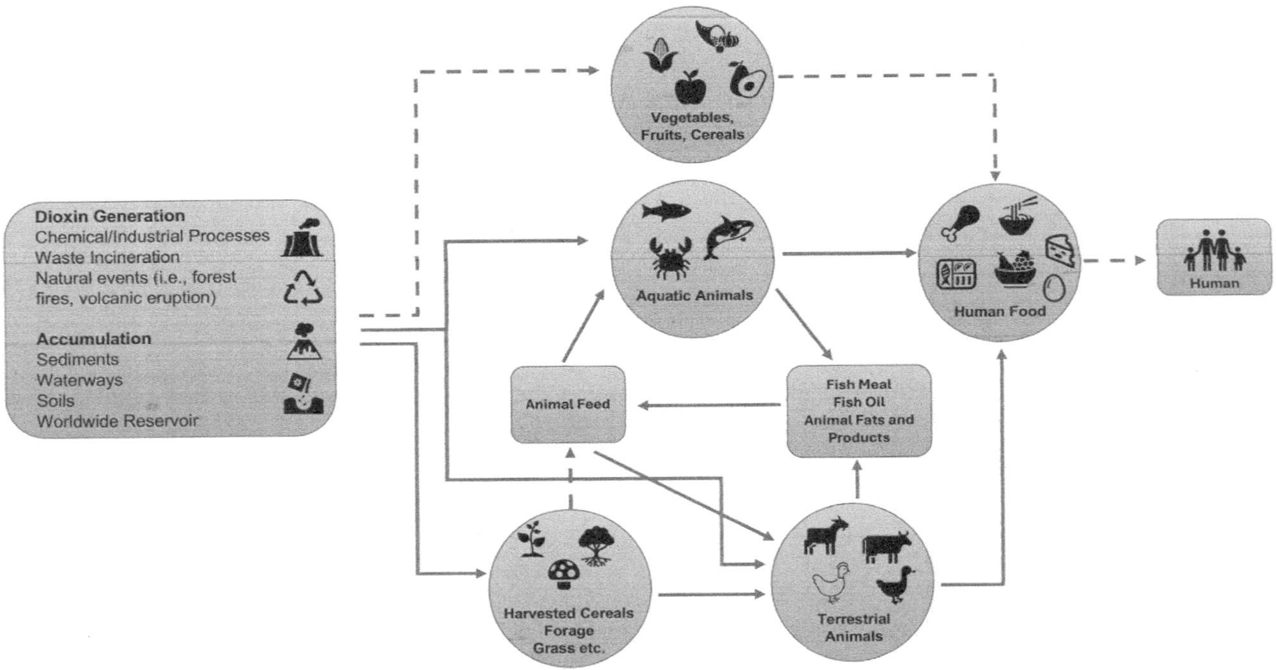

FIGURE 8.2 Routes of exposure to dioxins and dioxin-like compounds via the food supply chain. Solid lines represent pathways with a greater relative contribution of dioxin-like compounds, while dashed lines indicate pathways with lower contributions. Specific point sources are identified within these routes.

fertilizers that contained dioxins (Food and Agriculture Organization of The United Nations) (FAO, 2008).

Dioxin bioaccumulation in the food supply starts when animals consume contaminated plants, soil, or sediment, subsequently becoming food for humans or producing food products. Direct consumption of dioxin-contaminated plants by humans is a much less significant source (Yaktine et al., 2006). Dioxins are not added intentionally to food, nor are they created during the processing of food (Charnley & Doull, 2005). Human exposure to dioxins is predominantly through the consumption of contaminated food, which has accounted for more than 95% of the total environmental exposure. Other sources, such as water, air, and soil, contribute minimally. Besides, accidental air contamination in occupational settings and the contamination of animal feeds have sometimes been reported as sources of poisoning incidents involving human and livestock exposure (Parzefall, 2002; Patandin et al., 1999).

Combustion, incineration, smelting, and the pulp industry have been labeled as the key sources of dioxins by the U.S. EPA and other agencies. Incinerators that burned medical and municipal wastes were significant contributors to dioxin air emissions in the early 1990s, emitting several kilograms of toxic equivalents (TEQ) yearly. In 1995, those emissions had dropped dramatically due both to improved incinerator technology and more stringent regulations, banning certain chemicals and industrial practices. Even with such improvements, dioxins also emanate from various diffuse sources like home heating, diesel engines, and forest fires, which are more difficult to regulate (Huwe, 2002). Significantly, backyard burning of household waste contributed a great percentage and comprised 22% of dioxin emissions in North America from 1996 to 1997 (Commoner et al., 2000).

The mechanisms by which PCDD and PCDF compounds move through the environment are relatively well understood. As multimedia pollutants, once released, they spread across multiple environmental compartments (Fiedler, 2003a). PCDDs and PCDFs are by-products formed during industrial processes involving chlorine and organochlorines. They also result from incomplete combustion in incineration processes, including open burning of garbage and trash, especially wastes that are of electronic origin (Food and Agriculture Organization of The United Nations (FAO), 2008; Weber et al., 2008). These organic pollutants are also produced by natural events, such as forest fires and volcanic eruptions, but they are not as significant a source (Food and Agriculture Organization of the United Nations, 2024; Food and Agriculture Organization of The United Nations (FAO), 2008; World Health Organization (WHO, 2023). They result from a number of manufacturing processes, including metal smelting, paper pulp chlorine bleaching, and the manufacture of certain herbicides and pesticides. In most cases, the incineration of waste, including solid and hospital waste, results in considerable dioxin emissions due to incomplete combustion. However, modern technology makes controlled waste incineration possible that significantly minimizes dioxin emissions (World Health Organization (WHO, 2023). Although dioxin formation is local, their environmental distribution is global. Higher levels of dioxins are usually confined to soils, sediments, and foods, particularly dairy products, meat, fish, and shellfish, while plants, water, and air usually contain much lower levels (World Health Organization (WHO), 2023).

More than 10 million tons of PCB containing wastes remain available worldwide, mostly in countries which do not have the facilities to manage PCB wastes effectively. Most of the materials kept in any storage facility as a PCB-based waste oil is usually filled with high levels of PCDFs. Long-time improper storage and disposal of these materials can risk dioxin releases at quantities hence contaminating food supplies both in human and animal. Due to the environmental hazards, PCB-based waste should be treated as hazardous and disposed accordingly. The safest disposal is by high-temperature incineration in specially designed facilities intended for this purpose to avoid possible contamination (Melymuk et al., 2022; World Health Organization (WHO, 2023).

Due to efforts aimed at reducing dioxins, environmental levels have gone down. For instance, there was a decline in the levels of PCDD/PCDF by 50% in Lake Siskiwit between the 1980s to 1998 as stated by Baker and Hites (Baker & Hites, 2000). Dioxin levels in the eggs of birds in the Great Lakes and Baltic regions have declined. Human exposure to dioxins, mainly through diet, has also declined, with intake in Europe being reduced by about 60% between the 1980s and the mid-1990s (De Wit et al., 1994; Hebert et al., 1994). Current dioxin intake in the US is within the guidelines recommended by WHO, but exceeds the more stringent safety threshold of EPA and could cause health effects, especially on sensitive populations (Huwe, 2002).

8.6 FOOD AND DIOXINS

Once animals are exposed to dioxins, there are no rapid ways of reducing their dioxin burden effectively. Contaminated products may be removed from the market and destruction of the animals be considered. In an attempt to manage contamination and allow for the safety of the animal or its products, depuration is one of the common approaches. It is a practice where specific drugs or treatments are withheld for some period before allowing the animals or their products into the consumption market. This is a very important waiting period, as it ensures all the contaminants are cleared off the animals' systems for minimal risk to the consumers (Huwe, 2002).

Dioxins persist for a long period in animals; thus, this requires an extremely long withdrawal period to control contamination. For instance, the half-lives of PCDD/Fs in milk were estimated to range between 40 to 190 days (Bogdal et al., 2017; Firestone et al., 1979; Jensen & Hummel, 1982; Tuinstra et al., 1992), and between 90 to 200 days in the adipose tissue of beef cattle (Startin et al., 1994; Stephens et al., 1995; Thorpe et al., 2001). Although in chickens, half-lives are shorter in both adipose tissue and eggs, from 25 to 60 days (Stephens et al., 1995), hens generally accumulate less dioxins in the body as compared to males. This reduced accumulation could be the effect of the physiological functions in female, which, in general, deposit the dioxins in eggs, therefore "clearing" their body burdens of these toxic substances (De Vries et al., 2006).

Surprisingly enough, a study carried out in the Netherlands and other EU countries shows that organic eggs are generally more contaminated with dioxins than conventional ones, and some organic farms exceed the EU's maximum permissible levels. This is because organic hens, having more access to the outdoors, take on dioxins from several sources, including commercial and non-commercial feed, soil, plants, worms, and insects. The most important source is soil, followed by worms and insects, while plants and commercial feed contribute only negligible. Free-ranging chickens consume dioxin-contaminated soil, insects, and worms, passing these toxins into the egg yolk with remarkable efficiency. To make matters worse, environmental dioxin pollution due to historical waste burning and industrial processes adds to the problem of dioxin intake. Moreover, dioxins could cross-contaminate between chickens. Although this reduction in the amount of time hens spend outdoors would lower the dioxin levels of organic eggs, doing so may also eliminate some of the advantages thought to be derived from organic farming (De Vries et al., 2006; Huwe, 2002; Schoeters & Hoogenboom, 2006). Primary deputation methods are depuration itself and increasing carcass and fat weights to dilute the concentration. However, these methods often have one significant disadvantage: they are economically unfavorable because of the extended time required for effective reduction (Huwe, 2002).

Fish intake dioxins and PCBs into their body mainly through their gills and food, accumulating these contaminants primarily in their fatty tissues and liver. While bottom-dwelling and bottom-feeding fish species are generally more exposed to contaminated sediments than pelagic fish, the levels of dioxins and PCBs in bottom-dwelling species do not always exceed those found in pelagic species. The accumulation of the contaminants also related to the factors, including fish's size, diet, and physiological characteristics, fish's age, weight, lipid content, and the environmental conditions of their surroundings (Codex Alimentarius International Food Standards, 2018). Dioxin contaminants are found in higher concentrations in the fillet of the oily fish, like salmon, herring, mackerel, or trout, compared to lean fish, which tend to accumulate these contaminants more in their liver (Codex Alimentarius International Food Standards, 2018; Nøstbakken et al., 2021). The transfer of these contaminants to fillet is affected by the factors such as fish species, growth patterns, and the level of dioxins and related contaminants present in the environment (Codex Alimentarius International Food Standards, 2018).

The U.S. FDA conducted a market basket survey of 280 commonly consumed food products to monitor human exposure to contaminants, including dioxins. This survey also assesses pesticide residues, contaminants, and nutrient levels to protect public health. This study provided annual estimates of the average daily dioxin intake for the US population from 2000 to 2004 (U.S. Food and Drug Administration (FDA, 2007). Using these FDA estimates, Charnley and Doull calculated dietary dioxin intake across foods sources for the US population, attributing 32.2% of intake to meat, 16.4% to dairy, 10.8% to fruits and vegetables, 6.3% to poultry, 5.8% to fish, 2.4% to eggs, 1.5% to fats/oils, and 24.6% to other foods. It is worth noting that the food categories used in this study were not always precise; may not be perfectly generated, for example "vegetable beef soup" is grouped under "fruits and vegetables" despite containing meat, with its dioxin content likely coming from the beef fat (Charnley & Doull, 2005). Although fruits, vegetables, and grains contain much lower concentrations of dioxin and related compounds compared to meat, fish, poultry, eggs, and dairy, their high consumption volume means they still contribute to overall dioxin exposure, which may help to explain the higher intake levels (Institute of Medicine, 2003).

8.7 MITIGATION STRATEGIES IN FOOD PROCESSING

The Food and Agriculture Organization of the United Nations (FAO) and World Health Organization (WHO) recommend control measures and practices to reduce the dioxin and related compound's contamination risk in food and feed. These practices include implementing Good Agricultural Practices (Gaps), Good Manufacturing Practices (Gmps), Good Storage Practices (Gsps), Good Animal Feeding Practices (Gafps), and Good Laboratory Practices (Glps). Accordingly, in order to prevent environmental contamination by dioxins and related compounds, it is crucial to identify agricultural land affected by emissions, accidents, or illegal waste disposal to reduce risks to food and feed. If the agricultural area is under the risk of contamination, the agricultural practices should be avoided or limited. Sewage sludge is also a risk factor for dioxin contamination; therefore, the application of the sewage sludge contaminated with dioxins may result in these substances attaching to plants and increase the livestock exposure. It is essential to monitor sewage sludge used in agriculture for dioxins and should be treated if needed. Livestock, game, and poultry contacted with dioxin-contaminated soil can accumulate these contaminants if they consume the soil or the plants. Therefore, it is essential to identify these areas and regulate the access of the animals. Outdoor production in these areas should also be limited if needed. Reducing contamination in water and wild fish can take years, as dioxin and related compounds are very persistent in the environment. To reduce the expose through wild fish, it is important to identify polluted areas (sea, lake, river, etc.) and affected fish species. Fishing in these waters should be closely monitored and restricted if necessary (Codex Alimentarius International Food Standards, 2018).

As mentioned in the previous sections, the primary source of the dioxin exposure of human is animal-derived foods, such as poultry, fish, eggs, meat, and milk, as these substances tend to accumulate in the fat of those food products. Also, in lactating animals, these toxic substances can be present in milk fat, while in hens, they may concentrate in the egg yolk. To mitigate this transfer risk, implementing the control measures should be taken starting at the feed and feed ingredient levels. Reducing dioxin contamination in animal feed level will quickly lower contaminant levels in food products of animal origin. These control measures include identifying potentially contaminated areas in the feed supply chain, tracing sources of frequently contaminated feed or ingredients, following national guidelines, and monitoring feed and ingredient compliance with these guidelines. Buyers and consumers should seek assurances from suppliers that feed and its ingredients are sourced from certified facilities with certified production processes and quality assurance programs, such as HACCP principles, and that documentation verifying compliance with national guideline or maximum levels, if applicable, is provided (Codex Alimentarius International Food Standards, 2018).

REFERENCES

Aoki, Y. (2001). Polychlorinated biphenyls, polychloronated dibenzo-p-dioxins, and polychlorinated dibenzofurans as endocrine disrupters–What we have learned from Yusho disease. *Environmental Research*, *86*(1), 2–11. https://doi.org/10.1006/enrs.2001.4244

Arslan-Alaton, I., & Olmez-Hanci, T. (2013). Sources of environmental pollution: Persistent organic pollutants. In Suib, S. L. (ed), *New and Future Developments in Catalysis: Catalysis for Remediation and Environmental Concerns* (pp. 303–398). Elsevier B.V. https://doi.org/10.1016/B978-0-444-53870-3.00012-5

Baker, J. I., & Hites, R. A. (2000). Siskiwit lake revisited: Time trends of polychlorinated dibenzo-p-dioxin and dibenzofuran deposition at Isle Royale, Michigan. *Environmental Science and Technology*, *34*(14), 2887–2891. https://doi.org/10.1021/es991280p

Bergkvist, C., Öberg, M., Appelgren, M., Becker, W., Aune, M., Ankarberg, E. H., Berglund, M., & Håkansson, H. (2008). Exposure to dioxin-like pollutants via different food commodities in Swedish children and young adults. *Food and Chemical Toxicology*, *46*(11), 3360–3367. https://doi.org/10.1016/j.fct.2008.07.029

Bertazzi, P. A., Llaria, B., Brambilla, G., Consonni, D., & Pesatori, A. C. (1998). The seveso studies on early and long-term effects of dioxin exposure: A review. *Environmental Health Perspectives, 106*, 625–633. http://ehpnetl.niehs.nih.gov/docs/1998/Suppl-2/625-633bertazzi/abstract.html

Boersma, E. R. (2001). Environmental exposure to polychlorinated biphenyls (PCBs) and dioxins. *APMIS*, *109*(S103), 625–633. https://doi.org/10.1111/j.1600-0463.2001.tb05773.x

Bogdal, C., Züst, S., Schmid, P., Gyalpo, T., Zeberli, A., Hungerbühler, K., & Zennegg, M. (2017). Dynamic transgenerational fate of polychlorinated biphenyls and dioxins/furans in lactating cows and their offspring. *Environmental Science and Technology*, *51*(18), 10536–10545. https://doi.org/10.1021/acs.est.7b02968

Casey, D. K., Lawless, J. S., & Wall, P. G. (2010). A tale of two crises: The Belgian and Irish dioxin contamination incidents. *British Food Journal*, *112*(10), 1077–1091. https://doi.org/10.1108/00070701011080212

Charnley, G., & Doull, J. (2005). Human exposure to dioxins from food, 1999–2002. *Food and Chemical Toxicology*, *43*(5), 671–679. https://doi.org/10.1016/j.fct.2005.01.006

Codex Alimentarius International Food Standards. (2018). *Code of Practice for The Prevention and Reduction of Dioxins, Dioxin-Like PCBs and Non-Dioxin-Like PCBs in Food and Feed.* Codex Alimentarius.

Committee on Toxicity. (2001). *Committee on Toxicity of Chemicals in Food, Consumer Products and the Environment Statement on the Tolerable Daily Intake for Dioxins and Dioxin-Like Polychlorinated Biphenyls.* Committee on Toxicity.

Commoner, B., Woods, P., Eisl, B. H., & Couchot, K. (2000). *Long-Range Air Transport of Dioxin from North American Sources to Ecologically Vulnerable Receptors in Nunavut, Arctic Canada Final Report to the North American Commission for Environmental Cooperation.* North American Commission for Environmental Cooperation. www.cec.org

De Vries, M., Kwakkel, R. P., & Kijlstra, A. (2006). Dioxins in organic eggs: A review. *NJAS–Wageningen Journal of Life Sciences*, *54*(2), 207–221. https://doi.org/10.1016/S1573-5214(06)80023-0

DeVito, M. (2012). Toxicity equivalence factors for dioxin and related compounds. In A. Schecter (Ed.), *Dioxins and Health Including Other Persistent Organic Pollutants and Endocrine Disruptors* (3rd ed., pp. 37–52). Wiley.

De Wit, C. A., Jamberg, U. G., Asplund, L. T., Jansson, B., Olsson, M., Odsjo, T., Lindstedt, I. L., Andersson, O., Bergek, S., Hjelt, M., Rappe, C., Jansson, A., & Nygren, M. (1994). The Swedish dioxin survey: Summary of results from PCDD/F and coplanar PCB analyses in Biota. *Organohalogen Compounds*, *20*, 47–50.

Dopico, M., & Gómez, A. (2015). Review of the current state and main sources of dioxins around the world. *Journal of the Air and Waste Management Association*, *65*(9), 1033–1049. https://doi.org/10.1080/10962247.2015.1058869

Dyke, P., Coleman, P., & James, R. (1997). Dioxins in ambient air, Bonfire Night 1994. *Chemosphere, 34*, 1191–1201.

Esder, B., & Arkun, G. (2018). Contamination of dioxins in foods. *International Journal of Food Engineering Research*, *4*(1), 1–12.

Eskenazi, B., Warner, M., Brambilla, P., Signorini, S., Ames, J., & Mocarelli, P. (2018). The Seveso accident: A look at 40 years of health research and beyond. *Environment International*, *121*, 71–84. https://doi.org/10.1016/j.envint.2018.08.051

European Commission. (2001). *Fact Sheet on Dioxin in Feed and Food*. European Commission.

Fiedler, H. (2003a). Dioxins and Furans (PCDD/PCDF). In H. Fiedler (Ed.), *The Handbook of Environmental Chemistry Vol. 3, Part O Persistent Organic Pollutants*. Springer-Verlag.

Fiedler, H. (2003b). Dioxins in milk, meat, eggs and fish. In J. P. Felix D'Mello (Ed.), *Food Safety: Contaminants and Toxins* (pp. 153–174). CABI publishing. https://cabidigitallibrary.org

Firestone, D., Clower, M., Borsetti, A. P., Teske, R. H., & Long, P. E. (1979). Polychlorodibenzo-p-dioxin and pentachlorophenol residues in milk and blood of cows fed technical pentachlorophenol. *Journal of Agriculture Food Chemistry*, *27*(6), 1171–1177.

Food and Agriculture Organization of the United Nations. (2024). *Chapter 2. Main soil Contaminants and Their Fate in the Soil Environment.* Food and Agriculture Organization of the United Nations. https://openknowledge.fao.org/server/api/core/bitstreams/fe5df8d6-6b19-4def-bdc6-62886d824574/content/src/html/chapter-02-2.html

Food and Agriculture Organization of the United Nations (FAO). (2008). *Dioxins in the Food Chain Prevention and Control of Contamination What Are Dioxins?* Food and Agriculture Organization of the United Nations. www.fao.org/uploads/media/FAO_Fact_Sheet_020408.pdf

Gilpin, R. K., Wagel, D. J., & Solch, J. G. (2003). Production, distribution, and fate of polychlorinated dibenzo-p-dioxins, dibenzofurans, and related organohalogens in the environment. In A. Schecter & T. A. Gasiewicz (Eds.), *Dioxins and Health* (pp. 55–87). Wiley. https://doi.org/10.1002/0471722014.ch2

Hashimoto, S., Ito, H., & Morita, M. (1992). Elution of polychlorinated dibenzo-P-dioxins and dibenzofurans from coffee filter papers. *Chemosphere*, *25*(3), 297–305.

Hebert, E., Norstrom, I. R. J., Simon, M., Braune, B. M., Weseloh, D. V, & Macdonaldtgs, C. R. (1994). Temporal trends and sources of PCDDs and PCDFs in the great lakes: Herring gull egg monitoring, 1981-1991. *Environmental Science Technology*, *28*(7), 1268–1277.

Heres, L., Hoogenboom, R., Herbes, R., Traag, W., & Urlings, B. (2010). Tracing and analytical results of the dioxin contamination incident in 2008 originating from the Republic of Ireland. *Food Additives and Contaminants – Part A*, *27*(12), 1733–1744. https://doi.org/10.1080/19440049.2010.522598

Hoffman, A. J. (1995). An uneasy rebirth at love canal. *Environment*, *37*(2), 4–31. https://doi.org/10.1080/00139157.1995.9929215

Holt, E., Weber, R., Stevenson, G., & Gaus, C. (2012). Formation of dioxins during exposure of pesticide formulations to sunlight. *Chemosphere*, *88*(3), 364–370. https://doi.org/10.1016/j.chemosphere.2012.03.058

Homberger, E., Reggiani, G., Sambeth, J., & Wipf, H. K. (1979). The Seveso accident: Its nature, extent and consequences. *Annals of Occupational Hygiene, 22*, 327–370. http://annhyg.oxfordjournals.org/

Huang, H., Wang, A., Morello-Frosch, R., Lam, J., Sirota, M., Padula, A., & Woodruff, T. J. (2018). Cumulative risk and impact modeling on environmental chemical and social stressors. *Current Environmental Health Reports*, *5*(1), 88–99. https://doi.org/10.1007/s40572-018-0180-5

Huisman, M., Koopman-Esseboomb, C., Fidler, V., Hadders-Algrad, M., van der Paauwe, C. G., GMTh Tuinstraf, L., Weisglas-Kuperusb, N., Sauerb, P. J., Touweng, B. C., & Rudy

Boersmaa, E. (1995). Perinatal exposure to polychlorinated biphenyls and dioxins and its effect on neonatal neurological development. *Early Human Development, 41* (2), 111–127.

Huwe, J. K. (2002). Dioxins in food: A modern agricultural perspective. *Journal of Agricultural and Food Chemistry, 50*(7), 1739–1750. https://doi.org/10.1021/jf011265f

Inserm. (2000). *Dioxins in the Environment What are the Health Risks? Synthesis and Recommendations.* Institut national de la santé et de la recherche médicale (INSERM).

Institute of Medicine. (2003). Dioxins and dioxin-like compounds in the food supply: Strategies to decrease exposure. In *Dioxins and Dioxin-like Compounds in the Food Supply.* National Academies Press. https://doi.org/10.17226/10763

Jensen, D. J., & Hummel, R. A. (1982). Secretion of TCDD in milk and cream following the feeding of TCDD to lactating dairy cows. *Bulletin of Environmental Contamination and. Toxicology, 29*, 440–446.

Kulkarni, P. S., Crespo, J. G., & Afonso, C. A. M. (2008). Dioxins sources and current remediation technologies–A review. *Environment International, 34*(1), 139–153. https://doi.org/10.1016/j.envint.2007.07.009

Li, X., Shen, X., Jiang, W., Xi, Y., & Li, S. (2024). Comprehensive review of emerging contaminants: Detection technologies, environmental impact, and management strategies. *Ecotoxicology and Environmental Safety*, 278, 116420. https://doi.org/10.1016/j.ecoenv.2024.116420

Li, Y. F., Hao, S., Ma, W. L., Yang, P. F., Li, W. L., Zhang, Z. F., Liu, L. Y., & Macdonald, R. W. (2024). Persistent organic pollutants in global surface soils: Distributions and fractionations. *Environmental Science and Ecotechnology, 18*, 100311. https://doi.org/10.1016/J.ESE.2023.100311

Lida, T., HirakawaJ, H., Matsuedal, T., Takenakal, S., yu, M.-L., & Leon Guo, Y.-L. (1999). Recent trend of polychlorinated dibenzo-P-dioxins and their related compounds in the blood and Sebum of Yusho and Yu-Cheng patients. *Chemosphere, 38*(5), 981–993.

Marinković, N., Pašalić, D., Ferenčak, G., Gršković, B., & Rukavina, A. S. (2010). Dioxins and human toxicity. *Archives of Industrial Hygiene and Toxicology, 61*(4), 445–453. https://doi.org/10.2478/10004-1254-61-2010-2024

Melymuk, L., Blumenthal, J., Sáňka, O., Shu-Yin, A., Singla, V., Šebková, K., Pullen Fedinick, K., & Diamond, M. L. (2022). Persistent problem: Global challenges to managing PCBs. *Environmental Science and Technology, 56*(12), 9029–9040. https://doi.org/10.1021/acs.est.2c01204

Mitoma, C., Uchi, H., Tsukimori, K., Yamada, H., Akahane, M., Imamura, T., Utani, A., & Furue, M. (2015). Yusho and its latest findings–A review in studies conducted by the Yusho Group. *Environment International, 82*, 41–48. https://doi.org/10.1016/j.envint.2015.05.004

Nøstbakken, O. J., Rasinger, J. D., Hannisdal, R., Sanden, M., Frøyland, L., Duinker, A., Frantzen, S., Dahl, L. M., Lundebye, A. K., & Madsen, L. (2021). Levels of omega 3 fatty acids, vitamin D, dioxins and dioxin-like PCBs in oily fish; A new perspective on the reporting of nutrient and contaminant data for risk–benefit assessments of oily seafood. *Environment International*, 147, 106322. https://doi.org/10.1016/j.envint.2020.106322

Olson, K. R., & Morton, L. W. (2019). Long-term fate of agent orange and dioxin TCDD contaminated soils and sediments in Vietnam Hotspots. *Open Journal of Soil Science, 9*(1), 1–34. https://doi.org/10.4236/ojss.2019.91001

Parzefall, W. (2002). Risk assessment of dioxin contamination in human food. *Food and Chemical Toxicology, 40*, 1185–1189. www.elsevier.com/locate/foodchemtox

Patandin, S., Dagnelie, P. C., Mulder, P. G. H., Op De Coul, E., Van Der Veen, J. E., Weisgias-Kuperus, N., & Sauer1, P. J. J. (1999). Dietary exposure to polychlorinated biphenyls and dioxins from infancy until adulthood: A comparison between breast-feeding, toddler, and long-term exposure. *Environmental Health Perspectives, 107*(1), 45–51.

Pelclovä, D., Urban, P., Preiss, J., Lukää, E., Fenclovä, Z., Navrätil, T., Dubskä, Z., & Senholdovä, Z. (2006). Adverse health effects in humans exposed to 2,3,7,8-Tetrachlorodibenzo-p-Dioxin (TCDD). *Reviews on Environmental Health, 21*(2). www.pops.int/documents/convtext/convtext_en.pdf

Pesatori, A. C., & Bertazzi, P. A. (2012). The Seveso accident. In A. Schecter (Ed.), *Dioxins and Health: Including Other Persistent Organic Pollutants and Endocrine Disruptors* (3rd ed., p. 644). John Wiley & Sons.

Pesatori, A. C., Consonni, D., Bachetti, S., Zocchetti, C., Bonzini, M., Baccarelli, A., & Bertazzi, P. A. (2003). Short-and long-term morbidity and mortality in the population exposed to dioxin after the "Seveso accident". *Industrial Health, 41*, 127–138.

Ryan, J. J., Levesque, D., Panopio, L. G., Sun, W. F., Masuda, Y., & Kuroki, H. (1993). Environmental contamination and toxicology elimination of Polychlorinated Dibenzofurans (PCDFs) and Polychlorinated Biphenyls (PCBs) from human blood in the Yusho and Yu-Cheng rice oil poisonings. *Archives of Environmental Contamination. Toxicology, 24*, 504–512.

Schecter, A., Birnbaum, L., Ryan, J. J., & Constable, J. D. (2006). Dioxins: An overview. *Environmental Research, 101*(3), 419–428. https://doi.org/10.1016/j.envres.2005.12.003

Schecter, A., Dai, L. C., Thi, L., Thuy, B., Quynh, T., Minh, Q., Cau, H. D., Hoang Phiet, P., Thi, N., Phluong, N., Constable, J. D., Baughman, R., Ryan, J. J., Furst, P., & Raisanen, S. (1995). Agent orange and the vietnamese: The persistence of elevated dioxin levels in human tissues. *American Journal of Public Health, 85*(4), 516–522.

Schoeters, G., & Hoogenboom, R. (2006). Contamination of free-range chicken eggs with dioxins and dioxin-like polychlorinated biphenyls. *Molecular Nutrition and Food Research, 50*(10), 908–914. https://doi.org/10.1002/mnfr.200500201

Srogi, K. (2008). Levels and congener distributions of PCDDs, PCDFs and dioxin-like PCBs in environmental and human samples: A review. *Environmental Chemistry Letters, 6*(1), 1–28. https://doi.org/10.1007/s10311-007-0105-2

Startin, J. R., Wrigiit, C., & Kelly, M. (1994). TOX depletion rates of PCDDs in Bull Calf tissues. *Organohalogen Compounds, 21*, 347–350.

Stellman, S. D., & Stellman, J. M. (2004). Exposure opportunity models for Agent Orange, dioxin, and other military herbicides used in Vietnam, 1961–1971. *Journal of Exposure Analysis and Environmental Epidemiology, 14*(4), 354–362. https://doi.org/10.1038/sj.jea.7500331

Stephens, R. D., Petreas, M. X., & Hayward', D. G. (1995). Biotransfer and bioaccumulation of dioxins and furans from soil: Chickens as a model for foraging animals. *Science of the Total Environment, 175*, 253–273.

Sun, M. (1983). Missouri's costly dioxin lesson. *Science, 219*(4583), 367–369. https://doi.org/DOI:10.1126/science.6849139

Sun, M., Wang, Z., Cao, Z., & Dong, Z. (2024). Infants exposure to chemicals in diapers: A review and perspective. *Science of the Total Environment, 953*, 176072. https://doi.org/10.1016/j.scitotenv.2024.176072

Tai, P. T., Nishijo, M., Anh, N. T. N., Maruzeni, S., Nakagawa, H., Van Luong, H., Anh, T. H., Honda, R., Kido, T., & Nishijo, H. (2013). Dioxin exposure in breast milk and infant neurodevelopment in Vietnam. *Occupational and Environmental Medicine,* 70(9), 656–662. https://doi.org/10.1136/oemed-2012-101021

Thacker, N. P., Nitnaware, V. C., Das, S. K., & Devotta, S. (2007). Dioxin formation in pulp and paper mills of India. *Environmental Science and Pollution Research, 14*(4), 225–226. https://doi.org/10.1065/espr2007.02.386

Thorpe, S., Kelly, M., Startin, J., Harrison, N., & Rose, M. (2001). Concentration changes for 5 PCDD/F congeners after administration in beef cattle. *Chemosphere, 43*, 869–879.

Tuinstra, L. G. M. T., Roos, A. H., M Berende, P. L., van Rhijn, J. A., Traag, W. A., & Mengelers, M. J. B. (1992). Excretion of polychlorinated dibenzo-p-dioxins and-furans in milk of cows fed on dioxins in the dry period. *Journal of Agricultural Food Chemistry, 1092*(40), 1772–1776.

Tysklind, M., Fangmark, I., Marklund, S., Lindskog, A., Thaning, L., & Rappel, C. (1993). Atmospheric transport and transformation of polychlorinated dibenzo-p-dioxins and dibenzofurans. *Environmental Science Technology, 27*, 2190–2197. https://pubs.acs.org/sharingguidelines

United States Environmental Protection Agency (EPA). (2024a). *A Town, a Flood, and Superfund: Looking Back at the Times Beach Disaster Nearly 40 Years Later*. United States Environmental Protection Agency (EPA). www.epa.gov/mo/town-flood-and-superfund-looking-back-times-beach-disaster-nearly-40-years-later

United States Environmental Protection Agency (EPA). (2024b). *Tittabawassee River, Saginaw River & Bay Midland, MI Cleanup Activities*. United States Environmental Protection Agency (EPA). https://cumulis.epa.gov/supercpad/SiteProfiles/index.cfm?fuseaction=second.cleanup&id=0503250

U.S. Food and Drug Administration (FDA). (2007). *Dioxin Analysis Results/Exposure Estimates*. U.S. Food and Drug Administration (FDA). www.fda.gov/food/environmental-contaminants-food/dioxin-analysis-resultsexposure-estimates

Van Leeuwen, F. X. R., Feeley, M., Schrenk, D., Larsen, J. C., Farland, W., & Younes, M. (2000). Dioxins: WHO's tolerable daily intake (TDI) revisited. *Chemosphere, 40*(9–11), 1095–1101. https://doi.org/10.1016/S0045-6535(99)00358-6

Van Overmeire, I., Clark, G. C., Brown, D. J., Chu, M. D., Cooke, W. M., Denison, M. S., Baeyens, W., Srebrnik, S., & Goeyens, L. (2001). Trace contamination with dioxin-like chemicals: Evaluation of bioassay-based TEQ determination for hazard assessment and regulatory responses. *Environmental Science & Policy, 4*, 345–357. www.elsevier.com/locate/envsci

Wang, F., Xiang, L., Sze-Yin Leung, K., Elsner, M., Zhang, Y., Guo, Y., Pan, B., Sun, H., An, T., Ying, G., Brooks, B. W., Hou, D., Helbling, D. E., Sun, J., Qiu, H., Vogel, T. M., Zhang, W., Gao, Y., Simpson, M. J., … & Tiedje, J. M. (2024). Emerging contaminants: A one health perspective. *The Innovation, 5*(4), 100612. https://doi.org/10.1016/j.xinn.2024.100612

Weber, R., Tysklind, M., & Gaus, C. (2008). Dioxin–Contemporary and future challenges of historical legacies. *Environmental Science and Pollution Research, 15*(2), 96–100. https://doi.org/10.1065/espr2008.01.473

WHO European Centre for Environment and Health International Programme on Chemical Safety. (1998). *Executive Summary Assessment of the Health Risk of Dioxins: Re-Evaluation of the Tolerable Daily Intake (TDI)*. World Health Organization.

Wikoff, D., Fitzgerald, L., & Birnbaum, L. (2012). Persistent organic pollutants: An overview. In A. Schecter (Ed.), *Dioxins and Health Including Other Persistent Organic Pollutants and Endocrine Disruptors*. Wiley.

Woolf, A. D. (2022). Japan "Yusho" poisoning, 1968 and Taiwan "Yucheng" poisoning, 1979. In *History of Modern Clinical Toxicology* (pp. 121–135). Academic Press. https://doi.org/10.1016/B978-0-12-822218-8.00041-7

World Health Organization (WHO). (2023). *Dioxins*. www.who.int/news-room/fact-sheets/detail/dioxins-and-their-effects-on-human-health

Xiao, Q., Song, X., Li, W., Zhang, Y., & Wang, H. (2017). A primary estimation of PCDD/Fs release reduction from non-wood pulp and paper industry in China based on the investigation of pulp bleaching with chlorine converting to chlorine dioxide. *Chemosphere, 185*, 329–335. https://doi.org/10.1016/j.chemosphere.2017.06.119

Yaktine, A. L., Harrison, G. G., & Lawrence, R. S. (2006). Reducing exposure to dioxins and related compounds through foods in the next generation. *Nutrition Reviews, 64*(9), 403–409. https://doi.org/10.1301/nr.2006.sept.403-409

Furans and Safety *Food Processing and Manufacturing Related Effects*

9

Maria Tarapoulouzi, Georgios Stavroulakis, and Demetris Kafouris

9.1 INTRODUCTION

The thermal processing of food products typically gives rise to a diverse array of food processing contaminants. Furan, a particularly volatile compound within this category, has the potential to be generated in various thermally processed foods. Thus, it is essential to investigate the factors contributing to furan occurrence in different thermally processed foods, identify the primary sources of furan exposure, understand the influencing factors in its formation, and employ specific analytical methods for its detection. Furthermore, managing the formation of furan in processed foods at an industrial scale is a complex task, and ongoing research is actively addressing this challenge. Concurrently, gaining insights into the adverse effects of furan on human health at a molecular level is crucial for a comprehensive human risk assessment (Arisseto, 2016; Batool et al., 2023; Pawar et al., 2023).

While furans themselves are not intentionally added to food, they can be formed as byproducts during certain food processing, particularly those involving heat and certain precursors, as well as during cooking methods. Besides that, the formation of furans is often associated with thermal processes, also specific cooking conditions and ingredients used in food processing can influence their formation (Pawar et al., 2023).

 DOI: 10.1201/9781032712291-9

The presence of furans in food has raised concerns due to their potential toxicity, but it's essential to note that the levels found in food are generally low.

Coffee constitutes a significant dietary origin of furan in adults, whereas jarred or canned baby food emerges as the predominant contributor to furan exposure in the diet of infants and toddlers. In cereal-based products, especially in biscuits (wafer), crackers (cream cracker), breakfast cereals (corn flakes), snacks such as baked sweet potato. In addition, traditional Chinese liquor and black tea were found to have significant levels of furan. The levels of dietary exposure to furan exhibit variability across different countries or regions and among various age groups (Zhang and Zhang, 2023).

Furan exposure can lead to potential health risks. Upon ingestion, furan is rapidly absorbed and metabolized primarily in the liver, where its metabolites can bind to DNA and proteins, leading to cellular damage. Chronic exposure to furan has been linked to hepatotoxicity, characterized by liver cell necrosis, inflammation, and increased risk of liver cancer. Experimental studies in animals have demonstrated dose-dependent increases in liver tumors, prompting the classification of furan as a possible human carcinogen by regulatory bodies such as the International Agency for Research on Cancer (IARC) (IARC, 1995).

This chapter highlights existing gaps and challenges, serving as a guide for future research endeavors. Ongoing research should be focused on better understanding of the factors influencing furan formation and to develop strategies to mitigate their presence in food. Continuous monitoring and risk assessment of furan in food products is also important, to ensure consumer safety and public health protection.

9.2 STRUCTURAL INFORMATION ABOUT FURAN AND ITS DERIVATIVES PRODUCTION

9.2.1 Synthesis and Chemistry

Furans are a class of aromatic compounds containing a five-membered ring with four carbon atoms and one oxygen atom. They have unique properties and find applications in various fields, including organic synthesis and the pharmaceutical industry (Deepthi et al., 2019).

Furans can be synthesized by dehydrating diols in the presence of acid. For example, 1,4-butanediol can be dehydrated to yield tetrahydrofuran (THF), which can be further oxidized to furan. Additionally, the Diels-Alder reaction between a diene and a dienophile can be employed for furan synthesis. For instance, the reaction between 1,3-butadiene and maleic anhydride results in the formation of furan. Moreover, furans can also be obtained from biomass-derived precursors through various conversion processes. Finally, intramolecular cyclization of certain precursors can lead to furan formation. This method often involves the use of transition metal catalysts (Deepthi et al., 2019).

It is important to be noted that furans represent crucial flavor components within heterocyclic and aromatic compounds. They encompass lipophilic substances such as furaneol and furfural, characterized by a sweet, caramel-like scent. Thermal degradation reactions predominantly yield these compounds (Wang et al., 2021).

Furan can be transformed into furaneol through chemical reactions. Furaneol serves as a significant aroma constituent in various cheeses, contributing to caramel and nutty flavors, particularly in Cheddar cheese. One common method for synthesizing furaneol involves the acid-catalyzed dehydration of certain carbohydrates, such as glucose or fructose, which leads to the formation of other furan derivatives. These furan derivatives can then undergo further chemical reactions to yield furaneol. Initially, carbohydrates are treated with an acid catalyst under appropriate conditions, causing them to undergo dehydration reactions. This results in the formation of furan derivatives, including furfural. The furan derivatives obtained from the dehydration step can undergo additional chemical transformations, often involving reduction and other functional group modifications. These reactions ultimately lead to the formation of furaneol. Thus, furan itself is not directly converted into furaneol, it can serve as a starting material in the synthesis of furaneol through appropriate chemical reactions (Wang et al., 2021).

The formation pathways of compounds like furan may generate derivatives such as 2-furoic acid (FA), 5-hydroxymethylfurfural (5-HMF), and furfural by various processes. FA can be formed through the breakdown of polyphenols and carbohydrates, alongside oxidation. Phenolic compounds' thermal degradation, such as quinic acid, is also involved. Furan can arise from the decarboxylation of 2-furoic acid. 5-HMF and furfural are produced during the Maillard reaction's second phase, where hexoses or pentoses react with amino acids or proteins under heat. Caramelization, particularly under acidic and high-temperature conditions, is another source of 5-HMF and furfural. These pathways involve complex reactions like retro-aldol cleavage, α-dicarbonyl cleavage, and isomerization, leading to compound formation from their precursors (Meng et al., 2024).

9.2.2 Factors Influencing the Formation of Furan in Foods

Furan formation in foods can be influenced by various factors, and the main eight factors are presented in Figure 9.1. First, certain food ingredients, particularly those rich in carbohydrates and sugars, can contribute to furan formation and they are characterized as reaction precursors. Therefore, high levels of glucose, fructose, and ascorbic acid (vitamin C) can increase furan formation during processing or cooking (Shen et al., 2017). In addition, the presence of certain metals, such as iron and copper, can catalyze furan formation through metal-catalyzed oxidation reactions. These metals may be naturally present in food or introduced during processing. Second, prolonged processing times, such as extended heating or cooking durations, can lead to higher furan levels in foods. This is because furan formation tends to increase with exposure to heat over time. Third, higher processing temperatures can accelerate furan formation. For

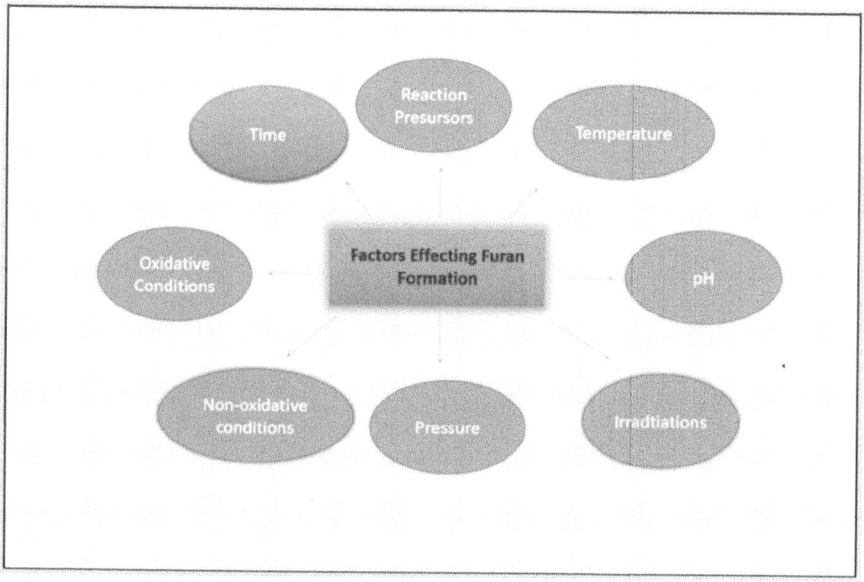

FIGURE 9.1 Eight main factors influencing furan production. (Batool et al., 2023)

instance, cooking methods like frying or baking at high temperatures can promote the formation of furan in foods. In addition, it has been shown that exposing fruit juices, which contain sugars, ascorbic acid, and amino acids, to high temperatures (90–105 °C) as well as high pressure (600 MPa) can also lead to furan formation. Furthermore, oxidative conditions, such as exposure to air or the presence of oxygen during food processing, can enhance furan formation. Lipid oxidation can contribute to furan formation in foods containing fats or oils. Similarly, non-oxidative conditions or reduced oxygen levels in food products that contain polyunsaturated fatty acids (PUFAs) can decrease furan formation during processing. Moreover, the pH of the food matrix can influence furan formation. Generally, acidic conditions promote higher furan formation compared to neutral or alkaline conditions. Therefore, foods with lower pH levels may exhibit increased furan levels. Lastly, irradiation of foods, such as γ-irradiation used for food preservation or sterilization, can induce furan formation, especially in products containing carbohydrates (Batool et al., 2020; Batool et al., 2023; Nie et al., 2013; Zhang and Zhang, 2023).

Overall, important parameters affecting the formation of furans during food cooking and processing include the presence of specific precursors, such as certain carbohydrates and amino acids, certain packaging materials, especially those containing recycled paper or board, can release furan precursors when in contact with food under certain conditions, and acid-catalyzed reactions, such as the dehydration of diols (Figure 9.2). Understanding and controlling these factors is crucial for mitigating furan formation in foods and ensuring food safety and quality. Techniques such as optimizing processing parameters, selecting appropriate ingredients, and minimizing exposure to oxidative conditions can help reduce furan levels in food products.

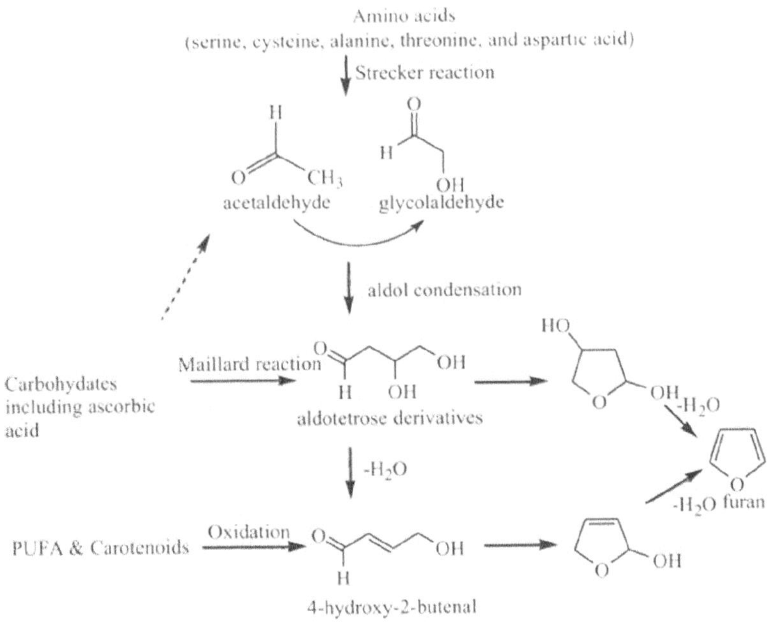

FIGURE 9.2 Possible pathways and precursors for furan formation. (Nie et al., 2013)

9.3 THE EFFECT OF FOOD PROCESSING TECHNIQUES AND MANUFACTURING ON FURANS

Furans can be formed during thermal processing techniques such as baking, roasting, frying, grilling, barbequing, canning and brewing. Table 9.1 summarizes the processing techniques, and the foodstuffs related to furans production. Thermal processing is a dynamic procedure that leads to numerous physical and chemical alterations, causing the deterioration of both color and flavor in foods and beverages (Huang et al., 2023).

The Maillard reaction, responsible for the browning of food, which occurs during baking and roasting, involves the reaction of amino acids with reducing sugars at elevated temperatures (Prata et al., 2021). However, it can also lead to the formation of undesirable derivatives, including furans. High-temperature frying can lead to the production of furans, specifically when using oils rich in unsaturated fatty acids. Cooking meat at high temperatures, such as grilling or barbecuing, can result in the formation of furans, particularly if there is fat accumulation onto open flames. Thermal processing during canning, especially in the presence of certain precursors, can also contribute to furan formation (Batool et al., 2023; Zheng et al., 2023).

It is important to be noted that the roasting of coffee beans is a significant source of furans. The high temperatures involved in the roasting process contribute to the formation of these compounds. Similar to coffee, the roasting of cocoa beans for chocolate

TABLE 9.1 Processing techniques and foodstuffs related to furans production

PROCESSING TECHNIQUES	*FOODSTUFF(S)*
Baking	Bread, cakes, and pastries
Roasting	Coffee, and nuts
Frying	Potato chips, French fries, fried meats, especially those with a coating or batter, fried dough-based products and other deep-fried snacks
Grilling – barbequing	Foods with fats or applying oils and marinades, such as meats with a higher fat content
Brewing	Malted barley or other grains

production can lead to furan formation. Furthermore, thermal treatments applied during the sterilization and pasteurization of food products, such as canned goods and dairy products, can contribute to furan formation (Barrios-Rodríguez et al., 2022; Alsafra et al., 2023).

9.3.1 Baking and Roasting

Furan formation during baking and roasting is primarily associated with the Maillard reaction, a complex series of chemical reactions that occur between amino acids and reducing sugars at elevated temperatures. The Maillard reaction involves amino acids (from proteins) and reducing sugars (e.g., glucose and fructose). It occurs at elevated temperatures, typically above 140°C (284°F). The Maillard reaction proceeds through a series of complex steps, including sugar fragmentation, dehydration, condensation, and polymerization. As a result of the Maillard reaction, various intermediate compounds are formed, some of which may contribute to the formation of furans (Kathuria et al., 2023).

During the Maillard reaction, dehydration reactions may occur, leading to the formation of furan rings. The cyclization of certain reaction intermediates can contribute to furan formation. The precursors to furan formation in the Maillard reaction include dehydrated sugars and amino acid derivatives. The extent of furan formation is influenced by the temperature and duration of the baking or roasting process. Higher temperatures and prolonged cooking times are generally associated with increased furan production (Kathuria et al., 2023; Shakoor et al., 2022).

The composition of ingredients, including the types of amino acids and sugars present, can influence furan formation. The moisture content in the food can impact the Maillard reaction and, subsequently, furan formation. The pH of the food matrix can affect the Maillard reaction, and the types of compounds formed (Batool et al., 2021; Shakoor et al., 2022).

Bread, cakes, and pastries can contain furans due to the Maillard reaction during baking. Coffee beans undergo roasting, a process that involves high temperatures and contributes to furan formation. Nuts that are roasted can also contain furans formed during the roasting process (Cha and Lee, 2020).

Understanding the mechanisms of furan formation during baking and roasting allows for the development of strategies to minimize their presence in food products while maintaining desirable sensory characteristics. Temperature control, ingredient selection, and ongoing research are key elements in managing furan levels during these cooking processes (Meng et al., 2024).

9.3.2 Frying

Furan formation during frying is a result of complex chemical reactions that occur at elevated temperatures. Amino acids from proteins and reducing sugars that are present in the food combined with high temperatures during frying initiate the Maillard reaction (Kathuria et al., 2023; Shakoor et al., 2022). Frying involves the exposure of fats and oils to high temperatures. Oxidation of unsaturated fatty acids can produce reactive compounds that contribute to furan formation. Thermal degradation of lipids can generate furan precursors, contributing to the overall furan content in the fried food (Kathuria et al., 2023; Chang et al., 2020).

Certain amino acids, notably those with reactive side chains, serve as precursors for furan formation. Sugars, such as glucose and fructose, participate in the Maillard reaction and contribute to furan formation. Oxidation products of unsaturated fatty acids can serve as precursors for furan formation. The extent of furan formation is influenced by the temperature of the frying oil. Higher temperatures can lead to increased furan production. Prolonged frying times can also contribute to the formation of furans (Chang et al., 2020; Jackson and Al-Taher, 2022).

Potato chips, French fries, and other deep-fried snacks can contain furans. Fried meats, especially those with a coating or batter, can contribute to furan formation. Fried dough-based products may also contain furans formed during frying (Dangal et al., 2024; Jackson and Al-Taher, 2022).

9.3.3 Grilling and Barbequing

Furan formation during grilling and barbecuing is associated with the cooking processes that involve high temperatures, open flames, and the breakdown of food components. Like other cooking methods, the Maillard reaction, lipid oxidation, and thermal degradation contribute to the formation of furans. Amino acids from proteins and reducing sugars in the food and high temperatures generated during grilling and barbecuing initiate the Maillard reaction. Dehydration and cyclization reactions within the Maillard reaction pathway can contribute to the formation of furans (Batool et al., 2021; Kathuria et al., 2023).

Grilling and barbecuing often involve cooking foods with fats or applying oils and marinades. Oxidation of unsaturated fatty acids in fats or cooking oils can produce reactive compounds that contribute to furan formation. Certain amino acids, especially those with reactive side chains, serve as precursors and in conjunction with sugars, such as glucose and fructose, contribute to furan formation during the Maillard reaction (Batool et al., 2021).

Oxidation products of unsaturated fatty acids in fats can serve as precursors for furan formation. The use of open flames and high temperatures during grilling and barbecuing can lead to increased furan formation. Prolonged exposure to high temperatures, especially when using open flames, can contribute to furan production (Batool et al., 2021; Shahidi and Hossain, 2022).

Grilling and barbecuing of meats, in particular those with a higher fat content, can result in the formation of furans. Marinating foods before grilling can introduce additional precursors that may contribute to furan formation. When wood or charcoal is used for grilling or barbecuing, the smoke produced may contain furan precursors that can be deposited on the food (Bassam et al., 2022).

9.3.4 Canning

Furan formation during canning is associated with the heat treatment involved in the canning process. The formation of furans is a result of various chemical reactions that occur under high-temperature conditions (Zheng et al., 2023).

The canning process involves the application of heat to food in sealed containers to destroy microorganisms and enzymes, ensuring the safety and preservation of the product. The temperatures used in canning are typically elevated, and the duration of heat exposure can vary depending on the type of food being processed (Owusu-Apenten and Vieira, 2022).

The Maillard reaction can occur during the heating phase of canning, leading to the formation of various reaction products, including furans. Certain amino acids in proteins, sugars, such as glucose and fructose, present in the food can participate in the Maillard reaction and contribute to furan formation. Depending on the type of food and the presence of acids, acid-catalyzed reactions may occur during canning, contributing to furan formation (Tsao et al., 2022).

Canning of meat and seafood products may contribute to furan formation, especially if these products contain proteins and reducing sugars. The canning of fruits and vegetables may also lead to furan formation, influenced by the presence of amino acids and sugars (Shen et al., 2021; Tsao et al., 2022). Proper control of processing parameters and storage conditions is essential to minimize furan formation in food products.

9.3.5 Brewing

Brewing can lead to the formation of furans in food and beverages (Pavesi Arisseto et al., 2011). Furans, including furan and its derivatives like 2-methylfuran, 2-ethylfuran, and furfural, are formed due to brewing processing of food and beverages. The Maillard reaction, which occurs between amino acids and reducing sugars under heat, is a primary pathway for furan formation. This reaction is a key process in brewing, contributing to the color, flavor, and aroma of the final product. Caramelization of sugars during the heating process can also lead to the formation of furans. This happens during the roasting of malted barley, or other grains used in brewing. Thermal degradation of fatty acids in the presence of oxygen can produce furans. Lipid oxidation can occur during the malting, mashing, and boiling stages of brewing (Prado et al., 2021; Pérez-Lucas et al., 2024).

During malting, which is the process of malt roasting, high temperatures can lead to Maillard reactions and caramelization, resulting in the formation of furans. Mashing is typically a lower-temperature process, yet any residual heat or local hotspots can contribute to minor furan formation. The boiling stage of brewing involves high temperatures, making it a critical stage for furan formation via Maillard reactions and caramelization. Although fermentation occurs at lower temperatures, furans formed during earlier stages can be carried through to the final product. Additionally, some microorganisms used in fermentation can produce or metabolize furan compounds (Elhalis et al., 2020; Pérez-Lucas et al., 2024).

9.3.6 Mitigation Strategies

Understanding the conditions under which furans are formed during food processing allows the development of strategies to minimize their presence while maintaining the desired quality and safety of food products. Ongoing research explores ways to mitigate furan formation through innovative processing methods and ingredient selection (Mariotti et al., 2013). Adhering to regulatory standards and guidelines regarding furan levels in food products is crucial for ensuring consumer safety.

Food manufacturers can optimize processing conditions by adjusting temperature, time, and other processing parameters that can help minimize furan formation. Modifying recipes or using different ingredients with lower furan precursors may constitute strategies to reduce furan levels in food. In addition, improving packaging materials to reduce the release of furan precursors into food can be another strategy (Kettlitz et al., 2019).

Using fresh and high-quality frying oil can reduce the formation of furan precursors. Optimizing food formulations and processing conditions can influence furan levels in fried products. Using marinades or rubs with minimal reducing sugars can decrease furan formation. If using wood for flavor, choosing wood chips carefully and using them sparingly may reduce furan content. Controlling the temperature and duration of the canning process can help minimize furan formation. Adjusting the pH of the food may influence the Maillard reaction and other acid-catalyzed reactions, potentially reducing furan levels.

9.4 ANALYTICAL TECHNIQUES FOR FURAN DETERMINATION

As accurate determination of furan levels is vital for food safety, there are considerations for furan analysis which are crucial to be considered. Sample preparation is very critical for accurate furan determination. Techniques like distillation, solvent extraction,

TABLE 9.2 The most widely used analytical methods to detect furans in food samples

NO	*ANALYTICAL METHOD*	*SAMPLE TYPE*	*REFERENCE*
1	SPME-GC-MS	Breakfast cereals	Lipinski et al. (2024)
2	HS-SPME-GC-MS	Baby food	Condurso et al. (2018)
3	SPME-GC-MS	Instant noodles	Lee et al. (2021)
4	UHPLC-UV	Coffee	Macheiner et al. (2021)
5	HS-GC-MS	Beer	Bogdanova et al. (2018)
6	HS-GC–MS	Coffee	Cao et al. (2022)
7	HS-SPME-GC-FID	Baby food, juices	Sarafraz-Yazdi et al. (2012)
8	GC-MS	Vegetable oils	Emektar et al. (2022)
9	HS-GC-MS	Coffee	Rahn and Yeretzian (2019)
10	HS-SPME-GC-MS	Cereals, coffee, and infant products	Frank et al. (2021)

Abbreviations: GC-MS: Gas Chromatography – Mass Spectrometry, HS-GC-MS: Headspace Gas Chromatography – Mass Spectrometry, HS-SPME-GC-FID: Headspace Solid-Phase Microextraction coupled with Gas Chromatography – Flame Ionization Detector, HS-SPME-GC-MS: Headspace Solid-Phase Microextraction coupled with Gas Chromatography – Mass Spectrometry, SPME-GC-MS: Solid-Phase Microextraction coupled with Gas Chromatography – Mass Spectrometry, and UHPLC-UV: Ultra-High Performance Liquid Chromatography with Ultraviolet Detection.

and clean-up procedures help in isolating furan from complex food matrices. In addition, calibration and validation are also important to be set up correctly, as the appropriate standards and rigorous validation ensures the reliability of analytical results. Consideration of matrix effects, thus the food matrix on furan extraction and detection is essential for accurate quantification (Batool et al., 2020).

Several analytical techniques are commonly employed to detect and quantify furan and its derivatives in several foodstuffs, as shown in Table 9.2. These methods vary in sensitivity, specificity, and the type of samples they can handle. Gas Chromatography (GC) with Flame Ionization Detection (FID) is very commonly used due to its sensitivity and specificity for volatile organic compounds. In addition, GC with Mass Spectrometry (GC-MS), is one of the most common and reliable methods for furan determination. It combines the separation capabilities of gas chromatography with the detection power of mass spectrometry. GC-MS offers high sensitivity, specificity, and the ability to provide both qualitative and quantitative data. It is widely used for detecting low levels of furan in complex matrices (Yi-Hsuan Huang et al., 2022). A very important category is the one of derivatization techniques. For instance, solid-phase derivatization involves reacting furan with a derivatizing agent on a solid phase, followed by GC analysis. This technique can simplify sample preparation and improve detection limits.

Furthermore, Liquid-Liquid Extraction (LLE) is a traditional extraction method that can be used prior to chromatographic analysis. It is a simple and cost-effective technique used for certain types of samples. Additionally, Solid Phase Extraction (SPE) is another sample preparation technique that can be used to concentrate and purify furan from complex matrices. It offers efficient cleanup, concentration of analytes, and reduction of matrix interferences.

Headspace Sampling Techniques, such as Static Headspace (SHS) GC-MS, involve heating of the sample in a sealed container and analyzing the vapor phase. This method is simple and effective for volatile compounds like furan, as well as Dynamic Headspace (DHS). It uses an inert gas to purge volatiles from the sample into a trap before GC analysis. The latter demonstrates higher sensitivity compared to SHS. Headspace GC is particularly the most useful technique for analyzing furan and its derivatives. It minimizes matrix effects and is suitable for complex food matrices. Moreover, in Solid-Phase Microextraction (SPME), a fiber coated with an extracting phase is exposed to the headspace above the sample, then desorbed in the GC inlet. SPME is efficient and solvent-free (Altaki et al., 2017; Drabińska et al., 2023; Sandjong-Sayon et al., 2024). GC-MS is often used in combination with sample preparation techniques like SPME or HS-GC and is widely preferred for its accuracy and reliability in detecting low levels of furan and derivatives (Frank et al., 2020; Condurso et al., 2018; Frank et al., 2021; Huang et al., 2023).

In terms of chromatography, High-Performance Liquid Chromatography with Ultraviolet detection (HPLC-UV) is used. It is less common for furan due to its volatility, yet used in specific contexts where non-volatile derivatives of furan are analyzed (Frank et al., 2020). Nowadays, researchers and scientists tend to use Fourier Transform Infrared Spectroscopy (FTIR) at a great frequency, due to its significant advantages such as simple and easy procedure, time-saving, and low cost compared to chromatographic techniques. FTIR Spectroscopy is used to identify and quantify furan by its characteristic absorption bands, and these are associated either with the C=C stretching of the furan ring (1400–1390 cm^{-1}) or with the in-plane C–H rocking vibration of the aldehyde functional group (1450–1325 cm^{-1}) (Lorenzo-Santiago et al., 2023; Teixeira dos Santos et al., 2024). Although it is less common than chromatography, this technique can be useful for specific applications, such as fast detection.

The choice of analytical technique for furan determination depends on factors such as the required sensitivity, specificity, available equipment, and the nature of the food matrix. Gas chromatography, particularly GC-MS with headspace sampling, is the most widely used due to its high sensitivity and accuracy. Other techniques, like HPLC and FTIR, are also employed based on specific requirements and contexts.

9.5 NEGATIVE EFFECTS OF FURANS ON HUMAN HEALTH AND THE RELEVANT RISK ASSESSMENT

The presence of furan in food products on the market is of high concern to many national public health agencies, as well as the European Food Safety Authority (EFSA) and the US Food and Drug Administration (FDA). This fact has pushed for research into the identification of risks caused by furan by several international organizations, including the IARC. Notably, IARC has classified furan as possibly carcinogenic to

TABLE 9.3 Types of toxicity in experimental studies

TOXICITY	EFFECT(S)	EXPERIMENTAL ANIMAL	REFERENCE
Hepatotoxicity	Cholangiofibrosis	Rat	Von Tungeln et al. (2017)
Carcinogenicity	Hepatocellular adenomas and carcinomas	Mouse	National Toxicology Program (1993)
Carcinogenicity	Cholangiocarcinoma	Rat	National Toxicology Program (1993)
Carcinogenicity	Mononuclear cell leukemia	Rat	National Toxicology Program (1993); Von Tungeln et al. (2017)
Carcinogenicity	Malignant mesothelioma of testes	Rat	Von Tungeln et al. (2017)
Endocrine toxicity	Increase of body weight Decreased testosterone, LH, and GH levels Increased cortisol levels Decreased sperm counts	Rat	Rehman et al. (2019)

humans (Group 2B). The toxicity of furan is well documented from animal studies, which are briefly described in the following paragraphs and summarized in Table 9.3.

Furan is a potent hepatotoxin. Oral exposure to furan leads to liver toxicity both in mice and rats, the latter being the most sensitive species. Specifically, cholangiofibrosis was observed in male Fischer 344 (F344) rats upon administration of 0.02, 0.044, 0.092, 0.2, 0.44, 0.92, and 2 mg furan per kg bw by gavage in corn oil 5 days per week. Notably, the incidence of cholangiofibrosis was significantly increased after 36 weeks at doses of 0.44 mg/kg bw and above (EFSA, 2017; Von Tungeln et al., 2017).

In addition to the oral toxicity of furan, there is well established evidence of furan carcinogenicity in mice and rats. Male and female B6C3F1 mice showed a dose-dependent increase in hepatocellular adenomas and carcinomas, which occurred at 104 weeks at doses of 4 mg/kg bw and above. Regarding male and female F344/N rats, significant increases in the incidence of cholangiocarcinomas were also observed in their liver at doses of 4 or 8 mg/kg bw. No relevant malignant tumors were observed at doses below or at 2 mg/kg bw up to 104 weeks (de Conti et al., 2017; National Toxicology Program, 1993; EFSA, 2017). Along with liver malignancies, exposure to furan lead to dose-dependent increases in mononuclear cell leukemia in both male and female F344/N rats (de Conti et al., 2017; National Toxicology Program, 1993; Von Tungeln et al., 2017) and malignant mesothelioma of the epididymis or testes in male F344/N rats (de Conti et al., 2017; Von Tungeln et al., 2017). Despite the wealth of existing data and evidence for the carcinogenicity of furan in experimental animals, there is a lack of data in relation to the carcinogenicity of furan in humans. Therefore, IARC has classified furan as "possibly carcinogenic to humans" (Group 2B) (IARC, 1995).

Furan and its compounds possess cytotoxic activity that leads to necrosis in their respective target organs, mostly in the liver, kidneys, and lungs. The cells (i.e. hepatocytes, tubule epithelial cells and Clara cells) of these target organs tend to express high levels of cytochrome P450 mixed-function oxidase activity (Kaya et al., 2019). Furan is rapidly absorbed by the digestive system and metabolized in the liver by the cytochrome P450, CYP2E1, which converts the molecule into a reactive metabolite, most likely cis-2-butene-1,4-dialaldehyde (BDA). This metabolite can react with DNA and induce mutagenic activity. BDA was found to be genotoxic in several *in vitro* tests. In the liver and kidney of rats and mice, furan induced very low levels of DNA adducts, whose chemical structures could not be defined and were not identical to those induced by BDA in *in vitro* model systems. Therefore, based on the available data, a genotoxic mechanism in furan-induced carcinogenesis cannot be ruled out and, therefore, the substance is assessed as a genotoxic carcinogen by European and national public health agencies (EFSA, 2017; Husøy et al., 2019).

The lack of evidence for furan genotoxicity in vivo suggests that non-genotoxic mechanisms may be involved in the pathogenesis of furan-induced hepatotoxicity and carcinogenicity. Exposure to furan at tumorigenic doses gives rise to a broad spectrum of molecular and cellular non-genotoxic aberrations, including among others epigenetic alterations, cell proliferation, oxidative stress, and apoptosis. Among these alterations, epigenetic mechanisms are of special interest due to their close correlation with the other non-genotoxic events. Epigenetic mechanisms consist of cytosine DNA methylation and hydroxymethylation, nucleosome positioning along DNA, and expression of non-coding RNAs (ncRNAs), particularly microRNAs (miRNAs) (de Conti et al., 2017). Of note, these miRNAs are directly or indirectly involved in the p53 pathway, a pathway intimately associated with carcinogenesis (Tivanello et al., 2023). The above information on the two main mechanisms, genotoxic and non-genotoxic, pertaining to furan liver toxicity and carcinogenicity is summarized in Figure 9.3.

Regarding endocrine toxicity, few animal studies have investigated the possible endocrine disrupting effects of furan. Rehman et al. (2019) found clear evidence that exposure to the two highest doses of furan (10 and 20 mg/kg per day) broadly and consistently affects the development of the neonatal male rats reproductive system, which ultimately lead to a reduction in fertility. The most prominent effects were dose-dependent increases in body weight, decreased plasma hormones (testosterone, LH, and GH), increased cortisol levels, and decreased sperm counts related to altered testicular

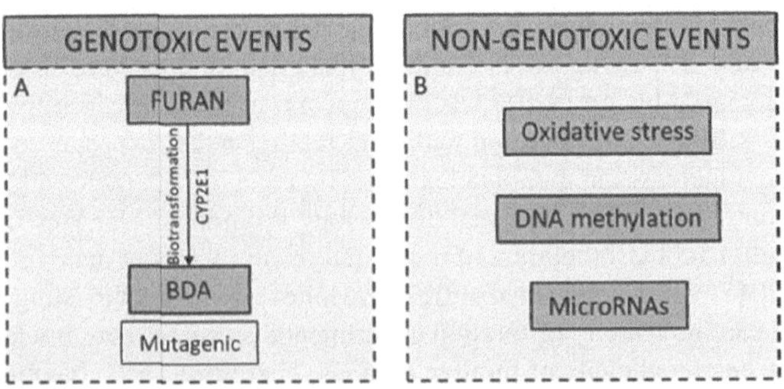

FIGURE 9.3 Events associated with furan toxicity. (Tivanello et al., 2023)

and epididymal histology. These results are in agreement with previous findings on furan exposure at the weaning stage and highlight that developmental exposure to furan can ultimately compromise male fertility in adulthood (Rehman et al., 2019).

Considering the various toxicities of furan, it is important to carry out an effective risk assessment of exposure to furan via food. To that end, EFSA used a margin of exposure (MOE) approach for the risk characterization, considering the potential contribution of genotoxic mechanisms in furan carcinogenicity. For non-neoplastic effects, the MOEs (calculated using the $BMDL_{10}$ of 0.064 mg/kg bw per day for the incidence of cholangiofibrosis in the rat as a reference point) were below 100 for several dietary surveys (EFSA, 2017). The MOEs for neoplastic effects (assessed using the $BMDL_{10}$ of 1.31 mg/kg bw per day for the incidence of hepatocellular adenomas and carcinomas in female mice as a reference point) were below 10,000 for most dietary surveys, notably for the 95th percentile of exposure across all age groups (EFSA, 2017). These MOEs indicate a public health concern for a potentially genotoxic carcinogen. Such findings were consistent with the risk assessment calculations from other public health agencies (Sirot et al., 2019; Husøy et al., 2019). However, due to the uncertainties associated with food consumption data and furan content in food that are used for the assessment of human exposure to furan, there is a need for alternative and possibly more accurate exposure assessment to inform risk assessment. Recently, Karlstetter and Mally (2020) developed a biomarker-based approach to monitor human exposure to furan. An important prerequisite for application of biomarker-based exposure assessment is to establish the correlation between external exposure and potential biomarkers of exposure. As mentioned before, the bioactivation of furan is realized via cytochrome P450 2E1 to form BDA, a chemically reactive α,β-unsaturated dialdehyde. BDA reacts with cellular nucleophiles to give rise to a range of metabolites that are excreted with urine and feces (Figure 9.4). In this study, two different BDA-derived metabolites were both excreted in a dose-related manner, thus serving as biomarkers of furan exposure. Furthermore, to perform a more precise and comprehensive risk assessment, it would be of interest to collect toxicological data and occurrence data in food of methylfurans, which are formed along with furan during thermal processing and are likely to undergo a metabolic fate similar to furan. Finally, the potential cumulative effects of furans and

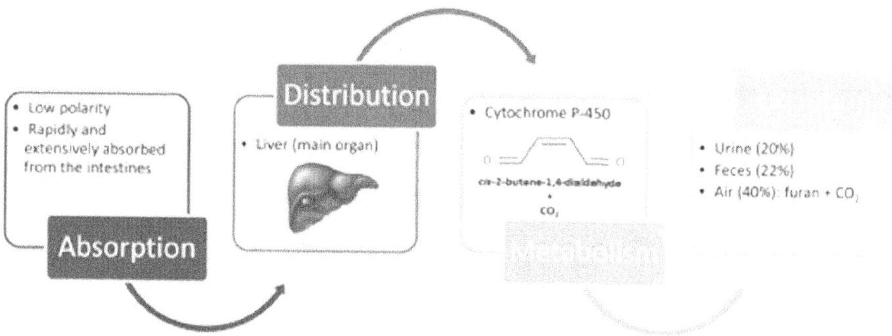

FIGURE 9.4 Metabolic pathway of furan leading to formation of cis-2-butene-1,4-dialaldehyde (BDA). (Arisseto, 2016)

methylfurans that could have the same biological target should also be considered for risk assessment (Sirot et al., 2019).

9.6 CONCLUSIONS

Furan is a chemical compound that can form in food during heat processing, particularly in products like canned and jarred foods, coffee, and some baby foods. It has raised concerns due to its potential carcinogenicity. Furan has been classified as a possible human carcinogen by the IARC, based on sufficient evidence of carcinogenicity in animals. Long-term exposure to high levels of furan can lead to liver toxicity and cancer. The typical dietary exposure to furan is generally low, but certain groups, such as infants consuming jarred baby foods, might have higher exposure levels relative to body weight. Currently, there are no specific regulatory limits for furan in foods in many regions, but it is monitored as a potential contaminant. Efforts to minimize furan formation during food processing and preparation are crucial, especially for vulnerable groups like infants. Consumers can also contribute to minimizing potential risks by adopting a balanced and varied diet and following recommended cooking and food storage practices.

REFERENCES

Alsafra, Z., Kuuliala, L., Scholl, G., Saegerman, C., Eppe, G. and De Meulenaer, B., 2023. Characterizing the formation of process contaminants during coffee roasting by multivariate statistical analysis. *Food Chemistry*, 427, pp. 136655.

Altaki, M.S., Santos, F.J., Puignou, L. and Galceran, M.T., 2017. Furan in commercial baby foods from the Spanish market: Estimation of daily intake and risk assessment. *Food Additives & Contaminants: Part A*, 34(5), pp. 728–739.

Arisseto, A.P., 2016. *Furan in Processed Foods*. Elsevier eBooks, pp. 383–396.

Barrios-Rodríguez, Y.F., Gutiérrez-Guzmán, N., Pedreschi, F. and Mariotti-Celis, M.S., 2022. Rational design of technologies for the mitigation of neo-formed contaminants in roasted coffee. *Trends in Food Science & Technology*, 120, pp. 223–235.

Bassam, S.M., Noleto-Dias, C. and Farag, M.A., 2022. Dissecting grilled red and white meat flavor: Its characteristics, production mechanisms, influencing factors and chemical hazards. *Food Chemistry*, 371, pp. 131139.

Batool, Z., Xu, D., Zhang, X., Li, X., Li, Y., Chen, Z., Li, B. and Li, L., 2021. A review on furan: Formation, analysis, occurrence, carcinogenicity, genotoxicity and reduction methods. *Critical Reviews in Food Science and Nutrition*, 61(3), pp. 395–406.

Batool, Z., Chen, J.H., Liu, B., Chen, F. and Wang, M., 2023. Review on furan as a food processing contaminant: Identifying research progress and technical challenges for future research. *Journal of Agricultural and Food Chemistry*, 71(13), pp. 5093–5106.

Batool, Z., Li, L., Xu, D., Wu, M., Weng, L., Jiao, W., Cheng, H., Roobab, U., Zhang, X., Li, X. and Liang, Y., 2020. Determination of furan and its derivatives in preserved dried fruits and roasted nuts marketed in China using an optimized HS-SPME GC/MS method. *European Food Research and Technology*, 246, pp. 2065–2077.

Bogdanova, E., Rozentale, I., Pugajeva, I., Emecheta, E.E. and Bartkevics, V., 2018. Occurrence and risk assessment of mycotoxins, acrylamide, and furan in Latvian beer. *Food Additives & Contaminants: Part B*, 11(2), pp. 126–137.

Cao, P., Zhang, L., Yang, Y., Wang, X., Liu, Z., Li, J., Wang, L., Chung, S., Zhou, M., Deng, K., Zhou, P. and Wu, P., 2022. Analysis of furan and its major furan derivatives in coffee products on the Chinese market using HS-GC–MS and the estimated exposure of the Chinese population. *Food Chemistry,* 387, p. 132823.

Cha, C.Y. and Lee, K.-G., 2020. Effect of roasting conditions on the formation and kinetics of furan in various nuts. *Food Chemistry,* 331, p. 127338.

Chang, C., Wu, G., Zhang, H., Jin, Q. and Wang, X., 2020. Deep-fried flavor: Characteristics, formation mechanisms, and influencing factors. *Critical Reviews in Food Science and Nutrition*, 60(9), pp. 1496–1514.

Condurso, C., Cincotta, F. and Verzera, A., 2018. Determination of furan and furan derivatives in baby food. *Food Chemistry,* 250, pp. 155–161.

Dangal, A., Tahergorabi, R., Acharya, D., Timsina, P., Rai, K., Dahal, S., Acharya, P. and Giuffrè, A.M., 2024. Review on deep-fat fried foods: Physical and chemical attributes, and consequences of high consumption. *European Food Research and Technology*, 250(6), pp. 1537–1550.

de Conti, A., Beland, F.A. and Pogribny, I.P., 2017. The role of epigenomic alterations in furan-induced hepatobiliary pathologies, *Food and Chemical Toxicology*, 109, pp. 677–682.

Deepthi, A., Babu, B.P. and Balachandran, A.L., 2019. Synthesis of furans–recent advances. *Organic Preparations and Procedures International*, 51(5), pp. 409–442.

Drabińska, N., Marcinkowska, M.A., Wieczorek, M.N. and Jeleń, H.H., 2023. Application of sorbent-based extraction techniques in food analysis. *Molecules*, 28(24), p. 7985.

EFSA, 2017. Scientific opinion on the risks for public health related to the presence of furan and methylfurans in food. *EFSA Journal*, 15(10), 5005, p. 142.

Elhalis, H., Cox, J., Frank, D. and Zhao, J., 2020. The crucial role of yeasts in the wet fermentation of coffee beans and quality. *International Journal of Food Microbiology*, 333, p. 108796.

Emektar, K., Kantekin-Erdogan, M.N. and Tekin, A., 2022. Furan formation in some vegetable oils during heat treatments. *Food Chemistry*, 386, p. 132744.

Frank, N., Dubois, M. and Pérez, J. F. H., 2020. Detection of furan and five alkylfurans, including 2-pentylfuran, in various food matrices. *Journal of Chromatography A*, 1622, p. 461119.

Frank, N., Dubois, M., Nguyen, K. H., Fromberg A. and Delatour, T., 2021. Quantification of furan and 5 alkylfurans with complete separation of 2-ethylfuran and 2,5-dimethylfuran isomers in cereals, coffee and Infant products by GC–MS. *Journal of Chromatography A*, 1179, p. 122765.

Huang, H., Chen, J., Zheng, M., Zhang, L., Ji, H., Cao, H., Dai, F. and Wang, L., 2023. Precursors and formation pathways of furfural in sugarcane juice during thermal treatment. *Food Chemistry*, 402, p. 134318.

Husøy, T., Arukwe, A., Bindrup, M.-L., Brantsæter, A.L., Kruse Fæste, C., Bogen Hetland, R., Alexander, J., Amlund, H., Helkås Dahl, K., Sundstøl Eriksen, G., Granum, B., Grave, K., Katrine Knutsen, H., Erik Paulsen, J., Ruus, A., Sanner, T., Steffensen, I.-L., Thomsen, C., Thrane, V. and Utne Skåre, J., 2019. Risk assessment of furan exposure in the Norwegian population. *European Journal of Nutrition & Food Safety,* 11(1), pp. 44–46.

IARC, 1995. IARC Monographs on the evaluation of carcinogenic risks to humans Volume 63, Dry cleaning, some chlorinated solvents and other industrial chemicals. IARC Publications. (Accessed: 30 July 2024).

Jackson, L.S. and Al-Taher, F., 2022. Processing issues: Acrylamide, furan, and trans fatty acids. In Martinović, A., Oh, S. & Lelieveld, H. (eds), *Ensuring Global Food Safety*, pp. 229–257. Academic Press.

Karlstetter D. and Mally A., 2020. Biomonitoring of heat-induced food contaminants: quantitative analysis of furan dependent glutathione- and lysine-adducts in rat urine as putative biomarkers of exposure. *Food and Chemical Toxicology,* 143, 111562.

Kathuria, D., Gautam, S. and Thakur, A., 2023. Maillard reaction in different food products: Effect on product quality, human health and mitigation strategies. *Food Control*, 153, pp. 109911.

Kaya, E., Yılmaz, S. and Ceribasi, S., 2019. Protective role of propolis on low and high dose furan-induced hepatotoxicity and oxidative stress in rats. *Journal of Veterinary Research*, 63(3), pp. 423–431.

Kettlitz, B., Scholz, G., Theurillat, V., Cselovszky, J., Buck, N.R., O' Hagan, S., Mavromichali, E., Ahrens, K., Kraehenbuehl, K., Scozzi, G., Weck, M., Vinci, C., Sobieraj, M. and Stadler, R.H., 2019. Furan and methylfurans in foods: An update on occurrence, mitigation, and risk assessment. *Comprehensive Reviews in Food Science and Food Safety*, 18, pp. 738–752.

Lee, H.Y. and Lee, K.G., 2021. Analysis of furan in various instant noodles by solid-phase microextraction–gas chromatography/mass spectrometry. *Food Control*, 126, p. 108047.

Lipinski, S., Lindekamp, N., Funck, N., Cramer, B. and Humpf, H.U., 2024. Determination of furan and alkylfuran in breakfast cereals from the European market and their correlation with acrylamide levels. *European Food Research and Technology*, 250, pp. 167–180.

Lorenzo-Santiago, M.A., Rodríguez-Campos, J., Rendón-Villalobos, R., García-Hernández, E., Vallejo-Cardona, A.A. and Contreras-Ramos, S.M., 2023. Thermal treatment to obtain 5-hydroxymethyl furfural (5-HMF), furfural and phenolic compounds from vinasse waste from agave. *Molecules*, 28(3), p. 1063.

Macheiner, L., Schmidt, A., Karpf, F. and Mayer, H.K., 2021. A novel UHPLC method for determining the degree of coffee roasting by analysis of furans. *Food Chemistry*, 341(1), p. 128165.

Mariotti, M. S., Granby, K., Rozowski, J. and Pedreschi, F., 2013. Furan: A critical heat induced dietary contaminant. *Food & Function*, 4(7), pp. 1001–1015.

Meng, D., Zhang, F., Jia, W., Jiao, J. and Zhang, Y., 2024. Understanding the health implications of furan and its derivatives in thermally processed foods: occurrence, toxicology, and mixture analysis. *Current Opinion in Food Science*, 57, p. 101151.

National Toxicology Program, 1993. Toxicology and carcinogenesis studies of Furan (CAS No. 110-00-9) in F344 rats and B6C3F1 mice (Gavage Studies). *National Toxicology Program Technical Report Series*, 402, pp. 1–286.

Nie, S., Huang, J., Hu, J., Zhang, Y., Wang, S., L, C., Marcone, M. and Xie, M., 2013. Effect of pH, temperature and heating time on the formation of furan in sugar–glycine model systems. *Food Science and Human Wellness*, 2(2), pp. 87–92.

Owusu-Apenten, R. and Vieira, E., 2022. Thermal processing and canning. In Dennis R. Heldman (ed), *Elementary Food Science,* pp. 253–288. Springer International Publishing.

Pavesi Arisseto, A., Vicente, E., Soares Ueno, M., Verdiani Tfouni, S.A. and De Figueiredo Toledo, M.C., 2011. Furan levels in coffee as influenced by species, roast degree, and brewing procedures. *Journal of Agricultural and Food Chemistry*, 59(7), pp. 3118–3124.

Pawar, B., Dhamne, S.A., Pingale, P.L., Sheth, A., Tekade, M., Mansour, R.S. and Tekade, R.K., 2023. Thermal processing and food-related toxicants. In Tekade, R. K. (ed), *Essentials of Pharmatoxicology in Drug Research, Volume 1*, pp. 343–368, Academic Press.

Pérez-Lucas, G., Navarro, G. and Navarro, S., 2024. Understanding how chemical pollutants arise and evolve in the beer brewing process: A scoping review. *Foods*, 13(11), p. 1709. https://doi.org/10.3390/foods13111709

Prado, R., Gastl, M. and Becker, T., 2021. Aroma and color development during the production of specialty malts: A review. *Comprehensive Reviews in Food Science and Food Safety*, 20(5), pp. 4816–4840.

Prata, R., Petrarca, M.H., Teixeira Filho, J. and Godoy, H.T., 2021. Simultaneous determination of furfural, 5-hydroxymethylfurfural and 4-hydroxy-2, 5-dimethyl-3 (2H)-furanone in baby foods available in the Brazilian market. *Journal of Food Composition and Analysis*, 99, p. 103874.

Rahn, A. and Yeretzian, C., 2019. Impact of consumer behavior on furan and furan-derivative exposure during coffee consumption. A comparison between brewing methods and drinking preferences. *Food Chemistry*, 272, pp. 514–522.

Rehman, H., Ullah, I., David, M., Ullah, A. and Jahan, S., 2019. Neonatal exposure to furan alters the development of reproductive systems in adult male Sprague Dawley rats. *Food and Chemical Toxicology*, 130, pp. 231–241.

Sandjong-Sayon, D.R., Fakih, A., Mercier, F., Kondjoyan, N., Meurillon, M., Ratel, J. and Engel, E., 2024. Targeted quantification and untargeted exploration of furan and derivatives in infant food by headspace extraction-gas chromatography-Q Exactive Orbitrap mass spectrometry. *Food Research International*, 191, Article 114614, 10.1016/j.foodres.2024.114614.

Sarafraz-Yazdi, A., Abbasian, M. and Amiri, A., 2012. Determination of furan in food samples using two solid phase microextraction fibers based on sol–gel technique with gas chromatography–flame ionisation detector. *Food Chemistry*, 131, pp. 698–704.

Shahidi, F. and Hossain, A., 2022. Role of lipids in food flavor generation. *Molecules*, 27(15), p. 5014.

Shakoor, A., Zhang, C., Xie, J. and Yang, X., 2022. Maillard reaction chemistry in formation of critical intermediates and flavour compounds and their antioxidant properties. *Food Chemistry*, 393, p. 133416.

Shen, M., Liu, L., Zhang, F., Jiang, Y., Huang, Z., Xie, J., Zhong, Y., Nie, S. and Xie, M., 2021. Effects of processing parameters on furan formation in canned strawberry jam. *Food Chemistry*, 358, p. 129819.

Shen, M., Zhang, F., Hong, T., Xie, J., Wang, Y., Nie, S. and Xie, M., 2017. Comparative study of the effects of antioxidants on furan formation during thermal processing in model systems. *LWT – Food Science and Technology*, 75, pp. 286–292.

Sirot, V., Rivière, G., Leconte, S., Vin, K., Traore, T., Jean, J., Carne, G., Gorecki, S., Veyrand, B., Marchand, P., Le Bizec, B., Jean-Pierre, C., Feidt, C., Vasseur, P., Lambert, M., Inthavong, C., Guérin, T. and Hulin, M., 2019. French infant total diet study: Dietary exposure to heat-induced compounds (acrylamide, furan and polycyclic aromatic hydrocarbons) and associated health risks. *Food and Chemical Toxicology*, 130, pp. 308–316.

Teixeira dos Santos, C.A., Páscoa, R.N.M.J., Pérez-del-Notario, N., González-Sáiz, J.M., Pizarro, C. and Lopes, J.A., 2024. Application of Fourier-transform infrared spectroscopy for the assessment of wine spoilage indicators: A feasibility study. *Molecules*, 29(8), p. 1882.

Tivanello, R.G., Lawrie, C.H. and Bragotto, A.P., 2023. The role of micrornas in the toxic mechanisms of furan and 3-MCPD and links to cancer risk–A review. *Food Additives & Contaminants: Part A*, 40(6), pp. 733–744.

Tsao, W.X., Chen, B.H. and Kao, T.H., 2022. Effect of sterilization conditions on the formation of furan and its derivatives in canned foods with different substrates. *Journal of Food and Drug Analysis*, 30(4), p. 614.

Von Tungeln, L.S., Walker, N.J., Olson, G.R., Mendoza, M.C.B., Felton, R.P., Thorn, B.T., Marques, M.M., Pogribny, I.P., Doerge, D.R. and Beland, F.A., 2017. Low dose assessment of the carcinogenicity of furan in male F344/N Nctr rats in a 2-year gavage study, *Food and Chemical Toxicology*, 99, pp. 170–181.

Wang, J., Yang, Z.J., Wang, Y.D., Cao, Y.P., Wang, B. and Liu, Y., 2021. The key aroma compounds and sensory characteristics of commercial Cheddar cheeses. *Journal of Dairy Science*, 104(7), pp. 7555–7571.

Yi-Hsuan Huang, Y.-H., Kao, T.-H., Inbaraj, B. S. and Chen, B.-H., 2022. Improved analytical method for determination of furan and its derivatives in commercial foods by HS-SPME arrow combined with gas chromatography–Tandem Mass Spectrometry. *Journal of Agricultural and Food Chemistry*, 70(25), pp. 7762–7772.

Zhang, Y. and Zhang, Y., 2023. A comprehensive review of furan in foods: From dietary exposures and in vivo metabolism to mitigation measures. *Comprehensive Reviews in Food Science and Food Safety*, 22(2), pp. 809–841.

Zheng, J., Tian, L. and Bayen, S., 2023. Chemical contaminants in canned food and can-packaged food: A review. *Critical Reviews in Food Science and Nutrition,* 63(16), pp. 2687–2718.

10 4(5)-Methylimidazole and Safety

Food Processing and Manufacturing Related Effects

Maomao Zeng and Peng Deng

10.1 OCCURRENCE, EXPOSURE ASSESSMENTS, AND TOXICITY

10.1.1 Occurrence in Foods

The Maillard Reaction (MR) constitutes a complex series of interactions that occur between reducing sugars and amino compounds under both ambient and elevated thermal conditions[1, 2]. This reaction is ubiquitous across numerous disciplines, including culinary baking, coffee processing, meat treatments, flavoring production, and the brewing industries, serving as a principal contributor to the development of rich hues and robust aromatics in processed foods[3]. The Maillard reaction is positioned at the vanguard of interdisciplinary research, spanning food science, chemistry, biology, and medicine[4, 5]. In recent years, the focus of academic inquiry has increasingly shifted to the underlying mechanisms of the Maillard reaction, protein cross-linking,

DOI: 10.1201/9781032712291-10

and the emergence of associated hazardous substances along with their toxicological characteristics, each of which has profound implications for human health[6, 7].

4(5)-Methylimidazole (4-MI) is recognized as a pernicious by-product arising from the Maillard reaction, chiefly produced during the production of ammonia-based caramel colorants. Caramel, a multifunctional additive imparting color adjustment and flavor enhancement, manifests as a spectrum from brown to dark brown, either as a viscous liquid or a hygroscopic powder[8]. The caramel is pervasively utilized within the food industry due to its quintessential color and flavor profile[9]. This additive is a complex mixture generated during the thermal treatment of carbohydrates (caramelization reaction), and caramel can be bifurcated into four distinct classifications contingent based on the feedstock employed: Cass I – plain caramel, Cass II – sulfite caramel, Cass III – ammonia caramel, and Cass IV – sulfite ammonium caramel[10]. The explanations of four class caramel are listed in Table 10.1.

Plain caramel, with a solid matter concentration ranging from 62% to 77%, is concocted devoid of any ammonium or sulfite additives. Moreover, plain caramel has the minimum content of nitrogen and sulfur, which can be added into products such as cereals, energy bars, rice cakes, and bread. Sulfite caramel has the solid content ranging from 65% to 72%, and the sulfites are added as catalytic agents during its fabrication. Sulfite caramel is mainly utilized in particular alimentary items, such as distilled spirits. Ammonia caramel has a solid content of 53–83%, and ammonia is employed as a catalyst during its process. This type of caramel is mainly used in soy sauce, stout beers, bread, cake mixes, cereals, and biscuits. Sulfite ammonium caramel has the solid content of 40–75%, and sulfite ammonium is utilized as a catalyst during its production. Sulfite ammonium caramel is extensively used in confectioneries, soft drinks, baked goods, and cereals. The applications of Class I, III, and IV caramel across a diverse food matrix are cataloged in Table 10.2.

The addition limits on 4(5)-MI concentrations within the four classes of caramel are specified in Table 10.3. European Union regulations stipulate that the levels of 4(5)-MI in ammonia caramel production should be below 200 mg/kg. Furthermore, 4(5)-MI can be detected in foods without adding caramel colorants, such as coffee and cookies[11]. The content of 4(5)-MI within a broad spectrum of food products, including coffee, biscuits, and soy sauce, are listed in Table 10.4.

10.1.2 Toxicity of Imidazole Compounds

2-MI ($C_4H_6N_2$) features a methyl substitution at the second carbon of imidazole, with its structural formula depicted in Figure 10.1. 2-MI has been detected in caramels and soft drinks in lots of studies. However, there was also research indicated that 2-MI cannot be detected with capillary electrophoresis in commercial caramel samples, whose detection limit was 0.22 mg/L (ppm) [12]. It has been reported that 2-MI also cannot be detected in 97 types of cola samples and 13 types of caramel by LC-MS/MS, with a detection limit of 0.2 mg/L. Consequently, these findings suggest the improbability presence of 2-MI in caramels and beverages, indicating that the risk can be neglected in food products containing caramels, hence omitting safety assessments of 2-MI[13].

TABLE 10.1 Classification and related characteristics of caramels

CLASS	*INTERNATIONAL NUMBER*	*TYPICAL APPLICATION*	*COLOR RATE*	*CHARGE*	*CONTAINING NITROGENOUS SUBSTANCES*	*CONTAINING SULFUR SUBSTANCES*
Plain caramel (I)	INS 150a EEC NO.E150a	baked products, ready-to-eat cereals, compound seasonings	1.7×10^4	negative (weak)	no	no
Sulfite caramel (II)	INS 150b EEC NO.E150b	alcoholic beverages	2.7×10^4	negative	no	yes
Ammonia caramel (III)	INS 150c EEC NO.E150c	soy sauce, beer, vinegar	$(3.2 \sim 5)\times10^4$	positive	yes	no
Sulfite ammonium caramel (IV)	INS 150d EEC NO.E150d	carbonated drinks, cooking wine, tea	$(2 \sim 8)\times10^4$	negative	yes	yes

TABLE 10.2 The addition amount of Class I, III, and IV caramel in various food products

	CLASS I CARAMEL (MG/KG)		*CLASS III CARAMEL (MG/KG)*		*CLASS IV CARAMEL (MG/KG)*	
TYPE	*TL*	*ML*	*TL*	*ML*	*TL*	*ML*
soy sauce	38571	95556	40000	85000	150000	200000
vinegar	2700	15250	22500	38333		
compound seasoning	18757	36266	13127	31525	14472	36359
cola				6240	1600	2000
fruit and vegetable juice	2325	46550	7500		1000	2000
tea drinks					161	1305
milk drinks					700	900
solid drinks					79	121
distilled spirits (brandy, whiskey)	2604	4200		5000	600	1000
Chinese rice wine			801	1487	600	850
cooking wine					1000	1177
cocoa products and chocolate			1160	1636	600	4267
cookies					3283	5600
puffed food	1000	2500			503	1000

TL stands for typical addition level, ML stands for maximum addition level.

Solid-phase extraction combined with pulsed amperometric detection was employed to identify 2-MI in sodas, while the concentration was found below the limit of quantification (0.03 mg/L).

4-MI, an isomer of 2-MI, presents as a colorless or slightly yellow nitrogen-containing compound. It has been confirmed that 4-MI and 5-MI are reciprocal isomers for each other, resulting in their existence in predominantly in mixtures[14]. Therefore, it is also denoted as 4(5)-MI, and the structural illustration is shown in Figure 10.1. 4-MI contains a heterocycle, produced from the interaction between ammonia and reducing sugars. These precursors are the prevalent components in foodstuffs, and thereby 4-MI can be formed in various food products during their baking and heating processes[15]. 4-MI also can be detected in Class III caramel, Class IV caramel, and caramel-containing products such as soy sauces, beers, and soft drinks, as well as these foods utilizing caramel colors during manufacturing (coffee and bread) [16]. The limited level of 4-MI made by FCC (Food Chemicals Codex), JECFA (the Joint FAO/WHO Expert Committee on Food Additives), and EU (European Union) was 250 mg/kg (ppm), while JECFA imposed a stricter limit of 4-MI in Class III caramel (200 mg/kg). Recent studies have investigated that the level of 4-MI in caramels below ppb. Previous studies have investigated the commercial caramel samples, and the content of 4-MI ranged from 88.2 to 168.6 mg/kg in Class III caramel, and 43.2 to 120.1 mg/kg in Class IV caramel. The

TABLE 10.3 International Technical Caramel Association / European Technical Caramel Association classification of caramel colors and related properties

	TYPE	EM (610 NM)	EBC单位(×103)	4(5)-MI (MG/KG)	TOTAL NITROGEN (%)	TOTAL SULFUR (%)	SO2(%)	PH
Plain	CP-1	5–35	2–12	<25	<0.1	<0.1	<0.1	3–4
caramel (I)	CP-1	40–80	15–25					2.5–4
Sulfite caramel (II)	CCS-1	40–80	16–24	<25	<0.1	0.15–2.5	<0.15	2.8–5.5
Ammonia caramel (III)	AC-1	60–90	16–24	<200	0.5–3.0	<0.7	<0.02	
	AC-2	100–140	27–37.5		0.5–5.0			
	AC-3	150–200	40–54	·	1.5–6.5			
Sulfite ammonium caramel (IV)	SAC-1	35–70	8–16	0.1–1.3	0.3–2.0	0.3	<0.1	
	SAC-2	75–100	17.5–23	0.5–2.8	0.8–3.2	0.5		2–3.5
	SAC-3	105–150	22.5–37	0.8–2.8	1.0–4.0	0.8		
	SAC-4	210–270	40–52	2.0–4.0	1.0–5.0	1.5		

TABLE 10.4 Content of 4(5)-MI in various food

TYPES	*4(5)-MI*
vinegar (ng/mL)	5 ~ 836
soy sauce (ng/g)	5 ~ 551
beverage (ng/mL)	31 ~ 790
beer (ng/mL)	19 ~ 224
bread (ng/g)	5 ~ 145
biscuit (ng/g)	5 ~ 376
coffee (ng/g)	649 ~ 5869

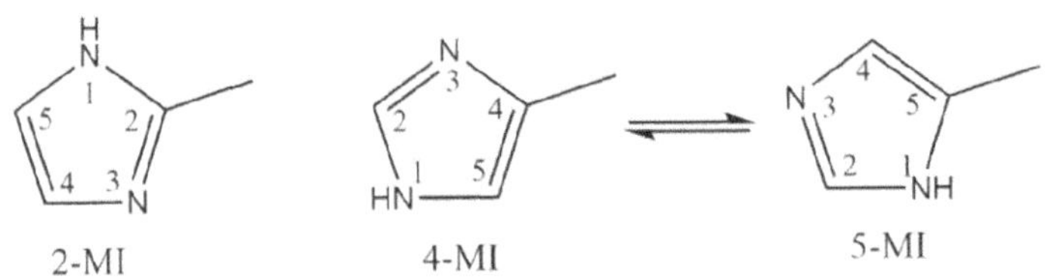

FIGURE 10.1 Chemical structure of 2-MI, 4-MI and 5-MI.

European Food Safety Authority (EFSA) reviewed 4-MI levels in production batches of Class III and IV caramel over two years in 2011, revealing the average level of 4-MI reached 41.6 mg/kg (5–140 mg/kg) in Class III caramel and 88.3 mg/kg (22.7–147 mg/kg) in Class IV caramel. Nowadays, manufacturers have improved the process of caramels, achieving the low level of 4-MI in caramels (<15–60 mg/kg).

10.1.3 Exposure Assessments

Exposure to specific dosages of 4(5)-MI can induce the state of hyperexcitability in various animals, including rodents, rabbits, chickens, and bovines, causing acquired damage to ruminants and manifesting in severe convulsive symptoms[17]. Over a span of approximately two years, Chan et al.[18] found that the consumption of varying dosages of imidazole caused the adverse symptoms in mice such as occurrences of excitement and convulsions. These results verified that imidazole was a neurotoxin with a potent convulsant effect, causing hepatocellular proliferation, chronic inflammation, monocellular pathologies, retarded growth, and predisposition towards bronchial tumors in mice[19]. According to the conclusive evidence of carcinogenicity in experimental animals but lacking in epidemiological data, the International Agency for Research on Cancer (IARC) has classified 4-MI as a Group 2B carcinogen, denoting it possibly carcinogenic to humans[20]. Considering the safety, JECFA and the EU have stipulated maximum permissible levels of 4-MI in Class III and IV caramels at 200–250 mg/kg. The limitation of 4-MI in caramels defined by GB8817-2001 in China was 200 mg/kg. The Office of Environmental Health Hazard Assessment (OEHHA) in California announced a critical risk level for 4(5)-MI in 2012, positing that its daily intake should not exceed 29 μg/day.

10.2 FORMATION MECHANISMS OF IMIDAZOLE COMPOUNDS

Imidazoles are by-products of the Maillard reaction, and thus it is necessary to make a comprehensive analysis of the Maillard reaction pathway to explore their formation. The Maillard reaction primarily imparts characteristic color and flavor to foods[21]. It begins with the condensation between reducing sugars and amino groups, leading to the formation of brown pigments and flavor compounds such as pyrazines, pyrroles, furans, oxazoles. Extensive research has been undertaken on the Maillard reaction, which can be divided into three reaction stages: initial, intermediate, and advanced[22]. In the initial stage of the Maillard reaction, predominant reactions involve the condensation of sugars with amino acids and the Amadori rearrangement. The nucleophilic nitrogen atom of the reactant's amino group adds to the carbonyl carbon of the sugar to form a hydroxylamine (N-glycosylamine), which subsequently attacks other sugar molecules to form disaccharide amines. These amines then dehydrate to form Schiff bases (imines), which are highly unstable and then undergo enol and ketone interconversion isomerization to yield Amadori rearrangement products. The intermediate stage of the Maillard reaction is characterized by the degradation of Amadori rearrangement products and amino acids[23]. The pathway for Amadori rearrangement products can change according to the pH of the reaction system. It can be found that the final products include 5-hydroxymethylfurfural and black essence when the pH ≤ 7, and when the pH > 7 and the temperature is relatively low, the primary products are reduced ketones and dehydroreduced ketones, which are unstable and likely to form α-amino ketones and aldehydes. The main products are dicarbonyl or reactive carbonyl compounds (butanedione and methylglyoxal) when the temperature reaches higher[24].

The advanced stage of the Maillard reaction involves complex reactions, notably the hydroxyaldol condensation of aldehydes and ketones catalyzed by amines, producing aldehydes (especially α,β-unsaturated aldehydes) and nitrogen-free polymers. Additionally, the rapid polymerization or copolymerization of amines at low temperatures results in high molecular weight nitrogenous substances, including melanoidins. See Figure 10.2.

The Maillard reaction can be influenced by various factors, including oxidizing and reducing agents. Notably, metal ions may form complexes with Maillard reaction products, facilitating the oxidation of Amadori compounds. Cu^{2+} is able to bind with Amadori compounds to form unstable complexes, which then degrade into hydroxyl radicals and dicarbonyl compounds. Furthermore, sulfites can inhibit the Maillard reaction in the presence of glucose, as they react with the glucosamine degradation product 3,4-dideoxyhexulose-3-ene, a crucial intermediate in the color formation process of the Maillard reaction.

4-MI is a nitrogen-containing heterocyclic compound, and its formation is closely associated with the presence of amine compounds. Hengel et al.[14] proposed that 4-MI is formed through the interaction between 2-aminopropanal and formamide. As shown in Figure 10.3, 2-aminopropanal is generated from the reaction between methylglyoxal and ammonium ions, and formamide is produced by the amination of methylglyoxal. Additionally, previous study revealed that the addition of formaldehyde

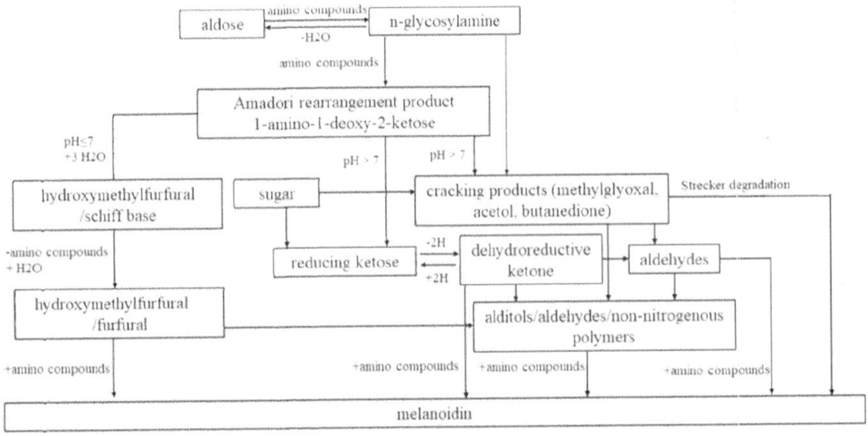

FIGURE 10.2 Maillard reaction path diagram.

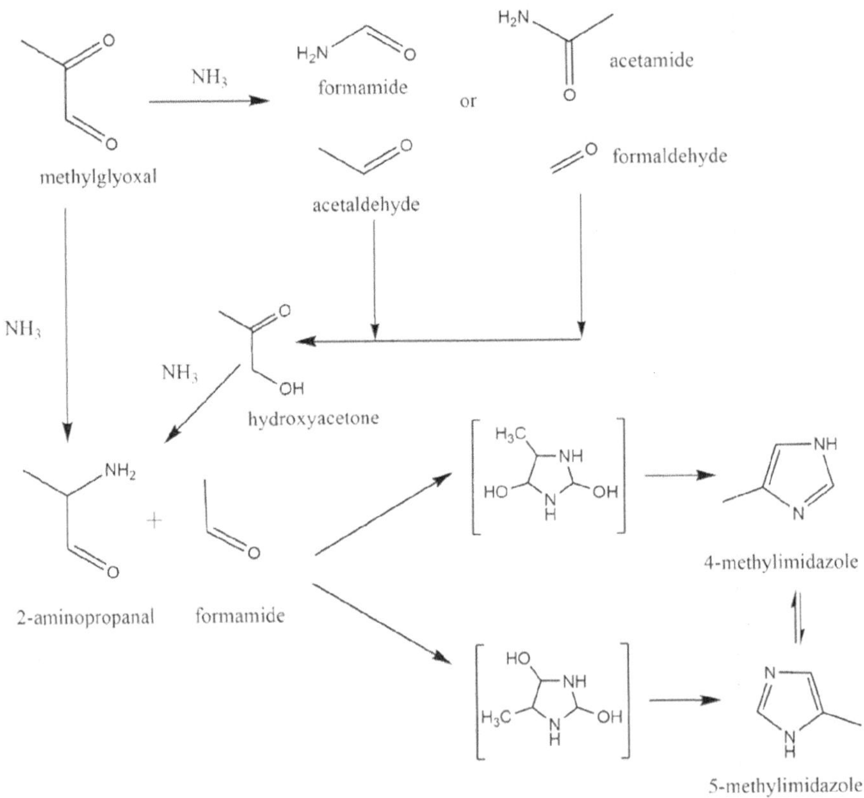

FIGURE 10.3 The formation pathway of 4(5)-MI.

into methylglyoxal-ammonia mixture had no influence on the formation of 4-MI. Therefore, it is speculated that formaldehyde cannot directly participate in the reaction process leading to the formation of 4-MI. Meanwhile, Wu et al. has investigated the formation of 4-MI by the reaction of α-dicarbonyl compounds and aldehydes (formaldehyde, acetaldehyde) in the presence of ammonium sulfite. Two formation pathways of 4-MI were proposed: (1) α-dicarbonyl compounds react with equimolar ammonia to form corresponding intermediate, and aldehyde reacts with equimolar ammonia to form another intermediate. The reaction of these two intermediates led to the formation of 4-MI. (2) α-dicarbonyl compounds react with diamine to form corresponding intermediate, which subsequently reacts with an aldehyde to form 4-MI. Although the mechanism of 4-MI formation is not entirely clear, evidence indicates that methylglyoxal is the key intermediate, primarily formed through the reverse aldol condensation of 3-deoxyglucosone or 1-deoxyglucosone.

10.3 ANALYSIS METHODS

The development of highly efficient, sensitive, and precise methodologies for detecting 4(5)-MI is crucial in the domain of food quality and safety. This significance is underscored by the potential carcinogenic and immunotoxin implications of 4(5)-MI, and thus the level of 4(5)-MI needs stringent quality control and oversight.

10.3.1 Extraction Technology

The initial step in the extraction of methylimidazole from foodstuffs involves selecting an optimal solvent to both homogenize the sample and solubilize the imidazole, encompassing a range of both inorganic and organic solvents. Typically, water-soluble imidazoles are extracted utilizing inorganic solvents to increase extraction efficiency. Nevertheless, certain food matrices, such as bread and cookies, which are rich in sugars, proteins, and fats, may precipitate the agglomeration of these macromolecules, potentially obstructing the solid-phase extraction (SPE) apparatus[25]. Consequently, organic solvents are preferred compared with their aqueous counterparts because the organic solvents can avoid this issue. Furthermore, acidification of the extraction solvent is advisable to facilitate imidazole ionization, thereby increasing extraction efficiency. To mitigate the interference from complex components in food matrix, subsequent filtration or centrifugation of the extract is recommended.

After the above extraction steps, for further isolation and purification of methylimidazole, pretreatment of the sample is also required before quantitative analysis. The most widely used method is SPE, and the general steps of SPE are activation of column, sample uptake, and elution[26]. In order to achieve a better adsorption effect, it is crucial to choose a suitable solid-phase extraction column. Referring to the solid-phase extraction method proposed by Gross et al.[27], the samples were hydrolyzed by alkali and extracted with organic solvents, and then extracted by solid-phase extraction columns such as PRS or C18. Imidazoles have weak alkalinity, and solid-phase extraction with ion-exchange function is mostly used for sample pretreatment in recent years. Waters Oasis MCX SPE columns have also been utilized in many studies, which are reversed-phase/strongly cation-exchange, water-wettable polymers. The MCX resin tends to be base selective and stable in organic solvents, acidifying the sample prior to SPE allows for ionic interactions to retain compounds. To eliminate the sample matrix effect, Jacobs et al. used Waters MCX column to clean up the samples and obtained good recovery results (2-MI: 90–113%; 4-MI and THI: 93–115%). Mottier et al.[28] used this method to quantify 2-MI, 4-MI and THI in thermally processed foods, and the obtained results showed that RSD below 17%, and recovery ranges from 89% to 102%. Although the quantitative results of SPE turned to be better, there are still many shortcomings such as the high consumption of organic solvents, slow purification speed, and high pretreatment cost, so it is still necessary to optimize the existing purification method or develop a new purification method to achieve the purpose of low-cost, miniaturization, and high extraction efficiency.

10.3.2 Instrument Analysis

The aqueous solubility of 4-MI presents significant challenges for its analytical determination[29]. Nevertheless, methodological advancements from conventional paper chromatography, gas chromatography (GC), and gas chromatography-mass spectrometry (GC-MS) have transitioned to cutting-edge techniques such as high-performance liquid chromatography-mass spectrometry (HPLC-MS) and ultra-high-performance liquid chromatography-mass spectrometry (UHPLC-MS). The detection of imidazoles predominantly employs methods such as GC, liquid chromatography (LC), GC-MS, high-performance liquid chromatography-tandem mass spectrometry (HPLC-MS/MS), and UHPLC-MS/MS[30]. Moreover, there are lots of research focusing on pursuing the development of other quantitative methods characterized by low detection limits and higher accuracy, including ion chromatography (IC), ion chromatography-mass spectrometry (IC-MS), two-dimensional chromatography (LC-LC), and enzyme-linked immunosorbent assay (ELISA), which currently lead the efforts in imidazolium compound analysis. The initial instrumental method for the analysis of 4-MI was GC, and thus the derivatization of 4-MI was needed for its effective chromatographic separation. With the broader application of mass spectrometry, recent studies have concentrated on the quantification of 4-MI using GC coupled with tandem mass spectrometry, aiming to increase detection sensitivity. Cunha et al.[9] reported a detection limit of 2.2 μg/L for 4-MI in stout beer and cola using GC-MS, underscoring the technique's efficacy in tracing minimal quantities of this compound.

In recent years, liquid chromatography-mass spectrometry (LC-MS) has become an effective method for the quantitative analysis of imidazoles, especially for beverages, since most beverages that can be injected directly into the LC after some simple preprocessing[31]. The quantitative analysis of imidazoles by isotopic internal standard method has been reported, which has the advantage of eliminating matrix effect, but if it is out of the working range of the method, a different dilution factor has to be used.

Moreover, the limit of quantification of this method is relatively high (80–180 ng/mL), which is not suitable for the analysis of most samples. For the quantitative analysis of imidazoles in samples, it is important to achieve a low LOQ. When suitable sample extraction and purification methods were selected, and the analytical methods were optimized to obtain suitable parameters to achieve the simultaneous determination of 4-MI, 2-MI and THI. The determination of THI, 2-MI, and 4-MI was performed using LC-MS/MS, and the determination of 4-MI was performed using the internal standard method, with matrix effects eliminated using the 4-MI-d_3 as isotope internal standard. The isotope-labeled standards were stable and had the properties of the target analytes, and they only differed in the molecular mass. It has been reported that SPE has been employed for the cleanup of food samples, and strong cation exchangers are the best adsorbents. The use of dispersive liquid-liquid microextraction (DLLME) for sample pretreatment in the determination of 4-MI in beverages has also been investigated, and this pretreatment method is able to meet the trace requirements for the determination of 4-MI in beverages with better reliability and accuracy compared with SPE SCX. However, this method has not been applied in a wide range of food products.

Liquid chromatography (HPLC) serves as a standard method in instrumental analysis for detecting substances of low to medium polarity, including 4-MI. While HPLC-UV can achieve high accuracy in the quantitative analysis of 4-MI, its practical application is limited by low sensitivity and interference from substances of similar polarity. Therefore, the accuracy can be affected when only HPLC is employed for 4-MI separation. Consequently, HPLC-MS/MS is typically utilized to overcome these challenges, enhancing both sensitivity and the avoidance of matrix interference.

Ion exchange chromatography (HPAEC) can achieve the sensitive quantitative analysis of food components, such as amino acids and carbohydrates. Its methodology is well-developed, with optimized parameters for the analysis of food component and established optimal gradient elution conditions. Therefore, it is applicable to the quantitative assessment of imidazoles in caramels. HPAEC can be combined with mass spectrometry or pulsed amperometry to achieve higher sensitivity, and thus HPAEC-PAD and HPAEC-MS exhibit high recovery rates in the analysis of 4-MI.

Despite the improved resolution of chromatography, there are still some shortcomings, such as time consumption, high operational costs, and the need for expensive equipment. ELISA has emerged as a favored analysis technique due to its sensitivity, accuracy, and rapidity, and ELISA has been increasingly applied in food safety for the detection of pesticides, veterinary drugs, and toxins. ELISA can be divided into double-antibody sandwich ELISA for large molecules and competitive ELISA for small molecules, which has been instrumental in detecting small molecule organic compounds. Indirect competitive ELISA (ic-ELISA) has the advantages of high sensitivity and low limit of detection.

The ultraviolet detector (UV) faces limitations in sensitivity and high detection thresholds, hindering accurate quantification. Liquid chromatography with a diode array detector (LC-DAD) also fails to meet the requirements for precise analysis. There are complex sample preparation and derivatization processes in the detection by GC-MS. HPLC-MS/MS and UPLC-MS/MS are preferred for their high selectivity, high sensitivity, and simple sample processing.

10.4 CONTROL OF IMIDAZOLES

10.4.1 Factors Affecting the Formation of Imidazoles

10.4.1.1 pH

The pH has a significant effect on the Maillard reaction, and the variety and quantity of resultant by-products are different under different pH[32]. The optimal pH range for the Maillard reaction is identified as 7.8 to 9.2, where alkaline conditions promote browning acceleration. Conversely, acidic environments (pH<2) prevent the reactions to form N-glucosamine. Furthermore, pH decides the subsequent reaction progressions, and thus the types of Amadori rearrangement product also changed. Evaluating the consumption of reactant and UV absorption spectra within the temperature of 100°C and pH range from 4.0 to 12.0, only part of lysine participates in this reaction below pH 8.0.

Imidazole has been confirmed as the by-product of the Maillard reaction, and their formation also depends on pH, indicating the importance in exploring the 4-MI formation across different pH levels. Methylglyoxal is the crucial intermediate responsible for the formation of 4-MI. The content of methylglyoxal is positively correlated with the pH in the reaction mixture of glucose, ammonium sulfite, and the reasons may be that the nucleophilic activity of $NH4^+$ increased when pH increased. The content of methylglyoxal produced in the glucose/ammonium sulfite model system is higher than that of glucose/ammonium sulfate model system under the same pH, indicating that sulfite can promote the formation of methylglyoxal. The level of 4-MI has the same trend as pH in glucose/ammonium sulfite model system; the content of 4-MI increased from 38.32 μmol/L to 97.94 μmol/L when pH increased from 4.9 to 6.9. However, the reduction is observed in the level of 4-MI when pH further elevated, and the promotion of sulfite in the formation of 4-MI is weakened at higher pH (>7). Therefore, the interplay between sulfite and pH crucially impacts 4-MI production by regulating the formation of methylglyoxal.

10.4.1.2 Temperature of heat treatment

It has been confirmed that the level of 4-MI formation is significantly increased at high temperatures (>120°C). The content of 4-MI reached 2.56–2.99 μM in the heating mixture of glucose and lysine at the temperature of 120°C, and the level of 4-MI increased to 7.02 μM and then decreased to 3.64 μM, when the temperature increased from 120°C to 140°C. Therefore, the temperature varying from 120°C to 140°C has a significant impact on the 4-MI formation. The crucial reactants contribute to the formation of 4-MI are methylglyoxal and $NH4^+$, and the content of methylglyoxal increased at high temperature, thus promoting the formation of 4-MI. A distinct linear relationship (R^2 = 0.986) was observed between the MGO and 4-MI at the temperature of 120°C. However, the positive correlation disappeared when the temperature reached

140°C, which further verified the significant effect of temperature on the formation of 4-MI. This phenomenon may be caused by the degradation of 4-MI at high temperature. Moon et al.[33] demonstrated that when the heating duration prolonged, the ammonium ion protonation was enhanced, and thus the nucleophilic reaction between $NH4^{+}$ and methylglyoxal was affected, interrupting the formation of 4-MI. On the other hand, the high heating temperature may cause the degradation of 4-MI, also reducing the content of 4-MI[34].

10.4.1.3 Types of reducing sugars

Maillard reaction is known as a carbonyl-ammonia interaction of reducing sugars with amino acid residues of proteins or free amino acids, which is significantly influenced by the type of reducing sugar. Predominantly utilized reducing sugars include arabinose, galactose, mannose, glucose, and fructose. The rate and final products of the Maillard reaction have significant difference, depending on the reducing sugar types utilized. Studies indicate that pentoses (five-carbon sugars) can rapidly promote the formation of Maillard reaction, followed by hexoses (six-carbon sugars), with disaccharides exhibiting the least effect. Benjakul et al.[3] investigated the reaction between porcine plasma protein and various reducing sugar, pH and the content of intermediate were closed with the sugar type. Chen et al.[35] heated the mixture of amino acids (glycine or lysine) and different types of reducing sugars (glucose, fructose, ribose), comparing the level of intermediates generated. Maillard reaction is known as a carbonyl-ammonia interaction of reducing sugars with amino acid residues of proteins or free amino acids, which is significantly influenced by the type of reducing sugar. Predominantly utilized reducing sugars include arabinose, galactose, mannose, glucose, and fructose. The rate and final products of the Maillard reaction have significant difference, depending on the reducing sugar types utilized. Studies indicate that pentoses (five-carbon sugars) can rapidly promote the formation of Maillard reaction, followed by hexoses (six-carbon sugars), with disaccharides exhibiting the least effect. Benjakul et al. investigated the reaction between porcine plasma protein and various reducing sugar, pH and the content of intermediate were closed with the sugar type. Chen et al. heated the mixture of amino acids (glycine or lysine) and different types of reducing sugars (glucose, fructose, ribose), comparing the level of intermediates generated. This study revealed significant differences in the formation of intermediates such as α-dicarbonyl compounds and 3-deoxyvalerose. α-Dicarbonyl compounds is formed during the intermediate stages of the Maillard reaction, having the high reactivity, which serves not only as the primary precursor for unique flavors and colors but also as the intermediates for the formation of imidazole hazards. Therefore, the types of reducing sugar may influence the formation of methylglyoxal, thus causing the changes in the level of 4-MI. It has been found that fructose has greater reactivity in the caramel-producing model system, resulting in a higher rate of Maillard reaction and higher content of 4-MI. The reason may be that fructose has the stronger capacity to form methylglyoxal.

10.4.2 Mitigation Strategies of Imidazoles

10.4.2.1 Ultrafiltration

4-MI is small molecular compounds and it can be significantly or entirely removed from the caramels utilizing ultrafiltration (UF) technology. This method uses ultrafiltration membranes featuring diverse molecular weight cut-offs to reduce the concentration of 4-MI in caramel colorants. These membranes are constructed from high-polymer films, which may originate from natural or synthetic sources. Based on the sieving principle, ultrafiltration uses membranes with specific molecular weight cut-offs for the separation. Propelled by the pressure differential, larger molecules are thereby retained and concentrated in the retentate.

Rufián-henares et al.[36] utilized ultrafiltration membranes to concentrate the intermediates of Maillard reaction, including α-dicarbonyl compounds and final products such as melanoidins. Guan et al.[37] identified that ultrafiltration can remove the substrates of Maillard reaction, particularly reducing sugars, substantially prevents Maillard browning. There are limited researches focusing on using ultrafiltration to reduce the Maillard reaction by-product 4-MI. By modifying the pH of the caramel aqueous solution and applying a specified voltage, ultrafiltration with a molecular weight cut-off exceeding 1 kDa effectively concentrated the caramel solution, causing that no imidazoles was detected in the concentrate.

10.4.2.2 Control of reaction conditions

There are detailed investigations on the process of formulating caramels with minimized level of 4-MI, and the ammonium carbonate was employed as the catalyst during the process of caramels. The methodology included a series of single-factor experiments to assess the effects of various parameters on the formation of 4-MI such as the addition amount of ammonium salt, reaction temperature, heating duration, pH of the reaction system (modulated by the concentration of ammonium salt), and the type of nitrogenous compounds. It has been confirmed that the addition amount of ammonium salt, heating duration, temperature, and pH are the critical determinants of 4-MI levels in caramels, and thus laying the groundwork for orthogonal experimental designs. Meticulous control over reaction conditions could significantly curtail the formation of 4-MI. The caramels with low level of 4-MI can be produced by analyzing the reaction parameters through response surface methodology.

The direct relationship of temperature and pH with the formation of 4-MI has been confirmed by investigating the level of 4-MI during different reaction conditions. The content of 4-MI increased when the heating duration was prolonged, while the effect of heating duration turned to be minor relative to the pH and temperature. The addition amount of ammonium salt seems to have the least impact. Within the framework of process for ammonia caramel, the degree of influence on 4-MI content was observed to decrease in the order of heating temperature, pH, heating duration, and finally, the addition amount of ammonium salt.

10.4.2.3 Addition of exogenous substances

In glucose-ammonia model reaction systems, the content of 4-MI was measured at 556 ± 1.3 μg/mL. The addition of 0.1 mol/L sodium sulfite achieved the increase in the level of 4-MI levels, with the specific value reaching 583 ± 2.6 μg/mL, indicating that low concentrations of sulfite ions can promote the formation of 4-MI. However, when the addition amount of sulfite was above 0.1 mol/L, the content of 4-MI decreased compared with that of 0.1 mol/L sulfite, but the level of 4-MI was still higher than that of the control sample. These findings indicated that sulfites can promote the formation of methylglyoxal, leading to the increase in the level of 4-MI. When the addition amount of sulfite exceeded 0.2 mol/L, the formation of 4-MI was inhibited. Lee et al.[23] reported that the inhibitory effect may be caused by the interaction between $SO3^{2-}$ and methylglyoxal, thus blocking the subsequent reactions required for the formation of 4-MI, which was contradictory to the facilitative effect at lower addition amount of sulfite.

Numerous studies have investigated the mitigation of Maillard reaction by-products by adding exogenous substances[38]. Divalent metal ions significantly influence the Maillard reaction, as they can react with Maillard reaction substrates and form relatively stable complexes with Maillard intermediates, thus changing the subsequent reaction pathways. These ions have the capability to form complexes with the Amadori rearrangement products and melanoidins generated during the initial stages of the Maillard reaction. The structural stability of complexes formed between different divalent metal ions and the Maillard reaction or its intermediates varies, potentially leading to varying degrees of influence on the Maillard reaction[39]. Experiments have demonstrated that upon the introduction of divalent metal ions into the caramel color Maillard model reaction system, the levels of 4-Methylimidazole (4-MI) production and reduction rates are as follows: 0.1M Fe^{2+} at 110±0.7 μg/mL (80% reduction); 0.1M Mg^{2+} at 483±2.0 μg/mL (13% reduction); 0.1M Zn^{2+} at 460±2.0 μg/mL (17% reduction). The efficacy in reducing 4-MI content follows the sequence: $Fe^{2+} > Zn^{2+} > Mg^{2+}$. Divalent metal ions ($Ca^{2+}$ and Mg^{2+}) have become a focal point in domestic and international research on the Maillard reaction. Studies indicate that the early products of the Maillard reaction form complexes with Ca^{2+} and Mg^{2+}, thereby inhibiting the formation of methylglyoxal (MGO). Furthermore, MGO interacts with divalent metal ions (Ca^{2+} and Mg^{2+}), reducing the generation of imidazoles. The mechanism for this interaction is illustrated in Figure 10.4.

It has been confirmed that when adding various compounds such as metal ions and amino acids into glucose/ammonia model system, the level of 4-MI was changed as follows: 110 ± 0.7 μg/mL (0.1 M Fe^{2+}) ~ 452 ± 2.9 μg/mL (0.01 M Fe^{2+}), 483 ± 2.0 μg/mL (0.1 M Mg^{2+}) ~ 576 ± 1.3 μg/mL (0.01 M Mg^{2+}), 656 ± 3.4 μg/mL (0.1 M Ca^{2+}) ~ 652 ± 2.6 μg/mL (0.01 M Ca^{2+}), 460 ± 2.0 μg/mL (0.1 M Zn^{2+}) ~ 581 ± 5.7 μg/mL (0.01M Zn^{2+}), 409 ± 4.4 μg/mL (0.1M tryptophan) ~ 556 ± 1.8 μg/mL (0.01M tryptophan). Generally, the addition of these metal ions has the dose-dependent manner in the inhibition of 4-MI formation, expect for Ca^{2+}. Notably, the addition of 0.01 M Ca^{2+}, Mg^{2+}, Zn^{2+}, tryptophan, and cysteine increased the content of 4-MI at 0.01 mol/L addition level. Therefore, these compounds can serve as food additives to inhibit the formation of 4-MI at optimal addition level. It has been reported that amino acids such as arginine, glycine, cysteine, and lysine also can inhibit the formation of imidazoles[40]. Many studies have indicated that the production of methylglyoxal depended on the reactivity of amino groups in Maillard reaction systems[41]. The reactivity of lysine is higher than that of glycine, thus exhibiting stronger inhibitory effect. However, the amino group in glycine showed higher reactivity to react with methylglyoxal than that of proline, and thus proline seems to be the more effective inhibitors. Arginine

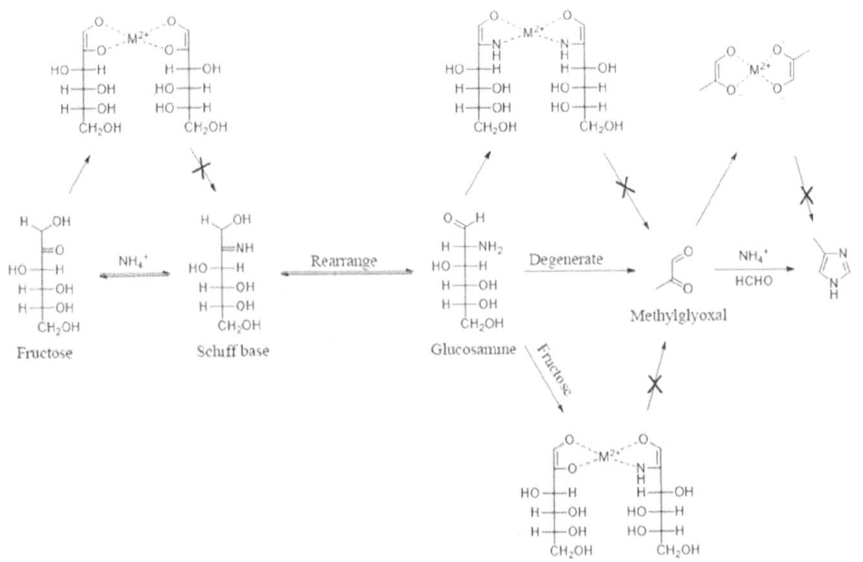

FIGURE 10.4 Possible inhibitory pathways of 4-MI by divalent metal ions.

interrupted the formation of imidazoles by controlling the level of methylglyoxal. Seo et al.[39] found that 0.1 mol/L tryptophan and cysteine reduced imidazoles by 26% and 29%, respectively. The formation of methylglyoxal also was affected by pH, indicating that the change in pH may cause the difference in the reactivity of amino acids, thus affecting the formation of methylglyoxal.

REFERENCES

[1] Göncüoğlu Taş N, Gökmen V. Maillard reaction and caramelization during hazelnut roasting: A multiresponse kinetic study. *Food Chemistry*, 2017, 221: 1911–1922.

[2] Tamanna N, Mahmood N. Food processing and Maillard reaction products: Effect on human health and nutrition. *International Journal of Food Science*, 2015, 2015: 526762.

[3] Benjakul S, Lertittikul W, Bauer F. Antioxidant activity of Maillard reaction products from a porcine plasma protein–sugar model system. *Food Chemistry*, 2005, 93(2): 189–196.

[4] Cui H, Yu J, Zhai Y, et al. Formation and fate of Amadori rearrangement products in Maillard reaction. *Trends in Food Science & Technology*, 2021, 115: 391–408.

[5] Şen D, Gökmen V. Kinetic modeling of Maillard and caramelization reactions in sucrose-rich and low moisture foods applied for roasted nuts and seeds. *Food Chemistry*, 2022, 395: 133583.

[6] Lund M N, Ray C A. Control of Maillard reactions in foods: Strategies and chemical mechanisms. *Journal of Agricultural and Food Chemistry*, 2017, 65(23): 4537–4552.

[7] Martins S I F S, Jongen W M F, van Boekel M A J S. A review of Maillard reaction in food and implications to kinetic modelling. *Trends in Food Science & Technology*, 2000, 11(9): 364–373.

[8] Li T-X, Ji L-B, Jiang Z-R, et al. Caramel products of glucose with water during heating process and their bioactivities. *International Journal of Food Properties*, 2020, 23(1): 971–978.

[9] Cunha S C, Barrado A I, Faria M A, et al. Assessment of 4-(5)methylimidazole in soft drinks and dark beer. *Journal of Food Composition and Analysis*, 2011, 24(4–5): 609–614.

[10] Sengar G, Sharma H K. Food caramels: A review. *Journal of Food Science and Technology*, 2014, 51(9): 1686–1696.

[11] Cunha S C, Senra L, Cruz R, et al. 4-Methylimidazole in soluble coffee and coffee substitutes. *Food Control*, 2016, 63: 15–20.

[12] Wu X, Huang M, Kong F, et al. Short communication: Study on the formation of 2-methylimidazole and 4-methylimidazole in the Maillard reaction. *Journal of Dairy Science*, 2015, 98(12): 8565–8571.

[13] Petruci J F d S, Pereira E A, Cardoso A A. Determination of 2-methylimidazole and 4-methylimidazole in caramel colors by capillary electrophoresis. *Journal of Agricultural and Food Chemistry*, 2013, 61(9): 2263–2267.

[14] Hengel M, Shibamoto T. Carcinogenic 4(5)-methylimidazole found in beverages, sauces, and caramel colors: Chemical properties, analysis, and biological activities. *Journal of Agricultural and Food Chemistry*, 2013, 61(4): 780–789.

[15] Moon J-K, Shibamoto T. Formation of carcinogenic 4(5)-methylimidazole in Maillard reaction systems. *Journal of Agricultural and Food Chemistry*, 2011, 59(2): 615–618.

[16] Vollmuth T A. Caramel color safety–An update. *Food and Chemical Toxicology*, 2018, 111: 578–596.

[17] Mousa R M A. Simultaneous mitigation of 4(5)-methylimidazole, acrylamide, and 5-hydroxymethylfurfural in ammonia biscuits by supplementing with food hydrocolloids. *Food Science & Nutrition*, 2019, 7(12): 3912–3921.

[18] Chan P C, Hill G D, Hills G D, et al. Toxicity and carcinogenicity studies of 4-methylimidazole in F344/N rats and B6C3F1 mice. *Archives of Toxicology*, 2008, 82(1): 55.

[19] Hong S-M, Park M, Lee K-G. Development of caramel colour with improved colour stability and reduced 4-methylimidazole. *Food Additives & Contaminants: Part A*, 2020, 37(7): 1110–1117.

[20] Hyong S, Chu M, Park H, et al. Analysis of α-dicarbonyl compounds and 4-methylimidazole in coffee made with various roasting and brewing conditions. *LWT*, 2021, 151: 112231.

[21] Jang H W, Jiang Y, Hengel M, et al. Formation of 4(5)-methylimidazole and its precursors, α-dicarbonyl compounds, in Maillard model systems. *Journal of Agricultural and Food Chemistry*, 2013, 61(28): 6865–6872.

[22] Karim F, Smith J S. Formation of 4(5)-methylimidazole in aqueous d-glucose-amino acids model system. *Journal of Food Science*, 2016, 81(1): T268–T274.

[23] Lee K-G, Jang H, Shibamoto T. Formation of carcinogenic 4(5)-methylimidazole in caramel model systems: A role of sulphite. *Food Chemistry*, 2013, 136(3–4): 1165–1168.

[24] Cömert E D, Gökmen V. Kinetic evaluation of the reaction between methylglyoxal and certain scavenging compounds and determination of their in vitro dicarbonyl scavenging activity. *Food Research International*, 2019, 121: 257–268.

[25] Wang Q, Chang H, Deng P, et al. Investigation on the simultaneous inhibition of advanced glycation end products, 4-methylimidazole and hydroxymethylfurfural in thermal reaction meat flavorings by liquiritigenin, liquiritin and glycyrrhizic acid and possible pathways. *Food Research International*, 2023, 173: 113414.

[26] Yamaguchi H, Masuda T. Determination of 4(5)-methylimidazole in Soy sauce and other foods by LC-MS/MS after solid-phase extraction. *Journal of Agricultural and Food Chemistry*, 2011, 59(18): 9770–9775.

[27] Gross G A, Grüter A. Quantitation of mutagegnic/carcinogenic heterocyclic aromatic amines in food products. *Journal of Chromatography A*, 1992.

[28] Mottier P, Mujahid C, Tarres A, et al. Process-induced formation of imidazoles in selected foods. *Food Chemistry*, 2017, 228: 381–387.

[29] Zhang L, Li C. Simultaneous analysis of 2-methylimidazole, 4-methylimidazole, and 5-hydroxymethylfurfural potentially formed in fermented soy sauce by "quick, easy, cheap, effective, rugged, and safe" purification and UHPLC with tandem mass spectrometry. *Journal of Separation Science*, 2019, 42(2): 501–508.

[30] Moretton C, Crétier G, Nigay H, et al. Quantification of 4-methylimidazole in Class III and IV caramel colors: Validation of a new method based on heart-cutting two-dimensional liquid chromatography (LC-LC). *Journal of Agricultural and Food Chemistry*, 2011, 59(8): 3544–3550.

[31] Wu C, Wang L, Li H, et al. Determination of 4(5)-methylimidazole in foods and beverages by modified QuEChERS extraction and liquid chromatography-tandem mass spectrometry analysis. *Food Chemistry*, 2019, 280: 278–285.

[32] Wu X, Kong F, Huang M, et al. Effects of pH on the formation of 4(5)-Methylimidazole in glucose/ammonium sulfate and glucose/ammonium sulfite caramel model reactions. *Food Research International*, 2015, 76: 661–665.

[33] Moon J-K, Shibamoto T. Formation of carcinogenic 4(5)-methylimidazole in Maillard reaction systems. *Journal of Agricultural and Food Chemistry*, 2010, 59(2): 615–618.

[34] Xue C, Li Y, Quan W, et al. Unraveling inhibitory effects of Alpinia officinarum Hance and curcumin on methylimidazole and acrylamide in cookies and possible pathways revealed by electron paramagnetic resonance. *Food Chemistry*, 2022, 389: 133011.

[35] Chen X-M, Kitts D D. Identification and quantification of α-dicarbonyl compounds produced in different sugar-amino acid Maillard reaction model systems. *Food Research International*, 2011, 44(9): 2775–2782.

[36] Rufian-henares J A, Morales F J. Effect of in vitro enzymatic digestion on antioxidant activity of coffee melanoidins and fractions. *Journal of Agricultural and Food Chemistry*, 2007, 55(24): 10016–10021.

[37] Guan Y G, Shi W H, Yu S J, et al. Ultrafiltration of caramel color solutions reduces 5-hydroxymethyl-2-furaldehyde. *Journal of Membrane Science*, 2011, 380(1–2): 9–12.

[38] Zhang Z, Chen Y, Deng P, et al. Research progress on generation, detection and inhibition of multiple hazards–acrylamide, 5-hydroxymethylfurfural, advanced glycation end products, methylimidazole–in baked goods. *Food Chemistry*, 2024, 431: 137152.

[39] Seo S, Ka M H, Lee K G. Reduction of carcinogenic 4(5)-methylimidazole in a caramel model system: Influence of food additives. *Journal of Agricultural and Food Chemistry*, 2014, 62(27): 6481–6486.

[40] Zhang Z, Deng P, Chen Y, et al. Exploration of the synergistic inhibition of 4-methylimidazole and 5-hydroxymethylfurfural in caramel colors by amino acid combinations and potential mechanisms. *Food Bioscience*, 2023, 57: 103471.

[41] Zheng J, Guo H, Ou J, et al. Benefits, deleterious effects and mitigation of methylglyoxal in foods: A critical review. *Trends in Food Science & Technology*, 2021, 107: 201–212.

Aromatic Nitro Compounds and Safety 11

Food Processing and Manufacturing Related Effects

Hansika Sati, Priyanka Kataria,
Devyani Shinde, and Sunil Pareek

11.1 INTRODUCTION

The burgeoning global food industry, driven by increasing population, urbanization, and changing dietary patterns, has led to a substantial rise in processed and manufactured food products. While this has undoubtedly increased food availability and convenience, it has also introduced new challenges to food safety (Ruel et al., 2017). One such concern is the presence of aromatic nitro compounds (ANCs) in the food supply. ANCs, a class of organic compounds characterized by the presence of a nitro group (-NO_2) attached to an aromatic ring, have been increasingly detected in various food products. Their formation is often linked to food processing and manufacturing operations, where complex chemical reactions occur (Deng et al., 2015). Some common occurring ANCs

DOI: 10.1201/9781032712291-11

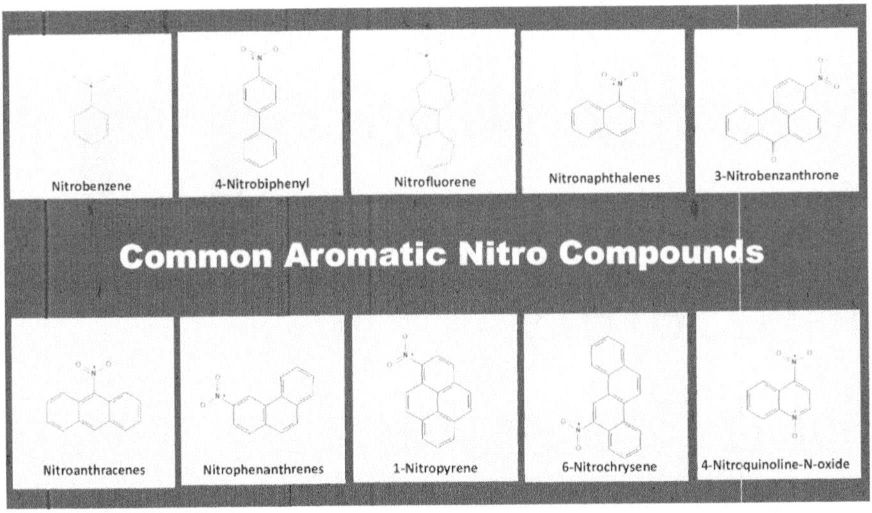

FIGURE 11.1 Common aromatic nitro compounds that occur in processed and manufactured food.

are shown in Figure 11.1. The potential health risks associated with ANC exposure have raised significant alarm among consumers, regulatory agencies, and researchers. Despite growing awareness of the issue, our understanding of ANCs in food remains limited.

While studies have focused on their toxicity, occurrence, and detection, multiple critical gaps appear. An extensive understanding of the mechanisms by which ANCs form during food processing is crucial for developing effective preventive strategies (Molognoni et al., 2019). Additionally, the oral toxicity of ANCs and their specific efforts on human health require further investigation. Another major challenge lies in the diversity of processed and manufactured food items. This heterogeneity complicates the assessment of ANC contamination, as different food matrices may have varying levels and types of ANCs (Ju & Parales, 2010). Moreover, matrix interferences can hinder their accurate detection and quantification. The ambiguity surrounding reported ANC levels, consumption data, and the application of different risk assessment methodologies further exacerbates the problem. Inconsistencies in these areas make it difficult to draw definite conclusions about the overall risk posed by ANCs in diet (Wollin & Dieter, 2005).

To address these challenges, this chapter aims to provide a comprehensive overview of ANCs in processed and manufactured food. It delves into their occurrence, formation, and health implications, while also discussing the limitations of current risk assessment approaches. Furthermore, ensuring the safety of processed and manufactured food needs a multidisciplinary approach involving food scientists, chemists, and regulatory authorities. By working collaboratively, it is possible to develop robust monitoring methods, implement preventive measures, and communicate risks effectively to consumers.

11.2 OCCURRENCE OF AROMATIC NITRO COMPOUNDS

Nitroaromatic compounds rarely occur in nature and have primarily been introduced into the environment through human activity (Ju & Parales, 2010). This significant category of industrial chemicals is extensively utilized in the production of a wide range of products, such as dyes, polymers, insecticides, and explosives. Unfortunately, the widespread utilization of these substances has resulted in the pollution of land and groundwater. The presence of the nitro group in these molecules not only enhances their chemical and functional variety, but also contributes to their resistance to biodegradation (Tiwari et al., 2019). The nitro group's electron-withdrawing property, together with the stability of the benzene ring, confers resistance to oxidative destruction onto nitroaromatic compounds. The presence of the nitro group significantly reduces the reactivity of the benzene ring towards electrophilic aromatic substitution (Kulkarni et al., 2007). The presence of nitro groups connected to aromatic rings leads to the development of partially positive charges at the ortho and para position. These positive charges prevent electrophiles, causing them to attack the open meta positions instead (Omelchenko et al., 2012).

In addition, the reaction between aromatic compounds containing numerous nitro groups and electrophiles can result in the formation of stable Meisenheimer complexes (Al-Kaysi et al., 2008). These attributes enhance the durability and resistance to decomposition of this category of compounds. These ubiquitous environmental contaminants are classically attributed to anthropogenic activities for their origins (Jiang et al., 2024). However, the contamination source can be divided into three significant routes into the food, viz., air, water, and different techniques and chemicals used in the food industry. Through these routes, these mutagens, carcinogenic ANCs enter into the food chain either at production, processing, or transporting food products, thereby putting the consumer's health at risk (Casado et al., 2024). The nitro group attached to this aromatic compound makes the molecule non-biodegradable, thus increasing complications in its environmental management.

11.2.1 Air

Air is the primary theater for all the pollutants occurring in our environment. Improper use and handling due to lack of awareness, skill, and knowledge contribute to unintentionally adding these compounds into the environment around us. Certain unavoidable chemicals in agriculture have nitroaromatics, like pesticides, which are intentionally applied to crops from where they get mixed into the atmosphere (Karić et al., 2022). The released hydrocarbons via industrial fuel consumption in urban settings serve as an underlayer for nitration reaction in combination with atmospheric nitrogen dioxide (NO_2) (Monteiro et al., 2023). This gave rise to a mixture of various nitro aromatics, such as nitrobenzene, nitrophenyl, 3-nitrotoluene, and 1- and 2-nitronapthalene via hydroxyl radical-initiated mechanism. The exhaust from different food industries has hydrocarbons that are prone to nitration, leading to the development of ANCs (Bandowe & Meusel, 2017; Sun et al., 2020). Also, diesel-powered engines are a significant source of these carcinogenic molecules in the environment. Among the various contaminants in the air, nitropyrenes are the most potent ones that come as airborne particulate matter (Hayakawa et al., 2021).

11.2.2 Water

The most regrettable consequence of the environmental presence of these ANCs is the contamination of the aqua sphere (Dai et al., 2014). In water sources, sunlight acts as a catalyst for nitration and halogenation reactions. Here, solar radiation dissolves the organics, nitrate, or nitrite, and generates hydroxyl radical, which in turn acts as a catalyst (Ju & Parales, 2010). The irradiation of the aqua sphere having phenols resulted in the synthesis of 2- and 4-nitro phenols and also chlorophenols and bromophenols. The ANCs also occur in the aqueous phase of the atmosphere where they can then be deposited on land by precipitation (Bandowe & Meusel, 2017).

11.2.3 Food

Ideally, raw food should not possess any of the ANCs, however, background contamination can be a case if they are detected at the farm level. This backend contamination could be the result of the natural production of polycyclic aromatic hydrocarbon (PAH) from volcanic eruptions or particulates traveling through the air. The ANCs occur in different food products under the cooking methods used or processing methods used. The principles used in food processing like smoking, baking, heating, roasting, grilling, frying, etc. add significantly to the formation of nitro aromatic compounds (Kulkarni et al., 2007). The free radicals generated in high-temperature processing interact with light aromatic compounds and form heavy compounds further, which are hydrophobic and finally retained in fat-rich foods (Palade et al., 2023).

11.3 CONTAMINATION OF DIFFERENT FOOD GROUPS DUE TO ANCS

ANCs, formed through various food processing techniques, represent a growing concern in food safety. These compounds, characterized by a nitro group attached to an aromatic ring, can emerge during processes such as cooking, curing, and smoking (Table 11.1). Their presence in a wide array of food groups, including meat, baked goods, and processed cereals, necessitates a comprehensive understanding of their formation, distribution, and potential health implications.

TABLE 11.1 Food products reported to generate aromatic nitro compounds during processing

FOOD AND FOOD PRODUCTS	*PROCESSING PROCEDURE GENERATING ANCS*	*ANCS GENERATED*	*REFERENCES*
Meat and meat products	Grilling	1-nitropyrene	(Kinouchi et al., 1986)
Bacon	Smoked	5-nitro-acenaphthene	(Sonego et al., 2022)
Fish	Smoked	2-nitro-naphthalene 5-nitro-acenaphthene 4-nitro-pyrene	(Sonego et al., 2022)
Coffee	Roasting	1-nitronaphthalene 2-nitronaphthalene 2-nitrofluorene 3-nitrofluoranthene 1-nitropyrene 6-nitrochrysene	(Ko et al., 2018)
Sausage	Grilling	2-nitronaphthalene 2-nitrofluorene l-nitropyrene	(Schlemitz & Pfannhauser, 1996)
Olive oil	Processed	2-nitronaphthalene 2-nitrofluorene l-nitropyrene	(Schlemitz & Pfannhauser, 1996)
Tea	Processed	2-nitronaphthalene 2-nitrofluorene l-nitropyrene	(Schlemitz & Pfannhauser, 1996)

ANCs: Aromatic nitro compounds.

11.3.1 Fresh Fruits and Vegetables

Fruits and vegetables are highly perishable commodities (Sati et al., 2023a, 2023b, 2023c). The food chain for fresh fruits and vegetables starts with the harvest at the field level to the consumer, where there is an active involvement of many stakeholders in the food supply chain (Gokarn & Choudhary, 2021). Fruits and vegetables are susceptible to ANC contamination through various pathways throughout their lifecycle. During growth, agricultural practices involving the use of fertilizers and pesticides may introduce aromatic compounds into the soil and subsequently into the plants. Additionally, pollution from sources such as industrial emissions and vehicular exhaust can deposit ANCs onto the surface of fruits and vegetables. Postharvest handling, transportation, and storage conditions can also contribute to ANC contamination. Washing fruits and vegetables thoroughly before consumption is essential to reduce ANC exposure, although it may not eliminate all contaminants, especially those absorbed into the produce.

Nitropyrene (NP) has been detected as the most frequent pollutant in the environment, which is carcinogenic (Hayakawa et al., 2021). NP is present as the particulate matter in the air that comes from the exhausts from industries as well as vehicle emissions. NP and nitrofluorene are the exhaust indicators and are ubiquitously present in the environment (Morgott, 2014). The densest concentrations of this particulate matter are found in urban settings as these urban areas have ample amounts of its source of origin. However, it is not necessary to encounter these contaminants only near urban settings and are also found in localities farther from urban areas. The air becomes the active carrier of this pollutant which carries it to fresh fruits and vegetables in local markets and vendors from there, the fresh produce gets contaminated at market sites (Lawal, 2017). Along with these, there are many other nitrated aromatic compounds which readily present in the environment and get into the food chain by air transfer and deposition in the produce.

The contamination of ANCs in food crops depends on many factors, like the physiology of fruit plants, location on the farm, contamination level of air, water, and soil. These are derivatives of PAHs and are lipophilic in nature and thus have low water solubility (Bandowe & Meusel, 2017; Han et al., 2019). The leaf can absorb the low molecular weight ANCs through lipid cuticles of the leaf surface making it difficult to wash off. They can settle themselves on the soil, thereby contaminating it, also they make the soil a huge sink of contamination and thereby enter into the food crop. ANCs can enter the food chain from contaminated water, air, and soil. The shoot can uptake these from the stomatal opening and also by adsorption of the contaminant on the surface (Manzetti, 2012, 2013). The root system takes these from contaminated water and soil and translocate these to all parts of the plant, thereby making it part of the plant itself. The main issue related to these compounds is, they are non-biodegradable and remain suspended in the atmosphere.

11.3.2 Thermally Processed Food

Thermally processed foods, including canned goods and packaged items, undergo heating during processing to ensure safety and extend shelf life (Amit et al., 2017). However, high temperatures used in processing can trigger chemical reactions, leading to the formation of the ANCs, particularly in foods rich in protein and fat. The Maillard reaction, which occurs between amino acids and reducing sugars during heating, can also contribute to the ANC formation. To mitigate the ANC contamination in thermally processed foods, precise control of processing temperatures and conditions is necessary. Another contributing factor to the widespread presence of the ANCs in the food chain is the omnipresence of PAHs in the surrounding air as particulate matter. These PAHs can undergo nitration in the presence of NO_2, further increasing the levels of ANCs in the environment and subsequently in the food chain. Nitro PAHs are formed at low concentrations as 1–33 ppm and 0–25 ppm (Dennis et al., 1984). These carcinogenic contaminants are produced in a wide range of processed food products due to their thermal processing, such as heating, roasting, grilling, smoking, etc.

11.3.3 Heating

Heating food, whether through conventional methods such as boiling, steaming, or microwaving, can result in the ANC formation (Molognoni et al., 2019). When food is exposed to elevated temperatures, precursor compounds present in the food, along with nitrogen oxides from various sources (e.g., combustion processes), can react to produce ANCs. The detection of ANCs in raw milk suggests contamination originating from veterinary residues or antibiotics, such as nitrofurans. Lactating cattle are susceptible to exposure from contaminated sources like air, water, and feed (Molognoni et al., 2021). Additionally, dermal absorption can facilitate the entry of nitro compounds into the food chain. Furthermore, boilers utilized in industries, particularly for plate heat exchangers or other purposes, may serve as sources of fuel-generated ANC, including nitrobenzene (Duan et al., 2015). Careful attention to cooking temperatures and techniques can help minimize ANC formation during food preparation at home or in food service establishments.

11.3.4 Roasting

Roasting food, whether in an oven, over an open flame, or on a grill, involves exposing the food to high temperatures. This cooking method is commonly used for meats, vegetables, coffee, and nuts (Sruthi et al., 2021). However, the direct heat can lead to charring and the formation of ANCs, especially in foods containing fat and protein. Coffee is the most consumed beverage worldwide and thus is a principal food commodity of economic value (Murthy & Madhava Naidu, 2012; Vegro & de Almeida, 2020). Coffee is made by roasting the coffee seeds, at a temperature range between 200–240°C with varying processing times, which could be from a few minutes to about half an hour. Roasting gives specific taste, flavor, aroma, and color to the coffee beans. The origin of fuel plays a significant role in the production of ANCs like nitropyrene and nitrobenzene. These compounds are prevalent in combustion contaminants, adhering to the surfaces of roasted coffee beans during the roasting process. Proper monitoring of roasting temperatures and avoiding overcooking or charring can help reduce ANC levels in roasted foods.

11.3.5 Toasting

Toasting bread and other grain products are a widespread culinary practice that involves exposing the food to elevated temperatures. While toasting is primarily done to improve flavor and texture, it can inadvertently lead to the formation of ANCs through a chemical process known as the Maillard reaction, especially concentrated in the crust of the bread (Ko et al., 2018). The Maillard reaction occurs between amino acids and reducing sugars when subjected to heat, resulting in the browning effect and the formation of various flavor compounds (Amaya-Farfan & Rodriguez-Amaya, 2021). However, at high temperatures, this reaction can also contribute to the formation of ANCs, which may have health implications.

To avoid ANC formation during toasting, it's essential to adjust toasting settings appropriately to achieve the desired level of browning without allowing excessive charring. By controlling the toasting process, it's possible to minimize ANC levels in toasted foods while still developing the desired flavor and texture enhancements. This careful adjustment helps to maintain the balance between achieving the desired sensory qualities of toasted foods and reducing potential health risks associated with ANC formation.

11.3.6 Frying

Frying food in oil at high temperatures is a popular cooking method used in both home and commercial kitchens (Spearpoint & Hopkin, 2021). However, the thermal degradation of fats and proteins during frying can result in the formation of ANCs. The interaction between the oil, food, and heat accelerates chemical reactions that lead to PAHs, which in turn form ANC (Singh et al., 2023). The temperature reaches up to 170–250°C during the frying of food items such as extruded products, and chips whereas in discontinuous frying of products like French fries, the temperature reaches around 160–185°C, respectively. Frying occurs in different products like potato, extruded products, port, ham, chicken, pork, hake, and fish (Singh et al., 2020). Frying facilitates the formation of PAHs and subsequent ANCs through the aromatization and dehydrocyclization processes of mono-unsaturated hydrocarbons present in the frying oil. Food product exhibits more contaminants in fried products and lesser levels of contamination in the same product but cooked by other methods than frying. This leads to the understanding that there forms PAHs and NPAHs in the cooking oil during heating, which contributes to the increased levels in the final product. Cooking fumes generated from heating oil are a salient source of indoor pollution which majorly impacts the aggregation of NPAHs to the final product 1-NP and 1,3-di-NP are found in cooking oil like peanut oil, soybean oil, and lard oil. The roasting process and its duration, along with factors such as the fat content of the food, direct contact with the heat source, and pyrolysis, all play significant roles in the release of fat drippings onto the heat source during cooking (Singh et al., 2020). This process favors the adsorption and absorption of NPAHs by the food, contributing to their presence in the final cooked product.

11.3.7 Ohmic-Infrared Cooking

Ohmic-infrared cooking represents a modern culinary technique that merges electrical resistance heating with infrared radiation to expedite and streamline the cooking process. This innovative method provides several advantages, including reduced cooking time and enhanced energy efficiency (Alkanan et al., 2021). However, it's crucial to acknowledge that the high temperatures associated with ohmic-infrared cooking may inadvertently contribute to the formation of ANCs. The key to reducing ANC formation during ohmic-infrared cooking lies in meticulous control of cooking parameters. This includes careful regulation of voltage, current, and cooking time (Adeyeye & Ashaolu, 2022). By precisely managing these factors, chefs and home cooks can minimize the risk of ANC formation while still harnessing the benefits of this advanced cooking method. Striking this balance ensures that food prepared using ohmic-infrared cooking maintains both its culinary excellence and safety standards.

11.3.8 Grilling

Grilling food over an open flame or on a heated surface is a popular outdoor cooking method, especially for meats and vegetables. However, the high temperatures and direct exposure to flames can lead to charring and the formation of ANCs, particularly in foods containing fat and protein (Vintilă, 2016). Meat and meat products, abundant in protein, are susceptible to contamination with PAHs when exposed directly to open flames. This exposure can lead to the conversion of PAHs into NPAHs. Given that non-vegetarian products already provide ample protein content, the formation of NPAHs from PAHs is particularly notable in meat and other non-vegetarian items. This process, where contaminants are deposited onto a product via smoke during cooking, is termed charbroiling (Alisson-Silva et al., 2016). The degree of contamination with both PAHs and NPAHs is influenced by the surface area of the food exposed to the heating source. Therefore, controlling the cooking process to minimize direct exposure to flames and reducing the surface area in contact with the heat source can help mitigate the formation of these harmful compounds. Additionally, marinating the food before grilling can serve as a protective barrier, preventing direct contact between the food and the flames (Dutta et al., 2022). Marinating with ingredients like citrus juices, vinegar, herbs, and spices not only enhances flavor but also reduces the formation of harmful compounds.

11.3.9 Barbecuing

Barbecuing involves cooking food over an open flame or hot coals, often in outdoor settings. While barbecuing imparts a distinct smoky flavor to food, it can also contribute to ANC contamination due to the combustion of organic matter such as wood or charcoal (Xu et al., 2023). The smoke produced during barbecuing contains nitrogen oxides, which can react with food components to form ANCs. The process of barbecuing can be optimized to minimize ANC contamination and ensure food safety. Proper ventilation is crucial to allow the smoke to disperse efficiently, reducing the concentration of nitrogen oxides and other contaminants around the food. Additionally, controlling cooking temperatures is essential to prevent excessive charring and minimize the production of harmful compounds (Sampaio et al., 2021). By maintaining moderate and consistent temperatures throughout the cooking process, the risk of ANC formation can be mitigated.

11.4 MECHANISM OF ANC FORMATION IN PROCESSED AND MANUFACTURED FOOD

A general mechanism of toxicity of ANCs is shown in Figure 11.2. The ANCs generally are emitted from the combustion of diesel engines or are formed from their parent PAHs by the reaction with atmospheric free radicals such as OH, HNO_3, NO, and NO_3.

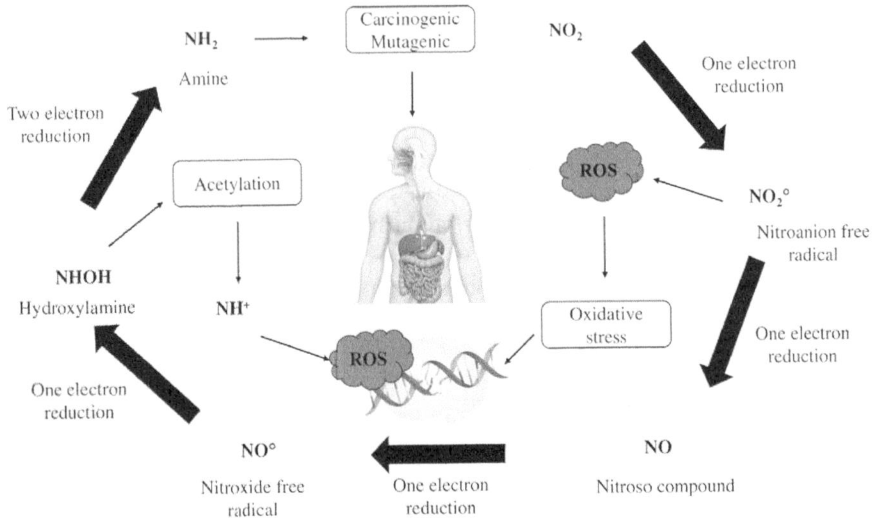

FIGURE 11.2 Major mechanisms of toxicity of aromatic nitro compounds.

This free radical-initiated reaction typically occurs in nitrogen oxide or dinitrogen oxide environments, where co-existing PAHs participate actively due to the high reactivity of their nitro groups. The levels of NPAHs discharged from diesel engines surpass those emitted by gasoline engines (Rhead & Hardy, 2003). The generation of NPAHs is directly linked to the temperature of combustion. Consequently, the ratios (1-NP/pyrene, 6-nitro chrysene/chrysene, 6-nitrobenzo[a]pyrene/benzo[a]pyrene, and 7-nitrobenzo[a] anthracene/benzo[a]anthracene) exhibit notable variations in emissions from diesel vehicles (2700–3000°C), wood-burning stoves (500–600°C), and coal-burning stoves (900–1200°C).

The primary source of NPAHs is attributed to secondary formation through photochemical reactions. However, nitro aromatics in food can result from heat treatments like grilling, smoking, frying, and roasting (Lin et al., 2016). Additionally, NPAHs can form in soil through the conversion of PAHs. PAHs are found ubiquitously in air, soil, and water, where they can undergo chemical reactions to form nitro aromatics (Lawal, 2017). A significant number of NPAHs containing two or more benzene rings are commonly formed through the gas-phase reaction of PAHs with free oxidants present in the atmosphere, such as OH, O_3, and NOx. The primary pathway for the formation of atmospheric ANCs is the OH free radical-initiated reaction (Q. Zhang et al., 2014). However, NPAHs also form through the heterogeneous oxidation of particle-bound PAHs. When ultraviolet (UV) light from the sun strikes O_3 in the presence of water vapor in the atmosphere, the ozone breaks down (photolyzes) and forms excited oxygen atoms. Some of these oxygen atoms then react with water vapor to produce two hydroxyl radicals. The concentration of hydroxyl radical is controlled by the balance between its chemical production and destruction. The hydroxyl radicals are formed by the reaction of excited atomic oxygen with water vapor. The production of excited atomic oxygen, which is produced by the photolysis of O_3, occurs over a very narrow spectral interval between 290 nm and 310 nm in the troposphere (Matsumi & Kawasaki,

2003). The production of excited atomic oxygen decreases with increasing latitude due to less available solar radiation for photolysis.

In the troposphere, the major mode of hydroxyl radical formation follows the photolysis of O_3 by sunlight in the spectral region longer than ~315 nm. This forms excited atoms, some of which then react with water to form two hydroxyl radicals (Gligorovski et al., 2015). About 90% of the $O(^1D)$ atoms are quenched by atmospheric gases back to the $O(^3P)$ ground state, which reforms O_3 through a reaction with O_2. The loss of OH is primarily due to reactions with the gases it oxidizes. The $^{\bullet}OH$ governs atmospheric chemistry during the day since its formation depends on solar radiation. The initial reaction involves the photolysis of ozone by solar radiation with wavelengths less than 310 nm, forming an oxygen atom that then reacts with water to form OH (Poulet, 2000). This mechanism highlights the importance of a small amount of ozone in the troposphere. Other sources of OH include the photolysis of nitrous acid (HONO), hydrogen peroxide (H_2O_2), or peroxy-methane (CH_3OOH), and reactions of nitrogen monoxide (NO) with the hydroperoxyl radical (HO_2) or alkenes with ozone.

11.5 ANCS DIETARY EXPOSURE AND ASSOCIATED HEALTH HAZARDS

11.5.1 Nitrobenzene

ANCs impose several health hazards, some of which are represented in Figure 11.3. Nitrobenzene can reduce blood capacity to carry oxygen (Soni et al., 2024). Frequent exposure to elevated concentrations of 1, 3-dinitrobenzene (1,3-DNB) and 1, 3, 5-trinitrobenzene (1,3,5-TNB) has been linked to anemia, bluish skin development, headaches, nausea, weakness, drowsiness, vomiting, irritability, and dizziness (Arora & Bae, 2014). A study *(in vivo)* conducted by Reddy et al. (2015) shown that 1,3-DNB and 1,3,5-TNB can harm male reproductive systems and potentially reduce sperm production (Reddy et al., 2015). Arora and Bae (2014) reported that 2-Chlorobenzene (2-CNB) is a mild hepatotoxicant, and 4-Chloronitrobenzene (4-CNB) is a strong hepatotoxicant (Arora & Bae, 2014). In a report by the International Agency for Research on Cancer (National Toxicology Program, 2011), nitrobenzene diazole (NBDs) was evaluated as probably carcinogenic to humans and are reasonably expected to be human carcinogens. Inhalation exposure to nitrobenzene has been associated with several tissue-specific tumors in experimental animals, including mice and rats (Hsu et al., 2007). These tumors encompass benign and malignant lung tumors (alveolar/bronchiolar adenoma and carcinoma), benign thyroid gland tumors (follicular cell adenoma), benign mammary gland tumors (adenocarcinoma), liver tumors (hepatocellular adenoma), and kidney tumors (renal adenoma). In rats and mice, oral exposure to 2-CNB led to the development of hepatocellular adenomas, carcinomas, and hepatoblastomas (Arora & Bae, 2014). Apart from causing chronic hepatotoxicity, 4-CNB induced vascular tumors in mice (Jones et al., 2006) and splenic and adrenomedullary tumors in rats (Arora &

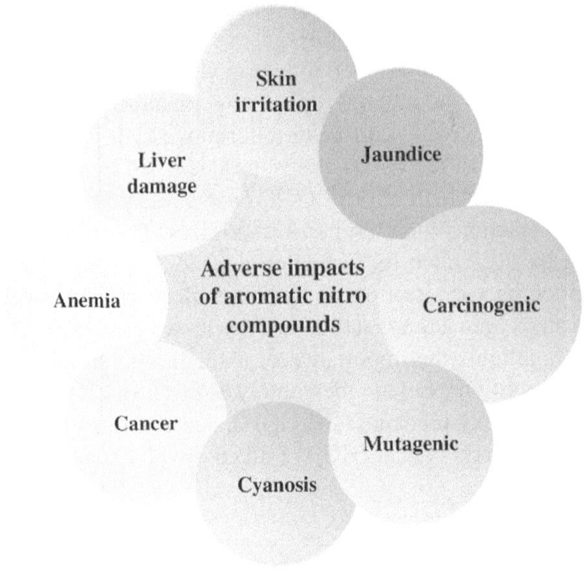

FIGURE 11.3 Health hazards caused by aromatic nitro compounds occurring in processed and manufactured food.

Bae, 2014). Another genotoxic chemical shown to induce mutations in bacteria is 2-CNB (Li et al., 2021).

Pentachloronitrobenzene (PCNB) is recognized as a fungicide capable of inducing allergic contact dermatitis and has been associated with skin sensitization and ocular irritation in individuals (O'Malley et al., 2020). Experimental tests revealed that individuals did not exhibit skin irritation following a 48-hour interaction, but 26% of test subjects experienced a reaction two weeks after a subsequent exposure. PCNB has the potential to harm the cornea and trigger conjunctivitis in the eye (Arora & Bae, 2014). Long-term effects of PCNB have been studied in various animals, although its impacts on human health regarding chronic, reproductive, developmental, or cancerous conditions remain unexplored. Dogs administered PCNB through their meals exhibited signs of liver injury (Foster & Engelhardt, 2023). Furthermore, mice given PCNB orally had fatal signs in their progeny, such as renal agenesis, cleft palates, and elevated fetal mortality (Reed et al., 2007). PCNB is highly harmful to fish and invertebrates and has the potential to bioaccumulate in the aquatic food chain (Tas & Pavlostathis, 2014). The US Environmental Protection Agency has designated PCNB as a priority pollution and a potential human carcinogen due to its extremely hazardous nature (Gibson, 2020). ANCs find extensive application in insecticides, explosives, medicines, and industrial synthesis as chemical intermediates (Tiwari et al., 2019). These substances are highly harmful or carcinogenic, posing a significant risk to human health. They are prevalent environmental pollutants present in workplaces (such as the chemical industry) and in emissions from gasoline and diesel engines (Kovacic & Somanathan, 2014).

Research has been done on these substances' toxicity, carcinogenicity, and metabolic pathways in living things. The synthesis of 2-methoxy aniline, an intermediary in producing several azo colors, begins with 2-methyl nitrobenzene. According to Esmaeili et al. (2006), 2-methy nitrobenzene has carcinogenic activity in rodents, resulting in neoplastic transformation in the kidneys, spleen, and urinary bladder. Xanthine oxidase products N-(2-methoxyphenyl) hydroxylamine and 2-anisidine, which form adducts with DNA, are responsible for the toxicity of 2-methyl nitrobenzene in mice (Kovacic & Somanathan, 2014). On the other hand, cytochrome P450 in the human liver detoxifies 2-methyl nitrobenzene to 2-nitrophenol and 2,6-hydroxy nitrobenzene (Gooch et al., 2017). The redox cycling of their free radicals (oxidative stress) and the production of nitroso and/or hydroxylamine metabolites are the primary causes of the toxicity of nitroaromatic explosives, which include trinitrotoluene (TNT) and 2,4,6-trinitrophenyl N-methyl-nitramine (tetryl) (Miliukiene & Čėnas, 2008).

11.5.2 4-Nitrobiphenyl

The carcinogenic potential of eleven alkyl-substituted derivatives of 4-(nitrobenzyl) pyridine (4-NBP) and two related nitroso compounds were evaluated in *Salmonella typhimurium* strains TA98 and TA100 (Purohit & Basu, 2000). With increasing steric demand of the alkyl substituents in both test strains, the mutagenicity of compounds substituted ortho to the nitro group 3-methyl-3, 3-ethyl-3, 3-isopropyl-3, 3-tertbutyl-3, 3,5-diethyl-3, 3,5-diisopropyl-3, and 3,5-ditertbutyl-4-NBP declined. Even at the highest doses, the most sterically hindered chemicals did not cause mutagenesis (Kulkarni et al., 2007). Deviations from the nitro group's coplanar orientation concerning the aromatic plane are associated with this decrease in mutagenicity. The compound 3-tertbutyl-4-hydroxyanisole did not exhibit a similar decrease in mutagenicity, it becomes essential to prevent the initial reduction step of non-planar nitro groups. The carcinogenic activity of 4-NBP 2-methyl-, 2-ethyl-, 2-isopropyl-, and 2,4,6-trimethyl-4NBP was also decreased with increasing size by the alkyl groups in the 2-position, and for the most sterically hindered species, 2-isopropyl-, 2,4,6-trimethyl-4-NBP, it was eliminated. The nitro group's co-planarity is unaffected in this instance (Klein et al., 2000). However, the two aromatic rings twisting is linked to a less successful charge delocalization of the nitrenium ion.

The mutagenicity of NBP, a canine bladder carcinogen, was examined for its catalytic activity by xanthine oxidase (Swaminathan & Hatcher, 1986). 4-Aminobiphenyl was shown to be the main product of NBP's in vitro enzymatic reduction. Compared to NBP, the reduction intermediate N-hydroxylaminobiphenyl exhibited equivalent mutagenesis activity (Kovacic & Somanathan, 2014). A report discusses the toxicity of chlorinated p-nitrophenyl ethers caused by their metabolites, nitroso, and amino derivatives, which interact with hemoglobin to produce methemoglobin (Fe^{2+} to Fe^{3+}). *Salmonella typhimurium* was investigated for mutagenicity activities in guinea pigs by metabolites of NBP ether (4-NO_2). The mutagenic activity of 4-NO and 4-NHOH metabolites was extreme, whereas that of 4-NO_2 was relatively moderate (Kovacic & Somanathan, 2014).

11.5.3 Nitrofluorene

There is limited research regarding the pure toxicological effects (toxicity) of 2-nitrofluorene (2-NF). Male Sprague-Dawley (SpD) rats administered intraperitoneal injections of 224 μmole 2-NF/kg body weight once daily for five days experienced a 10–15% loss in body weight without evident signs of toxicity. In feeding studies, both male and female rats given a diet containing 1.62 mM 2NF/kg for eight months, or until tumor formation occurred, exhibited no adverse effects of 2-NF on survival (Malejka-Giganti et al., 2008). The utilization of a synthetic diet when adding 2-NF to the diet somewhat complicates the interpretation of the rat condition in another feeding trial. However, compared to rats ingesting the carcinogen through a synthetic diet and control rats fed the synthetic food, all groups of rats treated with 2NF through skin painting and fed a regular diet lived longer (van Bekkum et al., 1999).

The 24-hour LD_{50} for 2-NF in male Swiss-Webster mice weighing between 25 and 30 g has been demonstrated to be 1.6 g/kg of body weight (Möller, 1994). The potential for teratogenic consequences of 2-NF has not been investigated. Nonetheless, teratogenicity has been demonstrated for a few of its putative metabolites. In a modified Ames experiment, 2-NF was shown to be non-toxic to *Salmonella* at dosages up to 333/~g/plate (Klein et al., 2000). In an incubation system, 2-NF was not harmed to *A. nidulans* up to 2000 ~g/mL. However, 2-NF has been shown to have harmful effects in mammalian cell cultures. Therefore, in Chinese hamster Don (CHD) cells, a 2-NF concentration of 9.48×10^{-4} M resulted in 50% mortality. In contrast, in mouse lymphoma cells, an incubation system containing 400 /~g 2-NF/mL reduced the survival around 16% (Beije & Möller, 1988).

11.5.4 Nitronaphthalenes

Sahu et al. (2024) reported that 1-Nitronaphthalene (1-NNT) is a carcinogenic substance (Sahu et al., 2024). Using mass spectrometry and proton nuclear magnetic resonance spectroscopy analyses, glutathionyl hydroxyl naphthalene products were detected in the liver and tracheobronchial airways of rats exposed to 1-NN and its metabolites (Watt & Buckpitt, 2000). As per the findings, liver metabolites were derived from C5 and C6 epoxides, whereas lung conjugates were derived from C7 and C8 epoxides. Similarly, Fanucchi et al. (2004) and Kovaic et al. (2014) reported that the newborn rats and mice were more susceptible to lung damage caused by 1-NNT (Fanucchi et al., 2004; Kovacic & Somanathan, 2014).

The existence of activating enzymes with high catalytic constant (kcat) in vulnerable species has been linked to the toxicity of lung toxicants. The mouse is susceptible to a range of lung toxicants that are activated metabolically. It has been demonstrated that recombinant CYP2F2 can metabolize the lung toxin naphthalene (Li et al., 2017). 1-nironaphthalene and 2-methylnaphthalene, lung toxicants, are quickly broken down into potentially cytotoxic intermediates with high kcat value.

Research using 1-NNT and naphthalene has shown how crucial cytochrome P450 monooxygenase is for producing reactive compounds that cause damage to Clara cells (Phimister et al., 2005). In most cases, metabolites of 1-NNT, particularly in the lower airways, were covalently bound. Interestingly, despite not needing metabolic activation, both 1-NNT and its isomer, 2-nitro naphthalene (2-NNT), demonstrated genotoxicity in vitro (Kovacic & Somanathan, 2014). Evidence suggests that 2-NNT can induce bladder and liver cancers (Kirkland & Beevers, 2006). While 1-NNT yielded negative results, 2-NNT provoked a slightly mutagenic reaction in the bladder and a mutagenic response in the liver. Unlike 1-NNT, 2-NNT acted as a direct mutagen in vitro (Kirkland & Beevers, 2006). An investigation into the mutagenicity, metabolism, and DNA adduct formation of 5-nitrobenzo[b]naphtha[2,1-d]thiophene was undertaken (Swartz et al., 2009). The data indicates that metabolism entails ring oxidation and nitro reduction, leading to the formation of DNA adducts and mutagenicity.

11.5.5 3-Nitrobenzanthrone

The nitrobenzanthrones (NBA) have four fused rings and are another class of nitrated polycyclic molecules (Ostojić et al., 2014). This genotoxic carcinogen has a very high mutagenesis activity which causes lung cancers in rats (Mizerovská et al., 2011; Stiborová et al., 2010). 3-NBA is thought to be carcinogenic in humans (Reshetnikova et al., 2016). According to Murata et al. (2006), the N-hydroxy metabolite causes oxidative damage in human cells by using H_2O_2. In addition to DNA adduct production, oxidative DNA damage may be a significant factor in the carcinogenesis of 3-NBA. It has been noted that the DNA of human hepatoma cells undergoes micronucleus production in response to NBA. 3-NBA is a mutagenic environmental pollutant that may cause human cancer. The amine derivative 3-abscisic acid (3-ABA) is the primary metabolite of 3-NBA. It has been shown that 3-NBA and 3-ABA can both boost ROS production (Hansen et al., 2007). The findings show that 3-ABA is genotoxic to lung cells. Both substances raise ROS and damage DNA, which results in cancer. Research indicates that 3-NBA can quickly decrease and activate in living cells (Chen et al., 2022). Previously reported inductions of hepatic DNA adduct align with the elevated mutant frequency values, which correspond to high levels of 3-NBA nitro reductase in the liver and colon.

11.5.6 Nitrated Polycyclic Aromatic Hydrocarbons

11.5.6.1 Nitroanthracenes

Nitroanthracenes are a class of organic compounds derived from anthracene with one or more nitro functional groups ($-NO_2$) attached. They are commonly encountered as environmental pollutants resulting from combustion processes involving organic materials. Nitroanthracenes are known to exhibit mutagenic and carcinogenic properties, posing significant health risks upon exposure. These compounds have been studied extensively due to their potential adverse effects on human health and the environment. A study evaluated the mutagenic activity and toxicity of PAHs and nitrated PAHs present in outdoor air samples from urban areas, primarily affected by traffic emissions (Ciganek et al., 2004). Nitronaphthalenes were the most abundant nitrated PAH derivatives and were exclusively detected in the vapor phase. Particulate matter was associated with 3-nitrofluoroanthene, 9-nitroanthracene (9-NA), and 9-nitrophenanthrene. PAH pollutants, particularly 9-NA, were commonly found in industrial anthraquinone samples. These impure substances were frequently utilized in toxicity research involving anthraquinones (Butterworth et al., 2004). Besides being mutagenic, the contaminated substance contained 9-NA and other impurities. Experiments using mammalian cells demonstrated the strong mutagenic potential of 9-NA.

11.5.6.2 Nitrophenanthrenes

Nitrophenanthrenes are a group of chemical compounds derived from phenanthrene by the addition of one or more nitro (-NO_2) groups. They are typically formed as byproducts of incomplete combustion processes involving organic materials, such as fossil fuels and biomass burning. These compounds are often found in urban air pollution, particularly in areas influenced by vehicular emissions and industrial activities (Bamford & Baker, 2003). Nitrophenanthrenes are a group of chemical compounds derived from phenanthrene by the addition of one or more nitro (-NO_2) groups. They are typically formed as byproducts of incomplete combustion processes involving organic materials, such as fossil fuels and biomass burning.

Nitrophenanthrenes can exist in various derivative forms, with different nitro substitutions at different positions on the phenanthrene molecule. Some common derivatives include 1-nitrophenanthrene, 3-nitrophenanthrene, and 9-nitrophenanthrene (Wadday & Ghafel, 2022). Exposure to nitrophenanthrenes and their derivatives has been associated with various adverse health effects. These compounds are known to exhibit mutagenic and carcinogenic properties, posing risks to human health. Inhalation or ingestion of air contaminated with nitrophenanthrenes can lead to respiratory issues, cardiovascular problems, and an increased risk of cancer, particularly lung cancer (Y. Zhang et al., 2023). Additionally, nitrophenanthrenes can contribute to environmental pollution and ecosystem disruption. Their presence in air and water can impact aquatic life and vegetation, leading to ecological imbalances (Manzetti, 2012). Overall, the formation and presence of nitrophenanthrenes in the environment raise concerns regarding public health and environmental quality, highlighting the importance of monitoring and controlling their emissions to mitigate potential risks.

11.5.7 1-Nitropyrene

Nitropyrene is a nitro derivative of pyrene. It is formed as a byproduct of incomplete combustion processes involving organic materials, such as fossil fuels and biomass burning. Nitropyrene is often found in urban air pollution, particularly in areas influenced by vehicular emissions and industrial activities. The cytotoxic and mutagenic characteristics of 1-nitropyrene (1-NP) and its reduced metabolite, 1-nitrosopyrene (1-NOP), were evaluated in diploid human fibroblasts. Conditions were established to ensure metabolic activation of the parent compound (1-NP) by human cells in culture (Su & Penning, 2023). The lethal effect of 1-NP on normal cells was compared to that of cells from a patient with hereditary cutaneous malignant melanoma (HCMM). The Xeroderma Pigmentosum (XP) cell survival curve had a 2.5-fold sharper slope than the normal cells; the HCMM cells survival curve had an intermediate slope. The XP cells' survival curve slope was 2.5 times steeper than normal when these cells were exposed to 1-NOP, but the HCMM cells responded normally, indicating that their problem is in activation rather than DNA adduct repair (Patton et al., 1986). In the human cell experiment, 1-NP and 1-NOP also showed mutagenic properties. In terms of concentration, 1-NOP was significantly more mutagenic than 1-NP (Su et al., 2022). However, the compounds were extremely similar when assessed based on the same number of DNA adducts initially attached to DNA or on equal cell death.

The cytotoxic or mutagenic effects of nitropyrenes are likely mediated by the cellular conversion of the parent chemical into a species that combines easily with DNA to create adducts, as they do not directly react with DNA (Yu, 2002). Since mutation tests using nitro reductase-deficient strains of *S. typhimurium* produce lower reversion frequencies than those obtained using nitro reductase proficient tester strains, it appears that reducing the nitro moiety is a crucial step in the metabolic activation of 1-NP (Grummt et al., 2006). Research demonstrated that the process for activating 1-NP progresses, with enzymatic reduction to 1-NOP serving as the crucial first step and rate-limiting step. After that, there is a reduction to the matching hydroxylamine and the easily reacted-with guanine to produce covalent bonded adducts at position 8 (N-deoxyguanosin-8-yl-l aminopyrene), which is formed by an acid-catalyzed intermediate. Ohnishi et al. (2001) reported that the DNA adduct produced by 1-NOP is identical to that of 1-NP (Ohnishi et al., 2001).

11.5.8 6-Nitrochrysene

6-Nitrochrysene is a nitro derivative of chrysene. It is formed as a byproduct of incomplete combustion processes involving organic materials, such as fossil fuels, biomass burning, and tobacco smoke. 6-nitrochrysene (6-NC) is often found in environmental samples, including air, soil, and water, particularly in areas affected by vehicular emissions and industrial activities. 6-NC is a mutagenic and carcinogenic substance (Kong et al., 2023; Peng et al., 2023). It is broken down by nitro reduction and ring oxidation. Trans-1,2-dihydro-N-hydroxy-6-aminochrysene (1,2-DHD 6-NHOH-C), produced via ring oxidation and nitro reduction, is the primary mutagenic metabolite (Guttenplan et al., 2007; Sun et al., 2009). The metabolite action resulting from each pathway in isolation showed characteristics like 6-NC mutation. Products of 1,2-DHD-6 NHOH-C with guanine and adenine generate the majority of DNA adducts in human cells treated with 6-NC; 1,2-DHD-6-NHOH-C seems to be the most genotoxic metabolite (Pande et al., 2020; Powell et al., 2020).

The potency of 6-NC is greater than that of benzo[a]pyrene, 2-amino-1-methyl-6-phenylimidazo[4,5-b] pyridine, and heterocyclic aromatic amines (Krzeminski et al., 2013). Simple nitroreduction activates 6-NC, forming 6-hydroxylaminochrysene, producing reduced amine-form DNA adducts. 6-NC is a potent breast carcinogen and mutagen in rats and an environmental pollutant (El-Bayoumy et al., 2002; Krzeminski et al., 2011; Y.-W. Sun et al., 2013). In a study, two metabolites of 6-nitrochrysene, namely 6-hydroxylamine and 1,2-dihydroxy-1,2-dihydro-6-nitrochrysenes, were found to demonstrate mutagenic properties (Y.-W. Sun et al., 2013).

11.5.9 Nitro Heteroaromatic Compounds

11.5.9.1 4-Nitroquinoline-N-oxide

4-Nitroquinoline-N-oxide(4-NQO) is a nitroaromatic compound that is often used in scientific research as a mutagen and carcinogen. 4-NQO induces DNA damage and mutations in a manner similar to other PAHs, making it a valuable tool for studying the mechanisms of carcinogenesis and DNA repair. It is frequently employed in laboratory experiments to simulate the effects of environmental pollutants and to investigate the molecular pathways involved in cancer development. Additionally, 4-NQO is utilized in dental research as a carcinogen for inducing oral cancer in animal models.

According to a research study, 4-NQO is a powerful inducer of the soxS gene, necessitating molecular oxygen for transcriptional activation (Gomes et al., 2005; Igonina et al., 2018; Kovacic & Somanathan, 2014; Baez & Shiloach, 2013). Two compounds related to 4-nitropyridine-4-N-oxide (4-nitropyridine-N-oxide, which is only slightly less effective than 4-NQO, and 4-hydroxylaminoquinoline N-oxide, which is less potent than 4-NQO) were also shown to regulate soxS. Consequently, 4-NQO induces substantial oxidative stress in cells, potentially enhancing its carcinogenic potency.

11.5.9.2 Other nitroheterocycles

Nitro heterocyclic chemicals have been utilized to treat bacterial and protozoan illnesses (Patterson & Wyllie, 2014; Wilkinson et al., 2011). There is a possibility that these drugs may induce cancer. A study re-evaluated the cytotoxic and genotoxic potential of metronidazole (MTZ), benzoic acid, and nifurtimox (NFX). NFX and benzoic acid exhibited higher mutagenicity in *Salmonella* compared to MTZ. Both NFX and benzoic acid induced DNA damage. MTZ, with its relatively low reduction potential, appears to act exclusively in anaerobic/hypoxic conditions. In addition to causing cellular damage, NFX and benzoic acid may potentially interact with DNA (Kovacic & Somanathan, 2014).

Similar to other nitro compounds, NFX, benzoic acid, and megazol are nitroheterocycles known for their efficacy against *Trypanosoma cruzi*, the parasite responsible for Chagas disease, which causes death of approximately 400,000 peoples annually in Central and South America. Their effectiveness is attributed, at least partially, to the sensitivity of trypanosomes, particularly *T. cruzi*, to oxidative stress (Sokolova et al., 2010). These compounds produce radical anions upon one electron reduction, which disrupt the metabolism of oxygen. It is also suggested that benzoic acid decreased reactive species that interact with parasite macromolecules. In contrast to host cells, *T. cruzi* lacks the aldehyde oxidase enzymes necessary to avert oxidative damage (Machado-Silva et al., 2016). The main enzyme that keeps *T. cruzi* safe from the oxidative stress is trypanothione reductase. Trypanothione reductase is depleted because of the radical anions produced by the reduction of the nitro compounds, which makes *T. cruzi* hazardous (Beltran-Hortelano et al., 2022; Vázquez et al., 2017).

11.5.9.3 *Other nitro heteroaromatics*

Nitro heteroaromatics encompass a diverse group of organic compounds featuring both nitro (-NO_2) groups and heteroaromatic structures. Nitro heteroaromatics find applications across various fields, including pharmaceuticals, agrochemicals, and materials science due to their diverse chemical properties. They are known for exhibiting biological activities such as antimicrobial, antiviral, and anticancer properties, making them essential in drug discovery e.g. nitrofurazone and nitrofurantoin (antibacterial) (Lewkowski et al., 2019), nitrofuran and nitrothiazole classes (antiprotozoal) (Ang et al., 2017), 2-nitroimidazoles (radiation sensitizers) (Bonnet et al., 2018), and MTZ and entramin (amebicide), acrintrazole, niridazole, and MTZ (anthelminths). Additionally, nitro heteroaromatics are studied for their environmental impact as contaminants in air, water, and soil.

11.6 RISK ASSESSMENT

ANCs, when found in food, may impose health problems because they have the potential to be poisonous and carcinogenic (Manzetti, 2012). To examine the possible dangers of ANCs in food and determine safe exposure limits, risk assessments usually entail multiple processes. This starts with identification of hazards. It determines the risks connected to food-borne ANCs. To do this, one must be aware of their toxicological characteristics, including the possibility that they could have harmful health impacts such toxicity, genotoxicity, and carcinogenicity. Identification of hazards involves understanding the chemical structure and properties (Koster et al., 2011). The solubility, stability, and reactivity of ANCs are determined by their chemical structure, and these characteristics can affect how they behave in food and the human body. Toxicological information from research examining the health impacts of ANCs on humans is used to identify hazards. These studies could include genotoxicity studies to evaluate potential DNA damage, carcinogenicity studies to investigate the possibility of cancer development, acute toxicity studies to ascertain the effects of short-term exposure, and chronic toxicity studies to assess the effects of long-term exposure (Kovacic & Somanathan, 2014).

Identification of hazards requires knowledge of the way that ANCs work. The body's metabolism can activate aromatic nitro molecules, resulting in the production of reactive intermediates that can harm cells and induce oxidative stress. The relationships between structure and activity (SAR) looks at the connection between an ANC's molecular structure and biological activity. Using their chemical structure, SAR investigations can predict the toxicity of related substances and uncover structural factors that contribute to toxicity. Determining the possible negative health consequences of eating food containing ANCs is also essential (Möller, 1994). These effects could be both acute, like headache, nausea, and irritation at high dosages, and chronic, such organ damage, neurotoxicity, reproductive toxicity, and carcinogenicity after long-term exposure to lower amounts. In addition to considering potential dietary intake of ANCs through contaminated food or water, occupational exposure during food processing or manufacturing, and environmental exposure from pollution or industrial sources, hazard identification also considers potential exposure scenarios. Estimating the amounts of ANCs in food and figuring out how much exposure humans have to these compounds are parts of exposure assessment (Zimmerli et al., 2001). This could entail checking food samples for these substances' existence and estimating dietary intake from consumption trends.

To ascertain the amounts of ANCs present, food samples are analyzed as the first step in the exposure assessment process. The quantities of these substances in diverse food matrices can be measured analytically using techniques like liquid chromatography-mass spectrometry (LC-MS) and gas chromatography-mass spectrometry (GC-MS) (Qu et al., 2020). Scientific literature, surveillance studies, and monitoring programs are the sources of information on the presence of ANCs in food. The frequency and concentrations of these substances in various food product kinds are disclosed by these statistics. Food consumption surveys are carried out to evaluate people's or populations' food intake. These surveys gather data on food types and quantities consumed, as well as patterns of consumption, dietary practices, and personal preferences. Consumer exposure to ANCs through diet is estimated using food consumption data. Exposure evaluations can consider a range of exposure scenarios, including high or average consumption levels, age groups, and vulnerable populations, including expectant mothers and children (Gushgari & Halden, 2018; Stuff et al., 2009).

To ascertain if exposure to ANCs in food is likely to be harmful to human health, exposure assessment results are compared to safety criteria that have been set, such as legal limits or reference dosages. In a comparative risk assessment, the degree of exposure in relation to these safety thresholds is evaluated, and possible health hazards related to aromatic nitro compound exposure are assessed (Suarez-Torres et al., 2021). To establish dose-response relationships, ascertain the potency of ANCs, and analyze their impacts on human health, toxicological data pertaining to these compounds are evaluated. Studies on both acute and chronic toxicity, as well as tests on laboratory animals for genotoxicity and carcinogenicity, may fall under this category.

Further, genotoxic research is conducted to investigate the potential for ANCs to damage DNA and result in chromosomal abnormalities or mutations (Purohit & Basu, 2000). The Ames test for gene mutations, the comet assay for DNA strand breakage, and the micronucleus assay for chromosomal damage are common genotoxicity assays. The process involves conducting carcinogenicity, reproductive toxicity, neurotoxicity research (Dixit & Kumar, 2018). Further, studies are conducted to study the mechanism of action. This could entail looking into the compound's metabolism, cellular targets, molecular pathways, and biological factors that contribute to its toxicity. The last step involves risk handling. To lessen or decrease exposure ANCs in food, risk management measures may be put into place based on the results of the risk assessment (Molognoni et al., 2019). This could entail regulating the maximum amounts of these substances that are permitted in food, putting control mechanisms in place throughout the manufacture and processing of food, and advising customers on safe food handling techniques.

11.7 CHALLENGES IN ANC RISK ASSESSMENT IN FOOD

The primary challenge in assessing the risk of ANCs in food is obtaining thorough and trustworthy data. Inadequate information regarding the frequencies, routes of exposure, toxicological characteristics, and impacts on human health can make it difficult to accurately evaluate the dangers connected to these substances. Additionally, the low amounts of ANCs, the complexity of food matrices, and the existence of interfering chemicals can make their analysis in food samples difficult (Molognoni et al., 2019). Although it can be technically challenging, developing sensitive and accurate analytical procedures for detecting and quantifying these chemicals is vital for risk assessment. The dearth of thorough toxicological data regarding aromatic nitro compounds is another difficulty. The toxicity of many of these substances has not been thoroughly investigated, and information regarding their genotoxicity, carcinogenicity, developmental toxicity, acute and long-term effects, and toxicity may be sparse or contradictory (Park et al., 2015).

The stability, bioavailability, and toxicity of other ingredients in food matrices may be impacted by interactions between ANCs and other ingredients. Accurate risk assessment requires an understanding of these substances' behavior in complex food matrices, which can be difficult given the complexity and variety of dietary compositions. There are several unknowns when estimating a person's dietary intake of ANCs, such as variations in food consumption habits, food processing techniques, cooking methods, and the compounds' bioavailability. Reliable exposure assessment techniques and data are necessary for precisely determining exposure levels (Wollin & Dieter, 2005). Due to data errors, model assumptions, and the extrapolation of findings from animal studies to human populations, it can be difficult to integrate information from hazard identification, exposure assessment, and toxicity assessment to define hazards associated with aromatic nitro compounds in food. It can be difficult to convey to decision-makers and stakeholders the scope and repercussions of risks. Risk assessment may face difficulties if novel ANCs are identified or if previously unknown health consequences are found. Researchers, regulators, and industry stakeholders must continue to monitor new chemicals and update their understanding of science, which calls for constant research, collaboration, and surveillance.

11.8 CONCLUSION AND FUTURE PROSPECTS

In conclusion, this chapter has conducted a comprehensive analysis of ANCs, elucidating their multifaceted chemical characteristics, biological effects, and environmental implications. Nitro heteroaromatics, including nitro benzenes, nitropyridines, nitroquinolines, and nitroimidazoles, exhibit considerable potential in the fields of medicines, agrochemicals, and materials science. Due to their remarkable antibacterial, antiviral, and anticancer properties, they are considered strong candidates for drug research and advancement.

Further study is necessary to fully exploit the potential of nitro heteroaromatics in medication development and environmental clean-up. Future research could focus on uncovering the methods by which nitro heteroaromatics affect biological systems, thereby improving their effectiveness and specificity. Furthermore, it would be advantageous to investigate innovative synthetic techniques for the synthesis of a wide range of nitroheteroaromatic compounds that exhibit enhanced characteristics. Evaluating the environmental destiny and harmfulness of nitro heteroaromatics is essential for devising efficient approaches to reduce and resolve their impact. Furthermore, the investigation of novel uses of nitro heteroaromatics in developing areas like nanotechnology and biotechnology shows potential for enhancing scientific understanding and practical implementations. By focusing on these study directions, we can maximize the potential of nitroheteroaromatic molecules, leading to progress in other fields.

ACKNOWLEDGMENT

Hansika Sati would like to thank the Department of Science and Technology, Government of India, for providing DST-INSPIRE fellowship [IF220107] for pursuing Ph.D. The authors are grateful to the National Institute of Food Technology Entrepreneurship and Management, Kundli, India for providing all necessary facilities.

ABBREVIATIONS

1-NNT	1-nitronaphthalene
1-NOP	1-nitrosopyrene
1-NP	1-nitropyrene
1,2-DHD 6-NHOH-C	1,2-dihydro-N-hydroxy-6-aminochrysene
1,3-DNB	1,3-dinitrobenzene
1,3,5-TNB	1,3,5-trinitrobenzene
2-CNB	2-chlorobenzene
2-NF	2-nitrofluorene
2-NNT	2-nitro naphthalene
3-ABA	3-abscisic acid
4-CNB	4-chlorobenzene
4-NBP	4-nitrobenzyl pyridine
4-NQO	4-nitroquinoline-N-oxide
6-NC	6-nitrochrysene
9-NA	9-nitroanthracene
CH_3OOH	Peroxy-methane
CHD	Chinese hamster Don
DNA	Deoxyribonucleic acid

GC-MS	Gas chromatography-mass spectrometry
H_2O_2	Hydrogen peroxide
HCMM	Hereditary cutaneous malignant melanoma
HO_2	Hydroperoxyl radical
HONO	Nitrous acid
LC-MS	Liquid chromatography-mass spectrometry
MTZ	Metronidazole
NBA	Nitrobenzanthrones
NBDs	Nitrobenzene diazole
NFX	Nifurtimox
NO	Nitrogen monoxide
NO_2	Nitrogen dioxide
NP	Nitropyrene
NPAH	Nitro-polycyclic aromatic hydrocarbon
O_3	Ozone
PAH	Polycyclic aromatic hydrocarbon
PCNB	Pentachloronitrobenzene
SAR	Structure and activity
SpD	Sprague-Dawley
TNT	Trinitrotoluene
UV	Ultraviolet
XP	Xeroderma pigmentosum

REFERENCES

Adeyeye, S.A.O. and Ashaolu, T.J. 2022. Polycyclic aromatic hydrocarbons formation and mitigation in meat and meat products. *Polycyclic Aromatic Compounds* 42, 3401–3411. https://doi.org/10.1080/10406638.2020.1866039

Alisson-Silva, F., Kawanishi, K. and Varki, A. 2016. Human risk of diseases associated with red meat intake: Analysis of current theories and proposed role for metabolic incorporation of a non-human sialic acid. *Molecular Aspects of Medicine* 51, 16–30. https://doi.org/10.1016/j.mam.2016.07.002

Alkanan, Z.T., Altemimi, A.B., Al-Hilphy, A.R.S., Watson, D.G. and Pratap-Singh, A. 2021. Ohmic heating in the food industry: Developments in concepts and applications during 2013–2020. *Applied Sciences* 11, 2507. https://doi.org/10.3390/app11062507

Al-Kaysi, R.O., Gallardo, I. and Guirado, G. 2008. Stable spirocyclic Meisenheimer complexes. *Molecules* 13, 1282–1302. https://doi.org/10.3390/molecules13061282

Amaya-Farfan, J. and Rodriguez-Amaya, D.B. 2021. The Maillard reactions. In: Rodriguez-Amaya, D.B. and Amaya-Farfan, J. (Eds.), *Chemical Changes During Processing and Storage of Foods,* Academic Press, Elsevier, London, UK, pp. 215–263. https://doi.org/10.1016/B978-0-12-817380-0.00006-3

Amit, S.K., Uddin, M.M., Rahman, R., Islam, S.M.R. and Khan, M.S. 2017. A review on mechanisms and commercial aspects of food preservation and processing. *Agriculture & Food Security* 6, 51. https://doi.org/10.1186/s40066-017-0130-8

Ang, C.W., Jarrad, A.M., Cooper, M.A. and Blaskovich, M.A.T. 2017. Nitroimidazoles: Molecular fireworks that combat a broad spectrum of infectious diseases. *Journal of Medicinal Chemistry* 60, 7636–7657. https://doi.org/10.1021/acs.jmedchem.7b00143

Arora, P.K. and Bae, H. 2014. Toxicity and microbial degradation of nitrobenzene, monochloronitrobenzenes, polynitrobenzenes, and pentachloronitrobenzene. *Journal of Chemistry* 2014, 265140. https://doi.org/10.1155/2014/265140

Baez, A. and Shiloach, J. 2013. Escherichia coli avoids high dissolved oxygen stress by activation of SoxRS and manganese-superoxide dismutase. *Microbial Cell Factories* 12, 1–9. https://doi.org/10.1186/1475-2859-12-23

Bamford, H.A. and Baker, J.E. 2003. Nitro-polycyclic aromatic hydrocarbon concentrations and sources in urban and suburban atmospheres of the mid-Atlantic region. *Atmospheric Environment* 37, 2077–2091. https://doi.org/10.1016/S1352-2310(03)00102-X

Bandowe, B.A.M. and Meusel, H. 2017. Nitrated polycyclic aromatic hydrocarbons (nitro-PAHs) in the environment–A review. *Science of the Total Environment* 581, 237–257. https://doi.org/10.1016/j.scitotenv.2016.12.115

Beije, B. and Möller, L. 1988. 2-Nitrofluorene and related compounds: Prevalence and biological effects. *Mutation Research/Reviews in Genetic Toxicology* 196, 177–209. https://doi.org/10.1016/0165-1110(88)90019-X

Beltran-Hortelano, I., Alcolea, V., Font, M. and Pérez-Silanes, S. 2022. Examination of multiple *Trypanosoma cruzi* targets in a new drug discovery approach for chagas disease. *Bioorganic & Medicinal Chemistry* 58, 116577. https://doi.org/10.1016/j.bmc.2021.116577

Bonnet, M., Hong, C.R., Wong, W.W., Liew, L.P., Shome, A., Wang, J., Gu, Y., Stevenson, R.J., Qi, W., Anderson, R.F., Pruijn, F.B., Wilson, W.R., Jamieson, S.M.F., Hicks, K.O. and Hay, M.P. 2018. Next-generation hypoxic cell radiosensitizers: Nitroimidazole alkylsulfonamides. *Journal of Medicinal Chemistry* 61, 1241–1254. https://doi.org/10.1021/acs.jmedchem.7b01678

Butterworth, B.E., Mathre, O.B., Ballinger, K.E. and Adalsteinsson, O. 2004. Contamination is a frequent confounding factor in toxicology studies with anthraquinone and related compounds. *International Journal of Toxicology* 23, 335–344. https://doi.org/10.1080/10915810490517072

Casado, N., Berenguer, C.V., Câmara, J.S. and Pereira, J.A.M. 2024. What are we eating? Surveying the presence of toxic molecules in the food supply chain using chromatographic approaches. *Molecules* 29, 579. https://doi.org/10.3390/molecules29030579

Chen, K.Y., Tseng, C.H., Feng, P.H., Sun, W.L., Ho, S.C., Lin, C.W., Van Hiep, N., Luo, C.S., Tseng, Y.H., Chen, T.T., Liu, W.T., Lee, K.Y. and Wu, S.M. 2022. 3-Nitrobenzanthrone promotes malignant transformation in human lung epithelial cells through the epiregulin-signaling pathway. *Cell Biology and Toxicology* 38, 865–887. https://doi.org/10.1007/s10565-021-09612-1

Ciganek, M., Neca, J., Adamec, V., Janosek, J. and Machala, M. 2004. A combined chemical and bioassay analysis of traffic-emitted polycyclic aromatic hydrocarbons. *Science of the Total Environment* 334, 141–148. https://doi.org/10.1016/j.scitotenv.2004.04.034

Dai, Y., Yan, H., Zhang, B., Wu, W., Yang, L., Li, S., Li, W., Li, H., Yan, L., Shan, D., Feng, Y., Terui, N. and Tanaka, S. 2014. A novel adsorbent obtained by caging activated carbon by konjac glucomannan gel for elimination of organic compounds. *Journal of Applied Polymer Science* 131, 40542. https://doi.org/10.1002/app.40542

Deng, K., Wong, T.Y., Wang, Y., Leung, E.M.K. and Chan, W. 2015. Combination of precolumn nitro-reduction and ultraperformance liquid chromatography with fluorescence detection for the sensitive quantification of 1-nitronaphthalene, 2-nitrofluorene, and 1-nitropyrene in meat products. *Journal of Agricultural and Food Chemistry* 63, 3161–3167. https://doi.org/10.1021/acs.jafc.5b00523

Dennis, M.J., Massey, R.C., McWeeny, D.J. and Knowles, M.E. 1984. Estimation of nitropolycyclic aromatic hydrocarbons in foods. *Food Additives & Contaminants* 1, 29–37. https://doi.org/10.1080/02652038409385820

Dixit, M. and Kumar, A. 2018. In vitro gene genotoxicity test methods. In: Dhawan, A. and Kwon, S. (Eds.), *In Vitro Toxicology*, Academic Press, Elsevier, London, UK, pp. 67–89. https://doi.org/10.1016/b978-0-12-804667-8.00004-3

Duan, F., Chyang, C.S., Zhang, L. and Chi, Y.T. 2015. Effect of the molecular structure of nitrogen compounds on the pollutant formation in a bubbling fluidized-bed combustor. *Energy* 83, 394–402. https://doi.org/10.1016/j.energy.2015.02.037

Dutta, K., Shityakov, S., Zhu, W. and Khalifa, I. 2022. High-risk meat and fish cooking methods of polycyclic aromatic hydrocarbons formation and its avoidance strategies. *Food Control* 142, 109253. https://doi.org/10.1016/j.foodcont.2022.109253

El-Bayoumy, K., Desai, D., Boyiri, T., Rosa, J., Krzeminski, J., Sharma, A.K., Pittman, B. and Amin, S. 2002. Comparative tumorigenicity of the environmental pollutant 6-nitrochrysene and its metabolites in the rat mammary gland. *Chemical Research in Toxicology* 15, 972–978. https://doi.org/10.1021/tx020019a

Esmaeili, A., Schlatterer, K., Demirhan, I., Schlatterer, B., Nauck, M., Chandra, P. and Chandra, A. 2006. Tumorigenic potential and the molecular mechanism of the carcinogenic effect exerted by 2-nitroanisole. *Anticancer Research* 26, 4203–4212. https://ar.iiarjournals.org/content/26/6b/4203

Fanucchi, M.V., Day, K.C., Clay, C.C. and Plopper, C.G. 2004. Increased vulnerability of neonatal rats and mice to 1-nitronaphthalene-induced pulmonary injury. *Toxicology and Applied Pharmacology* 201, 53–65. https://doi.org/10.1016/j.taap.2004.05.002

Foster, J.R. and Engelhardt, J.A. 2023. Assigning adversity to toxicologic outcomes. In: Haschek, W.M., Rousseaux, C.G., Wallig, M.A. and Bolon, B. (Eds.), *Toxicologic Pathology* 2, Academic Press, Elsevier, London, UK, pp. 567–615. https://doi.org/10.1016/b978-0-12-821047-5.00002-6

Gibson, J.C. 2020. Emerging persistent chemicals in human biomonitoring for populations in the Arctic: A Canadian perspective. *Science of the Total Environment* 708, 134538. https://doi.org/10.1016/j.scitotenv.2019.134538

Gligorovski, S., Strekowski, R., Barbati, S. and Vione, D. 2015. Environmental implications of hydroxyl radicals (•OH). *Chemical Reviews* 115, 13051–13092. https://doi.org/10.1021/cr500310b

Gokarn, S. and Choudhary, A. 2021. Modeling the key factors influencing the reduction of food loss and waste in fresh produce supply chains. *Journal of Environmental Management* 294, 113063. https://doi.org/10.1016/j.jenvman.2021.113063

Gomes, A.A., Silva-Júnior, A.C.T., Oliveira, E.B., Asad, L.M.B.O., Reis, N.C.S.C., Felzenszwalb, I., Kovary, K. and Asad, N.R. 2005. Reactive oxygen species mediate lethality induced by far-UV in *Escherichia coli* cells. *Redox Report* 10, 91–95. https://doi.org/10.1179/135100005x38833

Gooch, A., Sizochenko, N., Rasulev, B., Gorb, L. and Leszczynski, J. 2017. In vivo toxicity of nitroaromatics: A comprehensive quantitative structure–activity relationship study. *Environmental Toxicology and Chemistry* 36, 2227–2233. https://doi.org/10.1002/etc.3761

Grummt, T., Wunderlich, H.G., Chakraborty, A., Kundi, M., Majer, B., Ferk, F., Nersesyan, A.K., Parzefall, W. and Knasmüller, S. 2006. Genotoxicity of nitrosulfonic acids, nitrobenzoic acids, and nitrobenzylalcohols, pollutants commonly found in ground water near ammunition facilities. *Environmental and Molecular Mutagenesis* 47, 95–106. https://doi.org/10.1002/em.20172

Gushgari, A.J. and Halden, R.U. 2018. Critical review of major sources of human exposure to *N*-nitrosamines. *Chemosphere* 210, 1124–1136. https://doi.org/10.1016/j.chemosphere.2018.07.098

Guttenplan, J.B., Zhao, Z., Kosinska, W., Norman, R.G., Krzeminski, J., Sun, Y.W., Amin, S. and El-Bayoumy, K. 2007. Comparative mutational profiles of the environmental mammary carcinogen, 6-nitrochrysene and its metabolites in a lacI mammary epithelial cell line. *Carcinogenesis* 28, 2391–2397. https://doi.org/10.1093/carcin/bgm142

Han, M., Kong, J., Yuan, J., He, H., Hu, J., Yang, S., Li, S., Zhang, L. and Sun, C. 2019. Method development for simultaneous analyses of polycyclic aromatic hydrocarbons and their nitro-, oxy-, hydroxy- derivatives in sediments. *Talanta* 205, 120128. https://doi.org/10.1016/j.talanta.2019.120128

Hansen, T., Seidel, A. and Borlak, J. 2007. The environmental carcinogen 3-nitrobenzanthrone and its main metabolite 3-aminobenzanthrone enhance formation of reactive oxygen intermediates in human A549 lung epithelial cells. *Toxicology and Applied Pharmacology* 221, 222–234. https://doi.org/10.1016/j.taap.2007.03.003

Hayakawa, K., Tang, N., Matsuki, A., Inomata, Y., Toriba, A. and Nagato, E.G. 2021. Calculating source contributions to urban atmospheric polycyclic aromatic hydrocarbons and nitropolycyclic aromatic hydrocarbons using 1-nitropyrene and pyrene: An application to an Asian dust event. *Chemosphere* 280, 130662. https://doi.org/10.1016/j.chemosphere.2021.130662

Hsu, C.H., Stedeford, T., Okochi-Takada, E., Ushijima, T., Noguchi, H., Muro-Cacho, C., Holder, J.W. and Banasik, M. 2007. Framework analysis for the carcinogenic mode of action of nitrobenzene. *Journal of Environmental Science and Health Part C* 25, 155–184. https://doi.org/10.1080/10590500701399234

Igonina, E.V., Marsova, M.V. and Abilev, S.K. 2018. Lux biosensors: Screening biologically active compounds for genotoxicity. *Russian Journal of Genetics: Applied Research* 8, 87–95. https://doi.org/10.1134/s2079059718010082

Jiang, Y., Wang, X., Li, M., Liang, Y., Liu, Z., Chen, J., Guan, T., Mu, J., Zhu, Y., Meng, H., Zhou, Y., Yao, L., Xue, L. and Wang, W. 2024. Comprehensive understanding on sources of high levels of fine particulate nitro-aromatic compounds at a coastal rural area in northern China. *Journal of Environmental Sciences* 135, 483–494. https://doi.org/10.1016/j.jes.2022.09.033

Jones, C.R., Liu, Y.Y., Sepai, O., Yan, H. and Sabbioni, G. 2006. Internal exposure, health effects, and cancer risk of humans exposed to chloronitrobenzene. *Environmental Science & Technology* 40, 387–394. https://doi.org/10.1021/es050693p

Ju, K.S. and Parales, R.E. 2010. Nitroaromatic compounds, from synthesis to biodegradation. *Microbiology and Molecular Biology Reviews* 74, 250–272. https://doi.org/10.1128/mmbr.00006-10

Karić, N., Maia, A.S., Teodorović, A., Atanasova, N., Langergraber, G., Crini, G., Ribeiro, A.R.L. and Đolić, M. 2022. Bio-waste valorisation: Agricultural wastes as biosorbents for removal of (in)organic pollutants in wastewater treatment. *Chemical Engineering Journal Advances* 9, 100239. https://doi.org/10.1016/j.ceja.2021.100239

Kinouchi, T., Tsutsui, H. and Ohnishi, Y. 1986. Detection of 1-nitropyrene in yakitori (grilled chicken). *Mutation Research* 171, 105–113. https://doi.org/10.1016/0165-1218(86)90042-x

Kirkland, D. and Beevers, C. 2006. Induction of LacZ mutations in Muta™Mouse can distinguish carcinogenic from non-carcinogenic analogues of diaminotoluenes and nitronaphthalenes. *Mutation Research/Genetic Toxicology and Environmental Mutagenesis* 608, 88–96. https://doi.org/10.1016/j.mrgentox.2006.05.004

Klein, M., Voigtmann, U., Haack, T., Erdinger, L. and Boche, G. 2000. From mutagenic to non-mutagenic nitroarenes: Effect of bulky alkyl substituents on the mutagenic activity of 4-nitrobiphenyl in *Salmonella typhimurium*: Part I. Substituents ortho to the nitro group and in 2′-position. *Mutation Research/Genetic Toxicology and Environmental Mutagenesis* 467, 55–68. https://doi.org/10.1016/S1383-5718(00)00012-7

Ko, J.H., Das, G., Kim, J.E. and Shin, H.S. 2018. Study on formation of nitrated polycyclic aromatic hydrocarbons from different roasting condition in coffee. *Journal of Food Science and Technology* 55, 3991–4000. https://doi.org/10.1007/s13197-018-3324-6

Kong, J., Ma, T., Cao, X., Li, W., Zhu, F., He, H., Sun, C., Yang, S., Li, S. and Xian, Q. 2023. Occurrence, partition behavior, source and ecological risk assessment of nitro-PAHs in the sediment and water of Taige Canal, China. *Journal of Environmental Sciences* 124, 782–793. https://doi.org/10.1016/j.jes.2022.02.034

Koster, S., Boobis, A.R., Cubberley, R., Hollnagel, H.M., Richling, E., Wildemann, T., Würtzen, G. and Galli, C.L. 2011. Application of the TTC concept to unknown substances found in analysis of foods. *Food and Chemical Toxicology* 49, 1643–1660. https://doi.org/10.1016/j.fct.2011.03.049

Kovacic, P. and Somanathan, R. 2014. Nitroaromatic compounds: Environmental toxicity, carcinogenicity, mutagenicity, therapy and mechanism. *Journal of Applied Toxicology* 34, 810–824. https://doi.org/10.1002/jat.2980

Krzeminski, J., Kropachev, K., Kolbanovskiy, M., Reeves, D., Kolbanovskiy, A., Yun, B.H., Geacintov, N.E., Amin, S. and El-Bayoumy, K. 2011. Inefficient nucleotide excision repair in human cell extracts of the n-(deoxyguanosin-8-yl)-6-aminochrysene and 5-(deoxyguanosin-n2-yl)-6-aminochrysene adducts derived from 6-nitrochrysene. *Chemical Research in Toxicology* 24, 65–72. https://doi.org/10.1021/tx100284h

Krzeminski, J., Kropachev, K., Reeves, D., Kolbanovskiy, A., Kolbanovskiy, M., Chen, K.M., Sharma, A.K., Geacintov, N., Amin, S. and El-Bayoumy, K. 2013. Adenine-DNA adduct derived from the nitroreduction of 6-nitrochrysene is more resistant to nucleotide excision repair than guanine-dna adducts. *Chemical Research in Toxicology* 26, 1746–1754. https://doi.org/10.1021/tx400296x

Kulkarni, S.A., Moir, D. and Zhu, J. 2007. Influence of structural and functional modifications of selected genotoxic carcinogens on metabolism and mutagenicity–A review. *SAR and QSAR in Environmental Research* 18, 459–514. https://doi.org/10.1080/10629360701430090

Lawal, A.T. 2017. Polycyclic aromatic hydrocarbons. A review. *Cogent Environmental Science* 3, 1339841. https://doi.org/10.1080/23311843.2017.1339841

Lewkowski, J., Rogacz, D. and Rychter, P. 2019. Hazardous ecotoxicological impact of two commonly used nitrofuran-derived antibacterial drugs: Furazolidone and nitrofurantoin. *Chemosphere* 222, 381–390. https://doi.org/10.1016/j.chemosphere.2019.01.144

Li, L., Carratt, S., Hartog, M., Kovalchuk, N., Jia, K., Wang, Y., Zhang, Q.Y., Edwards, P., Van Winkle, L. and Ding, X. 2017. Human CYP2A13 and CYP2F1 mediate naphthalene toxicity in the lung and nasal mucosa of CYP2A13/2F1-humanized mice. *Environmental Health Perspectives* 125, 067004. https://doi.org/10.1289/ehp844

Li, T., Gao, Y.Z., Xu, J., Zhang, S.T., Guo, Y., Spain, J.C. and Zhou, N.Y. 2021. A recently assembled degradation pathway for 2,3-dichloronitrobenzene in *Diaphorobacter* sp. strain JS3051. *Mbio* 12, 10–1128. https://doi.org/10.1128/mbio.02231-21

Lin, Y., Qiu, X., Ma, Y., Wang, J., Wu, Y., Zeng, L., Hu, M., Zhu, T. and Zhu, Y. 2016. A novel approach for apportionment between primary and secondary sources of airborne nitrated polycyclic aromatic hydrocarbons (NPAHs). *Atmospheric Environment* 138, 108–113. https://doi.org/10.1016/j.atmosenv.2016.05.017

Machado-Silva, A., Cerqueira, P.G., Grazielle-Silva, V., Gadelha, F.R., Peloso, E. de F., Teixeira, S.M.R. and Machado, C.R. 2016. How *Trypanosoma cruzi* deals with oxidative stress: Antioxidant defence and DNA repair pathways. *Mutation Research/Reviews in Mutation Research* 767, 8–22. https://doi.org/10.1016/j.mrrev.2015.12.003

Malejka-Giganti, D., Parkin, D.R., Decker, R.W., Niehans, G.A., Bliss, R.L., Churchwell, M.I. and Beland, F.A. 2008. Tumorigenicity and genotoxicity of an environmental pollutant 2,7-dinitrofluorene after systemic administration at a low dose level to female rats. *International Journal of Cancer* 122, 1958–1965. https://doi.org/10.1002/ijc.23352

Manzetti, S. 2012. Ecotoxicity of polycyclic aromatic hydrocarbons, aromatic amines, and nitroarenes through molecular properties. *Environmental Chemistry Letters* 10, 349–361. https://doi.org/10.1007/s10311-012-0368-0

Manzetti, S. 2013. Polycyclic aromatic hydrocarbons in the environment: Environmental fate and transformation. *Polycyclic Aromatic Compounds* 33, 311–330. https://doi.org/10.1080/10406638.2013.781042

Matsumi, Y. and Kawasaki, M. 2003. Photolysis of atmospheric ozone in the ultraviolet region. *Chemical Reviews* 103, 4767–4782. https://doi.org/10.1021/cr0205255

Miliukiene, V. and Čėnas, N. 2008. Cytotoxicity of nitroaromatic explosives and their biodegradation products in mice splenocytes: Implications for their immunotoxicity. *Zeitschrift Für Naturforschung C* 63, 519–525. https://doi.org/10.1515/znc-2008-7-809

Mizerovská, J., Dračínská, H., Frei, E., Schmeiser, H.H., Arlt, V.M. and Stiborová, M. 2011. Induction of biotransformation enzymes by the carcinogenic air-pollutant 3-nitrobenzanthrone in liver, kidney and lung, after intra-tracheal instillation in rats. *Mutation Research/Genetic Toxicology and Environmental Mutagenesis* 720, 34–41. https://doi.org/10.1016/j.mrgentox.2010.12.003

Möller, L. 1994. In vivo metabolism and genotoxic effects of nitrated polycyclic aromatic hydrocarbons. *Environmental Health Perspectives* 102, 139–146. https://doi.org/10.1289/ehp.102-1566915

Molognoni, L. Daguer, H. and Hoff, R.B. 2021. Analysis of nitrofurans residues in foods of animal origin. In: Galanakis, C.M. (Ed.), *Food Toxicology and Forensics,* Academic Press, Elsevier, London, UK, pp. 379–419. https://doi.org/10.1016/b978-0-12-822360-4.00015-7

Molognoni, L., Daguer, H., Motta, G.E., Merlo, T.C. and Lindner, J.D.D. 2019. Interactions of preservatives in meat processing: Formation of carcinogenic compounds, analytical methods, and inhibitory agents. *Food Research International* 125, 108608. https://doi.org/10.1016/j.foodres.2019.108608

Monteiro, F.C., Guimaraes, I.D.L., de Almeida Rodrigues, P., da Anunciação de Pinho, J.V. and Conte-Junior, C.A. 2023. Degradation of PAHs using TiO2 as a semiconductor in the heterogeneous photocatalysis process: A systematic review. *Journal of Photochemistry and Photobiology A: Chemistry* 437, 114497. https://doi.org/10.1016/j.jphotochem.2022.114497

Morgott, D.A. 2014. Factors and trends affecting the identification of a reliable biomarker for diesel exhaust exposure. *Critical Reviews in Environmental Science and Technology* 44, 1795–1864. https://doi.org/10.1080/10643389.2013.790748

Murata, M., Tezuka, T., Ohnishi, S., Takamura-Enya, T., Hisamatsu, Y. and Kawanishi, S. 2006. Carcinogenic 3-nitrobenzanthrone induces oxidative damage to isolated and cellular DNA. *Free Radical Biology and Medicine* 40, 1242–1249. https://doi.org/10.1016/j.freeradbiomed.2005.11.015

Murthy, P.S. and Madhava Naidu, M. 2012. Sustainable management of coffee industry by-products and value addition—A review. *Resources, Conservation and Recycling* 66, 45–58. https://doi.org/10.1016/j.resconrec.2012.06.005

National Toxicology Program. 2011. NTP 12th report on carcinogens. *Report on Carcinogens: Carcinogen Profiles* 12, iii–499.

Ohnishi, S., Murata, M., Fukuhara, K., Miyata, N. and Kawanishi, S. 2001. Oxidative DNA damage by a metabolite of carcinogenic 1-nitropyrene. *Biochemical and Biophysical Research Communications* 280, 48–52. https://doi.org/10.1006/bbrc.2000.4095

O'Malley, M., Ngo, M.A. and Maibach, H.I. 2020. Pesticide-related dermatoses. In: John, S.M., Johansen, J.D., Rustemeyer, T., Elsner, P. and Maibach H.I. (Eds.), *Kanerva's Occupational Dermatology*, Springer, New York, pp. 1029–1057. https://doi.org/10.1007/978-3-319-68617-2_68

Omelchenko, I.V., Shishkin, O.V., Gorb, L., Hill, F.C. and Leszczynski, J. 2012. Properties, aromaticity, and substituents effects in poly nitro- and amino-substituted benzenes. *Structural Chemistry* 23, 1585–1597. https://doi.org/10.1007/s11224-012-9971-8

Ostojić, B.D., Stanković, B. and Đorđević, D.S. 2014. The molecular properties of nitrobenzanthrone isomers and their mutagenic activities. *Chemosphere* 104, 228–236. https://doi.org/10.1016/j.chemosphere.2013.11.057

Palade, L.M., Negoiță, M., Adascălului, A.C. and Mihai, A.L. 2023. Polycyclic aromatic hydrocarbon occurrence and formation in processed meat, edible oils, and cereal-derived products: A review. *Applied Sciences* 13, 7877. https://doi.org/10.3390/app13137877

Pande, P., Rebello, K.R., Chatterjee, A., Naldiga, S. and Basu, A.K. 2020. Site-specific incorporation of n-(2′-deoxyguanosine-8-yl)-6-aminochrysene adduct in DNA and its replication in human cells. *Chemical Research in Toxicology* 33, 1997–2005. https://doi.org/10.1021/acs.chemrestox.0c00197

Park, J., Seo, J., Lee, J. and Kwon, H. 2015. Distribution of seven N-nitrosamines in food. *Toxicological Research* 31, 279–288. https://doi.org/10.5487/tr.2015.31.3.279

Patterson, S. and Wyllie, S. 2014. Nitro drugs for the treatment of trypanosomatid diseases: Past, present, and future prospects. *Trends in Parasitology* 30, 289–298. https://doi.org/10.1016/j.pt.2014.04.003

Patton, J.D., Maher, V.M. and McCormick, J.J. 1986. Cytotoxic and mutagenic effects of 1-nitropyrene and 1-nitrosopyrene in diploid human fibroblasts. *Carcinogenesis* 7, 89–93. https://doi.org/10.1093/carcin/7.1.89

Peng, B., Dong, Q., Li, F., Wang, T., Qiu, X. and Zhu, T. 2023. A systematic review of polycyclic aromatic hydrocarbon derivatives: Occurrences, levels, biotransformation, exposure biomarkers, and toxicity. *Environmental Science & Technology* 57, 15314–15335. https://doi.org/10.1021/acs.est.3c03170

Phimister, A.J., Williams, K.J., Winkle, L.S.V. and Plopper, C.G. 2005. Consequences of abrupt glutathione depletion in murine clara cells: Ultrastructural and biochemical investigations into the role of glutathione loss in naphthalene cytotoxicity. *Journal of Pharmacology and Experimental Therapeutics* 314, 506–513. https://doi.org/10.1124/jpet.105.084533

Poulet, G. 2000. General description of atmospheric chemistry. In: Vovelle, C. (Ed.), *Pollutants from Combustion: Formation and Impact on Atmospheric Chemistry*, Springer, Netherlands, pp. 75–95. https://doi.org/10.1007/978-94-011-4249-6_5

Powell, B.V., Bacurio, J.H.T. and Basu, A.K. 2020. Translesion synthesis of 6-nitrochrysene-derived 2′-deoxyadenosine adduct in human cells. *DNA Repair* 95, 102935. https://doi.org/10.1016/j.dnarep.2020.102935

Purohit, V. and Basu, A.K. 2000. Mutagenicity of nitroaromatic compounds. *Chemical Research in Toxicology* 13, 673–692. https://doi.org/10.1021/tx000002x

Qu, L., Yu, H., Yin, S., Li, Y. and Sun, C. 2020. Solid-phase extraction combined with ultra-high-performance liquid chromatography-tandem mass spectrometry for the determination of 5 trace nitro-polycyclic aromatic hydrocarbons in barbecued foods. *Journal of AOAC International* 103, 1512–1520. https://doi.org/10.1093/jaoacint/qsaa062

Reddy, G. 2015. Wildlife toxicity assessment for 1,3-dinitrobenzene. In: Williams, M.A., Reddy, G., Quinn, M.J. and Johnson, M.S. (Eds.), *Wildlife Toxicity Assessments for Chemicals of Military Concern*, Elsevier, London, UK, pp. 161–182. https://doi.org/10.1016/b978-0-12-800020-5.00008-9

Reed, L., Büchner, V. and Tchounwou, P.B. 2007. Environmental toxicology and health effects associated with hexachlorobenzene exposure. *Reviews on Environmental Health* 22, 213–244. https://doi.org/10.1515/reveh.2007.22.3.213

Reshetnikova, G., Sidorenko, V.S., Whyard, T., Lukin, M., Waltzer, W., Takamura-Enye, T. and Romanov, V. 2016. Genotoxic and cytotoxic effects of the environmental pollutant 3-nitrobenzanthrone on bladder cancer cells. *Experimental Cell Research* 349, 101–108. https://doi.org/10.1016/j.yexcr.2016.10.003

Rhead, M.M. and Hardy, S.A. 2003. The sources of polycyclic aromatic compounds in diesel engine emissions. *Fuel* 82, 385–393. https://doi.org/10.1016/s0016-2361(02)00314-9

Ruel, M.T., Garrett, J., Yosef, S. and Olivier, M. 2017. Urbanization, food security and nutrition. In: de Pee, S., Taren, D. and Bloem, M.W. (Eds.), *Nutrition and Health in a Developing World*, Springer International, New Work, pp. 705–735. https://doi.org/10.1007/978-3-319-43739-2_32

Sahu, T.K., Shukla, P.C., Mondal, A., Gupta, S., Belgiorno, G. and Di Blasio, G. 2024. Assessment of particulate PAHs toxicity from alcohol-diesel blends fuelled high compression ratio CI engine. *Cleaner Engineering and Technology* 18, 100725. https://doi.org/10.1016/j.clet.2024.100725

Sampaio, G.R., Guizellini, G.M., da Silva, S.A., de Almeida, A.P., Pinaffi-Langley, A.C.C., Rogero, M.M., de Camargo, A.C. and Torres, E.A.F.S. 2021. Polycyclic aromatic hydrocarbons in foods: Biological effects, legislation, occurrence, analytical methods, and strategies to reduce their formation. *International Journal of Molecular Sciences* 22, 6010. https://doi.org/10.3390/ijms22116010

Sati, H., Chinchkar, A.V., Kataria, P. and Pareek, S. 2023a. Melatonin: A potential abiotic stress regulator. *Plant Stress* 10, 100293. https://doi.org/10.1016/j.stress.2023.100293

Sati, H., Kataria, P., Chinchkar, A.V. and Pareek, S. 2023b. Melatonin: A biomolecule for mitigating postharvest chilling injury in fruits and vegetables. *Crop Science* 63, 3175–3197. https://doi.org/10.1002/csc2.21126

Sati, H., Khandelwal, A. and Pareek, S. 2023c. Effect of exogenous melatonin in fruit postharvest, crosstalk with hormones, and defense mechanism for oxidative stress management. *Food Frontiers* 4, 233–261. https://doi.org/10.1002/fft2.180

Schlemitz, S. and Pfannhauser, W. 1996. Monitoring of nitropolycyclic aromatic hydrocarbons in food using gas chromatography. *Zeitschrift Für Lebensmittel-Untersuchung Und Forschung* 203, 61–64. https://doi.org/10.1007/bf01267771

Singh, L., Agarwal, T. and Simal-Gandara, J. 2020. PAHs, diet and cancer prevention: Cooking process driven-strategies. *Trends in Food Science & Technology* 99, 487–506. https://doi.org/10.1016/j.tifs.2020.03.030

Singh, L., Agarwal, T. and Simal-Gandara, J. 2023. Summarizing minimization of polycyclic aromatic hydrocarbons in thermally processed foods by different strategies. *Food Control* 146, 109514. https://doi.org/10.1016/j.foodcont.2022.109514

Sokolova, A.Y., Wyllie, S., Patterson, S., Oza, S.L., Read, K.D. and Fairlamb, A.H. 2010. Cross-resistance to nitro drugs and implications for treatment of human african trypanosomiasis. *Antimicrobial Agents and Chemotherapy* 54, 2893–2900. https://doi.org/10.1128/aac.00332-10

Sonego, E., Bhattarai, B. and Duedahl-Olesen, L. 2022. Detection of nitrated, oxygenated and hydrogenated polycyclic aromatic compounds in smoked fish and meat products. *Foods* 11, 2446. https://doi.org/10.3390/foods11162446

Soni, A., Mishra, S., Vaya, D. and Surolia, P.K. 2024. Role of Ag and g-C3N4 over CaTiO3 for effective photocatalytic degradation of nitrobenzene. *Inorganic Chemistry Communications* 159, 111862. https://doi.org/10.1016/j.inoche.2023.111862

Spearpoint, M. and Hopkin, C. 2021. Household cooking oil use and its bearing on fire safety. *Journal of Fire Sciences* 39, 265–284. https://doi.org/10.1177/07349041211012994

Sruthi, N.U., Premjit, Y., Pandiselvam, R., Kothakota, A. and Ramesh, S.V. 2021. An overview of conventional and emerging techniques of roasting: Effect on food bioactive signatures. *Food Chemistry* 348, 129088. https://doi.org/10.1016/j.foodchem.2021.129088

Stiborová, M., Martínek, V., Svobodová, M., Šístková, J., Dvořák, Z., Ulrichová, J., Šimánek, V., Frei, E., Schmeiser, H.H., Phillips, D.H. and Arlt, V.M. 2010. Mechanisms of the different DNA Adduct forming potentials of the urban air pollutants 2-nitrobenzanthrone and carcinogenic 3-nitrobenzanthrone. *Chemical Research in Toxicology* 23, 1192–1201. https://doi.org/10.1021/tx100052d

Stuff, J.E., Goh, E.T., Barrera, S.L., Bondy, M.L. and Forman, M.R. 2009. Construction of an N-nitroso database for assessing dietary intake. *Journal of Food Composition and Analysis* 22, S42–S47. https://doi.org/10.1016/j.jfca.2009.01.008

Su, A.L., Mesaros, C.A., Krzeminski, J., El-Bayoumy, K. and Penning, T.M. 2022. Role of human aldo–keto reductases in the nitroreduction of 1-nitropyrene and 1,8-dinitropyrene. *Chemical Research in Toxicology* 35, 2296–2309. https://doi.org/10.1021/acs.chemrestox.2c00271

Su, A.L. and Penning, T.M. 2023. Role of human aldo-keto reductases and nuclear factor erythroid 2-related factor 2 in the metabolic activation of 1-nitropyrene via nitroreduction in human lung cells. *Chemical Research in Toxicology* 36, 270–280. https://doi.org/10.1021/acs.chemrestox.2c00337

Suarez-Torres, J.D., Orozco, C.A. and Ciangherotti, C.E. 2021. The numerical probability of carcinogenicity to humans of some antimicrobials: Nitro-monoaromatics (including 5-nitrofurans and 5-nitroimidazoles), quinoxaline-1,4-dioxides (including carbadox), and chloramphenicol. *Toxicology in Vitro* 75, 105172. https://doi.org/10.1016/j.tiv.2021.105172

Sun, C., Qu, L., Wu, L., Wu, X., Sun, R. and Li, Y. 2020. Advances in analysis of nitrated polycyclic aromatic hydrocarbons in various matrices. *Trends in Analytical Chemistry* 127, 115878. https://doi.org/10.1016/j.trac.2020.115878

Sun, Y.W., Guttenplan, J.B., Cooper, T., Krzeminski, J., Aliaga, C., Boyiri, T., Kosinska, W., Zhao, Z.L., Chen, K.M., Berg, A., Amin, S. and El-Bayoumy, K. 2013. Mechanisms underlying the varied mammary carcinogenicity of the environmental pollutant 6-nitrochrysene and its metabolites (−)-[R,R]- and (+)-[S,S]-1,2-dihydroxy-1,2-dihydro-6-nitrochrysene in the rat. *Chemical Research in Toxicology* 26, 547–554. https://doi.org/10.1021/tx300478z

Sun, Y.W., Guttenplan, J.B., Khmelnitsky, M., Krzeminski, J., Boyiri, T., Amin, S. and El-Bayoumy, K. 2009. Stereoselective metabolism of the environmental mammary carcinogen 6-nitrochrysene to trans-1,2-dihydroxy-1,2-dihydro-6-nitrochrysene by aroclor 1254-treated rat liver microsomes and their comparative mutation profiles in a laci mammary epithelial cell line. *Chemical Research in Toxicology* 22, 1992–1997. https://doi.org/10.1021/tx9002897

Swaminathan, S. and Hatcher, J.F. 1986. Xanthine oxidase-mediated mutagenicity of the bladder carcinogen 4-nitrobiphenyl. *Mutation Research/Genetic Toxicology* 172, 37–45. https://doi.org/10.1016/0165-1218(86)90103-5

Swartz, C.D., King, L.C., Nesnow, S., Umbach, D.M., Kumar, S. and DeMarini, D.M. 2009. Mutagenicity, stable DNA adducts, and abasic sites induced in *Salmonella* by phenanthro[3,4-b]- and phenanthro[4,3-b]thiophenes, sulfur analogs of benzo[c]phenanthrene. *Mutation Research/Fundamental and Molecular Mechanisms of Mutagenesis* 661, 47–56. https://doi.org/10.1016/j.mrfmmm.2008.11.001

Tas, D.O. and Pavlostathis, S.G. 2014. Occurrence, Toxicity, and biotransformation of pentachloronitrobenzene and chloroanilines. *Critical Reviews in Environmental Science and Technology* 44, 473–518. https://doi.org/10.1080/10643389.2012.728809

Tiwari, J., Tarale, P., Sivanesan, S. and Bafana, A. 2019. Environmental persistence, hazard, and mitigation challenges of nitroaromatic compounds. *Environmental Science and Pollution Research* 26, 28650–28667. https://doi.org/10.1007/s11356-019-06043-8

van Bekkum, Y.M., van den Broek, P.H.H., Scheepers, P.T.J., Noordhoek, J. and Bos, R.P. 1999. Biological fate of [14C]-1-nitropyrene in rats following intragastric administration. *Chemico-Biological Interactions* 117, 15–33. https://doi.org/10.1016/S0009-2797(98)00095-7

Vázquez, K., Paulino, M., O. Salas, C., J. Zarate-Ramos, J., Vera, B. and Rivera, G. 2017. Trypanothione Reductase: A target for the development of anti- trypanosoma cruzi drugs. *Mini Reviews in Medicinal Chemistry* 17, 939–946. https://doi.org/10.2174/1389557517666170315145410

Vegro, C.L.R. and de Almeida, L.F. 2020. Global coffee market: Socio-economic and cultural dynamics. In: de Almeida, L.F. and Spers, E.E. (Eds.), *Coffee Consumption and Industry Strategies in Brazil,* Woodhead Publishing, Cambridge, UK, pp. 3–19. https://doi.org/10.1016/B978-0-12-814721-4.00001-9

Vintilă, I. 2016. Typical traditional processes: Cooking and frying. In: Prakash, V., Martín-Belloso, O., Keener, L., Astley, S., Braun, S., McMahon, H. and Lelieveld, H. (Eds.), *Regulating Safety of Traditional and Ethnic Foods*, Academic Press, Elsevier, London, UK, pp. 29–62. https://doi.org/10.1016/b978-0-12-800605-4.00003-7

Wadday, A.K. and Ghafel, S.T. 2022. Study effect of nitro group on some properties polycyclic aromatic hydrocarbons (Anthracene; Phenanthrene). In: *2022 International Congress on Human-Computer Interaction, Optimization and Robotic Applications (HORA)*, IEEE, Ankara, Turkey, pp. 1–4. https://doi.org/10.1109/hora55278.2022.9799939

Watt, K.C. and Buckpitt, A.R. 2000. Species differences in the regio- and stereoselectivity of 1-nitronaphthalene metabolism. *Drug Metabolism and Disposition* 28, 376–378. https://dmd.aspetjournals.org/content/28/4/376

Wilkinson, R.S., Bot, C., Kelly, M.J. and Hall, S.B. 2011. Trypanocidal activity of nitroaromatic prodrugs: Current treatments and future perspectives. *Current Topics in Medicinal Chemistry* 11, 2072–2084. https://doi.org/10.2174/156802611796575894

Wollin, K.M. and Dieter, H.H. 2005. Toxicological guidelines for monocyclic nitro-, amino- and aminonitroaromatics, nitramines, and nitrate esters in drinking water. *Archives of Environmental Contamination and Toxicology* 49, 18–26. https://doi.org/10.1007/s00244-004-0112-2

Xu, C., Chen, J., Zhang, X., Cai, K., Chen, C. and Xu, B. 2023. Emission characteristics and quantitative assessment of the health risks of cooking fumes during outdoor barbecuing. *Environmental Pollution* 323, 121319. https://doi.org/10.1016/j.envpol.2023.121319

Yu, H. 2002. Environmental carcinogenic polycyclic aromatic hydrocarbons: Photochemistry and phototoxicity. *Journal of Environmental Science and Health, Part C* 20, 149–183. https://doi.org/10.1081/gnc-120016203

Zhang, Q., Gao, R., Xu, F., Zhou, Q., Jiang, G., Wang, T., Chen, J., Hu, J., Jiang, W. and Wang, W. 2014. Role of water molecule in the gas-phase formation process of nitrated polycyclic aromatic hydrocarbons in the atmosphere: A computational study. *Environmental Science & Technology* 48, 5051–5057. https://doi.org/10.1021/es500453g

Zhang, Y., Hu, Q., Fu, J., Li, X., Mao, H. and Wang, T. 2023. Influence of exposure pathways on tissue distribution and health impact of polycyclic aromatic hydrocarbon derivatives. *Environment & Health* 1, 150–167. https://doi.org/10.1021/envhealth.3c00060

Zimmerli, B., Rhyn, P., Zoller, O. and Schlatter, J. 2001. Occurrence of heterocyclic aromatic amines in the Swiss diet: Analytical method, exposure estimation and risk assessment. *Food Additives & Contaminants* 18, 533–551. https://doi.org/10.1080/02652030119545

Trans-Fats and Safety

12

Food Processing and Manufacturing Related Effects

Monia Ennouri

12.1 INTRODUCTION

According to the World Health Organization (WHO), more than 278,000 deaths each year globally can be attributed to the intake of industrially produced trans-fat (WHO, 2024). It can be found in numerous products like vegetable shortening, margarines, fried foods, street food, and baked products, including biscuits, crackers, and pies. Trans fatty acids are dangerous to health as it enhances low-density lipoprotein (LDL) and decreases high-density lipoprotein (HDL) (Bhat et al., 2022). However, trans-fats have unique functional properties that are hard to replicate with other fats. They contribute to the texture, mouthfeel, and stability of many processed foods, such as baked goods, margarine, and fried foods. Finding alternatives that can mimic these properties without compromising product quality is a significant challenge (Temkov et al., 2021). Trans-fats often interact with other ingredients in complex ways (Manzoor et al., 2023). Some alternative fats and oils that are low in trans-fats, such as high-oleic oils, may be more expensive or less readily available than traditional fats. This can impact the cost and feasibility of reformulating products to reduce trans-fat content.

Addressing these challenges requires collaboration among food scientists, food manufacturers, regulatory agencies, and consumers. By combining all efforts, it is

DOI: 10.1201/9781032712291-12

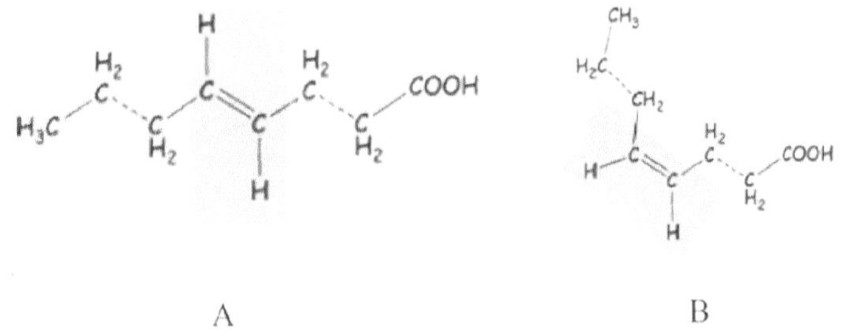

FIGURE 12.1 Configuration of *trans*-fatty acid (A) and configuration of *cis* fatty acid (B) (Madhujith and Sivakanthan, 2020).

possible to reduce the use of trans fats in processed foods and improve the overall healthfulness of the food supply.

12.2 ORIGIN OF TRANS-FATS

Two sources of trans fatty acids exist: industrial and natural. Natural ones are generated by animals as a result of the bio-hydrogenation process occurring in their digestive systems (Dhaka et al., 2011). The industrial trans fatty acids have two origins also: the heat treatment of oil (frying/ cooking/microwave) and the hydrogenation of oils. The configurations of trans fatty acid and cis fatty acid are presented in Figure 12.1 (Madhujith and Sivakanthan, 2020).

Hydrogenation is the principal source of trans-fatty acids, indeed the part of industrial hydrogenation affecting the increase in the level of trans fat is 10 to 50% against only 1 to 3% for thermal process (Shoaib et al., 2024).

12.3 EFFECTS OF TRANS FATTY ACIDS ON HEALTH

Clinical experiments have indicated the negative effects of trans-fats in food on human health and nutrition (Table 12.1). The most commonly known effect of trans fatty acids on health is cardiovascular disease since it increases the LDL level and reduces the HDL level (Chardigny et al., 2008; Matthan et al., 2004). Dashti et al. (2000) compared the long-term effects of palmitic (16:0), oleic (cis 18:1) and elaidic (trans 18:1) on hepatic lipoprotein production and found that trans fatty acids compared to cis fatty acids have more adverse effects on the composition and concentration of lipoproteins secreted by hepatic cells. The effects of trans-fats ingestion on cancer and pregnancy are less known than the effects of these fatty acids on heart disease. Okamura et al.

TABLE 12.1 Effects of trans-fats on human health

HUMAN HEALTH PROBLEM	TRANS-FATS INTAKE INDUCE	REFERENCE
Coronary heart diseases	Raise of the level of low-density lipoprotein and decrease of high-density lipoprotein level	Matthan et al. (2004) Chardigny et al. (2008) Dashti et al. (2000)
Cancer	Breast, prostate, and colon cancers	Smith et al. (2009) Ohmori et al. (2017)
Pregnancy	- Negative effect on fetal development, and fetal loss - Affect sperm concentration and sperm count	Mennitti et al. (2015) Chavarro et al. (2014)
Perturbation of cell functions	- Alteration of the fluidity of cell membrane and response of the receptors attached to it - Can participate in gene transcription by modeling the response of nuclear receptors	Takada et al. (2019)

(2021) noted that the intake of trans fatty acids induced intestinal inflammation and impaired glucose tolerance. Ohmori et al. (2017) reported that elaidic acid, a trans fatty acid, enhances the growth, survival, and invasion of the colorectal cancer cells, inducing the metastasis of the cancer cells. Takada et al. (2019) reported an alteration of the fluidity of cell membrane and response of the receptors attached to it with trans fatty acids intake, which can also participate in gene transcription by modeling the response of nuclear receptors. In the same vein, Mennitti et al. (2015) reported that maternal saturated fatty acids intake appears to provoke changes in liver and adipose tissue function correlated with insulin resistance and diabetes. They reported also that the trans fatty acids present in maternal plasma had adverse effects on the prenatal and postnatal development and growth. Chavarro et al. (2014) reported a study realized on a group of 209 healthy young university students (18–23 years) concerning the evaluation of trans fatty acids intake on semen quality, they found that trans fatty acids intake, which were derived from French potatoes and commercially baked items, was inversely related to total sperm count.

12.4 REGULATORY ACTIONS FOR TRANS-FATS

Several non-governmental and governmental bodies have worked out actions to restrict and remove trans fatty acids from food products. Regulatory actions have been taken up worldwide to eliminate the presence of trans-fats in the food supply (Figure 12.2).

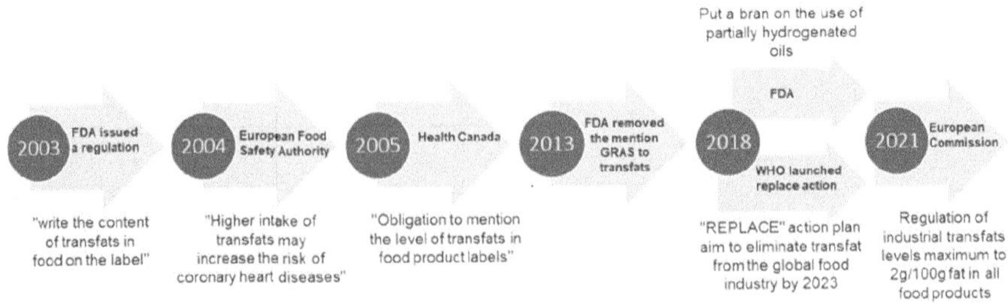

FIGURE 12.2 Evolution of regulatory actions for trans-fats.

The Food Drug Administration (FDA) in 2003 emitted a regulation requiring the writing of the content of trans-fats in food on the label (Parziale and Ooms, 2019). In 2004, the European Food Safety Authority declared that higher intakes of trans-fats may increase the risk of coronary heart diseases. In 2005, Health Canada imposed the mention of the level of trans-fats in food product labels. The FDA in 2013 deleted the mention Generally Recognized As Safe (GRAS) for trans fats, this action oblige industrials using oils containing trans fatty acids to demand a specific authorization to use them in food products and so limited their utilization, and in 2018 it imposes to put a ban on the use of partially hydrogenated oils. In the same year, the World Health Organization (WHO, 2020) has launched a "REPLACE" action plan to eliminate trans-fat from the global food industry by 2023. Since April 2021, the European Commission has put into effect a regulation for TFAs providing a maximum 2% of total fat in all industrially produced foods. In 2023 five countries: Lithuania, Saudi Arabia, Denmark, Poland, and Thailand were the first ones who received a validation certificate of trans-fat elimination in the new program of the WHO, which formally recognizes countries that have eliminated industrially produced trans-fats from their foods. These countries have demonstrated best practice trans-fat policy implementation, complete with systems to monitor and enforce the replacement of trans-fats with healthier alternatives. Exceedances still exist in several countries for example Sarwar et al. (2024) analyzed Composite soybean and palm oil samples from the retail market in Bangladesh and found that two-thirds of branded soybean oil samples exceeded WHO defined TFA limit of 2% and that soybean oil may contribute up to one-third of daily maximum limit of TFA intake.

12.5 FACTORS AFFECTING THE FORMATION OF TRANS FATTY ACIDS

Frequently processing and excessive cooking of food can have negative effects on the quality of oil, leading to various physiological and chemical changes. One such process is the isomerization of double bonds, which can result in the formation of trans fatty acids in food (Guo et al., 2023).

Effect of heat treatment of oils on trans fatty acids formation

Tripathi et al. (2022) reported that various thermal processing treatments of cooking oils having different degrees of unsaturation produce TFAs and their formation has been associated with time, temperature, and the number of processing cycles (Table 12.2). Hence, to maintain a safe level of TFA in the food, it is important to pay attention to the selection of cooking oil. Repeated processing/cooking of oil or oil-rich foods that may generate TFAs must be avoided. In the same vein, Tripathi and Chauhan (2022) tested the irradiation of soybean, mustard, and palm oil and concluded that the changes occurring after the irradiation are proportional to the linoleic and linolenic acid content of the oil. Changes in the oxidative parameters of oils tested were observed induced by the irradiation. Maximum difference after irradiation was observed in soybean oil followed by palm oil and then mustard oil (Table 12.2).

Most of the studies aiming to demonstrate the action of heating oils on the production of trans fatty acids have done so either by simply heating the oil or by putting potatoes in it to make fries, but it has been proven that the type of fried product can also influence the production of trans fatty acids differently and this is indeed what was highlighted in the study of Manzoor et al. (2023). These authors have reported that the proximate composition of food in addition to fatty acid oil composition and the frequency of frying affected significantly the trans fatty acid formation during deep frying. Cui et al. (2017) investigated the study of the effect of traditional Chinese cooking methods on fatty acid profiles of ten vegetable oils (peanut oil, soybean oil, rapeseed oil, corn oil, sunflower seed oil, rice bran oil, olive oil, sesame oil, linseed oil, and peony seed oil). The tested cooking methods were deep frying, stir frying, vegetable salad and pan frying. The result showed that *trans*-fatty acid (TFA) was produced during all processes. The highest TFA content was observed with rapeseed oil in vegetable salad oil with 2.88% of total fatty acid, whereas with sunflower seed oil no production of TFA was observed for the same preparation process. This study demonstrated also the superiority of olive oil compared to other oils, indeed the effect of four tested processes on olive oil was lower than other kinds of oils.

Table 12.3 reported the content of trans fatty acids of some margarines in different country, the lower contents were noted for Argentina and Indonesia.

12.6 POSSIBLE ALTERNATIVES TO TRANS-FATS

The global necessity to substitute trans fatty acids in food products has presented a challenge to food processors and manufacturers to discover innovative alternatives to trans-fats. Some technological approaches exist like oleogelation, interesterification, fractionation, and controlled hydrogenation.

Oleogelation

The oleogels are three-dimensional, thermo-reversible, anhydrous, viscoelastic, obtained by addition of an organo-gelling agent and/or a low molecular weight polymer, which can self-assemble into gel networks by non-covalent interactions, in an organic phase (oil) and by mixing and sometimes heating and cooling to trap the liquid phase.

TABLE 12.2 Effect of different treatments on trans fatty acids in oils and fats

		SOYBEAN OIL	*MUSTARD OIL*	*PALM OIL*	*HYDROGENATED FAT*	*CLARIFIED BUTTER*
Frying*	First cycle	0.88±0.11	1.17±0.18	1.15±0.25	10.13±0.57	3.64±0.32
	Fourth cycle	1.87±0.35	1.74±0.26	2.18±0.32	14.64±0.69	4.52±0.24
Heating (°C) **	100	0.65±0.09	0.80±0.06	0.93±0.12	9.40±0.47	3.21±0.28
	400	0.98±0.16	1.08±0.14	1.36±0.19	13.31±1.02	4.24±0.22
Microwave (min) ***	10	0.76±0.86	0.96±0.08	1.07±0.11	9.95±0.72	3.45±0.22
	40	1.18±0.20	1.28±0.23	1.44±0.15	11.95±1.06	3.98±0.31
Irradiation	2	0.67±0.12	0.91±0.13	0.94±0.14	nd	nd
(KGy)	6	0.89±0.11	1.03±0.14	1.07±0.11	nd	nd

*: each cycle is processed at 200–220°C during 10 min.
**: the heating is realized during 1h.
***: the application of microwave is at 2450Mhz and 150W power.
Sources: Tripahi and Chauhan, 2022; Tripahi et al., 2022.

TABLE 12.3 Contents in Trans fatty acids of some margarines from different countries

COUNTRY	*TFA (G/100G FOOD)*	*REFERENCE*
Iran	6.12	Abedi et al. (2016)
Poland	4.79	Monge Rojas et al. (2017)
Brazil	4.14	Monge Rojas et al. (2017)
China	1.68	Liu (2015)
Costa Rica	1.42	Monge Rojas et al. (2017)
Finland	0.34	Ritvanen et al. (2012)
Argentina	0	Monge Rojas et al. (2017)
Indonesia	0.03	Hartati et al. (2016)
Slovenia	5	Kusar et al. (2021)
Canada	0.3–16.3	Arcand et al. (2014)

Oleogel emulsions find their application in many food products in replacement to solid fat, like shortening and margarine. Oleogels can also be a vector of bioactive substances, such as antioxidants and polyphenols extracted from plants (Luo et al., 2019; Puscas et al., 2020).

Interesterification

Enzymatic interesterification is an innovative alternative for modifying the technological characteristics of natural edible oil. Due to the fatty acid positions change in the structure of glycerol resulting from interesterification, it participates in improving the physicochemical, functional, oxidative, and nutritional properties of fats and oil. For the production of the necessary lipids, enzymatic interesterification uses lipase as a biocatalyst with selectivity and specificity. The melting/dropping point, thermal characteristics, crystallization behavior, solid fat content, and stability against oxidation of triacylglycerol are all affected by changes in its molecular structure. This process is garnering more attention as a replacement for lipid modification because of its high acyl exchange reaction efficiency, easy reaction method, flexibility, environmentally friendly, and production of fewer by-products (Singh et al., 2022; Manzoor et al., 2024)

Fractionation

In opposition to interesterification and hydrogenation which are responsible for chemical change in the composition of the lipids, fractionalization is a reversible process. Fat fractionation works on the basis of separating the current TAG into two distinct products (stearin and olein) that have different melting points based on the degree of unsaturation and molecular weight. These products can then be processed further. Dry fractionation, solvent fractionation, and detergent fractionation are the three different forms of fractionation.

Without the use of any additives, the melt is used to create the crystals during the dry fractionation process. Based on the various solubility of TAG at a certain temperature, solvent fractionation uses solvents like hexane or acetone. However, it is important to pay attention to the quantity of stearin ingested because high intake rates of stearin fraction is associated with coronary diseases.

Controlled hydrogenation

The hydrogenation concerns the effect of the addition of hydrogen into the double bond of unsaturated fatty acid (Korver and Katan, 2006). The principal objective of edible oil hydrogenation is the extension of lipid shelf life and the modification of melting properties of oil. This process destroys the natural fatty acids and created a synthetic trans isomers offering the same texture as saturated fatty acids. The most used conditions for hydrogenation are 150°C to 225°C of temperature and 69 to 413 KPa of pressure. Other factors like the stirring rate and the quality and the type of oil influence the hydrogenation procedure. By increasing the pressure and the stirring speed, decreasing the temperature (40 to 60°C) of the hydrogenation process, it is possible to reduce the trans-fat formation.

12.7 CONCLUSIONS

There is no doubt that trans fatty acids are harmful to health causing various diseases, the scientific community continues to demonstrate the invasion of these fatty acids in food to raise awareness among consumers and educate them to demand from manufacturers working in the food sector to stop using these fatty acids. It is also important to note that international organizations are working to put pressure on manufacturers to impose low levels of fatty acids in food ready for consumption.

It is quite difficult to find a solution for the trans-fat problem. However, some alternatives exist to the use of hydrogenated oils rich in trans fatty acids. These alternatives must now be applied by manufacturers and tested to best meet consumer expectations in terms of taste, texture, and functionality.

REFERENCES

Abedi, A.-S., Hosseini, H., Mohammadi, A., Abdollahi, Z., Hajifaraji, M., Khaneghah, A.M., & Akbarirad, H. (2016). Fatty acid (FA) compositions and trans content of frequently consumed edible oils and fats from Iran' market. *Current Nutrition & Food Science*, *11*(1), 56–64.

Arcand, J., Scouboutakos, M.J., Au, J.T., & L'Abbe, M.R. (2014). Trans fatty acids in the Canadian food supply: An updated analysis. *The American Journal of Clinical Nutrition*, *100*, 1116–1123.

Bhat, S., Maganja, D., Huang, L., Wu, J.H.Y., & Marklund M. (2022). Influence of heating during cooking on trans fatty acid content of edible oils: A systematic review and meta-analysis. *Nutrients, 14*(7), 1489. https://doi.org/10.3390/nu14071489

Chardigny, J.M., Destaillats, F., Malpuech-Brugère, C., … & Sébédio, J.L. (2008). Do trans fatty acids from industrially produced sources and from natural sources have the same effect on cardiovascular disease risk factors in healthy subjects? Results of the trans fatty acids collaboration (TRANSFACT) study. *American Journal of Clinical Nutrition*, *87*(3), 558–566. https://doi.org/10.1093/ajcn/87.3.558.

Chavarro, J.E., Minguez-Alarcon, L., Mendiola, J., Cutillas-Tolin, A., Lopez-Espin, J.J., & Torres-Cantero, A.M. (2014). Trans fatty acid intake is inversely related to total sperm count in young healthy men. *Human Reproduction*, *29*(3), 429–440. http://doi.org/10.1093/humanrep/det464.

Cui, Y., Hao, P., Bingjie Liu, B., & Meng, X. (2017). Effect of traditional Chinese cooking methods on fatty acid profiles of vegetable oils. *Food Chemistry*, *233*, 77–84. https://doi.org/10.1016/j.foodchem.2017.04.084.

Dashti, N., Feng, Q., & Franklin, F.A. (2000). Long-term effects of cis and trans monounsaturated (18:1) and saturated (16:0) fatty acids on the synthesis and secretion of apolipoprotein A- and apolipoprotein B-containing lipoproteins in HepG2 cells. *Journal of Lipid Research*, *41*, 1980–1990.

Dhaka, V., Gulia, N., Ahlawat, K.S., & Khatkar, B.S. (2011). Trans fats sources, health risks and alternative approach–A review. *Journal of Food Science and Technology*, *48*, 534–541. https://doi.org/10.1007/s13197-010-0225-8.

Guo, Q., Li,T., Qu, Y., Liang, M., Ha, Y., Zhang, Y., & Wang, Q. (2023). New research development on trans fatty acids in food: Biological effects, analytical methods, formation mechanism, and mitigating measures. *Progress in Lipid Research*, *89*, 101199. https://doi.org/10.1016/j.plipres.2022.101199.

Hartati, S., Andarwulan, N., & Nuraida, L. (2016). Levels of trans-fatty acids present in processed food distributed in Indonesia. *Asian Journal of Chemistry*, *28*(9), 1955–1961. http://doi.org/10.14233/ajchem.2016.19849.

Korver, O., & Katan, M.B. (2006). The elimination of trans fats from spreads: How science helped to turn an industry around. *Nutrition Reviews*, *64*, 275–279.

Kusar, A., Hribar, M., Lavrisa, Z., Zupanic, N., Kupirovic, U.P., Hristov, H., & Pravst, I. (2021). Assessment of trans-fatty acid content in a sample of foods from the Slovenian food supply using a sales-weighting approach. *Public Health Nutrition*, *24*, 12–21.

Liu, A.D. (2015). Trans fatty acid levels in foods and intakes among population aged 3 years and above in Beijing and Guangzhou cities, China. *Biomedical and Environmental Sciences*, *28*(7), 477–485.

Luo, S.-Z., Hu, X.-F., Jia, Y.-J., Pan, L.-H., Zheng, Z., Zhao, Y.-Y., Mu, D.-D., Zhong, X.-Y., & Jiang, S.-T. (2019). Camellia oil-based oleogels structuring with tea polyphenol-palmitate particles and citrus pectin by emulsion-templated method: Preparation, characterization and potential application. *Food Hydrocolloids*, *95*, 76–87.

Madhujith, T., & Sivakanthan, S. (2020). Taking trans fats out of the food supply. In: De Silva, R.P., Pushpakumara, G., Prasada, P., & Weerahewa, J. (eds.), *Agricultural Research for Sustainable Food Systems in Sri Lanka*. Springer, Singapore. https://doi.org/10.1007/978-981-15-3673-1-13

Manzoor, S., Masoodi, F.A., & Rashid, R. (2023). Influence of food type, oil type and frying frequency on the formation of trans-fatty acids during repetitive deep-frying. *Food Control*, *147*, 109557. https://doi.org/10.1016/j.foodcont.2022.109557.

Manzoor, S., Masoodi, F.A., Akhtar, G., & Rashid, R. (2024). Production of trans-free shortening by lipase catalysed interesterification using mustard oil and palm stearin: optimisation and characterisation. *Biomass Conversion and Biorefinery*, *14*(12), 13027–13043.

Matthan, N.R., Welty, F.K., Barrett, P.H.R., … & Lichtenstein, A.H. (2004). Dietary hydrogenated fat increases high-density lipoprotein apo-I catabolism and decreases low-density lipoprotein apob-100 catabolism in hypercholesterolemic women. *Arteriosclerosis, Thrombosis, and Vascular Biology*, *24*(6), 1092–1097. https://doi.org/10.1161/01.ATV.000012410.23161.be.

Mennitti, L.V., Oliveira, J.L., Morais, C.A., Estadella, D., Oyama, L.M., Oller do Nacimento, C.M., & Pisani, L.P. (2015). Type of fatty acids in maternal diets during pregnancy and/

or lactation and metabolic consequences of the offspring. *The Journal of Nutritional Biochemistry*, *26*(2), 99–111. https://doi.org/10.1016/j.jnutbio.2014.10.001.

Monge-Rojas, R., Colon-Ramos, U., Jacoby, E., Alfaro, T., Tavares do Carmo, M., Das, G., Villalpando, S., & Bernal, C. (2017). Progress towards elimination of trans -fatty acids in foods commonly consumed in four Latin American cities. *Public Health Nutrition*, *20*(13), 2440–2449. https://doi.org/10.1017/S1368980017001227.

Ohmori, H., Fujii, K., Kadochi, Y., Mori, S., Nishiguchi, Y., … & Kuniyasu, H. (2017). Elaidic acid, a trans-fatty acid, enhances the metastasis of colorectal cancer cells. *Pathobiology*, *84*(3), 144–151. https://doi.org/10.1159/000449205.

Okamura, T., Hashimoto, Y., Majima, S., Senmaru, T., … & Fukui, M. (2021). Trans fatty acid intake induces intestinal inflammation and impaired glucose tolerance. *Frontiers in Immunology*, *12*, 669672. https://doi.org/10.3389/fimmu.2021.669672.

Parziale, A., & Ooms, G. (2019). The global fight against trans-fat: The potential role of international trade and law. *Globalization and Health*, *15*(1), 46. https://doi.org/10.1186/s12992-019-0488-4.

Puşcaş, A., Mureşan, V., Socaciu, C., & Muste, S. (2020). Oleogels in food: A review of current and potential applications. *Foods*, *9*, 70. https://doi.org/10.3390/foods9010070

Ritvanen, T., Putkonen, T., & Peltonen, K. (2012). A comparative study of the fatty acid composition of dairy products and margarines with reduced or substituted fat content. *Food and Nutrition Sciences*, *3*(9), 1189–1196. https://doi.org/10.4236/fns.2012.39156

Sarwar, S., Shaheen, N., Ashraf, M.M., Bahar, N., Akter, F., Hossain, M.M., ... & Shamim, A.A. (2024). Fatty acid profile emphasizing trans-fatty acids in commercially available soybean and palm oils and its probable intake in Bangladesh. *Food Chemistry Advances*, *4*, 100611. https://doi.org/10.1016/j.focha.2024.100611

Shoaib, H., Naz, S., Panhwar, A., Khaskheli, A.R., Sidhu, A.R., Siyal, F.J., ... & Almani, S. (2024). Review on health concern: Trans fatty acids and hydrogenation process: Review: Trans fatty acids impact on human health. *Biological Sciences-PJSIR*, *67*(1), 91–102.

Singh, P.K., Chopra, R., Garg, M., Dhiman, A., & Dhyani, A. (2022). Enzymatic interesterification of vegetable oil: A review on physicochemical and functional properties, and its health effects. *Journal of Oleo Science*, *71*(12), 1697–1709.

Smith, B.K., Robinson, L.E., Nam, R., & Ma, D.W.L. (2009). Trans-fatty acids and cancer: A mini-review. *British Journal of Nutrition*, *102*(9), 1254. https://doi.org/10.1017/s0007114509991437.

Takada, A., Shimizu, F., & Koba, S. (2019). Roles of trans and fatty acids in health; special references to their differences between Japanese and American old men. In: Tsipis, A. (ed.), *Visions of Cardiomyocete-Fundamental Concepts of Heart Life and Disease*, pp. 130, A. Tsipis, Athens, Greece.

Temkov, M., & Muresan, V. (2021). Tailoring the structure of lipids, oleogels and fat replacers by different approaches for solving the Trans-fat issue–A review. *Foods*, *10*, 1376. http://doi.org/10.3390/foods10061376.

Tripathi, P., Gupta, E., Purwar, S., & Pandey, A.T. (2022). Processing induced trans fatty acid quantification analysis of selected edible oils and their probable outcome. *Current Science*, *123*, 1455–1461. http://doi.org/10.18520/cs/v123/i12/1455-1461

Tripathi, P., & Chauhan, O.P. (2022). Irradiation induced changes in the trans fatty acid content and physicochemical properties of selected oils. *Carpathian Journal of Food Science and Technology*, *14*(2), 85–98. https://doi.org/10.34302/crpjfst/2022.14.2.8.

World Health Organization. (2020). *REPLACE Trans Fat: An Action Package to Eliminate Industrially Produced Trans-Fatty Acids. Module 2: Promote. How-to Guide for Determining the Best Replacement Oils and Interventions to Promote Their Use.* World Health Organization (WHO), Geneva.

World Health Organization. (2024). Trans fat. Knowledge action portal on NCDs. www.knowledge-action-portal.com/en/content/trans-fat

Food Additives and Safety

Food Processing and Manufacturing Related Effects

13

Teresa D'Amore, Slim Smaoui, and Sofia Agriopoulou

13.1 INTRODUCTION

> Food additive means any substance not normally consumed as a food by itself and not normally used as a typical ingredient of the food, whether or not it has nutritive value, the intentional addition of which to food for a technological (including organoleptic) purpose in the manufacture, processing, preparation, treatment, packing, packaging, transport or holding of such food results, or may be reasonably expected to result (directly or indirectly), in it or its by-products becoming a component of or otherwise affecting the characteristics of such foods. The term does not include contaminants or substances added to food for maintaining or improving nutritional qualities.

The authors would like to start this chapter with the definition of food additive introduced in the last version Codex Alimentarius Commission Procedural Manual and reported

DOI: 10.1201/9781032712291-13

in the Codex Standard 192-1995 – General Standard for Food Additives (GSFA) (Codex General Standard for Food Additives, 2023). These internationally adopted food standards were elaborated and refined in the years by the Codex Committee on Food Additives (CCFA) and the Joint Expert Committee on Food Additives (JECFA) of WHO (World Health Organization) and FAO (Food and Agriculture Organization of the United Nations), with the purpose to uniform and collect food requirements and specifications, and not the less to assure the best protection of consumers' health and ensure fair practices in the food trade. The worldwide international governments adopted and implemented these standards in their legislation. Consequently, when JECFA has deemed a food additive safe for use (safety assessment completed) and maximum permitted levels (MPLs) have been established in the CCFA, national food regulations must be implemented that allow the effective use of a food additive (WHO, FAO, 2020).

For this reason, although more than 4,000 active substances are added to food, feed, and beverages for additive-technological purposes, not all additives are permitted in specific regulations, while many of them were banned during the years, as a result of new evidences that changed the risk-benefit ratio. In fact, from the early twentieth century until today, the usage of chemical additives has increased, to extend shelf-life and maintain or improve the safety, freshness, taste, texture, and appearance of food (Carocho et al., 2014; Scotter & Castle, 2004). Several classifications, based on additives functions, labeling, and numbering, were developed worldwide. However, it is necessary to underline that it is difficult to schedule these molecules in one univocal list, because some agents may have more than one function in foods (Sun & Wang, 2017; Wu et al., 2022). The most-used classification is the Codex Alimentarius – GSFA – INS (International Numbering System), which divides all approved additives in categories, assigning a number code on the basis of technological purpose and functional class (27 classes), prefixed by "INS". In the European (EU) legislation, this classification is recognized by using the suffix "E" before the number code (WHO, FAO, 2024). All functional classes (44 in total) of additives with the Codex, EU, and USA classification number and a brief description are reported in Table 13.1.

The manufacturers are obliged to indicate which additives are in their products, also using this classification and in accordance with Regulation (EU) No 1169/2011 (European Commission, 2018). Monitoring and surveillance plans are carried out in all countries to ensure the observance of these procedures and regulations.

Over the years, the use of food additives has changed a lot, especially to ensure consumer safety and adapt to scientific and technological progress, but also due to a greater awareness and distrust of citizens towards processed and ultra-processed foods. However, the importance and role of these substances in fighting food insecurity and contrasting foodborne diseases and food waste should not be underestimated. Although food additives are generally safe when used within MPLs, some consumers prefer to avoid them for many reasons, including concerns about possible allergic reactions or a preference for unprocessed, minimally processed or natural foods (D'Amore et al., 2020; Sofos, 2008).

In this chapter, the authors would like to provide an overview of the current use, regulatory framework and occurrence of food additives in the global landscape. In

TABLE 13.1 Functional classes of food additives in foods

FUNCTIONAL CLASS	*DESCRIPTION*	*OTHER DEFINITION*	*CODEX CLASS NUMBER*	*EU CLASS NUMBER*	*USA CLASS NUMBER*	*INS NUMBER RANGE*
Acidity regulator (pH control agents)	substance added to change or maintain active acidity or basicity, including buffers, acids, alkalis, and neutralizing agents	acid, acidifier, alkali, base, buffer, buffering agent, pH adjusting agent	1	6–7	23	300–399 500–599
Anticaking agent and free-flow agent	substance added to finely powdered or crystalline food products to prevent caking, lumping, or agglomeration	anticaking agent, anti-stick agent, drying agent, dusting agent	2	8	1	500–599
Antifoaming agent	substance added to prevent or reduce foaming	defoaming agent	3	9	included in surface active-agent, 29	400–499
Antimicrobial agent or preservative	substance used to preserve food by preventing growth of microorganisms and subsequent spoilage	preservative, antimicrobial preservative, antimycotic agent, bacteriophage control agent, fungistatic agent, antimould and antirape agent, antimicrobial synergist	21	3	2	200–299

(*continued*)

TABLE 13.1 (Continued)

FUNCTIONAL CLASS	*DESCRIPTION*	*OTHER DEFINITION*	*CODEX CLASS NUMBER*	*EU CLASS NUMBER*	*USA CLASS NUMBER*	*INS NUMBER RANGE*
Antioxidant	substance used to preserve food by retarding deterioration, rancidity, or discoloration due to oxidation	antioxidant synergist, anti-browning agent	4	4	3	300–399
Bleaching agent	substance (non-flour use) used to decolourize food	-	5	-	-	915–930
Bulking agent	a food additive, which contributes to the bulk of a food without contributing significantly to its available energy value	filler	6	10	-	-
Carbonating agent	a food additive used to provide carbonation in a food	-	7	-	-	-
Carrier	a food additive used to dissolve, dilute, disperse or otherwise physically modify a food additive or nutrient without altering its function and without exerting any technological effect itself	carrier solvent, nutrient carrier, diluent for other food additives, encapsulating agent	8	5	-	1500–1522
Color	substance used to impart, preserve, or enhance the color or shading of a food	coloring agent, decorative pigment, surface colorant	9	2	4 (together with color retention agent)	100–199

Color retention agent	substance added to stabilize, retain or intensify the color of a food	color retention agent, color fixative, color stabilizer, color adjunct	10	2	4 (together with color)	100–199
Contrast enhancer	substance which, when applied to the external surface of fruit or vegetables following depigmentation of predefined parts (e.g. by laser treatment), helps to distinguish these parts from the remaining surface by imparting color following interaction with certain components of the epidermis	-	-	27	-	-
Curing agent	substance imparting a unique flavor and/or color to a food, usually producing an increase in shelf-life stability	pickling agent	included in preservative	included in preservative	5	249–252
Dough strengthener	substances used to modify starch and gluten, thereby producing a more stable dough	-	-	-	6	-
Drying agent	substance with moisture-absorbing ability, used to maintain an environment of low moisture	-	-	-	7	-

(*continued*)

TABLE 13.1 (Continued)

FUNCTIONAL CLASS	*DESCRIPTION*	*OTHER DEFINITION*	*CODEX CLASS NUMBER*	*EU CLASS NUMBER*	*USA CLASS NUMBER*	*INS NUMBER RANGE*
Emulsifier	substance able to modify surface tension in the component phase of an emulsion to establish a uniform dispersion or emulsion	plasticizer, dispersing agent, surface-active agent, crystallization inhibitor, density adjustment agent, suspension agent, clouding agent	11	11	8 (together with emulsifier salt)	400–499
Emulsifying salt	a food additive, which, in the manufacture of processed food, rearranges proteins in order to prevent fat separation	emulsifying salt synergist, melding salt	12	12	8 (together with emulsifier)	-
Enzyme	enzyme used to improve food processing and the quality of the finished food	-	-	included in specific regulation*	9	-
Firming agent	substance added to precipitate residual pectin, thus strengthening the supporting tissue and preventing its collapse during processing	-	13	13	10	300–399 400–499 500–599

Flavor enhancer	substance added to supplement, enhance, or modify the original taste and/or aroma of a food, without imparting a characteristic taste or aroma of its own	flavor synergist	14	14	11	600–699
Flavoring agent and adjuvant	substance added to impart or help impart a taste or aroma in food	-	-	-	12	600–699
Flour treating agent	substance added to milled flour, at the mill, to improve its color and/or baking qualities, including bleaching and maturing agents	flour bleaching agent, flour improver, dough conditioner, dough strengthening agent	15	26	13	915–930 1100–1101
Formulation aid	substance used to promote or produce a desired physical state or texture in food, including carriers, binders, fillers, plasticizers, film-formers, and tableting aids, etc.	-	-	-	14	-
Fumigant	volatile substance used for controlling insects or pests	fuming agent	-		15	220, 290 941–942

(*continued*)

TABLE 13.1 (Continued)

FUNCTIONAL CLASS	*DESCRIPTION*	*OTHER DEFINITION*	*CODEX CLASS NUMBER*	*EU CLASS NUMBER*	*USA CLASS NUMBER*	*INS NUMBER RANGE*
Gelling agent	a food additive, which gives a food texture through formation of a gel	-	17	16	-	400–499
Humectant	hygroscopic substance incorporated in food to promote retention of moisture, including moisture-retention agents and anti-dusting agents	moisture-retention agent, wetting agent	19	18	16	400–499
Leavening agent	substance used to produce or stimulate production of carbon dioxide in baked goods to impart a light texture, including yeast, yeast foods, and calcium salts	-	-	-	17	400–499
Lubricant and release agent	substance added to food contact surfaces to prevent ingredients and finished products from sticking to them	-	-	17 (only lubricant)	18	400–499

Modified starch	substance obtained by one or more chemical treatments of edible starches, which may have undergone a physical or enzymatic treatment, and may be acid or alkali thinned or bleached	-	-	19	-	1400–1452
Non-nutritive sweetener	substance having less than 2 percent of the caloric value of sucrose per equivalent unit of sweetening capacity	intense sweetener, bulk sweetener	26 (sweetener, generally)	1	19	650–699
Nutrient supplement	substance necessary for the body nutritional and metabolic processes	-	-		20	-
Nutritive sweetener	substance having greater than 2 percent of the caloric value of sucrose per equivalent unit of sweetening capacity	intense sweetener, bulk sweetener	26 (sweetener, generally)	1	21	650–699
Oxidizing and reducing agent	substance which chemically oxidizes or reduces another food ingredient, thereby producing a more stable product	oxidizer, reducer, red-ox agent	-	-	22	300–399

(*continued*)

TABLE 13.1 (Continued)

FUNCTIONAL CLASS	*DESCRIPTION*	*OTHER DEFINITION*	*CODEX CLASS NUMBER*	*EU CLASS NUMBER*	*USA CLASS NUMBER*	*INS NUMBER RANGE*
Packaging gas	a food additive gas, which is introduced into a container before, during or after filling with food with the intention to protect the food, for example, from oxidation or spoilage	-	20	20	-	930–949
Processing aid	substance used as manufacturing aids to enhance the appeal or utility of a food or food components	clarifying agents, clouding agents, catalysts, flocculants, filter aids, and crystallization inhibitors	-	-	24	-
Propellant	gas used to supply force to expel a product or used to reduce the amount of oxygen in contact with the food in packaging	aerating agent, and gas	22	21	25	942–949
Raising agent	a substance or a combination of food additives, which liberate(s) gas and thereby increase(s) the volume of a dough or batter	-	23	22	-	-

Sequestrant	substance which combines with polyvalent metal ions to form a soluble metal complex, to improve the quality and stability of products	-	24	23	26	-
Solvent	substance used to extract or dissolve another substance	vehicle solvent	-	-	27	-
Stabilizer	substance, which makes it possible to maintain a uniform dispersion of two or more components	foam stabilizer, colloidal stabilizer, emulsion stabilizer, stabilizer synergist, binder	25	24	28	400–499
Surface-active agent	substance used to modify surface properties of liquid food components for a variety of effects, other than emulsifiers, but including solubilizing agents, dispersants, detergents, wetting agents, rehydration enhancers, whipping agents, foaming agents, and defoaming agents, etc.	whipping agent, aerating agent	16 (only foaming agent)	15 (only foaming agent)	29	400–499 990–999
Surface-finishing agent	substance used to increase palatability, preserve gloss, and inhibit discoloration of foods, including glazes, polishes, waxes, and protective coatings	glazing agent, sealing agent, coating agent, surface-finishing agent, polishing agent, film forming agent	18 (only glazing agent)	17 (only glazing agent)	30	400–499 910–919

(*continued*)

TABLE 13.1 (Continued)

FUNCTIONAL CLASS	*DESCRIPTION*	*OTHER DEFINITION*	*CODEX CLASS NUMBER*	*EU CLASS NUMBER*	*USA CLASS NUMBER*	*INS NUMBER RANGE*
Synergist	substance used to act or react with another food ingredient to produce a total effect different or greater than the sum of the effects produced by the individual ingredients	catalytic agent, catalyst	-	-	31	-
Thickener	substance which affects the appearance or feel of the food (viscosity)	texturizer, thickener, bodying agent, binder, texturizing agent, thickener synergist	27	25	32	400–499

* Regulation (EC) No 1332/2008 on food enzymes

addition, a focus on toxicological studies and the derivation of reference values are presented. Finally, the fate of food additives in processed products and related case studies are critically discussed.

13.2 OCCURRENCE AND REGULATION

The regulatory framework of food additive usage, handling and monitoring is complex. Although global standards (GSFA and others, as previous described) give guidelines and JECFA plays a precious in the evaluation of safety these substances, at national, regional, and international levels there are strictly and often not harmonized regulations, standards, guidelines that establish authorization procedure, monitoring, usage conditions of food additives (EFSA Panel on Food Additives and Nutrient Sources added to Food (ANS), 2012). A summary of these standards is presented in Table 13.2.

13.2.1 Good Manufacturing Practice and Regulations in the Use of Food Additives

It is well-known that all food additives must be used following good laboratory practices (GMPs), as well as hygiene and handling practices. First of all, they must have a high degree of purity and must comply with identity and quality specifications. Some impurities, such as metals or contaminants, must be monitored following specific regulatory criteria and monitoring plans (FAO & WHO, 2023a). In accordance with GMP, the amount of additive added at all processing steps must be well documented and limited to the lowest possible amount to achieve the desired effect (FAO & WHO, 2023b).

In Europe, the so-called "General Food Law" (GFL, Regulation (EC) No 178/2002) lays down the common principles and requirements of food law and food safety procedures. Moreover, another basic regulation is the Reg. (EC) No 852/2004 on the hygiene of foodstuffs, followed by the two specific regulations on food additive purity criteria and manufacturing procedures (Reg. (EU) No 231/2012 and Reg. (EC) No 1333/2008). As established in GFL, the authorization process of food additives overseen by the European Commission (EC) is under of the independent scientific advice of the European Food Safety Authority (European Commission, 2002, 2004, 2023, 2024). The EC and EU Member States decide on the approval of food additives on the EU market and their conditions of use, such as labeling and MPLs. Moreover, EFSA tasks include also the re-evaluation of the safety of all food additives and to respond to specific requests on the subject proposed by EC. In 2010, a program to re-assess the safety of all the existing approved food additives, following Regulation (EU) No 257/2010, was started. This sequence of re-assessment opinions on all additives authorized before 20 January 2009 was brought to a close in 2020 (D'Amore et al., 2022; Dusemund et al., 2012). The studies comprise data relevant to the toxicological evaluations (metabolism and pharmacokinetic studies, special studies designed to investigate specific effects, such as the mechanism of toxicity, immune responses, or macromolecular binding) and

TABLE 13.2 International and national regulations on food additives

COUNTRY/ NATION	*REGULATION*	*NOTES*
Australia and New Zealand	FSANZ (Food Standards Australia New Zealand) – Food Standards Code 1.3.1. – Food Additives	It is part of Australia New Zealand Food Standards Code, 1991 Some natural additives recently approved (e.g., rosemary extract)
Brazil	ANVISA (Agência Nacional de Vigilância Sanitária) – Regulation n° 272 14 March 2019, Aditivos alimentares para uso em carnes e produtos cárneos, suas respectivas funções, limites máximos e condições de uso	Latest amendment: 2022
Canada	CFIA (Canadian Food Inspection Agency) – Food and Drug Regulation	Food additives inserted in Division 16 Under the Canadian Minister of Justice
China	NHFPC (National Health and Family Planning Committee of China) – GB 2760 National Food Safety Standard for Uses of Food Additives	Technical evaluations for the approval of a new food additive are carried out by the China National Center for Food Safety Risk Assessment
Europe	(EC) No. 1333/2008 of the European Parliament and of the Council on food additives. (EU) No. 1129/2011 amending Annex II to Regulation (EC) No. 1333/2008 of the European Parliament and of the Council by establishing a Union list of food additives	Last amendment: 2024
India	FSSAI (Food Safety and Standards Authority of India) – Food safety and standards Rules	Under the Ministry of Health and Family Welfare
Indonesia	NADFC (National Agency for Drug and Food Control of Indonesia) – Regulation of Head of Indonesian National Agency for Drug and Food Control Regulation No.36/2013 on the Maximum Limit of Preservative as Food Additive	Under the Minister of Health of Republic of Indonesia

TABLE 13.2 (Continued)

COUNTRY/ NATION	*REGULATION*	*NOTES*
Korea	MFDS (Ministry of Food and Drug Safety of Republic of Korea) – Food Additives Code 134-2019	Last amendment: 2019 For all additives also monographs are reported
Saudi Arabia	SFDA (Saudi Food and Drug Authority) – Additives Permitted for Use in Food Stuffs – GSO-FDS-05-2019	Last amendment: 2019 It uses EU regulations as model
South Africa	SAMH (South Africa Minister of Health) – Foodstuffs, Cosmetics and Disinfectants Act 54/1972	Last amendment: 2007
Taiwan	TFDA (Taiwan Food and Drug Administration) – Standards for Specification, Scope, Application and Limitation of Food Additives	Under the Ministry of Health and Family Welfare Last amendment: 2024
United States	US FDA (United States Food and Drug Administration) Code of Federal Regulations – Title 21 – Food and Drugs	The new inventory included in Title 21 replaced the EAFUS (Everything Added to Foods in the United States)

epidemiological studies. They are useful tools to control and re-confirm the levels of exposure and the reference values, which are associated with methods of risk evaluation.

In the United States, the regulations for food additives are described in the Code of Federal Regulations (CFR) – Title 21 – Food and Drugs (Alger et al., 2013; Neltner et al., 2011). The Part 170 contains the list of all additives permitted for direct addition to food for human consumption, replacing the previous epithet, "Everything Added to Food in the United States" (EAFUS). Similarly, in CFR – Title 9 – Part 424, all operations for food preparation and processing are explained. Some food additives in this regulation are food ingredients that are recognized as "GRAS" (Generally Recognized As Safe) (Burdock & Carabin, 2004; Frestedt, 2018). These substances intentionally added to foods, although subject to studies, approval, and review by the FDA, have been shown and have been deemed by assessors to be safe under the intended conditions of use. As an example, ascorbic acid and calcium alginate are included in this list. The eligibility criteria for GRAS classification are indicated in Title 21, Part 170.30.

13.2.2 Reference Value Calculation: Acceptable Daily Intake

Safety and toxicology studies are particularly important for the entire chain of regulated products. They are the basis for deriving toxicological reference values, which are values that define the level of safe exposure to a substance for a specific period (Gürtler, 2021).

Analytical and bioanalytical quality and safety controls along the entire production chain of a food, as well as maximum exposure limits, are determined on the basis of these values and consumption data.

The derivation of these values is always done by experts, institutions, and regulatory agencies that analyze toxicological studies, the most part long term *in vivo*, but also novel methodologies based on *in vitro*, *ex vivo*, *in silico* studies stand-alone or integrated (Ingenbleek et al., 2020). These studies are useful for deriving a reference point, the point of departure (PoD), or "the point on a dose–response curve established from experimental data used to derive a safe level" (EFSA glossary). Among them the most traditionally used are the NOAEL (no observed adverse effect level) and the LOAEL (low observed adverse effect level), which represent the highest concentration of a substance at which no measurable effect or the lowest measurable effect, respectively, occurs in a given population exposed to that substance. Other approaches, such as the benchmark dose – BMD, the minimum dose of a substance that produces a clear, low-level health risk, usually in the range of a 1–10% change in a specific toxic effect such as cancer induction, are recently introduce for overcoming some limitations of NOAEL (EFSA Scientific Committee et al., 2017).

These PoDs are used for the calculation of for estimating the health-based guidance values (HBGVs). For chemical substances intentionally added to foods such as food additives, pesticide residues, and veterinary drugs, the HBGV used is the acceptable daily intake (ADI), while for contaminants (e.g., toxic trace elements), which are not added deliberately to foods, tolerable daily intake (TDI) is adopted (Pompa et al., 2021). It is an estimate of the amount of a substance in food/feed and beverage that can be ingested daily over a lifetime without appreciable risk to health. The ADI is usually expressed as milligrams of the substance per kilogram of body weight per day (mg/kg/day; standard human = 60 kg) (D'Amore, Di Taranto, Berardi, et al., 2019).

The ADI is calculated as a ratio of the PoD, (NOAEL, LOAEL, BMDL, etc.) and safety factors or assessment factors (AFs). The AFs are dimensionless values, which include the contributions of all uncertainties arising from interspecies differences between humans and animals, but also interspecies differences within the human population. Other AFs, deriving from uncertainties in the toxicological studies conducted, such as the use of LOAEL instead of NOAEL or unavailability of long-term studies, can be added by risk assessors, based on expert judgment (EJ) and the weight of evidence available (WoE approach). Typically, the total safety factor is 100 (10 for species differences and 10 for human variability).

The resulting formula is:

$$ADI = \frac{PoD}{AFs} \tag{13.1}$$

As an example, for sodium nitrite, a common preservative agent used in several meat preparations, the established ADI is 0,07 mg/kg bw/day, expressed as nitrite ion. It is derived from a NOAEL of 10 mg/kg bw/day and the AFs of 100. The final calculation is corrected for the molecular weight of nitrite ion/sodium nitrite (MW_{NO2-}= 46.00 g/mol; MW_{NaNO2} = 68.99 g/mol), as follows:

$$ADI_{NaNO2} = \frac{NOAEL}{AF_{interspecies} \times AF_{intraspecies}} \times \frac{MW_{NO2-}}{MW_{NaNO2}} = \frac{10\ mg / kg\ bw / day}{10 \times 10} \times \frac{46.00\ g / mol}{68.99\ g / mol} \tag{13.2}$$

The ADIs of all food additives were periodically re-assessed by the EFSA, as described before, and re-confirmed or updated by the JECFA (WHO, FAO, 2020).

13.2.3 Monitoring of Food Additives: Analytical Methods

Monitoring of food additives is an essential process to ensure food safety and protect public health. Health authorities carry out regular controls on foods to verify compliance with the regulations on food additives, i.e. that the concentration of the additive in the finished product complies with the maximum limits established. Moreover, the illegal use of some concerning food additives (e.g., sulphites in fresh meat and meat preparations) is another mandatory activity of great importance to improve food information and protect consumers' health (D'Amore et al., 2020; Fernandes et al., 2022). Each regulation designates the competent authorities to take products from the market and carry out analytical controls, as well as the methods to be used, which must be internationally recognized and validated. The most common analytical methods for the determination of food additives include chromatographic, spectroscopic, electrophoretic, and enzymatic techniques.

The choice of method depends on the chemical-physical properties of the additive, since it is a heterogeneous class of compounds with very different characteristics and functional groups. Moreover, a key role is played also by the matrices and working range of the methods, as well as from the accuracy requested for the quantification of the analyte(s).

For colorants characterized by a high degree of conjugation and the presence of double bonds, conjugated system and chromophoric groups, such as azo-compounds or natural colorants based on carotenoids and anthocyanins, the method of choice is liquid chromatography with UV-visible detection (LC-UV-vis), or alternatively, with mass spectrometry detector (LC-MS). Also, spectrophotometric methods can also be used for these substances (Merino et al., 2017; Ntrallou et al., 2020).

Many additives, such as preservatives, pH regulators, antioxidants, and stabilizers are ionic compounds, so ion chromatography combined with conductometric detection (IC-CD) is the method of election, thanks to its high sensitivity, precision, possible automation, and low costs (D'Amore, Di Taranto, Vita, et al., 2019; D'Amore et al., 2021). Other detectors used in combination with ion chromatography include the pulsed amperometric detector (PAD) and the electrochemical detector (ED) (Kumar Meher et al., 2018; Rohrer & Avdalovic, 2021). Various sweeteners can also be determined with this technique as well as using separation by capillary electrophoresis (Martins et al., 2019; Wang et al., 2010). On the contrary, the application of gas chromatography coupled with several detectors is very limited, since many additives are not volatile molecules. Some applications using high-precision isotope dilution GC-MS are described in literature (Pagliano et al., 2019).

Most food additives are ionizable compounds. This makes them analyzable with spectrometric techniques, preceded by chromatographic separation. This technique is the most suitable for multi-analyte determinations of additives and provides a high number of identification points (Wu, 2021).

A panel of emerging techniques is represented by biosensors, which may offer significant advantages in terms of sensitivity, specificity, and speed of analysis, making them ideal tools for monitoring food additives. Several electrochemical, piezometric, thermal, optical and nanoparticle-based biosensors have been developed for the analysis of many additives, such as sulphites, benzoic acid, and azo colorants. However, challenges such as stability and reproducibility need to be addressed for their large-scale deployment (Bener et al., 2020; Turasan & Kokini, 2021).

Other methods such as distillation and titration (e.g., Monier-Williams method, Ripper method), enzymatic and microbiological assays have still a role in additive determination despite being dated. A key example is the analysis of sulphites in meat, shrimps, wine and other alcoholic beverages (Fernandes et al., 2022; Gajdár et al., 2022).

A comprehensive representation of the analytical procedures used for the determination of food additives is shown in Figure 13.1.

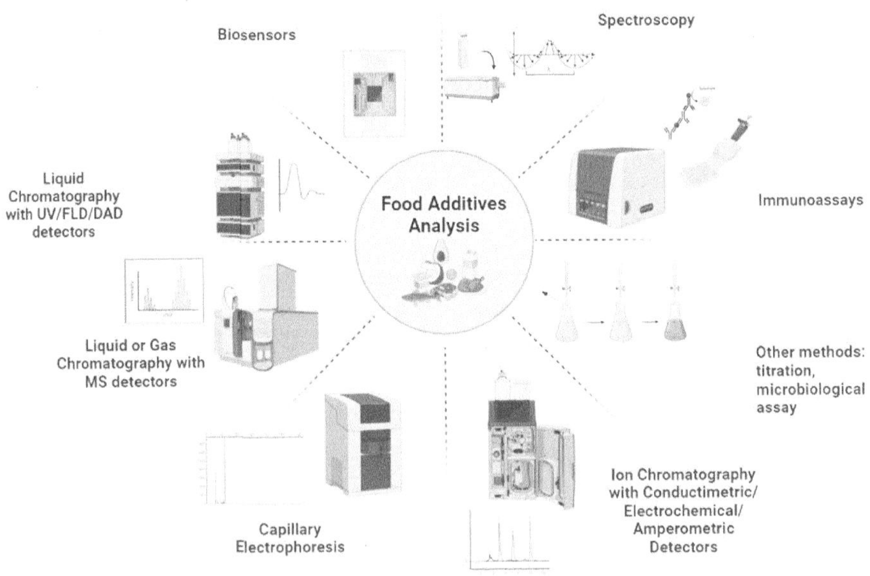

FIGURE 13.1 A comprehensive representation of the analytical procedures used for the determination of food additives.

13.3 TOXICITY, DIETARY EXPOSURE, AND RISK ASSESSMENT

Food additives are associated with a range of toxic effects on the human body as well as other organisms in food chains. There are four categories of acute toxicity for human hazard and precautionary statements with their corresponding labels: highly toxic (3), moderately toxic (2), slightly toxic (1), and practically nontoxic (0) (Shi et al., 2024). In a study conducted by Shi et al., which included 221,439 experimental toxicity records, it was found that food additives had the lowest toxicity in humans among other food-relevant chemicals to human health (pesticide residues, veterinary drug residues, food-related environmental pollutants, and natural toxins). Additionally, across all food additives including illegal additives, they had lower mean Lowest Published Toxic Dose (TD_{LO}: minimal toxic dose) values than pesticides, veterinary drugs, and environmental pollutants, indicating their higher potential to cause other adverse effects in humans (Shi et al., 2024).

According to the Rapid Alert System for Food and Feed, for the year 2021 food additives and flavorings were reported in 176 notifications, mainly due to unauthorized or very high levels of additive content (RASFF, 2022).

Salame and her co-workers found direct associations between the risk of type 2 diabetes and exposures to various food additive emulsifiers widely used in industrial foods, in a large prospective cohort of French adults. Actually of 104,139 participants, 1,056 were diagnosed with type 2 diabetes during follow-up (mean follow-up duration 6–8 years) (Salame et al., 2024). Similar associations were made by surveys conducted in the Netherlands, Spain, and the United Kingdom (Levy et al. 2021). Moreover, although there are experimental studies showing harmful effects of emulsifiers on intestinal microflora and metabolism, leading to chronic inflammation and increasing susceptibility to carcinogenesis, human epidemiological evidence does not associate them with cancer (Sellem et al., 2024).

Exposure assessment is the qualitative or quantitative evaluation of the likely intake of a chemical via food whose calculation is a function of consumption, concentration of the chemical hazard, and personal weight (FAO, 1997, EFSA, 2011). As food consumption data is the basis for assessing human exposure, comprehensive information is needed. In this respect, EFSA in the database FoodEx2's harmonized food consumption data across the European Union (EU). The information is divided by country, food category, age, and sex (EFSA, 2021) The FAO/WHO also provides a chronic individual food consumption database known as CIFOCOss (WHO, 2018). To assess dietary exposure to food chemicals such as food additives, official bodies have developed freely available dietary exposure assessment tools, such as FAIM (Food Additives Intake Model) for food additives. The purpose of the FAIM template is to provide a screening tool for estimating long-term exposure to food additives (Zenodo, 2013). It can be used as a first step in the dietary exposure assessment by applicants, risk assessors, and risk managers. It is based on summary food consumption statistics (in grams per kg body weight) from the EFSA Comprehensive European Food Consumption Database tailored to food additive categories. Based on FAIM, the mean and high-level exposure to food additives can be estimated for different population groups throughout several European countries and age groups (Gürtler, 2021).

13.4 FOOD PROCESSING METHODS AND THEIR EFFECTS: THE FATE OF ADDITIVES IN FOODS

Food processing and manufacturing can have a significant impact on all macromolecules contained in the food and on other ingredients, including additives, influencing their stability, efficacy, and organoleptic properties. Understanding these effects is crucial for the food industry to ensure the quality and safety of final products. Innovative techniques offer promising solutions to reduce dependence on synthetic additives, while maintaining quality and safety standards. Continuous research and development in this field is essential to optimize production processes and meet consumer needs (Cox et al., 2021).

It is also important to underline that most foods, whether fresh or processed, are dynamic matrices that can change during all stages of production, post-production, and consumption. Being active substances, additives can interact, bind, modify, and be modified by different ingredients of the food matrix. This can bring a technological benefit, such as avoiding the oxidation of lipids, the deterioration of proteins and stabilizing the color of a product. On the contrary, these reactions can lead to the formation of both potentially toxic intermediates and compounds, and of inactive derivates. These reactions can be enhanced by specific thermodynamic and kinetic factors (e.g., temperature, water activity, pressure, treatment time, pH, oxygen, light, and catalysis). For example, some additives can degrade or transform into less effective compounds when subjected to heat, radiation, and sudden changes in pressure, such as some colorants. Some secondary colorants produced may have the same wavelength of maximum absorption and do not lose color. In other cases, colorless breakdown products may be produced.

Nevertheless, the usage of several preservatives, added in traditional meat preparations, are often in conjunction with mild thermal treatments, leading to improvement in textural and organoleptic properties.

For these reasons also data relevant to the manufacturing and processing of additives need to be evaluated by manufacturers. The identification and characterization of products derived from additive-food interactions, or even from additive-additive interaction, despite being challenging, it is important for safety assessment of the final products (Mercer & Rodriguez-Amaya, 2021). Undesirable compounds formation should be avoided. As an example, sodium and potassium nitrite and nitrate are not used in combination with sulphites. In fact, it was shown that they can react, forming ammonium and hydroxylamine sulphonates. These products are inactive, so the antimicrobial activity of individual additives is lost.

Many examples of these cases are reported in literature, however, in this section some interesting case studies are analyzed in depth.

13.4.1 The Case of Nitrate and Nitrite Salts

Potassium and sodium salts of nitrate and nitrite (INS 249–252) are well-known antimicrobials, used for more than two centuries in several tradition meat preparations. However, these additives are characterized by a broad spectrum of functions and activities, having also antioxidant, color stabilizers and fixatives, flavorings, anti-browning, and texturing properties (EFSA Panel on Food Additives and Nutrient Sources added to Food (ANS), Mortensen, Aguilar, Crebelli, Di Domenico, Dusemund, Frutos, Galtier, Gott, Gundert-Remy, Lambré, Leblanc, Lindtner, Moldeus, Mosesso, Oskarsson, Parent-Massin, Stankovic, Waalkens-Berendsen, Woutersen, Wright, van den Brandt, Fortes, Merino, Toldrà, Arcella, Christodoulidou, Barrucci, et al., 2017; EFSA Panel on Food Additives and Nutrient Sources added to Food (ANS), Mortensen, Aguilar, Crebelli, Di Domenico, Dusemund, Frutos, Galtier, Gott, Gundert-Remy, Lambré, Leblanc, Lindtner, Moldeus, Mosesso, Oskarsson, Parent-Massin, Stankovic, Waalkens-Berendsen, Woutersen, Wright, van den Brandt, Fortes, Merino, Toldrà, Arcella, Christodoulidou, Cortinas Abrahantes, et al., 2017). Nitrate needs to be reduced to nitrite, which is the most active. Similarly, nitrite function is linked to the production of nitrosating agents, such as nitrogen oxide (NO) and nitrous acid (HNO_2). Moreover, in consequence of precise conditions and processing phases, several active species of nitrogen, the so-called reactive nitrogen species, may be formed. On the contrary, in low acid-matrices, nitrite may also be oxidized to nitrate (Hord et al., 2009).

During processed foods manufacturing (e.g., traditionally cured meat products and processed ripened cheeses) there are precise protocols to define their added amounts. Several commercial mixtures, containing also these salts, are also available. Many studies showed that after the manufacturing the initial amount of nitrite and nitrate dramatically decrease and continued to decrease also during the storage until the consumption. In fact, only 10–20 percent of the added nitrite/ nitrate remains at the end of the manufacturing process (Gómez et al., 2020). The residual amount may be in compliance with the regulations described above.

These losses depend on many processing conditions, including heat treatment period and temperature, pH, other additive addition, and the composition of food.

During the years, many researchers demonstrated that this dramatic decrease is only "apparent". In fact, these additives, in consequence of manufacturing and processing, may interact with all matrix constituents, including macromolecules and cofactors (Bonifacie et al., 2021; D'Amore et al., 2022).

Labeling the additives using the isotope ^{15}N, it was shown that most part of added nitrite remain in the final product bound to several molecular targets (Woolford & Cassens, 1977). Table 13.3 sums up the fate of nitrate when it is added to food and its molecular targets of nitrate and nitrite during manufacturing and processing.

C-, S-, O-, N- nitrosation reactions may be involved in the interaction of nitrite/ nitrate and proteins, amino acids, vitamins, and cofactors. Some of them, including S-nitrosothiols (R-S-NO) and S-nitrosoaminoacids (S-nitrosocysteine), exhibit also

TABLE 13.3 Molecular targets of nitrate and nitrite during manufacturing and processing

TARGET MOLECULES	*NITROSO-REACTIVE INTERMEDIATE*	*REACTIONS*	*PRODUCTS*	*ACTIVITY*	*REF*
cysteine	nitrite, nitrogen oxide	S- nitrosation	S-nitroso-cysteine	antioxidant, color-developer, and anti-clostridial	(Hammes, 2012)
enzymes	nitrite, nitrogen oxide	N-, O, S-, C- nitrosation	inactive or less active enzyme	antimicrobial, block of key microbial enzymes	(Hord et al., 2009)
glutathione	nitrogen oxide	S- nitrosation	S-nitrosoglutathione	-	(Shpaizer et al., 2018)
hemoglobin	nitrogen oxide	bond with Fe(II) of heme group	nitrosyl-hemoglobin	color fixative and stabilizer	(D'Amore et al., 2022)
iron sulfur-proteins	nitrite, nitrogen oxide	bond with Fe-S clusters		antimicrobial, texturizer	(Majou & Christieans, 2018)
lipids	nitrate, nitrite, nitrogen oxide	C-, O- nitrosation	nitro-fatty acids	antioxidant	(Kanner, 1994)
myoglobin	nitrogen oxide	bond with Fe(II) of heme group	nitrosyl-myoglobin	color fixative and stabilizer	(D'Amore et al., 2022)
nucleic acids	nitrite, nitrogen oxide	nitrosation of guanine, guanosine, and 2 – deoxyguanosine, either free or in DNA or RNA	8-hydroxydeoxy - guanosine 8-nitroguanidine	potentially toxic effects	(Shakil et al., 2022)
oxygen	nitrogen oxide	NO peroxidation	peroxynitrite	scavenging	(Kanner, 1994)
proteins	nitrate, nitrite, nitrogen oxide	N-, O, S-, C- nitrosation	inactive or misfolded proteins	color fixative and stabilizer	(Shakil et al., 2022)
secondary amines	nitrite, nitrogen oxide	N- nitrosation	N-nitrosamines	potentially toxic effects (carcinogenicity)	(Xie et al., 2023)
thiols	nitrite, nitrogen oxide	S- nitrosation	S-nitrosothiol, thionitrites	antioxidant	(Hammes, 2012)

antioxidant and flavor enhancer properties (Hammes, 2012; Shakil et al., 2022; Shpaizer et al., 2018). One of the most important manufacturing reactions is that with myoglobin and hemoglobin. NO may react with the ferrous ion of the heme group forming stable pigments, nitrosyl-myoglobin and nitrosyl-hemoglobin, respectively. In certain severe conditions, such as during heat-curing treatment or at low pH, it can be denatured, forming another adduct, nitroso-myochromogen, a pink pigment typical of heat-treated meat products, which consists in NO-porphyrin ring system alone. This reaction is responsible for the unique and particular color of cured meat products (D'Amore et al., 2022).

Other concerning reactions are also possible during heat-treatments. Nitrite can react with secondary amines (N-nitrosation) to form nitrosamines, compounds classified as carcinogenic to humans (De Mey et al., 2017; Xie et al., 2023). However, some recent studies suggested that in cured meat, the addition of nitrite promotes the formation of nitroso compounds without a dose–response effect. Moreover, no nitrosamine formation is observed in cooked ham up to 120 mg/kg of added nitrite (Bonifacie et al., 2021).

13.5 CONCLUSIONS AND FUTURE PERSPECTIVES

In this chapter, the importance and use of food additives has been described. However, despite their valuable technological functions, their use has raised concerns about safety and long-term health effects for years. In fact, some additives such as nitrites, sulphites, and aspartame have been associated with adverse health effects, including allergies, hyperactivity in children, and potential carcinogenic risks. Consequently, there is a growing interest in finding natural alternatives that can offer similar benefits without the potential associated risks. In the near future, these natural ingredients (plant total extracts, powders, oils, or purified bioactive molecules) may replace or reduce the use of conventional chemical additives. The adoption of natural additives can lead to a healthier and more sustainable future for the food system.

Similarly, the desire of the scientific community to replace and reduce synthetic food additives has led to the development of new processing and manufacturing techniques, from the more traditional, but optimized and improved drying techniques (both hot and freeze-drying) and thermal extrusion, to the massive increase in the use of microwaves assisted methods.

Among other procedures, the use of high-pressure processing methods can reduce the use of preservatives. In addition, the development of edible coatings manufactured with natural materials, such as polysaccharides, proteins, and lipids may protect foods from deterioration and oxidation, representing a promising alternative. The use of microorganisms, such as lactic acid bacteria, has been investigated for the same purposes. These microorganisms are able to produce various compounds that act as preservatives, antioxidants and at the same time improve the taste of various foods (e.g., dairy products and meat preparations).

The research on the development of active and intelligent packaging, which incorporates substances that can release or absorb compounds to extend the shelf-life of the food, is a very active field.

Although there are significant challenges in terms of stability, efficacy, and cost, continued research and innovation in the field of alternatives to conventional food additives promise to overcome these barriers. A renewed collaboration among industry, research, and regulator institutions will be crucial to overcome barriers and maximize the benefits of these new technologies.

REFERENCES

Alger, H. M., Maffini, M. V., Kulkarni, N. R., Bongard, E. D., & Neltner, T. (2013). Perspectives on how FDA assesses exposure to food additives when evaluating their safety: Workshop proceedings. *Comprehensive Reviews in Food Science and Food Safety*, *12*(1), 90–119. https://doi.org/10.1111/j.1541-4337.2012.00216.x

Bener, M., Şen, F. B., & Apak, R. (2020). Novel pararosaniline based optical sensor for the determination of sulfite in food extracts. *Spectrochimica Acta Part A: Molecular and Biomolecular Spectroscopy*, *226*, 117643. https://doi.org/10.1016/j.saa.2019.117643

Bonifacie, A., Promeyrat, A., Nassy, G., Gatellier, P., Santé-Lhoutellier, V., & Théron, L. (2021). Chemical reactivity of nitrite and ascorbate in a cured and cooked meat model implication in nitrosation, nitrosylation and oxidation. *Food Chemistry*, *348*, 129073. https://doi.org/10.1016/j.foodchem.2021.129073

Burdock, G. A., & Carabin, I. G. (2004). Generally recognized as safe (GRAS): History and description. *Toxicology Letters*, *150*(1), 3–18. https://doi.org/10.1016/j.toxlet.2003.07.004

Carocho, M., Barreiro, M. F., Morales, P., & Ferreira, I. C. F. R. (2014). Adding molecules to food, pros and cons: A review on synthetic and natural food additives. *Comprehensive Reviews in Food Science and Food Safety*, *13*(4), 377–399. https://doi.org/10.1111/1541-4337.12065

Codex General Standard for Food Additives. (2023). www.fao.org/fao-who-codexalimentarius/sh-proxy/en/?lnk=1&url=https%253A%252F%252Fworkspace.fao.org%252Fsites%252Fcodex%252FStandards%252FCXS%2B192-1995%252FCXS_192e.pdf

Cox, S., Sandall, A., Smith, L., Rossi, M., & Whelan, K. (2021). Food additive emulsifiers: A review of their role in foods, legislation and classifications, presence in food supply, dietary exposure, and safety assessment. *Nutrition Reviews*, *79*(6), 726–741. https://doi.org/10.1093/nutrit/nuaa038

D'Amore, T., Di Taranto, A., Berardi, G., Vita, V., & Iammarino, M. (2019). *Quantification of Nitrate Levels in Spinach and Simualtaneous Determination of Nitrite.* In Agricultural Research Updates - Volume 28, Chapter 5, (pp. 107–141). Nova Science Publisher.

D'Amore, T., Di Taranto, A., Berardi, G., Vita, V., & Iammarino, M. (2021). Going green in food analysis: A rapid and accurate method for the determination of sorbic acid and benzoic acid in foods by capillary ion chromatography with conductivity detection. *LWT, 141*, 110841. https://doi.org/10.1016/j.lwt.2020.110841

D'Amore, T., Di Taranto, A., Berardi, G., Vita, V., Marchesani, G., Chiaravalle, A. E., & Iammarino, M. (2020). Sulfites in meat: Occurrence, activity, toxicity, regulation, and detection. A comprehensive review. *Comprehensive Reviews in Food Science and Food Safety, 19*(5), 2701–2720. https://doi.org/10.1111/1541-4337.12607

D'Amore, T., Di Taranto, A., Vita, V., Berardi, G., & Iammarino, M. (2019). Development and validation of an analytical method for nitrite and nitrate determination in meat products by Capillary Ion Chromatography (CIC). *Food Analytical Methods, 12*(8), 1813–1822. https://doi.org/10.1007/s12161-019-01529-0

D'Amore, T., Taranto, A. D., Berardi, G., Vita, V., & Iammarino, M. (2022). Nitrate as food additives: Reactivity, occurrence, and regulation. In Christos Tsadilas (ed), *Nitrate Handbook*. CRC Press. DOI: 10.1201/9780429326806-17

De Mey, E., De Maere, H., Paelinck, H., & Fraeye, I. (2017). Volatile N-nitrosamines in meat products: Potential precursors, influence of processing, and mitigation strategies. *Critical Reviews in Food Science and Nutrition, 57*(13), 2909–2923. https://doi.org/10.1080/10408398.2015.1078769

Dusemund, B., Gilbert, J., Gott, D., Kenigswald, H., König, J., Lambré, C., Leblanc, J.-C., Mortensen, A., & Tobback, P. (2012). Food additives and nutrient sources added to food: Developments since the creation of EFSA. *EFSA Journal, 10*(10), s1006. https://doi.org/10.2903/j.efsa.2012.s1006

EFSA – European Food Safety Authority. (2011). Overview of the procedures currently used at EFSA for the assessment of dietary exposure to different chemical substances. *EFSA Journal, 9*(12), 2490, pp. 33. https://doi.org/10.2903/j.efsa.2011.2490

EFSA Panel on Food Additives and Nutrient Sources added to Food (ANS). (2012). Guidance for submission for food additive evaluations. *EFSA Journal, 10*(7), 2760. https://doi.org/10.2903/j.efsa.2012.2760

EFSA Panel on Food Additives and Nutrient Sources added to Food (ANS), Mortensen, A., Aguilar, F., Crebelli, R., Di Domenico, A., Dusemund, B., Frutos, M. J., Galtier, P., Gott, D., Gundert-Remy, U., Lambré, C., Leblanc, J.-C., Lindtner, O., Moldeus, P., Mosesso, P., Oskarsson, A., Parent-Massin, D., Stankovic, I., Waalkens-Berendsen, I., … & Younes, M. (2017). Re-evaluation of potassium nitrite (E 249) and sodium nitrite (E 250) as food additives. *EFSA Journal. European Food Safety Authority, 15*(6), e04786. https://doi.org/10.2903/j.efsa.2017.4786

EFSA Panel on Food Additives and Nutrient Sources added to Food (ANS), Mortensen, A., Aguilar, F., Crebelli, R., Di Domenico, A., Dusemund, B., Frutos, M. J., Galtier, P., Gott, D., Gundert-Remy, U., Lambré, C., Leblanc, J.-C., Lindtner, O., Moldeus, P., Mosesso, P., Oskarsson, A., Parent-Massin, D., Stankovic, I., Waalkens-Berendsen, I., … & Younes, M. (2017). Re-evaluation of sodium nitrate (E 251) and potassium nitrate (E 252) as food additives. *EFSA Journal. European Food Safety Authority, 15*(6), e04787. https://doi.org/10.2903/j.efsa.2017.4787

EFSA Scientific Committee, Hardy, A., Benford, D., Halldorsson, T., Jeger, M. J., Knutsen, K. H., More, S., Mortensen, A., Naegeli, H., Noteborn, H., Ockleford, C., Ricci, A., Rychen, G., Silano, V., Solecki, R., Turck, D., Aerts, M., Bodin, L., Davis, A., … & Schlatter, J. R. (2017). Update: Use of the benchmark dose approach in risk assessment. *EFSA Journal, 15*(1), e04658. https://doi.org/10.2903/j.efsa.2017.4658

EFSA – European Food Safety Authority. (2021). Food Consumption Data. Available online: www.efsa.europa.eu/en/data-report/food-consumption-data (accessed on 31 July 2024).

European Commission. (2002, January 28). *Regulation (EC) No 178/2002 of the European Parliament and of the Council of 28 January 2002 Laying Down the General Principles and Requirements of Food Law, Establishing the European Food Safety Authority and Laying Down Procedures in Matters of Food Safety*. European Commission, Brussels, Belgium. http://data.europa.eu/eli/reg/2002/178/oj/eng

European Commission. (2004). *Regulation (EC) No 852/2004 of the European Parliament and of the Council of 29 April 2004 on the Hygiene of Foodstuffs*. European Commission, Brussels, Belgium. https://eur-lex.europa.eu/eli/reg/2004/852/oj

European Commission. (2018). *Regulation (EU) No 1169/2011 of the European Parliament and of the Council of 25 October 2011 on the Provision of Food Information to Consumers, Amending Regulations (EC) No 1924/2006 and (EC) No 1925/2006 of the European Parliament and of the Council, and Repealing Commission Directive 87/250/EEC, Council Directive 90/496/EEC, Commission Directive 1999/10/EC, Directive 2000/13/EC of the European Parliament and of the Council, Commission Directives 2002/67/EC and 2008/5/EC and Commission Regulation (EC) No 608/2004 (Text with EEA relevance)Text with EEA Relevance.* European Commission, Brussels, Belgium. http://data.europa.eu/eli/reg/2011/1169/2018-01-01/eng

European Commission. (2023, October 29). *Commission Regulation (EU) No 231/2012 of 9 March 2012 Laying Down Specifications for Food Additives Listed in Annexes II and III to Regulation (EC) No 1333/2008 of the European Parliament and of the Council (Text with EEA Relevance) Text with EEA Relevance.* European Commission, Brussels, Belgium. http://data.europa.eu/eli/reg/2012/231/2023-10-29/eng

European Commission. (2024, April 23). *Regulation (EC) No 1333/2008 of the European Parliament and of the Council of 16 December 2008 on food additives (Text with EEA Relevance) Text with EEA Relevance.* European Commission, Brussels, Belgium. http://data.europa.eu/eli/reg/2008/1333/2024-04-23/eng

FAO & WHO. (1997). *FAO/WHO – Food and Agriculture Organization of the United Nations and the World Health Organization (). Food Consumption and Exposure Assessment of Chemicals: Report of a FAO/WHO Consultation, Geneva, Switzerland, 10–14 February 1997.* World Health Organization. Available online: https://apps.who.int/iris/bitstream/handle/10665/63988/WHO_FSF_FOS_97.5.pdf?sequence=1&isAllowed=y (accessed on 31 July 2024).

FAO, & WHO. (2023a). *Codex Alimentarius Commission Procedural Manual.* Twenty-eighth edition, revised. Rome.https://openknowledge.fao.org/handle/20.500.14283/cc5042en

FAO, & WHO. (2023b). *General Principles of Food Hygiene.* Codex Alimentarius Code of Practice, No. CXC 1-1969. Codex Alimentarius Commission. Rome. https://doi.org/10.4060/cc6125en

Fernandes, I., Rodrigues, J. A., & Almeida, C. M. M. (2022). Sulfites in fresh meat and meat preparations commercialized in Portugal. *Food Analytical Methods*, *15*(1), 172–184. https://doi.org/10.1007/s12161-021-02115-z

Frestedt, J. L. (2018). Chapter 16—Foods, food additives, and Generally Regarded as Safe (GRAS) food assessments. In A. M. Holban & A. M. Grumezescu (Eds.), *Food Control and Biosecurity* (pp. 543–565). Academic Press. https://doi.org/10.1016/B978-0-12-811445-2.00016-7

Gajdár, J., Herzog, G., & Etienne, M. (2022). Amperometric sensor for selective on-site analysis of free sulfite in wines. *ACS Sensors*, *7*(8), 2209–2217. https://doi.org/10.1021/acssensors.2c00611

Gómez, I., Janardhanan, R., Ibañez, F. C., & Beriain, M. J. (2020). The effects of processing and preservation technologies on meat quality: Sensory and nutritional aspects. *Foods*, *9*(10), 10. https://doi.org/10.3390/foods9101416

Gürtler, R. (2021). Risk assessment of food additives. In F.-X. Reichl & M. Schwenk (Eds.), *Regulatory Toxicology* (pp. 1323–1337). Springer International Publishing. https://doi.org/10.1007/978-3-030-57499-4_116

Hammes, W. P. (2012). Metabolism of nitrate in fermented meats: The characteristic feature of a specific group of fermented foods. *Food Microbiology*, *29*(2), 151–156. https://doi.org/10.1016/j.fm.2011.06.016

Hord, N. G., Tang, Y., & Bryan, N. S. (2009). Food sources of nitrates and nitrites: The physiologic context for potential health benefits. *The American Journal of Clinical Nutrition*, *90*(1), 1–10. https://doi.org/10.3945/ajcn.2008.27131

Ingenbleek, L., Lautz, L. S., Dervilly, G., Darney, K., Astuto, M. C., Tarazona, J., Liem, A. K. D., Kass, G. E. N., Leblanc, J. C., Verger, P., Bizec, B. L., & Dorne, J. L. C. M. (2020). Risk assessment of chemicals in food and feed: Principles, applications and future perspectives. In R. M. Harrison (ed), Book Environmental Pollutant Exposures and Public Health (pp. 1–38). The Royal Society of Chemistry. https://doi.org/10.1039/9781839160431-00001

Kanner, J. (1994). Oxidative processes in meat and meat products: Quality implications. *Meat Science*, *36*(1–2), 169–189. https://doi.org/10.1016/0309-1740(94)90040-X

Kumar Meher, A., Labhsetwar, N., & Bansiwal, A. (2018). An improved method for direct estimation of free cyanide in drinking water by Ion Chromatography-Pulsed Amperometry Detection (IC-PAD) on gold working electrode. *Food Chemistry, 240*, 131–138. https://doi.org/10.1016/j.foodchem.2017.07.041

Levy, R. B., Rauber, F., Chang, K., Louzada, M. L. D. C., Monteiro, C. A., Millett, C., & Vamos, E. P. (2021). Ultra-processed food consumption and type 2 diabetes incidence: A prospective cohort study. *Clinical Nutrition (Edinburgh, Scotland)*, *40*(5), 3608–3614. https://doi.org/10.1016/j.clnu.2020.12.018

Majou, D., & Christieans, S. (2018). Mechanisms of the bactericidal effects of nitrate and nitrite in cured meats. *Meat Science*, *145*, 273–284. https://doi.org/10.1016/j.meatsci.2018.06.013

Martins, F. C. O. L., Sentanin, M. A., & De Souza, D. (2019). Analytical methods in food additives determination: Compounds with functional applications. *Food Chemistry*, *272*, 732–750. https://doi.org/10.1016/j.foodchem.2018.08.060

Mercer, D. G., & Rodriguez-Amaya, D. B. (2021). Chapter 12—Reactions and interactions of some food additives. In D. B. Rodriguez-Amaya & J. Amaya-Farfan (Eds.), *Chemical Changes During Processing and Storage of Foods* (pp. 579–635). Academic Press. https://doi.org/10.1016/B978-0-12-817380-0.00012-9

Merino, L., Örnemark, U., & Toldrá, F. (2017). Analysis of nitrite and nitrate in foods: Overview of chemical, regulatory and analytical aspects. *Advances in Food and Nutrition Research*, *81*, 65–107. https://doi.org/10.1016/bs.afnr.2016.11.004

Neltner, T. G., Kulkarni, N. R., Alger, H. M., Maffini, M. V., Bongard, E. D., Fortin, N. D., & Olson, E. D. (2011). Navigating the U.S. Food additive regulatory program. *Comprehensive Reviews in Food Science and Food Safety*, *10*(6), 342–368. https://doi.org/10.1111/j.1541-4337.2011.00166.x

Ntrallou, K., Gika, H., & Tsochatzis, E. (2020). Analytical and sample preparation techniques for the determination of food colorants in food matrices. *Foods*, *9*(1), 1. https://doi.org/10.3390/foods9010058

Pagliano, E., Meija, J., Campanella, B., Onor, M., Iammarino, M., D'Amore, T., Berardi, G., D'Imperio, M., Parente, A., Mihai, O., & Mester, Z. (2019). Certification of nitrate in spinach powder reference material SPIN-1 by high-precision isotope dilution GC–MS. *Analytical and Bioanalytical Chemistry*, *411*(16), 3435–3445. https://doi.org/10.1007/s00216-019-01803-4

Pompa, C., D'Amore, T., Miedico, O., Preite, C., & Chiaravalle, A. E. (2021). Evaluation and dietary exposure assessment of selected toxic trace elements in durum wheat (Triticum durum) imported into the Italian market: Six years of official controls. *Foods*, *10*(4), 4. https://doi.org/10.3390/foods10040775

Rapid Alert System Feed and Food (RASFF). (2022). Available online: https://webgate.ec.europa.eu/rasff-window/screen/search (accessed on 31 July 2024).

Rohrer, J., & Avdalovic, N. (2021). Chapter 11—Ion chromatography methods that use electrochemical detection. In C. Pohl, N. Avdalovic, & K. Srinivasan (Eds.), *Separation Science and Technology* (Vol. *13*, pp. 287–301). Academic Press. https://doi.org/10.1016/B978-0-12-813075-9.00014-5

Salame, C., Javaux, G., Sellem, L., Viennois, E., de Edelenyi, F. S., Agaësse, C., De Sa, A., Huybrechts, I., Pierre, F., Coumoul, X., Julia, C., Kesse-Guyot, E., Allès, B., Fezeu, L. K., Hercberg, S., Deschasaux-Tanguy, M., Cosson, E., Tatulashvili, S., Chassaing, B., Srour, B., … & Touvier, M. (2024). Food additive emulsifiers and the risk of type 2 diabetes: Analysis of data from the NutriNet-Santé prospective cohort study. *The Lancet. Diabetes & Endocrinology*, *12*(5), 339–349. https://doi.org/10.1016/S2213-8587(24)00086-X

Scotter, M. J., & Castle, L. (2004). Chemical interactions between additives in foodstuffs: A review. *Food Additives & Contaminants*, *21*(2), 93–124. https://doi.org/10.1080/02652030310001636912

Sellem, L., Srour, B., Javaux, G., Chazelas, E., Chassaing, B., Viennois, E., Debras, C., Druesne-Pecollo, N., Esseddik, Y., Szabo de Edelenyi, F., Arnault, N., Agaësse, C., De Sa, A., Lutchia, R., Huybrechts, I., Scalbert, A., Pierre, F., Coumoul, X., Julia, C., Kesse-Guyot, E., … & Touvier, M. (2024). Food additive emulsifiers and cancer risk: Results from the French prospective NutriNet-Santé cohort. *PLoS Medicine*, *21*(2), e1004338. https://doi.org/10.1371/journal.pmed.1004338

Shakil, M. H., Trisha, A. T., Rahman, M., Talukdar, S., Kobun, R., Huda, N., & Zzaman, W. (2022). Nitrites in cured meats, health risk issues, alternatives to nitrites: A review. *Foods*, *11*(21), 3355. https://doi.org/10.3390/foods11213355

Shi, X. X., Wang, F., Wang, Z. Z., Huang, G. Y., Li, M., Simal-Gandara, J., Hao, G. F., & Yang, G. F. (2024). Unveiling toxicity profile for food risk components: A manually curated toxicological databank of food-relevant chemicals. *Critical Reviews in Food Science and Nutrition*, *64*(15), 5176–5191. https://doi.org/10.1080/10408398.2022.2152423

Shpaizer, A., Nussinovich, A., Kanner, J., & Tirosh, O. (2018). S-Nitroso-N-acetylcysteine generates less carcinogenic N-nitrosamines in meat products than nitrite. *Journal of Agricultural and Food Chemistry*, *66*(43), 11459–11467. https://doi.org/10.1021/acs.jafc.8b04549

Sofos, J. N. (2008). Challenges to meat safety in the 21st century. *Meat Science*, *78*(1), 3–13. https://doi.org/10.1016/j.meatsci.2007.07.027

Sun, B., & Wang, J. (2017). Food additives. In J.J. Jen & J. Chen (eds.), *Food Safety in China* (pp. 186–200). John Wiley & Sons, Ltd. https://doi.org/10.1002/9781119238102.ch12

Turasan, H., & Kokini, J. (2021). Novel nondestructive biosensors for the food industry. *Annual Review of Food Science and Technology*, *12*, 539–566. https://doi.org/10.1146/annurev-food-062520-082307

Wang, W., Wang, Y., Zhang, J., Chu, Q., & Ye, J. (2010). Simultaneous determination of electroactive and non-electroactive food preservatives by novel capillary electrophoresis with amperometric detection. *Analytica Chimica Acta*, *678*(1), 39–43. https://doi.org/10.1016/j.aca.2010.08.018

WHO – World Health Organization. (2018). Food Safety Collaborative Platform (CIFOCOss). Available online: https://apps.who.int/foscollab/Download/DownloadConso (accessed on 31 July 2024).

WHO, FAO. (2020). *Compendium of Food Additive Specifications. Joint FAO/WHO Expert Committee on Food Additives (JECFA), 87th Meeting June 2019.* https://openknowledge.fao.org/handle/20.500.14283/ca7513en

WHO, FAO. (2024). *GSFA Online Food Additive Functional Classes*. FAO and WHO. www.fao.org/gsfaonline/reference/techfuncs.html

Woolford, G., & Cassens, R. G. (1977). The fate of sodium nitrite in bacon. *Journal of Food Science*, *42*(3), 586–589. https://doi.org/10.1111/j.1365-2621.1977.tb12555.x

Wu, L. (2021). Chapter 7—Analysis of food additives. In C. M. Galanakis (Ed.), *Innovative Food Analysis* (pp. 157–180). Academic Press. https://doi.org/10.1016/B978-0-12-819493-5.00007-8

Wu, L., Zhang, C., Long, Y., Chen, Q., Zhang, W., & Liu, G. (2022). Food additives: From functions to analytical methods. *Critical Reviews in Food Science and Nutrition*, *62*(30), 8497–8517. https://doi.org/10.1080/10408398.2021.1929823

Xie, Y., Geng, Y., Yao, J., Ji, J., Chen, F., Xiao, J., Hu, X., & Ma, L. (2023). N-nitrosamines in processed meats: Exposure, formation and mitigation strategies. *Journal of Agriculture and Food Research*, *13*, 100645. https://doi.org/10.1016/j.jafr.2023.100645

Zenodo. (2013). Food Additives Intake Model (FAIM), Version 1.1. July 2013. Available online: https://zenodo.org/record/154725 (accessed on 31 July 2024).

Mycotoxins and Safety

14

Food Processing and Manufacturing Related Effects

Sofia Agriopoulou and Maria Tarapoulouzi

14.1 INTRODUCTION

Mycotoxins are already-known natural products of low molecular weight that are synthesized as secondary metabolites of filamentous fungi. Mycotoxins are a huge global food safety concern, with hundreds of identified species as they can be synthesized by fungi directly in the field, already on the plants during the growing season (if they remain in the agricultural products at the time of their collection), or produced during storage or technological processes (Nešić et al., 2023). The main producers of mycotoxins are species that belong to the genera *Aspergillus*, *Alternaria*, *Fusarium,* and *Penicillium* (Furlong et al., 2024). *Claviceps, Mucor, Trichoderma, Trichothecium, Myrothecium, Pyrenophora*, and *Stachybotrys* are also filamentous fungi that are produce mycotoxins (Khan, 2024). Mycotoxin contamination is common, with up to 80% of food-crops globally estimated to be contaminated (Eskola et al., 2020).

Currently, aflatoxins (AFs), trichothecenes (TCs) [such as deoxynivalenol (DON), T-2 toxin (T-2), HT-2 toxin (HT-2)], zearalenone (ZEN), fumonisins (FBs), ochratoxin A (OTA), ergot alkaloids (EAs), patulin (PAT), citrinin (CIT), and enniatins (ENs) are considered to be the mycotoxins of the greatest importance from food safety and regulatory viewpoints as they are extremely harmful to the health of humans and animals

DOI: 10.1201/9781032712291-14

(Agriopoulou et al., 2016; Agriopoulou 2016; Agriopoulou et al., 2020a; Agriopoulou et al., 2020b; Agriopoulou, 2021; Agriopoulou, 2023; Kumar et al., 2022; Smaoui et al., 2020; Smaoui et al., 2023a; Smaoui et al., 2023b; Smaoui et al., 2023c). Emerging mycotoxins such as beauvericin (BEA), moniliformin (MON), and fusaproliferin (FUS), have also been found in a number of agricultural products and animal feed (Llorens et al., 2024).

In order to produce safe agricultural raw material without mycotoxin contamination, pre-harvest and post- harvest factors must be controlled, although this is rather complicated (Lee et al., 2024). When managing mycotoxins in the food supply chain, it is important to explore methods to mitigate its adverse consequences. The term mitigation applied to mycotoxins means taking actions to minimize their toxic effect on consumers' health (Karlovsky et al., 2016; Liu et al., 2022; Owolabi et al., 2022). The term degradation is related to treatments that cause great and beneficial alteration (toxicity reduction) to mycotoxin structures (Furlong et al., 2024). There are three strategies to mitigate mycotoxins, namely, pre-harvest, post-harvest, and processing strategies (Llorens et al., 2024).

One strategy is to reduce the concentration of mycotoxins during food processing, which helps minimize the potential for consumer exposure to higher concentrations of mycotoxins. To mitigate the levels of mycotoxins in food products for human consumption, the food industry can consider two main approaches: nonthermal and thermal food processing techniques. Regardless of the approach chosen, it is important to choose a technically feasible, scalable, and cost-effective approach that does not produce harmful byproducts. In addition, it should not adversely affect other food quality elements such as flavor, taste, color and nutritional value. This chapter explores cutting-edge processing techniques that can be employed by the food industry to minimize mycotoxins, to mitigate its toxic effects.

14.2 OCCURRENCE OF MYCOTOXINS IN FOOD AND FEED

Mycotoxins exist in various agricultural commodities, worldwide, like peanuts, grapes, and wines, grains, nuts, dried fruit, coffee, cocoa, spices, oil seeds, fruits, fruit juices, beer, infant formulas, yogurt and other foodstuffs, and feed crops, both in the field and during transportation, posing a risk to food safety and human health (Figure 14.1) (Varzakas et al., 2016; Agriopoulou et al., 2020a; Wang et al., 2023). The frequency of mycotoxins increases at latitude between 26° and 35° north and south (Sun et al., 2023).

Mycotoxins present in contaminated feed might be transferred to animal products, thereby posing significant health risks to humans, especially infants (Bao et al., 2024). Since feed represents one of the most traded products, mycotoxin-contaminated feed may occur anywhere in the world. Aquatic feeds also derived from cereals are severely affected by mycotoxins, as cereals are a major category of products contaminated with mycotoxins (Smaoui et al., 2023b). Data of occurrence of mycotoxins in food and feed around the world are listed in Table 14.1.

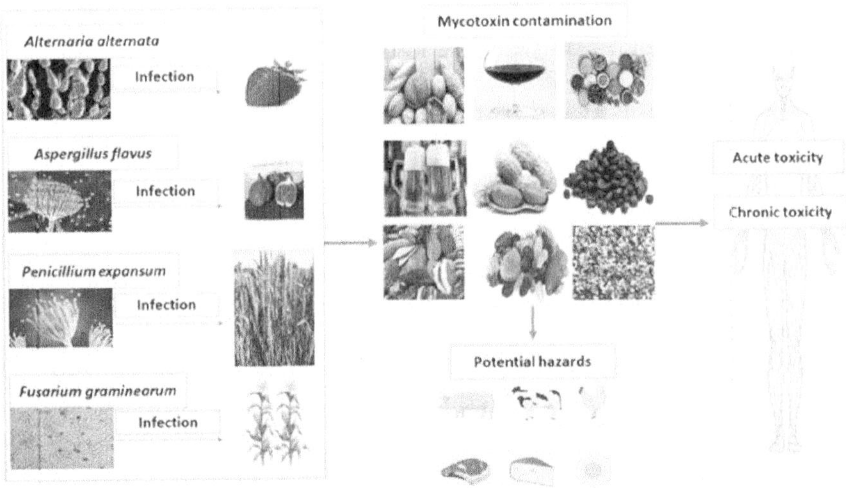

FIGURE 14.1 Mycotoxin contamination in food and feed and its hazard to human health.

TABLE 14.1 Occurrence of mycotoxins in food and feed around the world

FOOD/FEED MATRIX	*MYCOTOXIN*	*COUNTRY*	*RANGE (MG/KG OR MG/L)*	*DETECTION TECHNIQUE*	*REFERENCE*
Oat beverage	DON	Latvia	0.58–8.6	QuEChERS-LC-MS/MS	(Pavlenko et al., 2024)
Family cereal	CIT	Nigeria	1.2–151	LC-MS/MS	(Ojuri et al., 2019)
Almond	CIT	India	2.80–18.20	ELISA	(Kumar, 2019)
Coffee	OTA	Saudi Arabia	2.2–5.6	HPLC-FLD	(Almasoud et al., 2022)
Kalaharituber pfeilii	ZEA	Namibia	45–9680	ELISA	(Hainghumbi et al., 2022Z)
Sorghum	ZEA	Nigeria	<LOQ- 20	LC-MS/MS	(Olopade et al., 2021)
Fish feed	ZEA	Kenya	<38–757.9	HPLC	(Mwihia et al., 2020)
Rice	ZEA	Algeria	8.6–15.5	UHPLC-MS/MS	(Mahdjoubi et al., 2020)
Granola	ZEA	Nigeria	0.81–5.99	LC-MS/MS	(Ezekiel et al., 2020)
Cobs	ZEA	Côte d'Ivoire	38.61–234.87	HPLC	(Bamba et al., 2020)
Almond beverage	AFB1	Spain	<0.04–0.04	QuEChERS-LC-MS/MS	(Rodriguez Cañás et al., 2024)

TABLE 14.1 (Continued)

FOOD/FEED MATRIX	*MYCOTOXIN*	*COUNTRY*	*RANGE (MG/KG OR MG/L)*	*DETECTION TECHNIQUE*	*REFERENCE*
Agriculture maize fields	FB1	Portugal	426.7–5988.6	QuEChERS-LC-MS/MS	(Leite et al., 2024)
Plant-based meat alternatives	FB1	Italy	5.6–160.7	UHPLC-MS/MS	(Mihalache et al., 2024)

ELISA: enzyme-linked immunosorbent assay; UHPLC-MS/MS: Ultra-high-performance liquid chromatography tandem mass spectrometry; LC-MS/MS: liquid chromatography tandem mass spectrometry; QuEChERS: quick, easy, cheap, effective, rugged, and safe; CIT: Citrinin; AFB1: Aflatoxin B1; DON: Deoxynivalenol; FB1: Fumonisin B1; HPLC-FLD: high-performance liquid chromatography-fluorescence detector; ZEA:

14.3 FOOD PROCESSING METHODS AND THEIR EFFECTS: THE FATE OF MYCOTOXINS IN FOODS

14.3.1 Nonthermal Methods

Techniques such as sorting, cleaning, drying, and milling are employed to physically remove or reduce the concentration of mycotoxins. Sorting and cleaning help eliminate visibly contaminated grains, while drying and milling can reduce moisture content, thereby inhibiting fungal growth and mycotoxin production. These methods are often the first line of defense in post-harvest mycotoxin management (Llorens et al., 2024).

14.3.1.1 Separation and cleaning methods

After harvest, primary intervention methods such as the physical separation and cleaning of agricultural commodities can remove unwanted materials, such as dust, chaff, and small fragments, from agricultural commodities (Lee, H.J, et al., 2024). Mansouri-Nasrabadi et al., determined the effect of washing on OTA in rice and reported that about 26% of the OTA in rice was reduced by one-time washing and 39% of OTA was washed by two-time washing, and 43% of OTA decreased by three-time washing (Mansouri-Nasrabadi et al., 2018). Cleaning and sorting of damaged, discolored, and shriveled kernels reduced *Fusarium* toxins in grains. *Fusarium* toxins in maize, namely DON, ZEN and fumonisins B1 and B2 (FBs), was evaluated at an industrial-scale level by mechanical removal (sieving and density separation) of dust, coarse, small, broken, shriveled and low-density kernels and/or optical sorting of defected kernels. Depending on the contamination levels in raw incoming maize, the overall reduction rates ranged from 36 to 67% for DON, from 67 to 87% for ZEA and from 27 to 67% for FBs (Pascale et al., 2022). A low-cost sorting tool (Dropsort device) that was developed by Aoun et al., was able to separate grains based on bulk kernel density and 100 kernel weight, combined with size sorting, and was effective in reducing fumonisin concentration below 2 ppm (Aoun et al., 2020).

14.3.1.2 Milling

No operation can destroy mycotoxins in the milling process, but mycotoxin contamination, including OTA, can be redistributed and concentrated in some milled fractions such as bran and shorts (Lee et al., 2024). Massarolo et al. (Massarolo et al., 2022), and (Massarolo et al., 2020) studied the effects of parameters of corn processing and preparation on FB1 levels and bioacessibility. The authors found that, after dry milling, 23% of mycotoxins remain in the endosperm.

14.3.1.3 Brewing/Fermentation

Specific microbial strains can be used through fermentation processes in order to reduce mycotoxin levels in food. The most used microbial strains that are used in fermentation are lactic acid bacteria, which can bind to mycotoxins, and reduce their bioavailability (Adebo et al., 2019). Lactic acid bacteria, have been effective in binding and reducing aflatoxin and fumoniisin level in contaminated grains and ZEN and OTA through physical binding (Llorens et al., 2024). Yeasts, such as *Saccharomyces cerevisiae,* have inhibited the production of mycotoxins and specifically reduced patulin levels during fermentation (Yang et al., 2024). An effective reduction in the content of ZEA was observed in bread prepared by baking with the addition of yeast, ranging from 14.3% to 35.4% (Podgórska- Kryszczuk et al., 2022).

14.3.1.4 De-hulling

De-hulling is a widespread food processing technique used in Africa and has successfully been used to reduce mycotoxin concentrations in finished food (Hudu et al., 2024). In Malawi, Njombwa et al. evaluated occurrence levels of ZEA in dairy cattle concentrate feed and observed that 75% (83 out of 111) of corn bran tested positive for ZEA with minimum, maximum, and median concentration levels of 100, 2,400, and 240 μg/kg, respectively (Njombwa et al., 2020).

14.3.2 Thermal Methods

Mycotoxins are relatively stable compounds during conventional thermal food processing (80–121 °C) such as boiling, steaming, frying, and baking along with some relatively new approaches i.e., microwave heating. However, certain thermal food processing procedures with high temperatures and pressures, e.g., extrusion processing, can effectively reduce mycotoxin levels and their toxicities. In general, there are lots of important factors related to the degradation of and reduction in mycotoxins in food during food processing, such as the type of food matrix, the initial mycotoxin concentration, the heating temperature together with the time employed, the type of heat transfer, the degree of heat penetration, the pH, the moisture content, the nutritional composition such as fiber and fat levels, the type of additives, etc. (Lee et al., 2024).

14.3.2.1 Roasting

Roasting, in general, uses dry heat, often above 160°C, which can lead to a significant reduction of mycotoxins, including OTA. The thermal stability of OTA during roasting is of particular interest to the coffee industry due to the prevalence and high levels of contamination in coffee beans (Lee et al., 2024). Carbon and Lee (Carbon & Lee, 2022) reported that OTA was reduced less in brown rice (37%) compared to white rice (60%) when cooked at 160 °C and 200 °C for up to 30 min, suggesting that the outer fibrous layer or bran can provide insulation and prevent heat transfer.

In Daba et al., study where the impact of traditional conditions and processing methods on aflatoxin content of cooked maize was studied, it was found that roasting the soaked kernel resulted in significantly lower aflatoxin content to boiling the soaked maize kernel (Daba et al., 2024).

Table 14.2 summarizes the impacts of thermal food processing on different mycotoxins levels

14.3.2.2 Coffee brewing

Through coffee brewing, it is also possible to reduce OTA contamination levels even it is not significant (Lee et al., 2024). Malir et al. conducted a study demonstrating the reduction of OTA with different brewing methods, such as fake Turkish coffee, Turkish coffee, Lungo, Americano, Espresso, Doppio, and Ristretto, and found different degrees of reduction from 34% to 78% depending on the brewing method. In fact, the reduction of OTA in the coffee drink is not due to the reduction of OTA during the coffee brewing process and may simply be a transfer of OTA content from ground and roasted coffee beans to the coffee drink. It depends on several factors, such as the amount (or volume)

TABLE 14.2 The impacts of thermal food processing on different OTA levels

THERMAL PROCESSING	*MATRIX*	*REDUCTION (%)*	*REFERENCE*
Roasting	White/brown rice	15–60% (OTA)	(Carbon & Lee, 2022)
	Oat	2–18% (OTA)	(Lee, 2020)
Explosive puffing	Oat	38–52% (OTA)	(Lee et al., 2019)
	Rice	15–28% (OTA)	
Retorting	Oat	17% (OTA)	(Lee et al., 2020)
	Rice	54% (OTA)	
Direct steam injection	Oat	20–28% (OTA)	(Lee et al., 2019)
Indirect steaming	Oat	13%	(Lee et al., 2021)

of water and ground and roasted coffee beans used and the brewing time (or contact time between water and ground and roasted coffee beans) (Malir et al., 2014).

14.3.2.3 *Extrusion*

Extrusion processing combines different unit operations like cooking, kneading, shearing, shaping, and forming. During extrusion processing, high temperature, high pressure, and severe shear forces are generated that lead to chemical reactions and molecular modification of food components over a short period (Deng et al., 2021). Understanding the parameters affecting the efficacy of extrusion in mycotoxin removal is essential for the development of safe and high-quality cereal-based food products. This process requires the consideration of various extrusion parameters that influence mycotoxin reduction while preserving the quality and nutritional value of the final product, such as temperature, raw material moisture, screw speed, and pressure. The optimal temperature for reducing mycotoxins may vary depending on the type of mycotoxin and the raw material being processed. The pressure applied during extrusion is another crucial parameter affecting mycotoxin reduction (Janić Hajnal et al., 2024).

The mitigation of Alternaria toxins was investigated by Janić Hajnal, et al., through extrusion processing of whole grain red sorghum flour, and their research results showed a reduction of 61.4, 76.4, 12.1 and 50.8% for alternariol (AOH), alternariol monomethyl ether (AME), tenuazonic acid (TeA), and tentoxin (TEN) respectively (Janić Hajnal et al., 2024).

14.3.2.4 *Other innovative methods*

New detoxification strategies include the use of nanoparticles and the use of plant extracts (Awuchi et al., 2021). Nanoparticles such as magnetic carbon nonocomposites and chitosan-coated Fe_3O_4 nanoparticles have shown potential for mycotoxin removal, specifically patulin and AFB1, respectively (Zhou et al., 2020). Plant extracts, which can be incorporated into plant-based alternatives (PBMA) formulations, to inhibit the growth of mycotoxin-producing fungi during storage and shelf-life, such as essential oils exhibit antifungal properties and can inhibit mycotoxin production (Perczak et al., 2019).

Emerging green technologies, such as high pressure processing (HPP), pulsed electric field (PEF) and ultrasonic methods are effective for controlling fungi and mycotoxins in food. These technologies are applicable to a variety of food products, including canned products, meat products, fresh fruits and vegetables, and those that are sensitive to heat or contain bioactive compounds (Alizadeh et al., 2021), as well as PBMA, as they do not affect the texture and taste.

14.4 CONCLUSIONS AND FUTURE PERSPECTIVES

Mycotoxins are highly chemically stable, and once present in foods, they are difficult to remove, endangering both human and animal health. Ensuring the safety of crops from toxigenic fungi is of utmost importance to maintain a food supply free from mycotoxins contamination. This requires the implementation of preventive measures both in the field and throughout storage. Addressing this challenge is complex, as the fungi that produce mycotoxins are naturally occurring, making it difficult to prevent their contamination and the subsequent production of mycotoxins. It should be noted that the occurrence of toxigenic fungi does not always correspond to contamination levels, as external factors can inhibit mycotoxins production. Mycotoxins can be detected in raw agricultural products even without active mycotoxins producers. This can occur through various chemical processes or environmental conditions that can inactivate fungi while preserving mycotoxins within the product. Although nonthermal food processing technologies have exhibited their power to effectively eliminate a large number of different mycotoxins, in processed foods, regulations for its use in mycotoxin management are limited. More research will help evaluate the safety of these technologies (ozone, irradiation, dielectric barrier discharge) for mycotoxin management in food storage and processing and enhance consumer acceptance.

REFERENCES

Adebo, O.A., Kayitesi, E. and Njobeh, P.B. (2019). Reduction of mycotoxins during fermentation of whole grain sorghum to whole grain ting (a Southern African Food). *Toxins,* 11(3), 180.

Agriopoulou, S. (2016). Enniatins: An emerging food safety issue. *EC Nutrition*, 3, 1142–1146.

Agriopoulou, S. (2023). Ergot Alkaloids and Anti-Mycotoxin Strategies in Food and Feed. In *Anti-Mycotoxin Strategies for Food and Feed;* D. Nagaraju, S.M. Yanjarappa, P.N. Achar and A.M. Vaya., Eds.; Wiley. pp. 237–254. ISBN: 9781394160839. https://doi.org/10.1002/9781394160839.ch11.

Agriopoulou, S. (2021). Ergot alkaloids mycotoxins in cereals and cereal-derived food products: Characteristics, toxicity, prevalence, and control strategies. *Agronomy,* 11, 931.

Agriopoulou, S., Stamatelopoulou, E. and Varzakas, T. (2020a). Advances in occurrence, importance, and mycotoxin control strategies: Prevention and detoxification in foods. *Foods*, 9, 137

Agriopoulou, S., Stamatelopoulou, E. and Varzakas, T. (2020b). Advances in analysis and detection of major mycotoxins in foods. *Foods*, 9, 518.

Agriopoulou, S., Koliadima, A., Karaiskakis, G. and Kapolos, J. (2016). Kinetic study of aflatoxins' degradation in the presence of ozone. *Food Control,* 61, 221–226.

Alizadeh, A.M., Hashempour-Baltork, F., Khaneghah, A.M. and Hosseini, H. (2021). New perspective approaches in controlling fungi and mycotoxins in food using emerging and green technologies. *Current Opinion in Food Science*, 39, 7–15.

Almasoud, Y.F., Aldosari, Z.M., Almutairi, M.A. and Alowaifeer, A.M. (2022). Occurrence of ochratoxin A in green and roasted coffee beans from the Saudi market and a possible exposure assessment. *JSFA Reports,* 2, 255–260.

Aoun, M., Stafstrom, W., Priest, P., Fuchs, J., Windham, G.L., Williams, W.P. and Nelson, R.J. (2020). Low-cost grain sorting technologies to reduce mycotoxin contamination in maize and groundnut low-cost grain sorting technologies to reduce mycotoxin contamination in maize and groundnut. *Food Control,* 118, 107363.

Awuchi, C.G., Ondari, E.N., Eseoghene, I.J., Twinomuhwezi, H., Amagwula, I.O. and Morya, S. (2021). Fungal growth and mycotoxins production: Types, toxicities, control strategies, and detoxification. In *Fungal Reproduction and Growth;* S. Sultan and G. Kaur Surindar Singh, Eds.;. IntechOpen. P. 100207.

Bamba, S., Marius, H., Biego, G. and Ake, A.Y. (2020). Assessment of the health quality related to the presence of ochratoxin A, fumonisin B1 and zearalenone in maize (*Zea mays* L.) produced in Côte D' Ivoire. *Asian Journal of Research in Crop Science,* 5, 41–51.

Bao, Y., Dong, T., Shao, T., Wang, W., Varzakas, T., Agriopoulou, S. and Yuan, X. (2024). Antifungal and mycotoxin detoxification potentials of *Acetobacter tropicalis* AT7 and *Lactiplantibacillus plantarum* LP64 in whole-plant corn silage. *Animal Feed Science and Technology,* 313, 115987.

Carbon, H. and Lee, H. (2022). Varied reduction of ochratoxin A in brown and white rice during roasting. *World Mycotoxin Journal*, 15(4), 361–368.

Daba, H.G., Delele, M.A., Fanta, S.W., Habtu, N.G. and Abera, M.K. (2024). Impact of traditional processing conditions and methods on the aflatoxin content of cooked maize. *Cogent Food & Agriculture*, 10(1), 2297516.

Deng, L.Z., Sutar, P.P., Mujumdar, A.S., Tao, Y., Pan, Z., Liu, Y.H. and Xiao, H.W. (2021). Thermal decontamination technologies for microorganisms and mycotoxins in low-moisture foods. *Annual Review of Food Science and Technology,* 12(1), 287–305.

Eskola, M., Kos, G., Elliott, C. T., Hajšlová, J., Mayar, S. and Krska, R. (2020). Worldwide contamination of food-crops with mycotoxins: Validity of the widely cited 'FAO estimate' of 25%. *Critical Reviews in Food Science and Nutrition*, 60, 2773–2789.

Ezekiel, C.N., Oyedele, O.A., Kraak, B., Ayeni, K.I., Sulyok, M., Houbraken, J. and Krska, R. (2020). Fungal diversity and mycotoxins in low moisture content ready- to- eat foods in Nigeria. *Frontiers in Microbiology,* 11, 615.

Furlong, E.B., Buffon, J.G., Cerqueira, M.B. and Kupski, L. (2024). Mitigation of Mycotoxins in Food—Is It Possible? *Foods*, 13, 1112.

Hainghumbi, T.A., Embashu, W., Nantanga, K.K.M., Kadhila, N.P. and Iipumbu, L. (2022). Assessment of microbiological properties, my-cotoxins, and heavy metals in underprized raw Kalahari trufflessold in Namibia. *Journal of Food Quality and Hazards Control*, 9, 23–31. https://doi.org/10.18502/jfqhc.9.1.9687

Hudu, A.R., Addy, F., Mahunu, G.K., Abubakari, A.H. and Opoku, N. (2024). Zearalenone contamination in maize, its associated producing fungi, control strategies, and legislation in Sub-Saharan Africa. *Food Science and Nutrition,* 12, 4489–4512.

Janić Hajnal, E., Babič, J., Pezo, L., Banjac, V., Filipčev, B., Miljanić, J., Kos, J. and Jakovac-Strajn, B. (2024). Reduction of *Alternaria* toxins via the extrusion processing of whole-grain red Sorghum flour. *Foods,* 13(2), 255.

Karlovsky, P., Suman, M., Berthiller, F., De Meester, J., Eisenbrand, G., Perrin, I., Oswald, I.P., Speijers, G., Chiodini, A. and Recker, T. (2016). Impact of food processing and detoxification treatments on mycotoxin contamination. *Mycotoxin Research,* 32, 179–205.

Khan, R. (2024). Mycotoxins in food: Occurrence, health implications, and control strategies-A comprehensive review. *Toxicon*, 248, 108038.

Kumar, M. (2019). Quantitative estimation of aflatoxins, ochratoxin and citrinin in dried fruits and nuts samples from Indo-Gangetic region of Bihar. *International Journal of Advance Research and Innovative Ideas in Education*, 5, 2395–4396.

Kumar, P., Mahato, D.K., Gupta, A., Pandey, S., Paul, V., Saurabh, V. and Agriopoulou, S. (2022). Nivalenol mycotoxin concerns in foods: an overview on occurrence, impact on human and animal health and its detection and management strategies. *Toxins,* 14, 527.

Lee, C., Agriopoulou, S. and Varzakas, T. (2024). Pathways to Implementing Food Systems Capacity Building Programs for Smallholder Farmers: Major Constraints and Benefits. *Food Safety Magazine*, (June-July 2024 Issue) Management Column. Available online. www.food-safety.com/articles/9573-pathways-to-implementing-food-systems-capacity-building-programs-for-smallholder-farmers-major-constraints-and-benefits (accessed on 29 October 2024).

Lee, H.J., Gu, B.-J., Ganjyal, G. and Ryu, D. (2019). Reduction of ochratoxin A in direct steam-injected oat-based infant cereals with baking soda. *Food Control*, 96, 441–444.

Lee, H.J., Kim, H.D., Ryu, D. (2024). Practical Strategies to Reduce Ochratoxin A in Foods. *Toxins*, 16, 58.

Lee, H.J., Lee, C. and Ryu, D. (2020). Effects of baking soda and fructose in reduction of ochratoxin A in rice and oat porridge during retorting process. *Food Control*, 116, 107325.

Lee, H.J., Li, S., Gu, K. and Ryu, D. (2021). Reduction of ochratoxin A during the preparation of porridge with sodium bicarbonate and fructose. *Toxins*, 13, 224.

Leite, M., Freitas, A., Barbosa, J. and Ramos, F. (2024). Mycotoxins in Portuguese agricultural maize fields and dairy farms. *Toxins,* 16(8), 335.

Liu, W.-C., Pushparaj, K., Meyyazhagan, A., Arumugam, V.A., Pappuswamy, M., Bhotla, H.K., Baskaran, R., Issara, U., Balasubramanian, B. and Khaneghah, A.M. (2022). Ochratoxin A as an alarming health threat for livestock and human: A review on molecular interactions, mechanism of toxicity, detection, detoxification, and dietary prophylaxis. *Toxicon*, 213, 59–75.

Llorens, P., Juan-García, A., Juan, C., Moltó, J.C., Pena, A. and Silva, L.J.G. (2024). Mycotoxins contamination and their possible mitigation strategies in Plant-Based Meat Alternatives (PBMAs): A review. *Food Bioscience,* 62, 105230.

Mahdjoubi, C.K., Arroyo-Manzanares, N. and Hamini-Kadar, N. (2020). Multi- mycotoxin occurrence and exposure assessment approach in foodstuffs from Algeria. *Toxins,* 12, 1–18.

Malir, F., Ostry, V., Pfohl-Leszkowicz, A., Toman, J., Bazin, I., Roubal, T. (2014). Transfer of ochratoxin A into tea and coffee beverages. *Toxins,* 6, 3438–3453.

Mansouri-Nasrabadi, R., Milani, J.M., Nazari, S.S.J. (2018). Optimization of washing and cooking processes of rice for Ochratoxin A decrement by RSM. *Food Science and Nutrition,* 6, 2523–2529.

Massarolo, K.C., Rodrigues, P., Ferreira, C.F.J., Kupski, L., Badiale-Furlong, E. (2022). Simultaneous distribution of aflatoxins B1 and B2, and fumonisin B1 in corn fractions during dry and wet-milling. *Journal of Food Science and Technology,* 59, 3192–3200.

Massarolo, K.C., Ferreira, C.F., Collazzo, C.C., Bianchini, A., Kupski, L., Badiale-Furlong, E. (2020). Resistant starch and hydrothermal treatment of cornmeal: Factors in aflatoxins and fumonisin B1 reduction and bioaccessibility. *Food Control*, 114, 107274.

Mihalache, O. A., Torrijos, R. and Dall'Asta, C. (2024). Occurrence of mycotoxins in meat alternatives: Dietary exposure, potential health risks, and burden of disease. *Environment International,* 185, 108537.

Mwihia, E.W., Lyche, J.L., Mbuthia, P.G., Ivanova, L., Uhlig, S., Gathumbi, J.K., Maina, J.G., Eshitera, E.E. and Eriksen, G.S. (2020). Co-occurrence and levels of mycotoxins in fish feeds in Kenya. *Toxins,* 12, 1–31.

Nešić, K., Habschied, K. and Mastanjević, K. (2023). Modified mycotoxins and multitoxin contamination of food and feed as major analytical challenges. *Toxins*, 15, 511.

Njombwa, C.A., Hamie, J.C. and Banda, M. (2020). Occurrence of total aflatoxin and zearalenone in dairy cattle concentrate feeds in Malawi. *Research Square,* 14, 1–15.

Ojuri, O.T., Ezekiel, C.N., Eskola, M.K., Šarkanj, B., Babalola, A.D., Sulyok, M., Hajšlová, J., Elliott, C.T. and Krska, R. (2019). Mycotoxin co-exposures in infants and young children consuming household-and industrially-processed complementary foods in Nigeria and risk management advice. *Food Control*, 98, 312–322.

Olopade, B.K., Oranusi, S.U., Nwinyi, O.C., Gbashi, S. and Njobeh, P.B. (2021). Occurrences of deoxynivalenol, zearalenone and some of their masked forms in selected cereals from Southwest Nigeria. *Nutrition and Food Science Journal,* 23, 24–29.

Owolabi, I.O., Kolawole, O., Jantarabut, P., Elliott, C.T. and Petchkongkaew, A. (2022). The importance and mitigation of mycotoxins and plant toxins in Southeast Asian fermented foods. *NPJ Science of Food,* 6, 39.

Pascale, M., Logrieco, A.F., Lippolis, V., De Girolamo, A., Cervellieri, S., Lattanzio, V.M.T., Ciasca, B., Vega, A., Reichel, M., Graeber, M. and Slettengren, K. (2022). Industrial-scale cleaning solutions for the reduction of *Fusarium* toxins in maize. *Toxins,* 14, 728.

Pavlenko, R., Berzina, Z., Reinholds, I., Bartkiene, E. and Bartkevics, V. (2024). An occurrence study of mycotoxins in plant-based beverages using liquid chromatography–mass spectrometry. *Toxins,* 16(1), 53.

Perczak, A., Juś, K., Gwiazdowska, D., Marchwińska, K. and Waśkiewicz, A. (2019). The efficiency of deoxynivalenol degradation by essential oils under in vitro conditions. *Foods,* 8(9), 403.

Podgórska- Kryszczuk, I., Solarska, E. and Kordowska- wiater, M. (2022). Nivalenol and zearalenone by selected non-conventional. *Molecules,* 27, 1578.

Rodríguez-Cañás, I., González-Jartín, J.M., Alfonso, A., Alvariño, R., Vieytes, M.R., and Botana, L.M. (2024). Application of a multi-toxin detect method to analyze mycotoxins occurrence in plant-based beverages. *Food Chemistry,* 434, 137427.

Smaoui, S., Bra, O.B., Hlima, H.B. (2020). Mycotoxins analysis in cereals and related foodstuffs by liquid chromatography-tandem mass spectrometry techniques. *Journal of Food Quality,* 2020, 1–23.

Smaoui, S., D'Amore, T., Tarapoulouzi, M., Agriopoulou, S. and Varzakas, T. (2023a). Aflatoxins contamination in feed commodities: From occurrence and toxicity to recent advances in analytical methods and detoxification. *Microorganisms,* 11, 2614.

Smaoui, S., D'Amore, T., Agriopoulou, S. and Mousavi Khaneghah, A. (2023b). Mycotoxins in seafood: occurrence, recent development of analytical techniques and future challenges. *Separations,* 10, 217.

Smaoui, S., Agriopoulou, S., D'Amore, T., Tavares, L. and Mousavi Khaneghah, A. (2023c). The control of Fusarium growth and decontamination of produced mycotoxins by lactic acid bacteria. *Critical Reviews in Food Science and Nutrition,* 63, 11125–11152.

Sun, L., Li, R., Tai, B., Hussain, S., Wang, G., Liu, X. and Xing, F. (2023). Current status of major mycotoxins contamination in food and feed in Asia— a review. *ACS Food Science and Technology,* 3(2), 231–244.

Varzakas, T. (2016). Quality and safety aspects of cereals (Wheat) and their products. *Critical Reviews in Food Science and Nutrition,* 56, 2495–2510.

Wang, J., Zhang, Y.J., Yuan, J.J., Shao, W.Q., Peng, L.X., Chen, W.L. (2023). Contamination status and health risk assessment of aflatoxin B1 in foodstuffs in Yunfu City, 2019–2022 South China. *Journal of Preventive Medicine,* 50, 216–221.

Yang, C., Zhang, Z. and Peng, B. (2024). New insights into searching patulin degrading enzymes in Saccharomyces cerevisiae through proteomic and molecular docking analysis. *Journal of Hazardous Materials,* 463, 132806.

Zhou, Y., Wu, S., Wang, F., Li, Q., He, C., Duan, N. and Wang, Z. (2020). Assessing the toxicity *in vitro* of degradation products from deoxynivalenol photocatalytic degradation by using upconversion nanoparticles@ TiO2 composite. *Chemosphere*, 238, Article 124648.

15 Agrochemicals and Safety

Food Processing and Manufacturing Related Effects

Patroklos Vareltzis and Nikolaos Sidirokastritis

15.1 INTRODUCTION

As the global population continues to grow, the industrialization of agriculture and the expansion of livestock production to meet rising food demands present both opportunities and challenges for food safety. These challenges place greater responsibility on food manufacturers and processors to ensure food is safe and free from contamination before reaching consumers. Plant protection and veterinary formulations play an important role in the primary sector and in public health in general (Aktar et al., 2009). The ongoing development of agriculture has led to a significant increase in agrochemicals use worldwide, aimed at minimizing crop losses and improving productivity and quality. Large quantities of these formulations are produced and used annually to manage populations of pests, parasites, and diseases of both humans and living capital in the primary sector of the global economy. Currently, around 2 million tons of pesticides are applied to crops annually to combat pests and diseases. According to the United Nations Food and Agricultural Organization (FAO), the United States was the largest consumer of pesticides in 2020, followed by Brazil, China, Argentina, Russia, Canada, France, Australia, India, and Italy. However, in FAO's 2022 update, China's pesticide use had significantly declined, placing it third globally. Despite this, overall

DOI: 10.1201/9781032712291-15

pesticide use has increased by about 50% since the 1990s, though it has recently plateaued (Antonini & Argilés-Bosch, 2017; Kingsley Nwosu & John, 2022; Rajmohan et al., 2020).

For public health, the result of this use is the elimination or significant reduction of pandemics that were initiated by mosquito, flea, tick and rat populations (Kim et al., 2017; Lawler, 2017; Wenzel, 2012). In the primary sector, these formulations have led to a significant increase in production as well as an improvement in product quality (Aktar et al., 2009; Bortoli & Coumoul, 2018; Lamichhane, 2017). The reduction in crop losses caused by diseases and insects is largely attributed to the production and use of these formulations. Moreover, the response to third world food crises is precisely due to this increase in production (Aktar et al., 2009; Lawler, 2017). However, the extensive use of these formulations in agriculture and livestock production, violation of the legislated minimum time interval between application of formulations and harvest, and use outside the rules of good agricultural practice can pose a direct or indirect risk to consumer health. Wanwimolruk et al. (2016) in a study on fruit and vegetable contamination by pesticide formulations, reported that almost all fruits and vegetables collected from open-air markets in West Africa contained residues above MRLs. In fruit and vegetable samples from China, the percentage of samples found to exceed MRLs ranged from 35 to 70%, depending on the product. Lozowicka et al. (2015) analyzed vegetables from flea markets in Central Asia and reported that almost 60% of the samples contained 29 different active compounds and almost 30% of the samples were found above MRLs.

Agrochemical residues remaining in food materials after harvest, often beyond the control of consumers, can have harmful effects on human health. The presence of pesticide residues is a significant obstacle in international food trade. The distribution of these residues in food varies depending on the chemical nature of the pesticide, the type and part of the food, and environmental factors. Pesticide-treated crops typically contain unpredictable amounts of these chemicals, making it essential to explore alternatives for decontaminating foods. Methods such as washing with water, or soaking in solutions of salt, chlorine, chlorine dioxide, hydrogen peroxide, ozone, acetic acid, hydroxyperacetic acid, iprodione, and detergents have been shown to effectively reduce pesticide levels. Preparatory processes like peeling and trimming can remove residues from outer layers, while thermal treatments such as pasteurization, blanching, boiling, cooking, steaming, canning, and scrambling can degrade various pesticides, depending on the pesticide type and treatment duration. Conversely, preservation methods like drying, dehydration, and concentration can increase pesticide levels due to the concentration effect. Other processes, including refining, fermentation, and curing, affect pesticide residues to varying degrees. Techniques such as milling, baking, winemaking, malting, and brewing tend to reduce pesticide levels in the final products. Post-harvest treatments and cold storage have also proven effective. Many decontamination methods can reduce pesticide concentrations to below MRL, though the extent of reduction depends on the initial concentration at harvest, the food substrate, and the type of pesticide. The effect of classic and traditional pre- and post-harvest processing on agrochemical residues in food has been thoroughly reviewed in recent years (Bajwa & Sandhu, 2014; Munir et al., 2024; Nguyen et al., 2020).

This chapter focuses on the effects of novel food processing techniques on the fate of agrochemicals initially present in the raw materials. A thorough collection of scattered literature data and case studies has been conducted and presented in the Tables. Finally, the challenges and opportunities of the agrochemical processing methods in order to maximize the safety of the final product reaching consumers are discussed.

15.2 AGROCHEMICAL RESIDUES IN FOODSTUFF

Currently, pesticide residues in North America and in EU are thoroughly monitored. In general, market foods are compliant with maximum permissible concentrations (MPC) and percentages of samples with detected residues exceeding MPCs fortunately are in small number. In the latest EFSA report, among more than 87,000 food samples from Member States analyzed in 2021, 96.1% of samples were within legal limits, 3.9% exceeded this level, of which 2.5% were non-compliant, i.e. samples exceeding the MRL after accounting for the measurement uncertainty (Carrasco Cabrera et al., 2023). Notwithstanding, in plant products, 190 different substances were found in measurable concentrations including recent and old crop protection chemicals and, although the food authority EFSA assessed the risk to consumers as low, recommendations were deemed necessary to further improve food safety and abate consumers exposure through diet (Carrasco Cabrera et al., 2023).

Despite the controversy about their safety, local farmers and agro-wholesalers still predominantly use these chemicals in developing countries, especially Africa, to ensure an all-year supply of agriproducts. The average of active ingredients with an unacceptable risk was 4x higher than the corresponding risk in developed countries (Berlitz et al., 2023). These chemicals could have short- or long-term effects. Despite the state-of-the-art knowledge, factors, such as poor education and awareness, limited agricultural subventions, quests for cheap chemicals, over-dosage, and many more are the possible reasons for these toxic chemicals' setback and persistent use in developing countries.

Antibiotics are also widely used (Table 15.1) in animal husbandry and are mainly detected in dairy products (Bilandžić et al., 2011) while they are also present in honey (Molino et al., 2011; Saridaki-Papakonstadinou et al., 2006). Tetracyclines and sulfonamides are used in beekeeping mainly for the treatment of American and European brood rot (Al-Waili et al., 2012; Molino et al., 2011). The Saridaki-Papakonstadinou et al. (2006) reported that in 251 samples of Greek honey tetracyclines were detected in 29% of the samples. However, the presence of both antibiotics and insecticides in honey is increasing due to honey imports from third countries where the use of antibiotics is extensive. In several countries outside the European Union (Table 15.2), the use of formulations is problematic. Controls are inadequate and there are no MRLs or the limits are significantly higher than those in the European Union (Al-Waili et al., 2012; Molino et al., 2011). Wang et al. (2022) in 94 Chinese honey samples detected antibiotics in 84% of the samples. Sulfonamides were detected in

TABLE 15.1 Antibiotic residues in various foods

COUNTRY	*ANTIBIOTICS*	*FOOD*	*BIBLIOGRAPHY*
USA	Tetracyclines	Chicken	Leste et al., 2019
Brazil	Tetracyclines, fluoroquinolones	Chicken	Leste et al., 2019
Bangladesh	Amoxicillin	Milk, eggs	Chowdhury et al., 2015
Nigeria	Tetracyclines	Eggs	Ezenduka et al., 2011
India	Oxytetracycline, erythromycin	Honey	Kumar et al., 2020
Iran	Penicillin, gentamicin, chlormphenicol, tylosin	Honey	Mahmoudi et al., 2014
Zambia	Oxytetracycline, sulfamethazine	Beef and veal	Nchima et al., 2017
Turkey	Quinolones	Chicken, beef	Er et al., 2013
South Africa	Sulfanilamide, ciprofloxacin, streptomycin, tetracycline	Chicken, pork & beef	Ramatla et al., 2017
Cameroon	Tetracyclines	Chicken	Guetiya Wadoum et al., 2016
India	Tetracycline, oxytetracycline, sulfadimidine, sulfamethoxazole	Milk	Kumari Anjana & C. Jayachandran, 2017
India	Sulfamethaxazole, enrofloxacin, oxytetracycline, penicillin	Milk	Moudgil et al., 2019
China	Beta-lactams, tetracyclines, sulfonamides, quinolones	Milk	Zheng et al., 2013
China	Quinolones, tetracyclines, sulfonamides	Chicken, eggs	Yang et al., 2020
Ghana	Tetracyclines	Milk	Addo et al., 2011
Egypt	Tetracyclines	Beef and veal	Salama et al., 2011
Egypt	Tetracyclines	Chicken	Abdeta et al., 2024

TABLE 15.2 Food processes to manage residues in food

FOOD	*ACTIVE COMPOUNDS*	*CONDITIONS*	*PERCENTAGE REDUCTION (%)*	*BIBLIOGRAPHY*
Cold plasma				
Tomatoes	Chlorpyrifos	5 W, 6 min	89	Ranjitha Gracy et al., 2019
Mango	Chlorpyrifos Cypemethrin	8 kV, 5 min	74 63	Phan et al., 2018
Soybean	Chlorpyrifos	1kV, 6 min	41	Anbarasan et al., 2022
Blueberries	Boscalid Imidacloprid	80 kV, 5 min	80 76	Sarangapani et al., 2017
Lettuce	Malathion Chlorpyrifos	80kV, 3 min	53 51	Cong et al., 2021
Corn	Chlorpyrifos Carbaryl	20W, 1min, 1200Hz	86 66	Liu et al., 2021
Pulsed electric fields				
White wine	Cyprodinil Pyrimethanil Procymidone Vinclozolin	20 kV/cm, 2 ms	49 27 25 32	Delsart et al., 2016
Apple juice	Diazinon Dimethoate	20 kV/cm, 260 μs	48 35	Zhang et al., 2012
Apple juice	Methamidophos chlorpyrifos	20 kV/cm, 26 pulses	9.1 9	Chen et al., 2009
Cherry juice	pyraclostrobin chlorpyrifos ethyl cyprodinil malathion tau-fluvalinate	24.7 kV/cm, 655 μs	97 92 88 83 47	Akdemir Evrendilek et al., 2020
Radiation				
Strawberries	Azoxystrobin Carbendazim	4 kGy	56.70 27.99	Ciarrocchi et al., 2021
Peas	Profenosfos	32 kGy	47.90	Thihara Rodrigues et al., 2020
Ozonolysis				
Tomatoes	Azoxystrobin Chlorothalonil Difenoconazole	Growth in water with 3 ppm ozone for 30 minutes.	70–90	Rodrigues et al., 2019

(*continued*)

TABLE 15.2 (Continued)

FOOD	*ACTIVE COMPOUNDS*	*CONDITIONS*	*PERCENTAGE REDUCTION (%)*	*BIBLIOGRAPHY*
Grapes	Azoxystrobin	Exposure to ozone 0.64 mg/m^3	90.7	Karaca, 2019
	Bupirimate		63.4	
	Carbendazim		38.5	
	Penconazole		80.2	
	Triadimenol		61.4	
	Trifloxystrobin		51.8	
	Chlorpyrifos		98.0	
Cowpea	Emamectin benzoate	Ozone microbubbles	65	Li et al., 2023
	Carbendazim		44	
	Azoxystrobin		34	
	Boscalid		33	
	Difenoconazole		28	
Household practices				
Tomatoes	Dichlorvos	Wash with water for 15 minutes	26	Venkatachalapathy et al., 2020
	Dimethoate		38	
	Malathion		35	
	Chlorpyrifos		14	
Apples	Quinalphos	Wash with water for 15 minutes	34	T. Yang et al., 2017
	Diazinon		28	
	Fenitrothion		41	
Cucumbers	Diazinon	Peeling	67	Kwon et al., 2015
Pumpkins	Azoxystrobin	Peeling	100	Aguilera et al., 2012
	Acrinathrin		100	
	Kresoxim-methyl		90	
Household practices				
Amaranthus triocolor L.	Dimethoate	Wash with water for 10 minutes and cook for 10 minutes	76.94	Muralikrishna et al., 2016
	Malathion		84.86	
	Chlorpyrifos		61.15	
	Quinalphos		71.62	
	Profenophos		76.00	
	Ethion		62.30	

TABLE 15.2 (Continued)

FOOD	*ACTIVE COMPOUNDS*	*CONDITIONS*	*PERCENTAGE REDUCTION (%)*	*BIBLIOGRAPHY*
Okra	Acetamiprid Cypermethrin Chlorpyrifos Imidacloprid Profenofos Lambda-cyhalothrin	Steaming (100°C, 8 minutes). Boiling for 8 minutes. Baking (180°C, 8 minutes). Sautéing (80–100°C, 8 minutes)	11.11–55.55 8.57–37.14 12.50–100 12.50–50.0 11.11–100 28.57–67.85	Ishfaq et al., 2022
Heating				
Honey	Tetracycline Chlortetracycline Oxytetracycline Doxycycline	78°C, 1 minute	18.7 13.0 10.2 10.1	Molino et al., 2011
Honey	Chlorpyrifos	45°C, 1 hour	1.9	Q. Zhu et al., 2010
Milk	Dimethoate Malathion	72°C, 15 seconds	73.42 100	Abd-Rabo et al., 2016
Ultrasound				
Pakchoi	Chlorothalonil Pyrazophos Carbendazim	28–135 kHz, 20°C, 10 minutes	13.87–74.86 21.74–45.68 5.12–24.63	Y. Zhu et al., 2019
Water	isoproturon herbicide	power of 50W cm^2 1 hr	100%	Schieppati et al., 2019
Strawberries	Acetamiprid Alpha-cypemethrin Boscalid Bupirimate Chlorpyrifos Cyprodinil Deltamethrin Fenhexamid Folpet Iprodione Lambda-cyhalothrin Pirimicarb Pyraclostrobin Tetraconazole Trifloxystrobin	40 kHz, 5 minutes	45.1–91.2	Lozowicka et al., 2016

(*continued*)

TABLE 15.2 (Continued)

FOOD	*ACTIVE COMPOUNDS*	*CONDITIONS*	*PERCENTAGE REDUCTION (%)*	*BIBLIOGRAPHY*
Honey	HCL-Tetracycline Sulfathiazole	50kHz, 25°C, 2 hours	0 94	Sidirokastritis et al., 2022
Water	HCL-Tetracycline Sulfathiazole		0 0	
Milk	HCL-Tetracycline Sulfathiazole		29.8 4.1	
Cucumber	organophosphorus pesticides	Ultrasonic bath 20 minutes	85	Liang et al., 2012
Washes with chemical & natural plant extracts				
Apricots Nectarines Peaches	Tebuconazole	Wash with ClO_2 for 15 minutes	61.6 62.6 64.6	Calvo et al., 2019
Tomatoes	Diclorvos Dimethoate Malathion Chlorpyrifos	Immersion for 15 minutes in natural extracts of ***Albizia amara*** and ***Acacia concinna*** plants	87 84 83 64	Venkatachalapathy et al., 2020
High Hydrostatic Pressure (HHP)				
Tomatoes	Chlorpyrifos	Packaging with water as the surrounding liquid. Treatment with 0.1–400 MPa (5&25°C, 30 minutes)	75	Iizuka, Maeda, et al., 2013; Iizuka, Yahata, et al., 2013
Tomatoes & Brussels sprouts	Chlorpyrifos	Packed with 10% ethanol aqueous solution. Treatment with 75 MPa (5°C, 30 minutes)	100	Iizuka & Shimizu, 2014a, 2014b

TABLE 15.2 (Continued)

FOOD	*ACTIVE COMPOUNDS*	*CONDITIONS*	*PERCENTAGE REDUCTION (%)*	*BIBLIOGRAPHY*
Honey	malathion	Vacuum packed and processed at 580MPa (25°C, 6min)	30.6	Sidirokastritis et al., 2024
	chlorpyrifos		28.6	
	coumaphos		21	
	bromopropylate		15.7	
	tau-fluvalinate		10.6	
	HCL-tetracycline		0	Sidirokastritis et al., 2022
	Sulfathiazole		90	
Milk	HCl-tetracycline		78	Sidirokastritis et al., 2022
	Sulfathiazole		50	
Water	HCl-tetracycline		70	
	Sulfathiazole		83	
Filtration				
Honey	tau-fluvalinate	Filtration under vacuum with Whatman No 4 filter paper	100	Waliszewski et al., 2003
Honey	Chlorpyrifos	Filtration with a glass fiber filter	21.3	Q. Zhu et al., 2010
Adsorption				
Honey	Parathion	Adsorption resin immersion LS-803 for 90 minutes at 25°C	76.5	Xu et al., 2012
Photodisruption				
Honey	Coumaphos	250–750 W/m^2 for 15–60 minutes	90	Yuan et al., 2014
	methyl parathion		58.75–73.18	
	Fenitrothion		26.63–83.29	
Management with storage				
Honey	tau-fluvalinate	248 days at 35°C	Practically unchanged.	Tsigouri et al., 2001

(continued)

TABLE 15.2 (Continued)

FOOD	*ACTIVE COMPOUNDS*	*CONDITIONS*	*PERCENTAGE REDUCTION (%)*	*BIBLIOGRAPHY*
Honey	Tetracycline Chlortetracycline Oxytetracycline Doxycycline	Heating (78°C, 1 min) and stored in the dark at 20±5°C for 180 days	78–88	Molino et al., 2011

44.7% of samples while tetracyclines were detected in 26.6% of samples. Another important parameter of antibiotic use in beekeeping is the high half-life (Tillotson et al., 2006). Thompson et al. (2005) reported that the concentration of oxytetracycline in honey was 3.7 mg/kg eight weeks after its use in bees. Adams et al. (2009) detected lincomycin residues in honey up to 290 days after application to the hive. Martinello et al. (2013) detected sulfathiazole residues (180 mg/kg) in honey two weeks after application to the feeding syrup.

15.3 MAXIMUM RESIDUE LIMITS (MRLS)

A MRL is the maximum concentration of a plant protection or veterinary use compound that can be detected in a foodstuff so that it is considered safe for human health and there are no legal consequences for its production and disposal. The MRL of a compound is not always common for all foods, since the same compound may have a different MRL for each foodstuff detected. MRLs are regulated by the European Union (European Commission Regulation 396/2005; European Commission Regulation 37/2010) and the national legislation of many countries (Blasco et al., 2003) for certain acaricidal compounds, such as coumaphos. However, in several foods, such as honey, MRLs have not been legislated for many compounds, such as malathion. In these cases, the MRL is considered to be the minimum Limit Of Quantification (LOQ) of that compound in that food. The detection of insecticides in honey is more prevalent in countries outside the European Union, where several reports, often from honey imported into Europe, confirm their presence above the legislated MRLs (Eissa et al., 2014; A. Kumar et al., 2018; Rafique et al., 2018; Yavuz et al., 2010).

The European Union has also set MRLs for antibiotics in many foods (European Commission Regulation 37/2010). For example, sulfonamides and tetracyclines used in veterinary formulations have MRLs of 0.1 mg/kg in milk. The Food and Agriculture Organization of the United Nations (FAO) in 1998 established that the maximum acceptable daily intake (ADI) of tetracyclines is 0.03 mg/kg body weight. However, for tetracyclines and sulfonamides, the European Union has not established MRLs for certain foods, such as honey. In these cases, as with insecticides, MRLs are considered to be LOQs.

15.4 HEALTH EFFECTS

According to several studies, acute pesticide poisoning causes symptoms such as weakness, abdominal cramps, vomiting, muscle spasms, convulsions, irritation of the conjunctiva, headache, difficult breathing, loss of appetite, and nosebleed, among others, and each substance can be responsible for distinct symptoms or effects. Chronic poisoning also exhibits characteristic symptoms for each substance, including delayed neurotoxic effects, chromosomal changes, contact dermatitis, liver damage, cardiac arrhythmias, kidney damage, peripheral neuropathies, allergies, asthma, Parkinson's disease, cancers, teratogenic effects, and hearing loss, among others. A comprehensive review of the clinical studies for the effects of pesticides on human health has been done by Gerage et al. (2017).

The increasing use of antibiotics and insecticides in agriculture and livestock farming threatens human health, as several studies have reported direct and indirect effects on human health. Gastrointestinal problems have been reported even after consumption of food with very low concentrations of pesticide residues (Baù et al., 2020). Even when MRLs are respected, considering the cumulative nature of residues, it becomes evident that chronic consumption of residue-laden foods is hazardous to the health of consumers. Chronic exposure to residues of antibiotics and insecticides through food has been reported to have adverse effects on the nervous and reproductive systems (El-Nahhal, 2020; Giordano et al., 2007). However, the major problem of chronic exposure to antibiotic residues through food is the phenomenon of resistance of microorganisms to antibiotics. This phenomenon has increased in recent years and its enhancement has been linked to the continuous increase in the use of veterinary formulations in animal husbandry (Wu-Wu et al., 2023).

15.5 TRADITIONAL AND EMERGING FOOD PROCESSING TECHNIQUES

To reduce these negative effects of residues on human health, several food processing methods have been studied for their efficacy to decrease the concentrations of antibiotics and insecticides (Mir et al., 2022). These methods include a wide range of treatments (Table 15.2). Several simple household practices, such as communal washing, have been investigated by various researchers (Muralikrishna et al., 2016; Venkatachalapathy et al., 2020; T. Yang et al., 2017). Common processing methods for fruits and vegetables, such as drying, boiling, dehydration, canning, concentration, pasteurization have been studied in the past for their ability to decrease pesticide and contaminants in general. These methods have been thoroughly reviewed by researchers (Bajwa & Sandhu, 2014).

Ultrasound: A typical method for eliminating pesticide residue from food surfaces is washing the food in distilled water. It has been discovered that when sonication is employed during the washing process, pesticides are effectively removed by disrupting cell membranes, increasing the surface area of solvent and plant material interaction (Al-Taher et al., 2013). Ultrasound-assisted cleaning (UAC) is also thought to be an ecofriendly and effective pesticide removal technology that is superior to conventional approaches in its ability to remove pollutants. It is a time and energy-saving cleaning approach (Azam et al., 2020). The effectiveness of the methods depends on both the operating conditions (frequency, ratio of solvent to sample, intensity, time, temperature) and the food matrix. In the case of matrices with large surface area, ultrasonic treatments may have difficulty reaching insecticides on inner surfaces. Because ultrasonic energy degrades as it passes through different materials, less energy reaches the internal parts of the leaves. As a result, researchers discovered that using ultrasound to remove pesticides absorbed into the vine leaves may be more difficult than removing pesticides from the tomato surface. Several attempts have been made to investigate the effect of ultrasound treatment on pesticide residues, either as a stand-alone technique or in combination with others, as a pre-treatment and are summarized in Table 15.2.

Pulsed electric field: Pulsed electric field (PEF) technology, one of the promising methods of non-thermal preservation of food, has been used to address the biological hazards efficiently, as an alternative to pasteurization. In this technique, high intensity electric field with pulse amplitude from the range 100–300 V/cm to 20–80 kV/cm is generated between two electrodes and applies to a product for short duration (from milliseconds to microseconds, μs) to inactivate the microorganisms below the thermal processing temperatures, through the phenomenon of electroporation. At the same time, some useful chemical reactions occur, and electrochemical characteristics also change due to the complex situation of the food matrix (Q. Wang et al., 2018). As for pesticide residues, there is a suggestion that they are degraded by rotation and vibration of polar molecules dynamically or free radical oxidation during PEF (Chen et al., 2009; Zhang et al., 2012). However, the specific degradation pathway is still unclear. In contrast to conventional heat treatment, PEF treatment causes negligible deteriorative effect on color, flavor, taste, and nutrients of food materials.

Cold plasma: The ionized state of gas contains various species, including molecules, atoms, positive and negative ions, free radicals, photons, and other active components. Depending on the temperature of these species, plasma can be classified into two types: equilibrium (thermal) plasma and nonequilibrium (low-temperature) plasma. Cold plasma is generated by gas ionization at different pressures using instruments such as corona discharge (CD), dielectric barrier discharge (DBD), glow discharge (GD), radiofrequency discharge (RFD), plasma jet (PJ), gliding arc discharge (GAD), or microwave discharge (MWD). The initiation of cold plasma triggers inelastic collisions among reactive species, producing secondary electrons that contribute to pesticide dissociation, excitation, and interaction. Additionally, reactive oxygen species (ROS) and reactive nitrogen species (RNS) dominate among the active species and, under appropriate conditions, directly attack pesticide molecules, breaking chemical bonds and initiating various reactions that degrade pesticides into intermediates or less toxic compounds.

Bai et al. (2010) explored the degradation mechanism of dichlorvos during the treatment of cold plasma. There were two pathways that happened at the same time. One was the free radical reaction between pesticide molecules and reactive species initiated by cold plasma. Another was the electron energy reaction related to unsaturated bonds of molecules. As for different chemical bonds of pesticides, the attacking sequence shows differently. Generally, reactive species are more likely to attack lower chemical bond energy than higher ones. For example, ·OH radicals as dominated species firstly attacked P=S and then attacked P=O to degrade dimethoate to by-products and intermediates (Hu et al., 2013). The same result was shown in reference (Sarangapani et al., 2016).

High hydrostatic pressure: High Hydrostatic Pressure (HHP) has recently gained widespread use as a non-thermal processing method alternative to pasteurization, effectively inactivating microorganisms while causing minimal impact on the sensory and organoleptic properties of food products. One of the key advantages of HHP is the uniform and instantaneous application of pressure across the food matrix, regardless of its mass, size, or shape. This uniform pressure leads to reactions that compress biomolecules, reducing processing time, lowering the risk of over-processing, and preserving the sensory, nutritional, and functional qualities of the food. Mechanistically, high-pressure treatment disrupts hydrogen bonds between pesticide molecules adhered to food surfaces, weakening hydrophobic interactions. Through this mechanism (Iizuka, Yahata, et al., 2013) explained the transfer of pesticides from tomatoes to the surrounding medium (Iizuka, Maeda, et al., 2013). However, studies by Sidirokastritis et al. (2022, 2024) observed a significant reduction in certain pesticides and antibiotics within food matrices without transfer to the surrounding medium, which appeared to be dependent on the initial concentration of the substances. The precise mechanisms behind these findings have yet to be fully explored.

In addition to the above-mentioned methods, radiation (Ciarrocchi et al., 2021), ozonolysis (Rodrigues et al., 2019), heating (Abd-Rabo et al., 2016), chemical washes (Calvo et al., 2019) and plant extracts (Venkatachalapathy et al., 2020), filtration (Q. Zhu et al., 2010), adsorption (Xu et al., 2012), photodissociation (Yuan et al., 2014) and storage (Molino et al., 2011) have also been tested. However, several of these methods are either not sufficiently effective in reducing residues or their application depends on the nature of the food and, for example, cannot be applied to liquid foods such as honey. Moreover, they often degrade the quality characteristics of the product. For example, heating and pasteurization have been reported to degrade the legislated quality characteristics of honey. In particular, they increase the concentration of HMF (Sajtos et al., 2024) and negatively affect the enzyme diastase (Elamine et al., 2020). In addition, thermal processes affect the sensory profile of food matrices (Claeys et al., 2013), while reducing the concentrations of vitamins (Macdonald et al., 2011).

15.6 CHALLENGES AND FUTURE DEVELOPMENTS

One of the biggest challenges in the 21st century is how to feed the increasing population while decreasing the adverse consequences on the environment and human health that arose due to the continuous deployment of pesticides, fertilizers, and freshwater (Di Vaio et al., 2020; Tian et al., 2021). The European Green Deal and its Farm to Fork Strategy proposed targets to establish sustainable agriculture, considering nature conservation to ensure a fair, healthy, and environmentally friendly food system (European Union, 2020). Under the Farm to Fork Strategy, the EC plans to take actions to reduce by 50% the use and risk of chemical pesticides, including the use of more hazardous pesticides, until 2030. To achieve this goal, the EC will revise the SUD (European Parliament, 2009) and promote alternative practices, such as IPM, to protect harvests from pests and diseases. In this respect, strong support should be given to farmers and accelerate the transition toward sustainable agriculture. In addition, the Farm to Fork Strategy encourages organic farming, intending to have at least 25% of EU agricultural land under organic farming management by 2030 (European Union, 2020). However, some issues must be resolved to meet the proposed targets. As discussed, the risk assessment considering pesticide cocktails found in the major agricultural systems must be implemented since their effects on soil health and, consequently, on food safety are still unknown (Geissen et al., 2021). In addition, the issue of legacy pesticides that can persist in the environment for several decades after they were banned must be resolved (McGinley et al., 2023). Thus, besides reducing pesticide usage, to ensure food safety, it is necessary to implement novel approaches to detect pesticide residues, assess the real risk of combined exposure to multiple residues, and degrade them into non-toxic products to safeguard consumer health.

In the context of pesticide screening and detection, it is essential to prioritize new methods developed under the principles of green chemistry (Iammarino et al., 2022). Recent advancements have introduced analytical techniques that leverage nanosystems for precise, environmentally friendly, and highly sensitive detection of pesticide residues in both food and the environment (Chaudhary et al., 2023; Kaur et al., 2021). Furthermore, the integration of innovative, smartly designed nanomaterials alongside advanced instrumentation can improve the detection of pesticides in complex food matrices, making the process more efficient, cost-effective, and faster (Kaur et al., 2021).

When examining the impact of food processing on chemical levels, it is important to consider the fate of substances that are lost. If the chemical is removed through washing, sorting, or discarding cooking liquids, the process is straightforward. However, if the compound is broken down, it becomes crucial to identify any potential decomposition products that may remain in the final food product and assess their concentrations. Such findings could necessitate a new risk evaluation for any newly formed chemical entities in the food (Picó et al., 2008).

As previously mentioned, pesticide residues can degrade or deteriorate during cooking. Compounds containing functional groups like carbamate, amide, urea, thiocarbonyl, and imino are prone to hydrolysis in the presence of small amounts of acid or base during heating or boiling (Hamilton et al., 2004). Consequently, it is more appropriate to assess the health risks of pesticides based on data from monitoring their levels in cooked foods. For example, a study by Lee et al. (2009) demonstrated the hydrolysis of 44 different pesticides in trace amounts of acid or base when examining cooked samples of five types of vegetables.

Although degradation of pesticides in environmental and aqueous systems is well documented, studies about the development of breakdown products and metabolite identification during cooking or processing food are scarce. Some studies show an increase of ethylenethiourea (ETU) formation when cooking plant materials treated with ethylene bisdithiocarbamates (EDBCs) fungicides (U. Kumar & Agarwal, 1991; Zhang, 1991). Knio et al. (2000) also observed the formation of three metabolites, resulting from ETU decomposition during storage of the tomato paste, the ethylene urea being the main degradation product. Hence, while the cooking process reduces the level of EDBC residues and increased the ETU content, storing processed food containing ETU did not result in rapid degradation of the latter. In addition, the formation of primary (oxon) and secondary (alcohol) metabolites resulting from the pH- and temperature-catalyzed degradation of the parent organophosphate pesticides (OPs) were identified in cooked meat (Coulibaly & Smith, 1994).

Degradation of chlorfenvinphos in carrots during storage was also studied by (Uygun, 1997). On the first day, stored carrot samples did not exhibit any residue other than unchanged chlorfenvinphos, as was expected. On the fifteenth day of storage, the GC-MS chromatogram showed the presence of diethyl phosphate, which is one of the primary metabolites of the pesticide. 2,4-Dichlorobenzoic acid, 1-(2′,4′-dichlorophenyl)-ethan-1-ol and 2,2dichloro- 1-(2′,4′-dichlorophenyl)-vinyl alcohol were found as secondary metabolites in carrots on the fortieth day of storage.

Pesticide metabolites were also detected during the olive oil extraction process. Specifically, fenthion sulfoxide and endosulfan sulfate, the primary metabolites of fenthion and endosulfan, respectively, were found in laboratory-produced olive oils (Amvrazi & Albanis, 2008). Additionally, the concentration of fenthion sulfoxide increased with both the amount of water used during olive oil extraction and the initial concentration of fenthion in the olives. Notably, these metabolites exhibit greater toxicity than the parent compounds.

Furthermore, other processing methods can result in the formation of new chemical entities in foods. Some of these changes may occur incidentally, and detailed information about the newly formed components might not always be available. This is particularly relevant for traditional processing techniques applied to foods that have not previously undergone such treatment. In contrast, industrial processing, which is generally conducted under tightly controlled conditions, often provides more information on the types and levels of newly formed chemicals. This is not always the case with home cooking, where foods that are undercooked or overcooked may introduce unknown compounds. In-vitro hydrolysis procedures are commonly used to assess how processing affects residue composition and to identify alteration products. These hydrolysis studies help confirm the definition of residues in processed products or highlight additional breakdown products for further analysis. A well-studied example of this is the ethylene bisdithiocarbamate (EBDC) fungicides, which degrade into products of toxicological concern.

REFERENCES

Abd-Rabo, F. H. R., Elsalamony, H., & Sakr, S. S. (2016). Reduction of pesticide residues in egyptian buffalo milk by some processing treatments. *International Journal of Dairy Science*, *11*(2), 75–80. https://doi.org/10.3923/ijds.2016.75.80

Abdeta, D., Tafesse, M., & Bacha, B. (2024). Detection of selected antimicrobial residues in red meat and kidney of beef cattle slaughtered at Nekemte municipal abattoir, Ethiopia. *Veterinary Medicine and Science*, *10*(3), 1–9. https://doi.org/10.1002/vms3.1459

Adams, S. J., Fussell, R. J., Dickinson, M., Wilkins, S., & Sharman, M. (2009). Study of the depletion of lincomycin residues in honey extracted from treated honeybee (Apis mellifera L.) colonies and the effect of the shook swarm procedure. *Analytica Chimica Acta*, *637*(1–2), 315–320. https://doi.org/10.1016/j.aca.2008.09.013

Addo, K. K., Mensah, G. I., Aning, K. G., Nartey, N., Nipah, G. K., Bonsu, C., Akyeh, M. L., & Smits, H. L. (2011). Microbiological quality and antibiotic residues in informally marketed raw cow milk within the coastal savannah zone of Ghana. *Tropical Medicine and International Health*, *16*(2), 227–232. https://doi.org/10.1111/j.1365-3156.2010.02666.x

Aguilera, A., Valverde, A., Camacho, F., Boulaid, M., & García-Fuentes, L. (2012). Effect of household processing and unit to unit variability of azoxystrobin, acrinathrin and kresoxim methyl residues in zucchini. *Food Control*, *25*(2), 594–600. https://doi.org/10.1016/j.foodcont.2011.11.038

Akdemir Evrendilek, G., Keskin, E., & Golge, O. (2020). Interaction and multi-objective effects of multiple non-thermal treatments of sour cherry juice: Pesticide removal, microbial inactivation, and quality preservation. *Journal of the Science of Food and Agriculture*, *100*(4), 1653–1661. https://doi.org/10.1002/jsfa.10178

Aktar, W., Sengupta, D., & Chowdhury, A. (2009). Impact of pesticides use in agriculture: Their benefits and hazards. *Interdisciplinary Toxicology*, *2*(1), 1–12. https://doi.org/10.2478/v10102-009-0001-7

Al-Taher, F., Chen, Y., Wylie, P., & Cappozzo, J. (2013). Reduction of pesticide residues in tomatoes and other produce. *Journal of Food Protection*, *76*(3), 510–515. https://doi.org/10.4315/0362-028X.JFP-12-240

Al-Waili, N., Salom, K., Al-Ghamdi, A., & Ansari, M. J. (2012). Antibiotic, pesticide, and microbial contaminants of honey: Human health hazards. *The Scientific World Journal*, 2012(Table 1). https://doi.org/10.1100/2012/930849

Amvrazi, E. G., & Albanis, T. A. (2008). Multiclass pesticide determination in olives and their processing factors in olive oil: Comparison of different olive oil extraction systems. *Journal of Agricultural and Food Chemistry*, *56*(14), 5700–5709. https://doi.org/10.1021/jf703783u

Anbarasan, R., Jaspin, S., Bhavadharini, B., Pare, A., Pandiselvam, R., & Mahendran, R. (2022). Chlorpyrifos pesticide reduction in soybean using cold plasma and ozone treatments. *Lwt*, *159*(December 2021), 113193. https://doi.org/10.1016/j.lwt.2022.113193

Antonini, C., & Argilés-Bosch, J. M. (2017). Productivity and environmental costs from intensification of farming. A panel data analysis across EU regions. *Journal of Cleaner Production*, *140*, 796–803. https://doi.org/10.1016/j.jclepro.2016.04.009

Azam, S. M. R., Ma, H., Xu, B., Devi, S., Siddique, M. A. B., Stanley, S. L., Bhandari, B., & Zhu, J. (2020). Efficacy of ultrasound treatment in the and removal of pesticide residues from fresh vegetables: A review. *Trends in Food Science and Technology*, *97*(301), 417–432. https://doi.org/10.1016/j.tifs.2020.01.028

Bai, Y., Chen, J., Yang, Y., Guo, L., & Zhang, C. (2010). Degradation of organophosphorus pesticide induced by oxygen plasma: Effects of operating parameters and reaction mechanisms. *Chemosphere*, *81*(3), 408–414. https://doi.org/10.1016/j.chemosphere.2010.06.071

Bajwa, U., & Sandhu, K. S. (2014). Effect of handling and processing on pesticide residues in food- A review. *Journal of Food Science and Technology*, *51*(2), 201–220. https://doi.org/10.1007/s13197-011-0499-5

Baù, M., Moretti, A., Bertoni, E., Vazzoler, V., Luini, C., Agosti, M., & Salvatore, S. (2020). Risk and protective factors for gastrointestinal symptoms associated with antibiotic treatment in children: A population study. *Pediatric Gastroenterology, Hepatology and Nutrition*, *23*(1), 35–48. https://doi.org/10.5223/pghn.2020.23.1.35

Berlitz, F., Elias, S., Fraga, G., & Bender, R. (2023). Risk assessment of agrochemical residues in food: A systematic review and meta-analysis. doi: 10.21203/rs.3.rs-3286624/v1.

Bilandžić, N., Kolanović, B. S., Varenina, I., Scortichini, G., Annunziata, L., Brstilo, M., & Rudan, N. (2011). Veterinary drug residues determination in raw milk in Croatia. *Food Control*, *22*(12), 1941–1948. https://doi.org/10.1016/j.foodcont.2011.05.007

Blasco, C., Fernández, M., Pena, A., Lino, C., Silveira, M. I., Font, G., & Picó, Y. (2003). Assessment of pesticide residues in honey samples from Portugal and Spain. *Journal of Agricultural and Food Chemistry*, *51*(27), 8132–8138. https://doi.org/10.1021/jf034870m

Bortoli, S., & Coumoul, X. (2018). Impact of pesticides on human health. *Pratiques En Nutrition*, *14*(53), 18–24. https://doi.org/10.1016/j.pranut.2017.12.005

Calvo, H., Redondo, D., Remón, S., Venturini, M. E., & Arias, E. (2019). Efficacy of electrolyzed water, chlorine dioxide and photocatalysis for disinfection and removal of pesticide residues from stone fruit. *Postharvest Biology and Technology*, *148*(October 2018), 22–31. https://doi.org/10.1016/j.postharvbio.2018.10.009

Carrasco Cabrera, L., Di Piazza, G., Dujardin, B., & Medina Pastor, P. (2023). The 2021 European Union report on pesticide residues in food. *EFSA Journal*, *21*(4). https://doi.org/10.2903/j.efsa.2023.7939

Chaudhary, V., Rustagi, S., & Kaushik, A. (2023). Bio-derived smart nanostructures for efficient biosensors. *Current Opinion in Green and Sustainable Chemistry*, *42*, 100817. https://doi.org/10.1016/j.cogsc.2023.100817

Chen, F., Zeng, L., Zhang, Y., Liao, X., Ge, Y., Hu, X., & Jiang, L. (2009). Degradation behaviour of methamidophos and chlorpyrifos in apple juice treated with pulsed electric fields. *Food Chemistry*, *112*(4), 956–961. https://doi.org/10.1016/j.foodchem.2008.07.016

Chowdhury, S., Hassan, M. M., Alam, M., Sattar, S., Bari, M. S., Saifuddin, A. K. M., & Hoque, M. A. (2015). Antibiotic residues in milk and eggs of commercial and local farms at Chittagong, Bangladesh. *Veterinary World*, *8*(4), 467–471. https://doi.org/10.14202/vetworld.2015.467-471

Ciarrocchi, I. R., Mendes, K. F., Pimpinato, R. F., Spoto, M. H. F., & Tornisielo, V. L. (2021). The effect of radiation in the degradation of carbendazim and azoxystrobin in strawberry. *Radiation Physics and Chemistry*, *179*, 109269. https://doi.org/10.1016/j.radphyschem.2020.109269

Claeys, W. L., Cardoen, S., Daube, G., De Block, J., Dewettinck, K., Dierick, K., De Zutter, L., Huyghebaert, A., Imberechts, H., Thiange, P., Vandenplas, Y., & Herman, L. (2013). Raw or heated cow milk consumption: Review of risks and benefits. *Food Control*, *31*(1), 251–262. https://doi.org/10.1016/j.foodcont.2012.09.035

Cong, L., Huang, M., Zhang, J., & Yan, W. (2021). Effect of dielectric barrier discharge plasma on the degradation of malathion and chlorpyrifos on lettuce. *Journal of the Science of Food and Agriculture*, *101*(2), 424–432. https://doi.org/10.1002/jsfa.10651

Coulibaly, K., & Smith, J. S. (1994). Effect of pH and cooking temperature on the stability of Organophosphate pesticides in beef muscle. *Journal of Agricultural and Food Chemistry*, *42*(9), 2035–2039. https://doi.org/10.1021/jf00045a038

Delsart, C., Franc, C., Grimi, N., de Revel, G., Vorobiev, E., & Peuchot, M. M. (2016). Effects of pulsed electric fields on four residual fungicides in white wines. *IFMBE Proceedings*, *53*(Wc 2015), 124–127. https://doi.org/10.1007/978-981-287-817-5_27

Di Vaio, A., Boccia, F., Landriani, L., & Palladino, R. (2020). Artificial intelligence in the agri-food system: Rethinking sustainable business models in the COVID-19 scenario. *Sustainability (Switzerland)*, *12*(12). https://doi.org/10.3390/SU12124851

Eissa, F., El-sawi, S., & Zidan, N. E. (2014). Determining pesticide. *Journal of Bioprocessing & Biotechniques*, *23*(5), 1573–1580.

El-Nahhal, Y. (2020). Pesticide residues in honey and their potential reproductive toxicity. *Science of the Total Environment*, *741*, 139953. https://doi.org/10.1016/j.scitotenv.2020.139953

Elamine, Y., Anjos, O., Estevinho, L. M., Lyoussi, B., Aazza, S., & Miguel, M. G. (2020). Effect of extreme heat processing on the Moroccan Zantaz' honey antioxidant activities. *Journal of Food Science and Technology*, *57*(9), 3323–3333. https://doi.org/10.1007/s13197-020-04365-x

Er, B., Kaynak Onurdă, F., Demirhan, B., Özgen Özgacar, S., Bayhan Öktem, A., & Abbasoğlu, U. (2013). Screening of quinolone antibiotic residues in chicken meat and beef sold in the markets of Ankara, Turkey. *Poultry Science*, *92*(8), 2212–2215. https://doi.org/10.3382/ps.2013-03072

European Commission. (2005). Regulation (EC) No 396/2005, Maximum residue levels of pesticides in/on food and feed of plant and animal. *Official Journal of the European Union*, *L70(48)*, 1–16. www.fsai.ie/uploadedFiles/Legislation/Food_Legisation_Links/Pesticides_Residues_in_food/Regulation_EC_No_396_2005.pdf

European Parliament. (2009). Directive 2009/128/EC of the European Parliament and the Council of 21 October 2009 establishing a framework for Community action to achieve the sustainable use of pesticides. *October*, *309*, 71–86.

European Parliament and the Council of the European Union. (2010). Commission Regulation (EU) No 37/2010. *Official Journal of the European Union, L* 15/1.

European Union. *Farm to Fork Strategy. For a Fair, Healthy and Environmentally-Friendly Food System*; European Union: Brussels, Belgium, 2020.

Ezenduka, E. V., Oboegbulem, S. I., Nwanta, J. A., & Onunkwo, J. I. (2011). Prevalence of antimicrobial residues in raw table eggs from farms and retail outlets in Enugu State, Nigeria. *Tropical Animal Health and Production*, *43*(3), 557–559. https://doi.org/10.1007/s11250-010-9730-z

Food and Agriculture Organization of the United Nations (FAO). (2022). FAOSTAT Database. Pesticides Use. 2022. Available online: www.fao.org/faostat/en/#data/RP (accessed on 29 May 2023)

Geissen, V., Silva, V., Lwanga, E. H., Beriot, N., Oostindie, K., Bin, Z., Pyne, E., Busink, S., Zomer, P., Mol, H., & Ritsema, C. J. (2021). Cocktails of pesticide residues in conventional and organic farming systems in Europe – Legacy of the past and turning point for the future. *Environmental Pollution*, *278*, 116827. https://doi.org/10.1016/j.envpol.2021.116827

Gerage, J., Meira, A. P., & Vieira da Silva, M. (2017). Food and nutrition security: Pesticide residues in food. *Nutrire*, 42. doi: 10.1186/s41110-016-0028-4.

Giordano, G., Afsharinejad, Z., Guizzetti, M., Vitalone, A., Kavanagh, T. J., & Costa, L. G. (2007). Organophosphorus insecticides chlorpyrifos and diazinon and oxidative stress in neuronal cells in a genetic model of glutathione deficiency. *Toxicology and Applied Pharmacology*, *219*(2–3), 181–189. https://doi.org/10.1016/j.taap.2006.09.016

Guetiya Wadoum, R. E., Zambou, N. F., Anyangwe, F. F., Njimou, J. R., Coman, M. M., Verdenelli, M. C., Cecchini, C., Silvi, S., Orpianesi, C., Cresci, A., & Colizzi, V. (2016). Abusive use of antibiotics in poultry farming in Cameroon and the public health implications. *British Poultry Science*, *57*(4), 483–493. https://doi.org/10.1080/00071668.2016.1180668

Hamilton, D., Ambrus, Á., Dieterle, R., Felsot, A., Harris, C., Petersen, B., Racke, K., Wong, S. S., Gonzalez, R., Tanaka, K., Earl, M., Roberts, G., & Bhula, R. (2004). Pesticide residues in food – Acute dietary exposure. *Pest Management Science*, *60*(4), 311–339. https://doi.org/10.1002/ps.865

Hu, Y., Bai, Y., Li, X., & Chen, J. (2013). Application of dielectric barrier discharge plasma for degradation and pathways of dimethoate in aqueous solution. *Separation and Purification Technology*, *120*, 191–197. https://doi.org/10.1016/j.seppur.2013.10.005

Iammarino, M., Palermo, C., & Tomasevic, I. (2022). Advanced analysis techniques of food contaminants and risk assessment—Editorial. *Applied Sciences (Switzerland)*, *12*(10). https://doi.org/10.3390/app12104863

Iizuka, T., Maeda, S., & Shimizu, A. (2013). Removal of pesticide residue in cherry tomato by hydrostatic pressure. *Journal of Food Engineering*, *116*(4), 796–800. https://doi.org/10.1016/j.jfoodeng.2013.01.035

Iizuka, T., & Shimizu, A. (2014a). Removal of pesticide residue from Brussels sprouts by hydrostatic pressure. *Innovative Food Science and Emerging Technologies*, *22*, 70–75. https://doi.org/10.1016/j.ifset.2014.01.009

Iizuka, T., & Shimizu, A. (2014b). Removal of pesticide residue from cherry tomatoes by hydrostatic pressure (Part 2). *Innovative Food Science and Emerging Technologies*, *26*, 34–39. https://doi.org/10.1016/j.ifset.2013.04.011

Iizuka, T., Yahata, M., & Shimizu, A. (2013). Potential mechanism involved in removal of hydrophobic pesticides from vegetables by hydrostatic pressure. *Journal of Food Engineering*, *119*(1), 1–6. https://doi.org/10.1016/j.jfoodeng.2013.05.006

Ishfaq, B., Sameen, A., Pasha, I., Shahid, M., & Abrar, M. (2022). Evaluation of cookery methods in reduction of pesticide residues and quality attributes of okra fruit (Abelmoschus esculentus L.). *Food Science and Technology (Brazil)*, *42*, 1–9. https://doi.org/10.1590/fst.86021

Karaca, H. (2019). The effects of ozone-enriched storage atmosphere on pesticide residues and physicochemical properties of table grapes. *Ozone: Science and Engineering*, *41*(5), 404–414. https://doi.org/10.1080/01919512.2018.1555449

Kaur, N., Khunger, A., Wallen, S. L., Kaushik, A., Chaudhary, G. R., & Varma, R. S. (2021). Advanced green analytical chemistry for environmental pesticide detection. *Current Opinion in Green and Sustainable Chemistry*, *30*, 100488. https://doi.org/10.1016/j.cogsc.2021.100488

Kim, K. H., Kabir, E., & Jahan, S. A. (2017). Exposure to pesticides and the associated human health effects. *Science of the Total Environment*, *575*, 525–535. https://doi.org/10.1016/j.scitotenv.2016.09.009

Kingsley Nwosu, O., & John, A. (2022). Chemical Pesticides and Food Safety. In Ramón Eduardo Rebolledo Ranz (ed), *Insecticides – Impact and Benefits of Its Use for Humanity*, IntechOpen. April. https://doi.org/10.5772/intechopen.102395

Knio, K. M., Dagher, S., & Saad, A. (2000). The fate and persistence of zineb, maneb, and ethylenethiourea on fresh and processed tomatoes. *Food Additives and Contaminants*, *17*(5), 393–398. https://doi.org/10.1080/026520300404806

Kumar, A., Gill, J. P. S., Bedi, J. S., Chhuneja, P. K., & Kumar, A. (2020). Determination of antibiotic residues in Indian honeys and assessment of potential risks to consumers. *Journal of Apicultural Research*, *59*(1), 25–34. https://doi.org/10.1080/00218839.2019.1677000

Kumar, A., Gill, J. P. S., Bedi, J. S., & Kumar, A. (2018). Pesticide residues in Indian raw honeys, an indicator of environmental pollution. *Environmental Science and Pollution Research, 25*(34), 34005–34016. https://doi.org/10.1007/s11356-018-3312-4

Kumar, U., & Agarwal, H. C. (1991). Degradation of dithane M-45 residues in brinjals during cooking. *Bulletin of Environmental Contamination and Toxicology, 47*(5), 725–731. https://doi.org/10.1007/BF01701141

Kumari Anjana, R. K. N., & C. Jayachandran, K. G. M. (2017). Persistence of antibiotic residue in milk under region of Bihar, India. *International Journal of Current Microbiology and Applied Sciences, 6*(3), 2296–2299. https://doi.org/10.20546/ijcmas.2017.603.262

Kwon, H., Kim, T. K., Hong, S. M., Se, E. K., Cho, N. J., & Kyung, K. S. (2015). Effect of household processing on pesticide residues in field-sprayed tomatoes. *Food Science and Biotechnology, 24*(1), 1–6. https://doi.org/10.1007/s10068-015-0001-7

Lamichhane, J. R. (2017). Pesticide use and risk reduction in European farming systems with IPM: An introduction to the special issue. *Crop Protection, 97*, 1–6. https://doi.org/10.1016/j.cropro.2017.01.017

Lawler, S. P. (2017). Environmental safety review of methoprene and bacterially-derived pesticides commonly used for sustained mosquito control. *Ecotoxicology and Environmental Safety, 139*(December 2016), 335–343. https://doi.org/10.1016/j.ecoenv.2016.12.038

Lee, S. J., Park, S., Choi, J. Y., Shim, J. H., Shin, E. H., Choi, J. H., Kim, S. T., El-Aty, A. M. A., Jin, J. S., Bae, D. W., & Shin, S. C. (2009). Multiresidue analysis of pesticides with hydrolyzable functionality in cooked vegetables by liquid chromatography tandem mass spectrometry. *Biomedical Chromatography, 23*(7), 719–731. https://doi.org/10.1002/bmc.1176

Leste, T., Mahatmi, H., Costa, M., & Puja, I. K. (2019). Residues of antibiotic in chicken meat imported from Brazil and United States of America through quarantine stations in Dili,Timor Leste. *IOSR Journal of Agriculture and Veterinary Science, 12*(6), 39–43. https://doi.org/10.9790/2380-1206023943

Li, X., Liu, C., Liu, F., Zhang, X., Peng, Q., Wu, G., Lin, J., & Zhao, Z. (2023). Accelerated removal of five pesticide residues in three vegetables with ozone microbubbles. *Food Chemistry, 403*(April 2022). https://doi.org/10.1016/j.foodchem.2022.134386

Liang, Y., Lin, X., Yamada, S., Zhou, M., Inoue, M., & Inosako, K. (2012). Cucumber productivity and soil degradation in recropping system in greenhouse. *Communications in Soil Science and Plant Analysis, 43*(13), 1743–1748. https://doi.org/10.1080/00103624.2012.684820

Liu, H., Guo, D., & Feng, X. (2021). Plasma degradation of pesticides on the surface of corn and evaluation of its quality changes. *Sustainability (Switzerland), 13*(16). https://doi.org/10.3390/su13168830

Lozowicka, B., Abzeitova, E., Sagitov, A., Kaczynski, P., Toleubayev, K., & Li, A. (2015). Studies of pesticide residues in tomatoes and cucumbers from Kazakhstan and the associated health risks. *Environmental Monitoring and Assessment, 187*(10). https://doi.org/10.1007/s10661-015-4818-6

Lozowicka, B., Jankowska, M., Hrynko, I., & Kaczynski, P. (2016). Removal of 16 pesticide residues from strawberries by washing with tap and ozone water, ultrasonic cleaning and boiling. *Environmental Monitoring and Assessment, 188*(1), 1–19. https://doi.org/10.1007/s10661-015-4850-6

Macdonald, L. E., Brett, J., Kelton, D., Majowicz, S. E., Snedeker, K., & Sargeant, J. M. (2011). A systematic review and meta-analysis of the effects of pasteurization on milk vitamins, and evidence for raw milk consumption and other health-related outcomes. *Journal of Food Protection, 74*(11), 1814–1832. https://doi.org/10.4315/0362-028X.JFP-10-269

Mahmoudi, R., Norian, R., & Pajohi-Alamoti, M. (2014). Antibiotic residues in iranian honey by elisa. *International Journal of Food Properties*, *17*(10), 2367–2373. https://doi.org/10.1080/10942912.2013.809539

Martinello, M., Baggio, A., Gallina, A., & Mutinelli, F. (2013). Distribution of sulfathiazole in honey, beeswax, and honeybees and the persistence of residues in treated hives. *Journal of Agricultural and Food Chemistry,* 61(38): 9275–9279. doi: 10.1021/jf4027052. Epub 2013 Sep 10. PMID: 23978000.

McGinley, J., Healy, M. G., Ryan, P. C., O'Driscoll, H., Mellander, P. E., Morrison, L., & Siggins, A. (2023). Impact of historical legacy pesticides on achieving legislative goals in Europe. *Science of the Total Environment*, *873*(February), 162312. https://doi.org/10.1016/j.scitotenv.2023.162312

Mir, S. A., Dar, B. N., Mir, M. M., Sofi, S. A., Shah, M. A., Sidiq, T., Sunooj, K. V., Hamdani, A. M., & Mousavi Khaneghah, A. (2022). Current strategies for the reduction of pesticide residues in food products. *Journal of Food Composition and Analysis*, *106*(April 2021), 104274. https://doi.org/10.1016/j.jfca.2021.104274

Molino, F., Lázaro, R., Pérez, C., Bayarri, S., Corredera, L., & Herrera, A. (2011). Effect of pasteurization and storage on tetracycline levels in honey. *Apidologie*, *42*(3), 391–400. https://doi.org/10.1007/s13592-011-0015-y

Moudgil, P., Bedi, J. S., Aulakh, R. S., & Gill, J. P. S. (2019). Antibiotic residues and mycotoxins in raw milk in Punjab (India): A rising concern for food safety. *Journal of Food Science and Technology*, *56*(11), 5146–5151. https://doi.org/10.1007/s13197-019-03963-8

Munir, S., Azeem, A., Sikandar Zaman, M., & Zia Ul Haq, M. (2024). From field to table: Ensuring food safety by reducing pesticide residues in food. *Science of the Total Environment*, *922*(February), 171382. https://doi.org/10.1016/j.scitotenv.2024.171382

Muralikrishna, P., Mathew, T. B., Sreelakshmi, P., Koshy, B. A., Paul, A., & Rajith, R. (2016). Evaluation of different household practices to decontaminate organophosphate insecticide residues from amaranthus tricolor l. *Entomon*, *41*(3), 195–202.

Nchima, G., Choongo, K., Muzandu, K., Nalubamba, K., Muma, J., Bumbangi, F., Monga, G., & Kangwa, H. (2017). Determination of oxytetracycline and sulphamethazine residues in marketed beef from selected parts of Zambia to assess compliance with maximum residual limits. *American Journal of Research Communication*, *5*(9), 43. www.usa-journals.com www.usa-journals.com

Nguyen, T. T., Rosello, C., Bélanger, R., & Ratti, C. (2020). Fate of residual pesticides in Fruit and Vegetable Waste (FVW) processing. *Foods*, *9*(10). https://doi.org/10.3390/foods9101468

Phan, K. T. K., Phan, H. T., Boonyawan, D., Intipunya, P., Brennan, C. S., Regenstein, J. M., & Phimolsiripol, Y. (2018). Non-thermal plasma for elimination of pesticide residues in mango. *Innovative Food Science and Emerging Technologies*, 48(December 2017), 164–171. https://doi.org/10.1016/j.ifset.2018.06.009

Picó, Y., Farré, M. L., Tokman, N., & Barceló, D. (2008). Rapid and sensitive ultra-high-pressure liquid chromatography-quadrupole time-of-flight mass spectrometry for the quantification of amitraz and identification of its degradation products in fruits. *Journal of Chromatography A*, *1203*(1), 36–46. https://doi.org/10.1016/j.chroma.2008.07.018

Rafique, N., Nazir, S., Akram, S., Ahad, K., Gohar, A., Abbasi, S. T., Ahmed, I., & Rafique, K. (2018). Screening of multiclass pesticide residues in honey by SPE-GC/MSD: A pilot study. *Environmental Monitoring and Assessment*, *190*(11). https://doi.org/10.1007/s10661-018-7041-4

Rajmohan, K. S., Chandrasekaran, R., & Varjani, S. (2020). A review on occurrence of pesticides in environment and current technologies for their remediation and management. *Indian Journal of Microbiology*, *60*(2), 125–138. https://doi.org/10.1007/s12088-019-00841-x

Ramatla, T., Ngoma, L., Adetunji, M., & Mwanza, M. (2017). Evaluation of antibiotic residues in raw meat using different analytical methods. *Antibiotics*, *6*(4), 1–17. https://doi.org/10.3390/antibiotics6040034

Ranjitha Gracy, T. K., Gupta, V., & Mahendran, R. (2019). Influence of low-pressure nonthermal dielectric barrier discharge plasma on chlorpyrifos reduction in tomatoes. *Journal of Food Process Engineering*, *42*(6), 1–16. https://doi.org/10.1111/jfpe.13242

Rodrigues, A. A. Z., Queiroz, M. E. L. R. de, Neves, A. A., Oliveira, A. F. de, Prates, L. H. F., Freitas, J. F. de, Heleno, F. F., & Faroni, L. R. D. A. (2019). Use of ozone and detergent for removal of pesticides and improving storage quality of tomato. *Food Research International*, *125*(August), 108626. https://doi.org/10.1016/j.foodres.2019.108626

Sajtos, Z., Ragyák, Á. Z., Hódi, F., Szigeti, V., Bellér, G., & Baranyai, E. (2024). Hydroxymethylfurfural content of old honey samples – Does the sticky treat really last forever? *Lwt*, 193(October 2023). https://doi.org/10.1016/j.lwt.2024.115781

Salama, N. A., Abou-Raya, S. H., Shalaby, A. R., Emam, W. H., & Mehaya, F. M. (2011). Incidence of tetracycline residues in chicken meat and liver retailed to consumers. *Food Additives and Contaminants: Part B Surveillance*, *4*(2), 88–93. https://doi.org/10.1080/19393210.2011.585245

Sarangapani, C., Misra, N. N., Milosavljevic, V., Bourke, P., O'Regan, F., & Cullen, P. J. (2016). Pesticide degradation in water using atmospheric air cold plasma. *Journal of Water Process Engineering*, *9*, 225–232. https://doi.org/10.1016/j.jwpe.2016.01.003

Sarangapani, C., O'Toole, G., Cullen, P. J., & Bourke, P. (2017). Atmospheric cold plasma dissipation efficiency of agrochemicals on blueberries. *Innovative Food Science and Emerging Technologies*, *44*, 235–241. https://doi.org/10.1016/j.ifset.2017.02.012

Saridaki-Papakonstadinou, M., Andredakis, S., Burriel, A., & Tsachev, I. (2006). Determination of tetracycline residues in Greek honey. *Trakia Journal of Sciences*, *4*(1), 33–36. www.uni-sz.bg

Schieppati, D., Galli, F., Peyot, M. L., Yargeau, V., Bianchi, C. L., & Boffito, D. C. (2019). An ultrasound-assisted photocatalytic treatment to remove an herbicidal pollutant from wastewaters. *Ultrasonics Sonochemistry*, *54*(December 2018), 302–310. https://doi.org/10.1016/j.ultsonch.2019.01.027

Sidirokastritis, N. D., Karazafeiris, E., Tananaki, C., & Vareltzis, P. (2024). High hydrostatic pressure effect on the concentration of selected pesticides residues in spiked honey samples. *Innovative Food Science and Emerging Technologies*, 91(February 2023), 103527. https://doi.org/10.1016/j.ifset.2023.103527

Sidirokastritis, N. D., Tsiantoulas, I., Tananaki, C., & Vareltzis, P. (2022). The effect of high hydrostatic pressure on tetracycline hydrochloride and sulfathiazole residues in various food matrices–comparison with ultrasound and heat treatment. Food Additives and Contaminants – Part A Chemistry, Analysis, *Control, Exposure and Risk Assessment*, *39*(4), 687–698. https://doi.org/10.1080/19440049.2022.2036820

Thihara Rodrigues, F., Marchioni, E., Lordel-Madeleine, S., Kuntz, F., Casañas Haasis Villavicencio, A. L., & Julien-David, D. (2020). Degradation of profenofos in aqueous solution and in vegetable sample by electron beam radiation. *Radiation Physics and Chemistry*, 166(August 2019), 108441. https://doi.org/10.1016/j.radphyschem.2019.108441

Thompson, H. M., Waite, R. J., Wilkins, S., Brown, M. A., Bigwood, T., Shaw, M., Ridgway, C., & Sharman, M. (2005). Effects of European foulbrood treatment regime on oxytetracycline levels in honey extracted from treated honeybee (Apis mellifera) colonies and toxicity to brood. *Food Additives and Contaminants*, *22*(6), 573–578. https://doi.org/10.1080/02652030500089986

Tian, Z., Wang, J. W., Li, J., & Han, B. (2021). Designing future crops: Challenges and strategies for sustainable agriculture. *Plant Journal*, *105*(5), 1165–1178. https://doi.org/10.1111/tpj.15107

Tillotson, G. S., Doern, G. V., & Blondeau, J. M. (2006). Optimal antimicrobial therapy: The balance of potency and exposure. *Expert Opinion on Investigational Drugs*, *15*(4), 335–337. https://doi.org/10.1517/13543784.15.4.335

Tsigouri, A. D., Menkissoglu-Spiroudi, U., & Thrasyvoulou, A. (2001). Study of tau-fluvalinate persistence in honey. *Pest Management Science*, *57*(5), 467–471. https://doi.org/10.1002/ps.303

Uygun, U. (1997). Degradation of chlorfenvinphos in carrots during storage. *Food Chemistry*, *60*(4), 479–487. https://doi.org/10.1016/S0308-8146(96)00274-9

Venkatachalapathy, R., Anoop Chandra, I. R., Das, S., Vajiha Aafrin, B., Lalitha Priya, U., Peter, M. J., Karthikeyan, S., & Sukumar, M. (2020). Effective removal of organophosphorus pesticide residues in tomatoes using natural extracts. *Journal of Food Process Engineering*, *43*(2), 1–10. https://doi.org/10.1111/jfpe.13351

Waliszewski, S. M., Infanzón, R. M., Carvajal, O., & Hart, M. M. (2003). Removing τ-fluvalinate residues from press-extracted honey. *Journal of the Science of Food and Agriculture*, *83*(12), 1225–1227. https://doi.org/10.1002/jsfa.1530

Wang, Q., Li, Y., Sun, D. W., & Zhu, Z. (2018). enhancing food processing by pulsed and high voltage electric fields: Principles and applications. *Critical Reviews in Food Science and Nutrition*, *58*(13), 2285–2298. https://doi.org/10.1080/10408398.2018.1434609

Wang, Y., Dong, X., Han, M., Yang, Z., Wang, Y., Qian, L., Huang, M., Luo, B., Wang, H., Chen, Y., & Jiang, Q. (2022). Antibiotic residues in honey in the Chinese market and human health risk assessment. *Journal of Hazardous Materials*, *440*(August), 129815. https://doi.org/10.1016/j.jhazmat.2022.129815

Wanwimolruk, S., Phopin, K., Boonpangrak, S., & Prachayasittikul, V. (2016). Food safety in Thailand 4: Comparison of pesticide residues found in three commonly consumed vegetables purchased from local markets and supermarkets in Thailand. *PeerJ*, *2016*(9). https://doi.org/10.7717/peerj.2432

Wenzel, S. E. (2012). Asthma phenotypes: The evolution from clinical to molecular approaches. *Nature Medicine*, *18*(5), 716–725. https://doi.org/10.1038/nm.2678

Wu-Wu, J. W. F., Guadamuz-Mayorga, C., Oviedo-Cerdas, D., & Zamora, W. J. (2023). Antibiotic resistance and food safety: Perspectives on new technologies and molecules for microbial control in the food industry. *Antibiotics*, *12*(3). https://doi.org/10.3390/antibiotics12030550

Xu, R., Cheng, N., Huang, W., Gao, H., Deng, J., & Cao, W. (2012). Effects of the processing steps on parathion levels during honey production and parathion removal by macroporous adsorption resins. *Food Control*, *23*(1), 234–237. https://doi.org/10.1016/j.foodcont.2011.07.020

Yang, T., Doherty, J., Zhao, B., Kinchla, A. J., Clark, J. M., & He, L. (2017). Effectiveness of commercial and homemade washing agents in removing pesticide residues on and in apples. *Journal of Agricultural and Food Chemistry*, *65*(44), 9744–9752. https://doi.org/10.1021/acs.jafc.7b03118

Yang, Y., Qiu, W., Li, Y., & Liu, L. (2020). Antibiotic residues in poultry food in Fujian Province of China. *Food Additives and Contaminants: Part B Surveillance*, *13*(3), 177–184. https://doi.org/10.1080/19393210.2020.1751309

Yavuz, H., Guler, G. O., Aktumsek, A., Cakmak, Y. S., & Ozparlak, H. (2010). Determination of some organochlorine pesticide residues in honeys from Konya, Turkey. *Environmental Monitoring and Assessment*, *168*(1–4), 277–283. https://doi.org/10.1007/s10661-009-1111-6

Yuan, Z., Yao, J., Liu, H., Han, J., & Trebše, P. (2014). Photodegradation of organophosphorus pesticides in honey medium. *Ecotoxicology and Environmental Safety*, *108*, 84–88. https://doi.org/10.1016/j.ecoenv.2014.06.032

Zhang, L. Z. (1991). *Radiotracer Studies of Fungicide Residues in Food Punts.* Proceedings of a Final Research Co-Ordination Meeting Organized By the Joint Fao/Iaea Division of Nuclear Techniques in Food and Agriculture and Held in Ankara. March, 13–17.

Zhang, Y., Hou, Y., Zhang, Y., Chen, J., Chen, F., Liao, X., & Hu, X. (2012). Reduction of diazinon and dimethoate in apple juice by pulsed electric field treatment. *Journal of the Science of Food and Agriculture*, *92*(4), 743–750. https://doi.org/10.1002/jsfa.4636

Zheng, N., Wang, J., Han, R., Xu, X., Zhen, Y., Qu, X., Sun, P., Li, S., & Yu, Z. (2013). Occurrence of several main antibiotic residues in raw milk in 10 provinces of China. *Food Additives and Contaminants: Part B Surveillance*, *6*(2), 84–89. https://doi.org/10.1080/19393210.2012.727189

Zhu, Q., Cao, W., Gao, H., Chen, N., Wang, B., & Yu, S. (2010). Effects of the processing steps on chlorpyrifos levels during honey production. *Food Control*, *21*(11), 1497–1499. https://doi.org/10.1016/j.foodcont.2010.04.022

Zhu, Y., Zhang, T., Xu, D., Wang, S., Yuan, Y., He, S., & Cao, Y. (2019). The removal of pesticide residues from pakchoi (Brassica rape L. ssp. chinensis) by ultrasonic treatment. *Food Control*, *95*, 176–180. https://doi.org/10.1016/j.foodcont.2018.07.039

16 Regulatory Issues in Terms of Food Processing and Manufacturing Related Effects and Food Toxicology

Maria Tarapoulouzi and Sofia Agriopoulou

16.1 INTRODUCTION

In the context of regulatory framework, the EU's Registration, Evaluation, Authorisation, and Restriction of Chemicals (REACH) specifically includes provisions for chemicals with endocrine-disrupting properties. Chemicals identified as EDCs may be restricted or require authorization. REACH Regulation also covers heavy metals in industrial applications that might contaminate food. European Food Safety Authority (EFSA) assesses the risk of chemicals used in food packaging and food contact materials, focusing on potential endocrine-disrupting properties, as well as heavy metals, acrylamide, PAHs based on new evidence (EFSA CONTAM et al., 2024; EFSA, 2008). Plant Protection Products Regulation and Biocidal Products Regulation include specific criteria for identifying and regulating endocrine-disrupting pesticides and biocides (Lindeman and Ritz, 2020).

In the US, the Environmental Protection Agency (EPA) has set up the Endocrine Disruptor Screening Program (EDSP) focuses on identifying chemicals that may have

DOI: 10.1201/9781032712291-16

endocrine-disrupting properties, especially in pesticides and industrial chemicals (Maffini and Vandenberg, 2022). The EPA establishes drinking water standards for heavy metals and provides guidelines for heavy metal contaminants in agricultural land, which could affect food production (Wong et al., 2022). Also, Food and Drug Administration (FDA) regulates food additives, food packaging, acrylamide, and contact materials, but its approach to endocrine disruptors is often risk-based, meaning the focus is on the level of exposure rather than inherent hazard. FDA sets action levels (similar to maximum permissible levels) for heavy metals in food, focusing on specific metals and food types where contamination is a known issue (Hembrom et al., 2020). In terms of acrylamide, the FDA has not established formal limits on acrylamide (AA) in food but has issued guidance to help food manufacturers reduce its levels (Pietropaoli et al., 2022). Unlike the EU, the US regulatory approach for AA is more focused on voluntary measures rather than enforceable limits. In addition, Toxic Substances Control Act (TSCA) is a law that allows the EPA to regulate industrial chemicals, including those that may migrate into food from packaging, but action on endocrine disruptors has been slower compared to the EU (Duh-Leong et al., 2023).

In Japan, the Food Safety Commission of Japan (FSCJ) assesses the risk of chemicals in food packaging and food contact materials, including potential endocrine disruptors, AA and heavy metals in food (Blair and Regenstein, 2015). The Ministry of Health, Labour and Welfare (MHLW) regulates food safety and has set standards for food additives and contact materials based on toxicological assessments, including endocrine-disrupting properties, heavy metals (Rani et al., 2024). Japan has strict standards for food contaminants, including heavy metals, especially due to its history with mercury contamination (e.g., Minamata disease) (FSCJ, 2005; Suzuki et al., 2022). The Food Sanitation Law sets limits for heavy metals in food, while the Water Supply Act regulates heavy metal levels in drinking water. The Japanese Agricultural Standard (JAS) for honey includes testing for HMF to ensure freshness and safety, aligning with international standards, particularly for honey imported into Japan (Kamiloglu et al., 2021).

Health Canada and the Canadian Food Inspection Agency (CFIA) regulate chemicals in food, food contact materials, and additives, focusing on both toxicology and endocrine disruption. Canada uses a risk-based approach to limit heavy metal exposure through food and drinking water (Smythe, 2024). The CFIA monitors imported and domestic honey for compliance with general food safety regulations, which include the assessment of HMF levels, particularly for ensuring honey freshness (Chaven, 2023). Furthermore, the Canadian Environmental Protection Act (CEPA) allows for regulation of industrial chemicals, including those with endocrine-disrupting properties, that may migrate into food. CEPA regulates industrial emissions of heavy metals, preventing contamination of agricultural land and water sources (Rachkowski, 2024).

In China, the National Health Commission (NHC) regulates food safety and has established standards for food additives, acrylamide, and contact materials (Yang et al., 2023), and the Ministry of Ecology and Environment (MEE) oversees industrial chemicals, but endocrine disruptor-specific regulations are still emerging. MEE manages industrial emissions that can contaminate soil and water (Zhang et al., 2020). China has set maximum permissible levels for heavy metals in food, water, and food packaging materials (Han et al., 2022). In China, the regulation of acrylamide, is managed by the NHC and the SAMR under the Food Safety Law of the People's Republic of China. The SAMR oversees food safety, including monitoring HMF in honey, baked goods, and fruit juices as part of general quality control measures (Zhang et al., 2014).

In Australia, the Food Standards Australia New Zealand (FSANZ) is responsible for food safety, including assessing the risks associated with endocrine disruptors in food contact materials and additives (Tang et al., 2023). FSANZ sets maximum levels for contaminants in food, including heavy metals and PAHs through the Australia New Zealand Food Standards Code. On the other hand, FSANZ has not set maximum limits for AA in food but has issued guidelines for food manufacturers to reduce its presence in food products (Croft et al., 2004). Australia's Department of Health sets drinking water standards for heavy metals under the Australian Drinking Water Guidelines (Huston et al., 2012). The regulation of industrial chemicals and pesticides falls under NICNAS (National Industrial Chemicals Notification and Assessment Scheme), but endocrine disruptor-specific regulations are still developing (Australia, 1991).

16.2 ENDOCRINE DISRUPTORS

Regulatory frameworks for endocrine disruptors (Bisphenol A (BPA), pesticides and phthalates and plasticizers) vary significantly across countries, reflecting differences in scientific, economic, and political considerations. The EU leads with its precautionary and hazard-based approach, particularly in food safety, while the United States and other countries like Canada and Japan take more of a risk-based approach. Regulatory systems worldwide continue to evolve as scientific evidence of endocrine disruption becomes clearer, but challenges remain in harmonizing regulations, particularly in relation to food safety and toxicological assessments.

The EU has set very low tolerable daily intake (TDI) levels for BPA and has banned its use in baby bottles and other food contact materials aimed at infants. In addition, the EU uses a hazard-based approach to regulate pesticides with endocrine-disrupting properties. This means that if a pesticide is identified as an endocrine disruptor, it may be banned regardless of exposure level. Moreover, the EU has strict migration limits for phthalates in food contact materials (Vom Saal et al., 2024).

The FDA in the US has reviewed the safety of BPA in food contact materials multiple times and maintains that BPA is safe at current exposure levels, though it has banned BPA in baby bottles and sippy cups (Siddique et al., 2021). This contrasts with stricter European regulations. Furthermore, the FDA has set migration limits, but the regulatory process has been slower in addressing new scientific evidence on their endocrine-disrupting potential. The EPA regulates pesticides under the Federal Insecticide, Fungicide, and Rodenticide Act (FIFRA), but uses a risk-based assessment that includes evaluating the likelihood of exposure in food. Endocrine-disrupting pesticides remain in use in the US under controlled conditions (Yang et al., 2024).

Japan took early action to reduce BPA use in food contact materials, particularly in baby bottles and other products for infants. Alternatives to BPA are increasingly used in food packaging (Dehdashti et al., 2023). Moreover, Japan regulates pesticides under the Agricultural Chemicals Regulation Law, which includes consideration of endocrine-disrupting effects (Fujimori, 2004). However, the regulation of endocrine-disrupting pesticides has been slower compared to the EU. In terms of plasticizers and phthalates, food packaging materials are subject to migration limits, and Japan has been proactive in testing for potential leaching of endocrine disruptors (Ong et al., 2022).

Canada was the first country to declare BPA as toxic and banned its use in baby bottles. Health Canada continues to monitor BPA levels in food packaging and has recommended the reduction of BPA in all food contact materials (Dubeau et al., 2023). Health Canada regulates pesticides under the Pest Control Products Act (PCPA), and this agency assesses pesticides for endocrine-disrupting effects, but the regulatory system remains focused on risk assessment rather than a strict hazard-based approach (Ghamari and Helfinger, 2019). Also, Canada has implemented restrictions on phthalates in children's products but has less stringent regulations on their use in food packaging compared to the EU (Navaranjan et al., 2021).

China has not implemented a nationwide ban on BPA, though its use in baby products is restricted. Food packaging and containers still widely use BPA, but there are increasing efforts to transition to safer alternatives. The regulation of pesticides with endocrine-disrupting properties is evolving. China is becoming more aligned with international standards, but gaps remain in enforcement and risk assessment. Phthalates are still commonly used in food packaging, though concerns over their endocrine-disrupting properties have prompted tighter regulations and migration limits (Dueñas-Moreno et al., 2023).

In Australia, FSANZ has reviewed BPA and concluded that current exposure levels from food contact materials do not pose a significant risk to human health. However, it continues to monitor scientific evidence and consumer products (Charkiewicz et al., 2024). Also, Australia assesses pesticides for endocrine-disrupting properties during the registration process. While endocrine disruptors are considered in risk assessments, Australia has not adopted the same strict hazard-based bans as the EU. Furthermore, FSANZ sets migration limits for phthalates in food packaging, but has not banned their use outright (Adenuga, 2023).

16.3 HEAVY METALS

Heavy metals like lead, mercury, arsenic, and cadmium are toxic and persistent environmental contaminants that pose significant health risks when they enter the food chain. Regulatory frameworks for heavy metals in food vary across different countries, driven by differences in risk assessments, economic contexts, and scientific standards. European Commission Regulation (EC) No 2023/915 sets maximum levels for contaminants, including heavy metals, in various food categories (e.g., cereals, fish, vegetables, meat) (Commission Regulation (EC) No 2023/915). The EU and Japan have some of the strictest regulations, with precautionary limits and comprehensive monitoring systems. The US, Canada, Australia, and China also have strong regulatory frameworks, but tend to rely more on risk-based assessments.

The EU has established very low maximum levels for lead in baby food, drinking water, and cereals. Lead exposure can cause developmental issues, particularly in children, and the EU has progressively reduced permissible levels in food. In terms of mercury, fish is a primary source of mercury, and the EU has set maximum allowable levels for mercury in seafood, especially in species like tuna and swordfish. Methylmercury, a toxic organic form, is closely monitored. In addition, the EU has set maximum levels for arsenic in rice and rice-based products, a common source of arsenic contamination. The EU regulates cadmium in cereals, vegetables, and shellfish, setting limits based on its bioaccumulation and potential to cause kidney damage and bone demineralization (Koch et al., 2022).

In the US, FDA has set action levels for lead in food products, including bottled water, juices, and candy. The FDA's **closer to zero** initiative aims to further reduce lead exposure in foods, especially those consumed by infants and children (Hoffman-Pennesi et al., 2024). The FDA and EPA have joint guidelines on mercury levels in fish, with recommendations for safe consumption levels, especially for pregnant women and children. The focus is on methylmercury, a highly toxic form that accumulates in fish (Spiller et al., 2023). Arsenic, particularly in rice and rice-based products, is closely monitored. The FDA has set action levels for inorganic arsenic in infant rice cereals and is working to reduce exposure to arsenic through guidelines for water and food, as well as cadmium in shellfish, cereals, and leafy vegetables, setting limits to reduce its bioaccumulation in human tissues, which can lead to kidney damage and other health problems over time (Genchi et al., 2022).

Japan has strict limits on lead in food products, particularly in drinking water, rice, and infant food, given the harmful effects of lead on neurological development. Also, Japan strictly regulates mercury levels in fish and seafood due to its significant seafood consumption. Following the Minamata disease outbreak, which was caused by methylmercury poisoning, Japan closely monitors mercury levels in marine products (Zafar et al., 2024). Arsenic levels in water and food are regulated to prevent exposure to both organic and inorganic forms, particularly in seafood and rice. Japan has set stringent maximum permissible levels for arsenic in drinking water. In addition, Japan has established cadmium limits for rice and leafy vegetables, which can absorb cadmium from contaminated soils, to prevent kidney and bone damage in its population (Suzuki et al., 2022).

Lead contamination in food and water is a significant concern in China, given the country's historical issues with environmental pollution. China has established limits for lead in food, particularly in children's products, cereals, and drinking water. Mercury in seafood, especially freshwater fish, is a key concern due to pollution from industrial activities. China has set limits for mercury in fish, and efforts are underway to reduce emissions of mercury from coal combustion. Inorganic arsenic in rice is a major issue, as rice is a staple food in China. China has established maximum permissible levels for arsenic in rice and drinking water, following incidents of arsenic poisoning from contaminated groundwater. Soil contamination from industrial pollution is a concern, and cadmium has been found in rice and other crops. China has set limits for cadmium in food, especially rice, which is the main route of cadmium exposure for its population (Zhang et al., 2021).

Canada has set maximum levels for lead in food products, drinking water, and infant food. Lead contamination is closely monitored in water and food contact materials to prevent exposure, particularly in vulnerable populations. Health Canada issues advisories for fish consumption, particularly for pregnant women and children, due to mercury contamination in seafood. The focus is on methylmercury in predatory fish like tuna and swordfish. Canada has set stringent limits on inorganic arsenic in rice and rice products, as well as in drinking water. The country's guidelines aim to minimize exposure to this carcinogen. Cadmium is regulated in food products like shellfish and grains. Canada's limits on cadmium aim to reduce the risk of kidney damage and bone health issues due to chronic exposure (Chan et al., 2021).

Australia has strict lead limits in food and drinking water. FSANZ monitors lead contamination in imported products and drinking water, and the country has gradually reduced lead levels in the environment. FSANZ provides guidelines for mercury in fish, particularly focusing on safe consumption levels for at-risk populations. Australia's seafood consumption guidelines aim to minimize mercury exposure. Australia regulates arsenic levels in drinking water, rice, and seafood. FSANZ has established limits for inorganic arsenic in rice and rice-based products. Cadmium is regulated in vegetables, grains, and shellfish, with strict limits to protect consumers from its long-term toxic effects on the kidneys and bones (James, 2012).

16.4 ACRYLAMIDE

Regulatory approaches to AA vary across countries but share some common elements. The EU has the most formalized framework, with benchmark levels and mandatory mitigation efforts. The US, Japan, Canada, China, and Australia rely primarily on voluntary measures and industry cooperation, encouraging food manufacturers to adopt practices that lower AA levels. In all regions, the focus is on reducing exposure, particularly for vulnerable populations like infants and children, due to acrylamide's potential carcinogenic effects. Although there is no global consensus on maximum limits, the trend is toward mitigating AA formation through improved food processing methods, consumer education, and continuous monitoring.

The EU has a well-developed regulatory approach to AA in food, primarily governed by Commission Regulation (EU) 2017/2158, which sets benchmark levels for AA in various food categories, such as potato-based products, bread, and coffee (Koszucka et al., 2020).

In the US, AA has been classified by the EPA as a probable human carcinogen based on animal studies (Bušová et al., 2020). The FDA focuses on reducing exposure by encouraging industry to adopt practices that limit AA formation. Although AA levels in processed foods like potato products and coffee are a concern, the FDA has not set mandatory limits but encourages reducing its levels wherever feasible (Rifai and Saleh, 2020).

Japan does not have specific legal limits on AA in food but relies on general food safety regulations under the Food Sanitation Law, administered by the MHLW. In line with international assessments, Japan acknowledges the potential carcinogenic risks of acrylamide, focusing on research and monitoring of AA levels in foods like snacks, instant coffee, and baked goods. AA is currently monitored in various foods, but Japan's approach largely mirrors that of the US, relying on voluntary compliance by food manufacturers rather than setting mandatory limits (Zha et al., 2020).

In Canada, Health Canada and the CFIA monitor AA levels in food, with Health Canada providing guidelines for industry on reducing AA formation. Canada has not set legal maximum limits for AA in food but has worked with the food industry to implement mitigation strategies. Health Canada issued a Fact Sheet and guidance on AA reduction in 2012. Similar to the US, Canada's approach is based on voluntary industry cooperation rather than strict regulatory controls, although monitoring and risk assessments are regularly conducted (Sá and House, 2024).

While China does not have specific legal limits for AA in food, authorities monitor food safety through regular inspections and testing of food products. China has adopted guidelines similar to those used internationally, encouraging the reduction of AA through voluntary industry measures.

Public awareness campaigns in Australia stress the importance of adjusting home-cooking practices to lower AA formation in foods like toast and French fries. Research into AA levels in Australian foods continues, with FSANZ focusing on high-risk food categories for future risk assessments.

16.5 HYDROXYMETHYLFURFURAL (HMF)

Globally, regulatory approaches to HMF in food are still developing. The EU has taken the lead in regulating HMF in honey, with strict limits, while other regions like the United States, Canada, Japan, and China primarily focus on monitoring and voluntary industry compliance rather than enforcing specific legal limits across all food categories. Most countries follow the Codex Alimentarius standards for honey, which limits HMF to 40 mg/kg (and 80 mg/kg in tropical regions) (Bogdanov et al., 2002; Yap et al., 2024). However, in food categories like baked goods, coffee, and fruit juices, HMF levels are generally not subject to formal regulation, though governments encourage good manufacturing practices to reduce its formation.

In the EU, EFSA has reviewed the toxicological profile of HMF and found evidence of its genotoxic and mutagenic effects in animal studies, though clear evidence of carcinogenicity in humans is lacking. HMF is considered a potentially harmful byproduct of thermal processing, especially at high concentrations, but its presence in moderate amounts, as commonly found in food, is not considered an acute risk. Foods like honey, dried fruits, nuts, juices, and coffee are monitored for HMF content. While no formal restrictions are in place for many food categories, continuous monitoring of levels and public awareness are emphasized (Liu et al., 2023; Yap et al., 2024).

While no strict regulatory limits have been placed in the US, the FDA encourages minimizing HMF formation through better food processing practices. The focus is on monitoring HMF levels in processed foods such as honey, caramelized products, and baked goods.

Japan has no specific legal limits for HMF in food, but the MHLW oversees food safety, including the presence of HMF in imported and domestic honey and other processed foods.

China does not have specific regulations governing the maximum HMF content in most food categories, but there are limits for honey, in line with international standards. Honey producers are required to meet a maximum HMF limit of 40 mg/kg to ensure product quality (Bogdanov et al., 2002).

For honey, Canada and Australia (particularly the Australian Honey Bee Industry Council (AHBIC) as well as other regulatory bodies) recommend compliance with the international standard of 40 mg/kg for HMF, aligning with practices in the EU and other countries.

16.6 POLYCYCLIC AROMATIC HYDROCARBONS (PAHS)

The EU has one of the most comprehensive regulatory frameworks for PAHs in food. Commission Regulation (EC) No. 2023/915 sets maximum levels for PAHs in certain foodstuffs. Specifically, the regulation establishes limits for benzo(a)pyrene and the sum of four specific PAHs (benzo(a)pyrene, benzo(a)anthracene, chrysene, and benzo(b)fluoranthene). The key **maximum levels** defined for different foodstuffs are the following:

1. Benzo(a)pyrene
   - Smoked meats and smoked meat products: 2.0 μg/kg.
   - Smoked fish and smoked fishery products: 2.0 μg/kg.
   - Infant formulae and follow-on formulae: 1.0 μg/kg.
   - Infant foods and processed cereal-based foods for infants and young children: 1.0 μg/kg.
2. Sum of benzo(a)pyrene, benzo(a)anthracene, chrysene, and benzo(b) fluoranthene
   - Smoked meats and smoked meat products: 12.0 μg/kg.
   - Smoked fish and smoked fishery products: 12.0 μg/kg.
   - Infant formulae and follow-on formulae: 1.0 μg/kg.
   - Infant foods and processed cereal-based foods for infants and young children: 1.0 μg/kg.
3. Other foodstuffs
   - Cocoa beans and derived products (such as cocoa butter): 5.0 μg/kg for benzo(a)pyrene and 30.0 μg/kg for the sum of benzo(a)pyrene, benzo(a) anthracene, chrysene, and benzo(b)fluoranthene.
   - Coconut oil, palm oil, and similar oils: 5.0 μg/kg for benzo(a)pyrene and 30.0 μg/kg for the sum of benzo(a)pyrene, benzo(a)anthracene, chrysene, and benzo(b)fluoranthene.
4. Specific smoking or drying methods
   - Foods processed using traditional smoking or drying methods, especially in certain regions, are subject to specific derogations. However, these foods are still monitored to ensure compliance with food safety standards.

While there are no specific federal limits for PAHs in food in the US, there is a strong emphasis on reducing PAH formation during food preparation, especially in home-cooking and food processing industries. The EPA set a legal limit of 0.2 μg/L for BaP in drinking water as part of the National Primary Drinking Water Regulations (Sampaio et al., 2021).

Japan has conducted extensive research into PAH contamination in grilled fish and other common Japanese foods, however there are no specific PAH limits for most food categories, but PAHs are included in food safety monitoring programs (Gao et al., 2022). PAH levels in foods like fish, soy sauce, and grilled meats have been a subject of research, as these are staples in the Japanese diet. The government encourages food processors to adopt methods that reduce PAH formation. Public health recommendations advise limiting consumption of heavily grilled or smoked foods, but no strict regulatory limits on PAHs are in place.

Also, for China there are no specific legal limits for PAHs.

While there are no specific federal limits for PAHs in all foods in Canada, there are guidelines for monitoring PAHs, especially in smoked meats and seafood. Canada's focus is primarily on imported products such as smoked fish and grilled foods known to contain elevated PAH levels. Health Canada has created a list of regulatory limits for certain contaminants and adulterants in specific foods, including a BaP limit of 3.0 μg/kg in olive pomace oils, 0.04 μg/L in drinking water and the maximum allowable level of BaP in food has been set at 5 μg/kg in specific categories like fish and smoked meats, aligning with international standards (Sampaio et al., 2021).

In Australia and New Zealand, PAH levels are regulated under general **food contaminants** legislation. Therefore, FSANZ has set the daily exposure to BaP at 0.024 ng/kgbw (Barzegar et al., 2023).

16.7 N-NITROSO COMPOUNDS, AROMATIC AMINES, AMIDES, AND NITROSAMINES

The EU has developed some of the most comprehensive regulations, setting specific limits on nitrosamine levels in processed foods and limiting the use of nitrates and nitrites in preservation. The US, Japan, China, Canada, Australia, and New Zealand focus on monitoring, guidelines, and public health campaigns to reduce the consumption of processed meats and high-temperature cooking methods that promote nitrosamine formation. The toxicological concerns center around the mutagenic and carcinogenic potential of these compounds, particularly in cured, smoked, and grilled foods, and countries are working to minimize dietary exposure through regulation, food safety monitoring, and public health advice.

The EU has developed stringent regulations for controlling N-nitrosamines in food, primarily focusing on processed meats, dairy products, and cured foods, where nitrites are commonly used as preservatives. Commission Regulation (EC) No. 2023/915 sets maximum allowable limits for nitrosamines and related compounds in certain food categories. Some examples of these limits include:

1. Processed meat products
    - Cured meats (such as sausages, bacon, ham)
        - N-nitrosodimethylamine (NDMA): 2.5 μg/kg.
        - N-nitrosamines: 10 μg/kg.
    - Smoked meats
        - N-nitrosamines: 3.0 μg/kg for NDMA and total N-nitrosamines.
2. Dairy products
    - Cheese (especially processed and cured cheeses where nitrates may be used)
        - N-nitrosamines: 1.0 μg/kg.
    - Milk-based infant formulas
        - NDMA: 0.2 μg/kg, ensuring special protection for vulnerable populations, such as infants.
3. Fish and fish products
    - Cured and smoked fish products (including salted and smoked fish, like salmon, herring)
        - N-nitrosamines: 4.0 μg/kg.
4. Other cured or nitrite-preserved foods
    - Canned meats and fish
        - NDMA: 2.0 μg/kg.
    - Cured or pickled vegetables (where nitrites are sometimes used in the brining process)
        - total N-nitrosamines: 2.5 μg/kg.

In the US, the EPA regulates N-nitrosodimethylamine (NDMA) as a contaminant in drinking water, with the maximum allowable concentration set at 0.0007 μg/L (Szczuka et al., 2020). In 2020, the FDA issued warnings and recalls related to NDMA contamination in pharmaceuticals, which highlighted concerns about NDMA in food as well. Although specific regulatory limits for nitrosamines in most foods are limited, monitoring programs are in place for bacon, hot dogs, and other processed meats (Paustenbach et al., 2024).

Grilled fish and smoked meats are traditional in Japan, and research has shown elevated levels of nitrosamines in these products. However, Japanese authorities have not imposed specific limits, focusing instead on monitoring and public health guidance to minimize exposure (Barra et al., 2011).

Similar to Japan, in China no specific limits are set for these compounds in food, but they are monitored in products like cured meats and smoked foods. The government promotes improved food processing techniques to reduce nitrosamine formation, especially in traditional foods like pickled vegetables and salted meats (Li et al., 2021).

Canada does not have specific limits for nitrosamines in all food categories, but limits on nitrates and nitrites in processed foods are enforced to control their formation. Canada follows a similar approach to the EU in limiting the use of nitrites in cured meats, such as bacon, ham, and sausages, and common cooking practices like barbecuing and grilling, and this is also adopted by Australia and New Zealand.

16.8 DIOXINS OR DIOXIN-LIKE COMPOUNDS

Regulatory frameworks vary across countries, with the European Union setting some of the strictest maximum levels for dioxins in food, followed by the United States, Japan, Canada, and Australia/New Zealand, which regulate dioxins through both environmental controls and food safety measures. Despite ongoing efforts to reduce dioxin exposure, dioxins remain a major food safety concern, particularly in fatty animal products such as fish, meat, and dairy, where they tend to accumulate.

The EU has some of the most stringent regulations on dioxins in food. Commission Regulation (EC) No. 2023/951 sets maximum levels for dioxins and dioxin-like PCBs in a variety of foods, including:

1. Dairy products
   - Milk and dairy products (including butter, cheese, and cream)
     - Dioxins: 2.5 pg/g fat.
     - Dioxins and dioxin-like PCBs (sum): 5.5 pg/g fat.
2. Meat and meat products
   - Beef, pork, poultry, and derived meat products
     - Dioxins: 1.0 pg/g fat.
     - Dioxins and dioxin-like PCBs (sum): 1.5 pg/g fat.
3. Fish and fishery products
   - Fish and fishery products, especially fatty fish, like salmon, mackerel, and herring
     - Dioxins: 3.5 pg/g fat.
     - Dioxins and dioxin-like PCBs (sum): 6.5 pg/g fat.
   - Eel
     - Dioxins: 3.5 pg/g fat.
     - Dioxins and dioxin-like PCBs (sum): 10.0 pg/g fat.
4. Eggs and egg products
   - Eggs and egg products (e.g., powdered eggs)
     - Dioxins: 2.5 pg/g fat.
     - Dioxins and dioxin-like PCBs (sum): 5.0 pg/g fat.
5. Infant formula and baby foods
   - Infant formula, follow-on formula, and other baby foods
     - Dioxins and dioxin-like PCBs (sum): 1.0 pg/g fat.
     - Special consideration is given to protect vulnerable populations, such as infants and young children.
6. Animal feed
   - Feed for farm animals, including feed for cattle, poultry, and fish
     - Maximum levels are set for dioxins and dioxin-like PCBs in animal feed to prevent contamination in the food chain. For example, the limit for feed fat is typically 0.75 pg/g fat for dioxins.

In the US, the EPA has set a reference dose (RfD) for dioxins at 0.7 picograms per kilogram of body weight per day (0.7 pg/kg bw/day), based on non-cancer effects (Buckley, 2010).

Japan regulates dioxins under the Dioxin Law (1999), which aims to reduce dioxin emissions from industrial activities (Kawamoto and Weber, 2021). Food safety regulations are enforced by the MHLW. Japan has established maximum limits for dioxins and dioxin-like compounds in foods, particularly in fish, meat, and dairy products, as these are the primary sources of dioxin exposure in the Japanese diet (Takahashi et al., 2024).

In China, there are no specific national limits for dioxins in all food products, but China follows international guidelines, such as those from the Codex Alimentarius, which set recommended limits for dioxins in food and feed (De Witte et al., 2022).

Canada has established maximum allowable levels for dioxins and dioxin-like compounds in food, particularly in meat, fish, and dairy. Canada also follows Codex Alimentarius guidelines for dioxins in food and works to minimize exposure through a combination of environmental regulations and food safety controls (De Witte et al., 2022).

Similarly, in Australia and New Zealand dioxin levels in food are monitored by Codex Alimentarius standards, and maximum residue limits are set for dioxins and dioxin-like compounds in various foods, including fish, meat, and dairy products (De Witte et al., 2022).

16.9 FURANS

Furans are low-molecular-weight compounds with high volatility found in heat-treated commercial foods and produced through thermal degradation of natural food precursors such as ascorbic acid, amino acids, carbohydrates, unsaturated fatty acids, and carotenoids. Several of these compounds cause necrosis of target cells within certain organs, including the liver, the kidneys, and the lungs (Pérez-Lucas et al., 2024).

A number of maximum tolerances for contaminants, natural plant toxicants as well as for process contaminants such as 3-monochloropropane-1,2-diol are currently laid down in Commission Regulation (EC) No 1881/2006. Furan in food is currently not covered by this EU Regulation. Commission Recommendation 2007/196/EC recommended that member states should perform monitoring of the presence of furan in foodstuffs that have undergone heat treatment during the years 2007 and 2008. Results of this monitoring were published in an EFSA report (EFSA, 2011), and included data for commercial foodstuffs as purchased without any further preparation and commercial foodstuffs analyzed as consumed after further preparation in the laboratory. Both 2-methylfuran and 2, 5-dimethylfuran were on the Union list of flavorings and source materials for use in food and their conditions of use. However, they were removed from the Union list by Commission Regulation (EU) 246/2014, since their applications had been withdrawn by the industry. "The Good Scents Company" (a website maintained by the flavorings industry) reported that 3-methylfuran is not for use in flavorings (EFSA, 2017).

16.10 4(5)-METHYLIMIDAZOLE

4(5)-Methylimidazole (4(5)-MEI) occurs in food matrices as a result of the Maillard reaction that takes place between amino acids and sugars when food is thermally processed. Ammonia caramel colorant additives also constitute a significant source of 4(5)-MEI in food products. In the process of creating ammonia caramel and sulfite ammonia caramel color, the caramelization procedure causes ammonia to interact with reducing sugars, producing 4(5)-MEI (Revelou et al., 2021).

According to the Regulation (EC) No 1333/2008 the use of caramel color has been restricted to E150 a-d only in vegetables and red fruits, excluding olives, vinegar, oil, and brine. The same regulation limits the use of E150 a-d only in beers and malt/Tafelbier/Table beer (original wort content less than 6%), brown ale, porter and stout, and old ale. The use of E150a-d is banned in fruit spirits obtained by maceration and distillation of Geist, London Gin, Sambuca, Maraschino, Marrasquino or Maraskino, and Mistra. Whisky: whiskey can only contain E 150a. 4-MEI does not have a harmonized classification under Regulation (EC) No 1272/2008. In the inventory of the European Chemicals Agency (ECHA), almost 240 notifiers have classified the substance, but not as a mutagen, whereas more than 60% of the notifiers classify 4-MEI as a carcinogen category 2 (suspected of causing cancer) (Tzatzarakis et al., 2017)

The European Commission Regulation 231/2012 stipulates that the maximum level of 4(5)-MEI in ammonia caramel and ammonia sulfate caramel should not exceed 200 and 250 mg/kg, respectively (European Commission, 2012).

16.11 TRANS FATS

In the EU Regulation (EU) 2019/649 imposes a maximum limit of 2 grams of industrially produced trans fats per 100 grams of fat in foodstuffs intended for the final consumer and retail (Hishamuddin et al., 2022).

In 2015, the US FDA determined that Partially Hydrogenated Oils (PHOs), the primary source of industrial trans fats, are no longer "generally recognized as safe" (GRAS) (Kiralan et al., 2021). By 2018, PHOs were effectively banned from all processed foods. Any product containing trans fats must list them on the Nutrition Facts label, with a threshold where trans fats below 0.5 grams per serving can be labeled as "0 grams" (Golomb et al., 2021).

In 2018, Health Canada banned the use of PHOs in all foods, including imported products and those made in restaurants. The regulation stipulates that foods must not contain more than 2% trans-fat content in fats and oils (Franco-Arellano et al., 2020).

Denmark was the first country to implement strict limits on trans fats in 2003 (Nagpal et al., 2021). The regulation limits industrially produced trans fats to a maximum of 2 grams per 100 grams of total fat (Aued-Pimentel et al., 2021). This regulation is considered a model for other countries, as it has successfully led to a significant reduction in trans-fat consumption and related heart disease rates.

Argentina has Resolution No. 137/2010 that set a limit of 2% trans fats in vegetable oils and margarines and 5% in all other foods (World Health Organization, 2023). These standards align with recommendations from the Pan American Health Organization (PAHO).

Food Safety and Standards Authority of India (FSSAI), in 2021, reduced the trans-fat limit to 2% of total fat content in all foods, including bakery products and oils, aligning with the WHO's recommendations. Earlier, the limit was set at 5%, but it has been progressively reduced (Chopra et al., 2021; Firdaus and Mishra, 2021).

In general, the World Health Organization (WHO) has called for a global elimination of industrially produced trans fats by 2023, advocating for a maximum of 2 grams per 100 grams of total fat (World Health Organization, 2023). Many countries are aligning with the WHO's REPLACE initiative to ensure a coordinated effort towards the elimination of industrial trans fats, which contributes to reduced global health disparities related to diet and nutrition.

16.12 FOOD ADDITIVES

Food additives are increasingly used in processed foods worldwide, resulting in rising health exposure. The harmonization of regulations on food additives is a difficult goal due to the large differences in global food use patterns. In addition, the existence of various definitions of food additives and current regulations, makes harmonization extremely complex (Pressman et al., 2017). Magnusson et al. have compared the regulations, definitions, and approval processes for food additives intentionally added or unintentionally present in human food in the following specific countries/jurisdictions: Argentina, Australia, Brazil, Canada, China, European Union, Japan, Mexico, New Zealand, and the United States. This includes direct food additives, food ingredients, flavoring agents, food enzymes and/or processing aids, food contact materials, new food, and nanoscale materials for food applications. Although each regulatory authority uses its own regulatory framework and there may be differences in definitions, regulations, and approval processes, there are also many similarities. The main objective of any regulatory authority is to create a regulatory framework and succeed in providing consumers with safe food by maintaining and enforcing these regulations (Magnusson et al., 2013). Among the most significant differences is that the only country with a GRAS list is the United States. This means that compounds that are considered GRAS in the United States need other official approvals in order to be used as such in other countries (Pressman et al., 2017).

In the EU, the European Parliament and Council have established one of the world's most stringent legislative frameworks for food additives. Regulation (EC) No. 1333/

2008 established the common authorization steps for food additives, food flavorings, and food enzymes. Once an additive has been authorized by the EU, it is added to the EU list of allowed food additives outlined in regulation EU 1333/2008, which also specifies the conditions of use. The authorized food additives must also meet the purity criteria outlined in regulation (EC) No. 231/2012 (European Commission, 2012). Moreover, the safety of most additives approved in regulation (EC) No 1333/2008 was re-evaluated by the European Commission such as (EC) No. 257/2010 (European Commission, 2010) which set up a program for the re-evaluation of the authorized food additives in compliance with Regulation (EC) No 1333/2008 of the European Parliament and Council on food additives). The use of titanium dioxide (E171) was fully banned as a food additive by the European Commission (EC) after August 2022 based on the Commission Regulation (EC) No. 2022/63 (European Commission, 2022). Following the publication of this rule, Annex II and III of rule (EC) No. 1333/2008 were amended.

16.13 MYCOTOXINS

Due to a broad range of toxic effects on both humans and animals, some mycotoxins have been recognized by European regulatory authorities as emerging risks in food safety, thus, efficient regulatory frameworks and monitoring approaches have been developed globally relying on the latest scientific knowledge based mainly on new analytical tools and approaches. In this sense, maximum levels (MLs) have been established for the main classes of mycotoxins in several core commodities intended for food and feed by the EU, including cereals, nuts, fruits, and derived products, such as milk (Vila-López et al., 2023). Aflatoxins are ubiquitous in nature threatening the health of both consumers and animals when they consume mycotoxin-contaminated feed (Oyebamiji et al., 2024, Agriopoulou et al., 2020). Regulatory standards for aflatoxin M1 in milk and dairy products in countries worldwide appear to vary significantly from country to country, as well as by dairy product (liquid milk vs. milk powder vs. cheese, etc.). In general, dairy products intended for infants, such as infant formula, have more stringent AFM1 standards, while milk powder expected to be reconstituted by dilution with water tends to be subject to less stringent AFM1 standards (Hsu et al., 2024).

In terms of the EU, the maximum levels of the most toxicologically important mycotoxins permitted in foodstuffs have been established based on the opinions of the European Food Safety Authority (EFSA). EU Regulation 1881/2006 and its amendments (1126/2007, 105/2010, 165/2010, 594/2012 1058/2012, 2015/1137, and 2019/1901) established the maximum acceptable limits for AFs (in nuts, dried fruits, cereals and cereal products, spices, and milk), ochratoxin A (in cereals and cereal products, dried fruits, coffee, grape juice and wine, spices, licorice), patulin (in fruit juices, spirit drinks, cider), fumonisins, deoxynivalenol, zearalenone (in cereals and cereal products, maize and maize-based products, pasta, bread), and citrinin (in food supplements based on rice fermented with *Monascus purpureus* (Piotrowska et al., 2021). Commission Recommendation (EU) 2022/553 set MLs in Alternaria toxins in food (Commission Recommendation, 2022).

Those MLs were compiled in EU Commission Regulation (EC) No 1881/2006 and its amendments. In 2006, EU set a safe limit for total aflatoxins (AFs) and AFB1 at 4 g/kg and 2 g/kg, respectively, for human intake. In 2010, the limit was revised and set at 5 g/kg and 10 g/kg for AFB1 and total aflatoxin, respectively (European Commission, 2006). In 2023 the new Commission Regulation (EC) No. 2023/915, repealed Regulation (EC) No. 1881/2006 and added new food categories (Commission Regulation, 2023).

The US and Canada have set their limits for AFs and AFB1 at 20 g/kg and 15 g/kg, respectively (Ali, 2019). The United States Food and Drug Administration (FDA) has set an action level for AFM1 in milk of 0.5 μg/kg, while the EU has a very strict action level for milk of 0.05 μg/kg. The US standards were set on the basis of how much AFM1 is produced from AFB1 in dairy animal feed (U.S. FDA, 2005), while the EU standard was based on the as low as reasonably achievable (ALARA) principle (European Commission, 2006). For the sum of ergot alkaloids in milling products, the regulatory limit of 50–150 μg/kg (as of 1st July 2024) was set in Commission Regulation 2023/91521 (Commission Regulation, 2023).

Bosnia and Herzegovina and Switzerland have the lowest limit for AFB1, which is 1 g/kg (Ali, 2019).

In Japan, any concentration of aflatoxin in crops is prohibited (Dadzie et al., 2019).

In Malaysia, based on the 1985 Malaysia Food Regulation, the maximum limit of all mycotoxins was set at 35 μg/kg initially. Later, it was reviewed and set at 5 g/kg for all mycotoxins, including aflatoxin (Sabran et al., 2013).

Egypt set its limit at 4–15 g/kg and 2–12 g/kg for total aflatoxin and AFB1, respectively, while 0.01 g/kg was set as the limit for AFB1 in processed cereal-based foods and baby foods for infants and young children (Marrez and Ayesh, 2022).

Furthermore, the Food and Agriculture Organization (FAO) has also set the safe limit for AFB1 to be 1–20 g/kg in food (Ndagijimana et al., 2020).

However, different drawbacks exist nowadays, as the current MLs do not consider the exposure to multiple mycotoxins, and they are either based on the risk assessment of a single compound or their sum and no MLs have been established for other kinds of foods like pseudo-cereal grains (More et al., 2009; Palumbo et al., 2020; Ramos-Diaz et al., 2021).

16.14 AGROCHEMICALS

In the EU, Regulation (EC) No. 396/2005 establishes MRLs for pesticide residues in or on food and feed of plant and animal origin. MRLs are set for individual pesticides and are harmonized across all member states. As MRLs vary significantly depending on the pesticide and the type of food, and some of them are set as follows:

In the EU, Glyphosate in wheat is set at 10 mg/kg (Masci et al., 2024 and Mazuryk et al., 2023), and for Chlorpyrifos is reduced to near-zero levels due to its potential neurotoxicity, particularly affecting child development (Hites, 2023). In the US, MRLs for Permethrin in vegetables is typically 0.05 mg/kg (Amjad et al., 2020), and Glyphosate in wheat is set at 30 mg/kg (Masci et al., 2024). Canada has set an MRL for Glyphosate in wheat at 5 mg/kg (Masci et al., 2024). China has set up an MRL for Imidacloprid in fruits and vegetables in the range of 0.05–5 mg/kg (Zhu et al., 2020). In addition, Azoxystrobin for rice is set at 0.5 mg/kg (Cheng et al., 2022). In India for Chlorpyriphos has set a limit in vegetables at 0.2 mg/kg (Gowda and Ramesh, 2020). Regulatory bodies continue to adapt MRLs based on evolving toxicological data and scientific research to protect consumer health.

REFERENCES

Adenuga, M.D. (2023). Occurrence and dietary exposure to phthalates in the US population – a contextual review. *Food Additives and Contaminants: Part A*, *40*(1), 169–179.

Agriopoulou, S., Stamatelopoulou, E. and Varzakas, T. (2020). Advances in occurrence, importance, and mycotoxin control strategies: Prevention and detoxification in foods. *Foods*, *9*(2), 137.

Ali, N. (2019). Aflatoxins in rice: worldwide occurrence and public health perspectives. *Toxicology Reports, 6*, 1188–1197.

Amjad, A., Randhawa, M.A., Javed, M.S., Muhammad, Z., Ashraf, M., Ahmad, Z. and Murtaza, S. (2020). Dietary intake assessment of pyrethroid residues from okra and eggplant grown in peri-urban areas of Punjab, Pakistan. *Environmental Science and Pollution Research, 27*, 39693–39701.

Aued-Pimentel, S. and Kus-Yamashita, M.M. (2021). Analysis of the fat profile of industrialized food in Brazil with emphasis on trans-fatty acids. *Journal of Food Composition and Analysis*, *97*, 103799.

Australia, W. (1991). *National industrial chemicals notification and assessment scheme.* M390B Unsaturated Polyester Resin, Full Public Report: NA/166, Sydney Australia (July 1994).

Bara, V., Bara, C. and Bara, L. (2011). Nitrosamines occurrence in some food products. *Ecotoxicologie, Zootehnie si Tehnologii de Industrie Alimentară, 5*, 27–34.

Barzegar, G., Rezaei Kalantary, R., Bashiry, M., Jaafarzadeh, N., Ghanbari, F., Shakerinejad, G., Khatebasreh, M. and Sabaghan, M. (2023). Measurement of polycyclic aromatic hydrocarbons in edible oils and potential health risk to consumers using Monte Carlo simulation, southwest Iran. *Environmental Science and Pollution Research, 30*(2), 5126–5136.

Blair, R. and Regenstein, J.M. (2015). International regulations. In *Genetic Modification and Food Quality: A Down to Earth Analysis* (p. 4), First Edition. Blair, R. and Regenstein, J.M. Eds. Oxford, UK: Wiley.

Bogdanov, S., Martin, P., & Lullmann, C. (2002). Harmonised methods of the international honey commission. Swiss Bee Research Centre, FAM, Liebefeld, *5*(1), 1-62.

Buckley, D.D. (2010). Dr. Timothy Buckley, Chair Dioxin Review Panel Science Advisory Board Environmental Protection Agency (EPA), Washington, DC.

Bušová, M., Bencko, V., Laktičová, K.V., Holcátová, I. and Vargová, M. (2020). Risk of exposure to acrylamide. *Central European Journal of Public Health*, *28*, S43–S46.

Chan, H.M., Singh, K., Batal, M., Marushka, L., Tikhonov, C., Sadik, T., Schwartz, H., Ing, A. and Fediuk, K. (2021). Levels of metals and persistent organic pollutants in traditional foods consumed by First Nations living on-reserve in Canada. *Canadian Journal of Public Health*, *112*(Suppl 1), 81–96.

Charkiewicz, A.E., Omeljaniuk, W.J. and Nikliński, J. (2024). Bisphenol A—What do we know? A global or local approach at the public health risk level. *International Journal of Molecular Sciences*, *25*(11), 6229.

Chaven, S. (2023). Confectionery and bakery products; including honey. In Andersen, V., Lelieveld, H., and Motarjemi, Y. (Eds.) *Food Safety Management* (pp. 291–303). London, United Kingdom: Academic Press.

Cheng, C., Liang, Y. and Hu, J. (2022). Estimation of residue levels and dietary risk assessment of cyproconazole and azoxystrobin in cucumber after field application in China. *Environmental Science and Pollution Research*, *29*(3), 1–8.

Chopra, S., Arora, C., Malhotra, A. and Khurana, S.C. (2021). Industrially produced trans fat: Usage, health implications, global and indian regulations. *Indian Journal of Public Health*, *65*(1), 71–75.

Commission Recommendation (EU) 2022/553 of 5 April 2022 on monitoring the presence of Alternaria toxins in food. *Official Journal of European Union L* 107, 90–92.

Commission Regulation (EU) 2023/915 of 25 April 2023 on maximum levels for certain contaminants in food and repealing Regulation (EC) No 1881/2006. *Official Journal of European Union*, 119, 103–157.

Commission Regulation (EU) No 231/2012. Commission Regulation (EU) No 231/2012 of 9 March 2012 laying down specifications for food additives listed in Annexes II and III to Regulation (EC) No 1333/2008 of the European Parliament and of the Council Text with EEA relevance. Available online: https://eur-lex.europa.eu/legal content/EN/TXT/?uri= celex: 32012R0231 (accessed on 12 November 2024).

Croft, M., Tong, P., Fuentes, D. and Hambridge, T. (2004). Australian survey of acrylamide in carbohydrate-based foods. *Food Additives and Contaminants*, *21*(8), 721–736.

Dadzie, M., Oppong, A., Ofori, K., Eleblu, J., Ifie, E., Blay, E., Obeng-Bio, E., Appiah-Kubi, Z. and Warburton, M. (2019). Distribution of *Aspergillus flavus* and aflatoxin accumulation in stored maize grains across three agroecologies in Ghana. *Food Control, 104,* 91–98.

Dehdashti, B., Nikaeen, M., Amin, M.M. and Mohammadi, F. (2023.) Health risk assessment of exposure to bisphenol A in polymeric baby bottles. *Environmental Health Insights*, *17*, 11786302231151531.

De Witte, B., Coleman, B., Bekaert, K., Boitsov, S., Botelho, M.J., Castro-Jiménez, J., Duffy, C., Habedank, F., McGovern, E., Parmentier, K. and Tornero, V. (2022). Threshold values on environmental chemical contaminants in seafood in the European Economic Area. *Food Control*, *138*, 108978.

Dubeau, C., Aker, A., Caron-Beaudoin, É., Ayotte, P., Blanchette, C., McHugh, N.G.L. and Lemire, M. (2023). Perfluoroalkyl acid and bisphenol-A exposure via food sources in four First Nation communities in Quebec, Canada. *Public Health Nutrition*, *26*(1), 106–121.

Dueñas-Moreno, J., Mora, A., Kumar, M., Meng, X.Z. and Mahlknecht, J. (2023). Worldwide risk assessment of phthalates and bisphenol A in humans: The need for updating guidelines. *Environment International*, *181*, 108294.

Duh-Leong, C., Maffini, M.V., Kassotis, C.D., Vandenberg, L.N. and Trasande, L. (2023). The regulation of endocrine-disrupting chemicals to minimize their impact on health. *Nature Reviews Endocrinology*, *19*(10), 600–614.

EFSA (European Food Safety Authority), 2011. Update on furan levels in food from monitoring years 2004-2010 and exposure assessment. EFSA Journal, 2011;9(9):2347, 33 pp. https://doi.org/10.2903/j.efsa.2011.2347

EFSA Panel on Contaminants in the Food Chain (CONTAM), Knutsen, H.K., Alexander, J., Barregård, L., Bignami, M., Brüschweiler, B., Ceccatelli, S., Cottrill, B., Dinovi, M., Edler, L. and Grasl-Kraupp, B. (2017). Risks for public health related to the presence of furan and methylfurans in food. *EFSA Journal*, *15*(10), e05005.

EFSA Panel on Contaminants in the Food Chain (CONTAM), Schrenk, D., Bignami, M., Bodin, L., Chipman, J.K., del Mazo, J., Grasl-Kraupp, B., Hogstrand, C., Hoogenboom, L.,

Leblanc, J.C. and Nebbia, C.S. (2024). Update of the risk assessment of inorganic arsenic in food. *EFSA Journal*, *22*(1), e8488.

European Commission Regulation (EC) No 257/2010 of 25 March 2010 setting up a programme for the re-evaluation of approved food additives in accordance with Regulation (EC) No 1333/2008 of the European Parliament and of the Council on food additives.

European Commission Regulation (EC) No. 231/2012 laying down specifications for food additives listed in Annexes II and III to Regulation (EC) No. 1333/2008 of the European Parliament and of the Council.

European Commission Regulation (EC) 2022/63 of 14 January2022 amending Annexes II and III to Regulation (EC) No 1333/2008 of the European Parliament and of the Council as regards the food additive titanium dioxide (E 171).

European Commission Regulation (EC) No. 1881/2006 setting maximum levels for certain contaminants in foodstuffs. 2006. Available online: https://eur-lex.europa.eu/LexUriServ/LexUriServ.do?uri=OJ:L:2006:364:0005:0024:EN:PDF (accessed on 1 October 2024).

European Food Safety Authority (EFSA). (2008). Polycyclic aromatic hydrocarbons in food-scientific opinion of the panel on contaminants in the food chain. *EFSA Journal*, *6*(8), 724.

Firdaus, M.A. and Mishra, S. (2021). Industrial trans-fatty acid intake associated with coronary heart disease risk: a review. *Journal of Nutrition and Food Sciences*, *11*, 806.

Franco-Arellano, B., Arcand, J., Kim, M.A., Schermel, A. and L'Abbé, M.R. (2020). Progress towards eliminating industrially produced trans-fatty acids in the Canadian marketplace, 2013–2017. *Public Health Nutrition*, *23*(13), 2257–2267.

Fujimori, K. (2004). 15 The Regulatory System in Japan. *In Marrs, T. C. and Ballantyne, B. (Eds) Pesticide Toxicology and International Regulation. Wiley, Chichester, West Sussex, England.*

Genchi, G., Lauria, G., Catalano, A., Carocci, A. and Sinicropi, M.S. (2022). Arsenic: a review on a great health issue worldwide. *Applied Sciences*, *12*(12), 6184.

Ghamari, D.M. and Helfinger, M. (2019). Ghamari, D.M. and Helfinger, M. (2019). Re: Proposed changes to regulation of pesticides in Ontario, under Bill 132, Prevent Cancer Now, 1-9.

Golomb, B.A., Pavlovsky, A. and Koslik, H.J. (2021). Trans fats consumption and body mass index in cross sectional analysis. *medRxiv*, 2021–01. https://doi.org/10.1101/2021.01.13.21249701

Gowda, A. and Ramesh, H.L. (2020). Analysis of pesticide residues in carrot retailed in vegetable markets in five districts of Karnataka, India. *GSC Biological and Pharmaceutical Sciences*, *10*(3), 027–038.

Guo, X., Dai, Z., & Zhang, W. (2022). Pollution, exposure and risk of biogenic amines in canned sea fish: Classification of analytical methods based on carbon spheres QuEChERS extraction combined with HPLC. Molecules, 27(19), 6243.

Han, Y., Cheng, J., An, D., He, Y. and Tang, Z. (2022). Occurrence, potential release and health risks of heavy metals in popular take-out food containers from China. *Environmental Research*, *206*, 112265.

Hembrom, S., Singh, B., Gupta, S.K. and Nema, A.K. (2020). A comprehensive evaluation of heavy metal contamination in foodstuff and associated human health risk: a global perspective. In Pooja Singh, Rajeev Pratap Singh, Vaibhav Srivastava (Eds.) *Contemporary Environmental Issues and Challenges in Era of Climate Change* (pp. 33–63). Singapore: Springer.

Hishamuddin, E., Abd Razak, R.A., Yeoh, C.B. and Ahmad Tarmizi, A.H. (2022). Assessment of trans fatty acid levels in refined palm-based oils and commercial vegetable oils in the malaysian market. *Journal of Oil Palm Research*, *34*, 129–138.

Hites, R.A. (2021. The rise and fall of chlorpyrifos in the United States. *Environmental Science and Technology*, *55*(3), 1354–1358.

Hoffman-Pennesi, D., Winfield, S., Gavelek, A., Santillana Farakos, S.M. and Spungen, J. (2024). Infants' and young children's dietary exposures to lead and cadmium: FDA total diet study 2018–2020. *Food Additives and Contaminants*: Part A, *41*, 1–16.

Hsu, P., Pokharel, A., Scott, C.K. and Wu, F. (2024). Aflatoxin M1 in milk and dairy products: the state of the evidence for child growth impairment. *Food and Chemical Toxicology*, *193*, 115008.

Huston, R., Chan, Y.C., Chapman, H., Gardner, T. and Shaw, G. (2012). Source apportionment of heavy metals and ionic contaminants in rainwater tanks in a subtropical urban area in Australia. *Water Research*, *46*(4), 1121–1132.

James, A. (2012). Heavy metals in fish and shellfish. *EOS*, *20*, 07–12.

Japan Food Safety Commission. (2005). Food safety risk assessment related to methylmercury in seafood. www.fsc.go.jp/english/index.data/methylmercury_risk_assessment.pdf (accessed on 12 November 2024).

Kamiloglu, S., Ozdal, T. and Capanoglu, E., 2021. Regulatory aspects. In Galanakis, C. M. (Ed.) *Food Authentication and Traceability* (pp. 303–330). London, United Kingdom: Academic Press. https://doi.org/10.1016/B978-0-12-821104-5.00004-.

Kawamoto, K. and Weber, R. (2021). Dioxin sources to the aquatic environment: re-assessing dioxins in industrial processes and possible emissions to the aquatic. *Emerging Contaminants*, *7*, 52–62.

Kiralan, M., Ketenoglu, O. and Kiralan, S.S. (2021). Trans fatty acids—occurrence, technical aspects, and worldwide regulations. *Studies in Natural Products Chemistry*, *70*, 313–343.

Koch, W., Czop, M., Iłowiecka, K., Nawrocka, A. and Wiącek, D. (2022). Dietary intake of toxic heavy metals with major groups of food products—results of analytical determinations. *Nutrients*, *14*(8), 1626.

Koszucka, A., Nowak, A., Nowak, I. and Motyl, I. (2020). Acrylamide in human diet, its metabolism, toxicity, inactivation and the associated European Union legal regulations in food industry. *Critical Reviews in Food Science and Nutrition*, *60*(10), 1677–1692.

Li, X., Bei, E., Qiu, Y., Xiao, H., Wang, J., Lin, P., Zhang, X. and Chen, C. (2021). Intake of volatile nitrosamines by Chinese residents in different provinces via food and drinking water. *Science of the Total Environment*, *754*, 142121.

Lindeman, B. and Ritz, V. (2020). Regulation and risk management of endocrine disruptors: current status and future perspectives. In Mantovani, A and Fucic, A., (Eds) *Challenges in Endocrine Disruptor Toxicology and Risk Assessment* (pp. 495–511). Cambridge, UK: Royal Society of Chemistry. Doi: https://doi.org/10.1039/9781839160738.

Liu, Y., Peng, X., Huang, Y., Hu, H., Li, C., Chen, Y., Yu, Q. and Wang, Y., 2023. Recent advances in the occurrence, mechanisms, influence factors and control strategies of process contaminants in nuts: a comprehensive review. *Food Control*, *159*, 110265.

Maffini, M.V. and Vandenberg, L.N. (2022). Failure to launch: the endocrine disruptor screening program at the US Environmental Protection Agency. *Frontiers in Toxicology*, *4*, 908439.

Magnuson, B., Munro, I., Abbot, P., Baldwin, N., Lopez-Garcia, R., Ly, K., McGirr, L., Roberts, A. and Socolovsky, S. (2013). Review of the regulation and safety assessment of food substances in various countries and jurisdictions. *Food Additives and Contaminants: Part A*, *30*(7), 1147–1220.

Marrez, D.A. and Ayesh, A.M. (2022). Mycotoxins: the threat to food safety. *Egyptian Journal of Chemistry*, *65*(1), 1–2.

Masci, M., Caproni, R. and Nevigato, T. (2024). Glyphosate as Contaminant in Cereals: Chromatographic Methods for Its Determination. Discussion about Occurrence, Accumulation, Fate, Degradation, and Regulatory Status. Preprints.org. doi:10.20944/preprints202403.0921.v1

Mazuryk, J., Klepacka, K., Kutner, W. and Sharma, P.S. (2023). Glyphosate separating and sensing for precision agriculture and environmental protection in the era of smart materials. *Environmental Science and Technology*, *57*(27), 9898–9924.

More, S.J., Hardy, A., Bampidis, V., Benford, D., Hougaard Bennekou, S., Bragard, C., Boesten, J., Halldorsson, T.I., Hernández-Jerez, A.F., Jeger, M.J., et al. (2009). Guidance on harmonised methodologies for human health, animal health and ecological risk assessment of combined exposure to multiple chemicals. *EFSA Journal*, 17, 5634.

Nagpal, T., Sahu, J.K., Khare, S.K., Bashir, K. and Jan, K. (2021). Trans fatty acids in food: A review on dietary intake, health impact, regulations and alternatives. *Journal of Food Science*, *86*(12), 5159–5174.

Navaranjan, G., Diamond, M.L., Harris, S.A., Jantunen, L.M., Bernstein, S., Scott, J.A., Takaro, T.K., Dai, R., Lefebvre, D.L., Azad, M.B. and Becker, A.B. (2021). Early life exposure to phthalates and the development of childhood asthma among Canadian children. *Environmental Research*, *197*, 110981.

Ndagijimana, R., Shahbaz, U. and Sun, X. (2020) Aflatoxin B1 in food and feed: an overview on prevalence, determination and control tactics. *Journal of Academia and Industrial Research*, *8*(8), 144–152.

Ong, H.T., Samsudin, H. and Soto-Valdez, H. (2022). Migration of endocrine-disrupting chemicals into food from plastic packaging materials: an overview of chemical risk assessment, techniques to monitor migration, and international regulations. *Critical Reviews in Food Science and Nutrition*, *62*(4), 957–979.

Oyebamiji, Y.O., Adebayo, I., Umar, O.B., Mohd Zaini, N.A. and Ismail, M.N. (2024). Occurrence, health implications, and management of aflatoxin in cereal: A current review. *Egyptian Journal of Botany*, *64*(4), 1–15.

Palumbo, R., Crisci, A., Venâncio, A., Cortiñas Abrahantes, J., Dorne, J.L., Battilani, P. and Toscano, P. (2020). Occurrence and co-occurrence of mycotoxins in cereal-based feed and food. *Microorganisms*, *8*, 74.

Paustenbach, D.J., Brown, S.E., Heywood, J.J., Donnell, M.T. and Eaton, D.L. (2024). Risk characterization of N-nitrosodimethylamine in pharmaceuticals. *Food and Chemical Toxicology*, *186*, 114498.

Pérez-Lucas, G., Navarro, G. and Navarro, S. (2024). Understanding how chemical pollutants arise and evolve in the brewing supply chain: a scoping review. *Foods*, 13, 1709.

Pietropaoli, F., Pantalone, S., Cichelli, A. and d'Alessandro, N. (2022). Acrylamide in widely consumed foods–a review. *Food Additives and Contaminants: Part A*, *39*(5), 853–887.

Piotrowska, M. (2021). Microbiological decontamination of mycotoxins: opportunities and limitations. *Toxins*, *13*(11), 819.

Pressman, P., Clemens, R., Hayes, W. and Reddy, C. (2017). Food additive safety: a review of toxicologic and regulatory issues. *Toxicology Research and application*, *1*, 2397847317723572.

Rachkowski, D. (2024). *Strengthening environmental protection for a healthier Canada Act: the right to a healthy environment & accountability in chemicals management* (Doctoral dissertation, University of British Columbia).

Ramos-Diaz, J.M., Sulyok, M., Jacobsen, S.E., Jouppila, K. and Nathanail, A.V. (2021). Comparative study of mycotoxin occurrence in Andean and cereal grains cultivated in South America and North Europe. *Food Control*, *130*, 108260.

Rani, M., Sharma, S., and Shanker, U. (2024). Legal and safety issues of ECs. In Occurrence, Distribution and Toxic Effects of Emerging Contaminantsx (pp. 259-282). Shanker, U., and Rani, M. Eds, CRC Press. Roca Boca, FL, USA.

Regulation (EC) No 1333/2008, of the European Parliament and of the Council, 16 December 2008, on food additives,(OJ L 354, 31.12.2008, p. 16), 2008R1333-EN21.04.2015-022.001-1.

Revelou, P.-K., Xagoraris, M., Alissandrakis, E., Pappas, C.S. and Tarantilis, P.A. (2021). A review of the analytical methods for the determination of 4(5)-methylimi dazole in food matrices. *Chemosensors*, *9*, 322.

Rifai, L. and Saleh, F.A. (2020). A review on acrylamide in food: occurrence, toxicity, and mitigation strategies. *International Journal of Toxicology*, *39*(2), 93–102.

Sá, A.G. and House, J.D. (2024). Protein quality of cereals: digestibility determination and processing impacts. *Journal of Cereal Science*, *117*, 103892.

Sabran, M.R., Jamaluddin, R. and Ahmad, Z. (2013). A mini review on aflatoxin exposure in Malaysia: past, present and future. *Frontiers in Microbiology*, *4*, 334.

Sampaio, G.R., Guizellini, G.M., da Silva, S.A., de Almeida, A.P., Pinaffi-Langley, A.C.C., Rogero, M.M., de Camargo, A.C. and Torres, E.A. (2021). Polycyclic aromatic hydrocarbons in foods: biological effects, legislation, occurrence, analytical methods, and strategies to reduce their formation. *International Journal of Molecular Sciences*, *22*(11), 6010.

Siddique, S., Zhang, G., Coleman, K. and Kubwabo, C. (2021). Investigation of the migration of bisphenols from baby bottles and sippy cups. *Current Research in Food Science*, *4*, 619–626.

Smythe, E.A. (2024). Industrial meat in Canada, growth promoters and the struggle over international food standards. *Canadian Food Studies/La Revue canadienne des études sur l'alimentation*, *11*(1), 111–130.

Spiller, P., van Wijngaarden, E., Adams, H.R., Strain, J.J., McSorley, E.M., Mulhern, M.S., Conway, M.C., Yeates, A.J., Carrington, C., Bolger, P.M. and Morgan, K.M. (2023). Net effects explains the benefits to children from maternal fish consumption despite methylmercury in fish. *Neurotoxicology*, *99*, 195–205.

Suzuki, Y., Kondo, M., Akiyama, H. and Ogra, Y. (2022). Presence of nano-sized mercury-containing particles in seafoods, and an estimate of dietary exposure. *Environmental Pollution*, *307*, 119555.

Suzuki, Y., Kondo, M., Harimoto, M., Kitayama, I. and Akiyama, H. (2022). Dietary exposure to arsenic species in Japan in 2019 using a total diet study based on composite sample with market basket approach at the national level. *Journal of Food Composition and Analysis*, *108*, 104384.

Szczuka, A., Huang, N., MacDonald, J.A., Nayak, A., Zhang, Z. and Mitch, W.A. (2020). N-Nitrosodimethylamine formation during UV/hydrogen peroxide and UV/chlorine advanced oxidation process treatment following reverse osmosis for potable reuse. *Environmental Science and Technology*, *54*(23), 15465–15475.

Takahashi, M., Suzuki, Y., Aoyagi, M., Toda, E., Ito, K., Fukumitsu, T., Hagio, M., Hayashi, T., Shintaku, S., Ihara, S. and Nakashima, A. (2024). Estimated daily intake of residual agricultural chemicals across general Japanese people based on the total diet study from 2019 to 2021. *Japanese Journal of Food Chemistry and Safety*, *31*, 65–75.

Tang, C., Ramos, M.J.G., Heffernan, A., Kaserzon, S., Rauert, C., Lin, C.Y., Mueller, J.F. Wang, X. (2023). Evaluation and identification of chemical migrants leached from baby food pouch packaging. *Chemosphere*, *340*, 139758.

Tzatzarakis, M.N., Vakonaki, E., Moti, S., Alegakis, A., Tsitsimpikou, C., Tsakiris, I., Goumenou, M., Nosyrev, A.E., Rizos, A.K. and Tsatsakis, A.M. (2017). Quantification of 4-methylimidazole in soft drinks, sauces and vinegars of Greek market using two liquid chromatography techniques. *Food Chemical Toxicology*, *107*, 565–571.

U.S. FDA (U.S. Food and Drug Administration). (2005). *Whole Milk, Low Fat Milk, Skim Milk—Aflatoxin M1*. www.fda.gov/regulatory-information/search-fda-guidance-documents/cpg-sec-527400-whole-milk-lowfat-milk-skim-milk-aflatoxin-m1 (accessed on 12 November 2024).

Vila-López, M.V., Pallarés, N., Ferrer, E. and Tolosa, J. (2023). Mycotoxin determination and occurrence in pseudo-cereals intended for food and feed: a review. *Toxins*, *15*, 379.

Vom Saal, F.S., Antoniou, M., Belcher, S.M., Bergman, A., Bhandari, R.K., Birnbaum, L.S., Cohen, A., Collins, T.J., Demeneix, B., Fine, A.M. and Flaws, J.A.. (2024). The conflict between regulatory agencies over the 20,000-fold lowering of the tolerable daily intake (TDI) for bisphenol A (BPA) by the European Food Safety Authority (EFSA). Environmental health perspectives, 132(4), p.045001.

Wong, C., Roberts, S.M. and Saab, I.N. (2022). Review of regulatory reference values and background levels for heavy metals in the human diet. *Regulatory Toxicology and Pharmacology, 130*, 105122.

World Health Organization. (2023). *Countdown to 2023: WHO Report on Global Trans-Fat Elimination 2022*. Geneva, Switzerland: World Health Organization.

Yang, D., Yang, H., Shi, M., Jia, X., Sui, H., Liu, Z. and Wu, Y. (2023). Advancing food safety risk assessment in China: development of new approach methodologies (NAMs). *Frontiers in Toxicology, 5*, 1292373.

Yang, M., Wang, Y., Yang, G., Wang, Y., Liu, F. and Chen, C. (2024). A review of cumulative risk assessment of multiple pesticide residues in food: Current status, approaches and future perspectives. *Trends in Food Science and Technology, 144*, 104340.

Yap, S.K., Chin, N.L., Niranjan, K., Shah, N.N.A.K. and Than, L.T.L. (2024). An antioxidant-and hydroxymethylfurfural-based index for health impact grading of honey. *Journal of Food Measurement and Characterization, 18*(4), 2609–2617.

Zafar, A., Javed, S., Akram, N. and Naqvi, S.A.R. (2024). Health risks of mercury. In Nitish Kumar (Ed.) *Mercury Toxicity Mitigation: Sustainable Nexus Approach* (pp. 67–92). Cham: Springer Nature Switzerland.

Zha, L., Sobue, T., Kitamura, T., Kitamura, Y., Ishihara, J., Kotemori, A., Liu, R., Ikeda, S., Sawada, N., Iwasaki, M. and Tsugane, S. (2020). Dietary acrylamide intake and the risk of liver cancer: the Japan public health center-based prospective study. *Nutrients, 12*(9), 2503.

Zhang, H., Wei, L., Liu, J., Lin, S. and Yuan, Y. (2014). Detection of 5-hydroxymethyl-2-furfural levels in selected Chinese foods by ultra-high-performance liquid chromatograph analytical method. *Food Analytical Methods, 7*, 181–188.

Zhang, J., Liu, L., Zhao, Y., Li, H., Lian, Y., Zhang, Z., Huang, C. and Du, X. (2020). Development of a high-resolution emission inventory of agricultural machinery with a novel methodology: a case study for Yangtze River Delta region. *Environmental Pollution, 266*, 115075.

Zhang, X., Wang, Z., Liu, L., Zhan, N., Qin, J., Lu, X. and Cheng, M. (2021). Assessment of the risks from dietary lead exposure in China. *Journal of Hazardous Materials, 418*, 126134.

Zhu, W., Zhao, L., Jin, J. and Song, Y. (2022). Preparation of core-shell rare earth-doped upconversion nanomaterials and simultaneous detection of two pesticides in food. *Foods, 11*(10), 1485.

Index

Note: Page locators in **bold** and *italics* represents tables and figures, respectively.